The negative regulation of hematopoiesis
from fundamental aspects to clinical applications

Régulation négative de l'hématopoïèse
des aspects fondamentaux à l'application clinique

British Library Cataloguing in Publication Data
A catalogue record for this book
is available from the British Library

ISBN 2-7420-0015-1
ISSN 0768-3154

First published in 1993 by

Editions John Libbey Eurotext
6 rue Blanche, 92120 Montrouge, France. (33) (1) 47 35 85 52
ISBN 2-7420-0015-1

John Libbey and Company Ltd
13 Smiths Yard, Summerley Street, London SW18 4HR,
England.
(44) (81) 947 27 77

Institut National de la Santé et de la Recherche Médicale
101 rue de Tolbiac, 75654 Paris Cedex 13, France.
(33) (1) 44 23 60 00
ISBN 2-85598-541-2

ISSN 0768-3154

© 1993 Colloques INSERM/John Libbey Eurotext Ltd,
All rights reserved
Unauthorized publication contravenes applicable laws

The negative regulation of hematopoiesis
from fundamental aspects to clinical applications

Régulation négative de l'hématopoïèse des aspects fondamentaux à l'application clinique

Proceedings of the Third International Conference on Negative Regulation of Hematopoiesis, Paris (France), 18-22 April 1993

Organized by the Department of Hematology of the Faculté de Médecine Saint-Antoine, Université Pierre et Marie Curie, Paris, France

Under the auspices of Monsieur le Ministre Chargé de la Recherche et de l'Enseignement Supérieur

Sponsored by the Institut National de la Santé et de la Recherche Médicale (INSERM), the National Institutes of Health (NIH) and the Assistance Publique-Hôpitaux de Paris (AP-HP)

Edited by
Martine Guigon
François M. Lemoine
Nicholas Dainiak
Alan Schechter
Albert Najman

LES EDITIONS INSERM

Preface

Since 1987, at the time of the first Symposium that we have organized at the Pasteur Institute in Paris, the field of inhibitory factors involved in hematopoiesis has considerably increased. This evolution was already noticeable at the 2nd Conference in 1990 in Providence (RI, USA). Putative molecules are now produced by genetic engineering or by chemical synthesis. Furthermore, negative regulators of hematopoiesis are now studied not only at the cellular level but also at the molecular level at each step of the proliferation/differentiation: gene expression, transcription, transduction signalling, modulation of surface receptors. Based on the concept of negative regulation, new tools and new therapeutic strategies are now being developed. This wide variety of approaches involved the participation of scientists and clinicians working with different models. The invited speakers of this conference have been selected for their high competence in the most representative fields of the present ways of research. The pluridisciplinarity of this meeting allowed fruitful discussions and reflected the necessary opening for hematology to various aspects of Biology. In this book, we have found it more convenient for the reader to present the papers according to their topic, whatever a plenary lecture, a brief communication or a poster summary.

Martine Guigon
François M. Lemoine
Nicholas Dainiak
Alan Schechter
Albert Najman

Préface

Depuis 1987, date de la première rencontre que nous avons organisée à l'Institut Pasteur de Paris, le domaine des phénomènes inhibiteurs impliqués dans l'hématopoïèse s'est beaucoup élargi. Cette évolution avait déjà été remarquée lors de la 2e Conférence qui s'est tenue à Providence (USA) en 1990. De nouvelles molécules sont maintenant produites par génie génétique ou par synthèse chimique. En outre, les facteurs intervenant dans la régulation négative de l'hématopoïèse ne sont plus seulement analysés à l'échelle cellulaire, mais également à l'échelon moléculaire : expression des gènes, transcription, voies de transduction et modulation des récepteurs de surface. Le concept de régulation négative a ouvert la voie à de nouvelles technologies et débouchera dans l'avenir sur de nouvelles stratégies thérapeutiques. Cette grande diversité d'approches fait appel à des chercheurs et à des médecins travaillant sur des modèles les plus variés. Les orateurs de cette 3e Conférence ont été choisis pour leur extrême compétence dans les domaines les plus représentatifs des avancées actuelles. L'aspect pluridisciplinaire de cette réunion a permis des discussions fructueuses et traduit l'ouverture obligatoire de l'hématologie aux disciplines les plus variées de la Biologie. Dans cet ouvrage sont réunis à la fois les textes des conférences plénières, des communications brèves et des communications affichées qui ont été regroupés par thème.

Martine Guigon
François M. Lemoine
Nicholas Dainiak
Alan Schechter
Albert Najman

Acknowledgments

We wish to thank the following sponsoring Institutions
> Institut National de la Santé et de la Recherche Médicale (INSERM)
> The National Institutes of Health (National Institute of Diabetes and Digestive and Kidney Diseases, National Heart, Lung and Blood Institute, National Cancer Institute)
> Ligue Nationale contre le Cancer
> Assistance Publique-Hôpitaux de Paris

for providing financial support.

We also thank the following pharmaceutical companies:
> Ipsen-Biotech, Roche and Amgen (France), Wellcome, Sarget, Smith Kline Beecham, Schering Plough, Immunotech, Coultronics France SA, Biosys, Immunex Research and Development Corporation, Amgen (USA), Hoechst, ERIA Diagnostics Pasteur, Cell Science Products-Becton Dickinson, Upjohn

who have contributed by their help to the success of the meeting.

We are grateful to Pr Jean-Claude Imbert, Dean of the Faculté de Médecine Saint-Antoine and to Mr Yves Thomas and his team who were very helpful to organize the Conference in the Faculté de Médecine. We want to thank the members of the Laboratory of Hematology, especially Claude Baillou and also Judith Brami-Najman who was in charge of the Secretariat during the conference and Frédéric Alran for their efficient and nice help.

We would like to thank Pr Jacques Picard (Faculté de Médecine Saint-Antoine), Mr Dominique Deroubaix (Assistance Publique-Hôpitaux de Paris), Pr Jean Rosa (INSERM) and Pr Henri Rochant (Société Française d'Hématologie) for introducing the Conference and also the Chairpersons for their help to respect a busy schedule and organizing fruitful discussions (C. Bréchot, H. Broxmeyer, J. Caen, Y. Cayre, N. Dainiak, M. Dexter, C. Eaves, C. Hesdorffer, E. Gluckman, N.C. Gorin, M. Guigon, F.M. Lemoine, H. Lodish, B. Lowenberg, A. Najman, I. Pragnell, J. Rosa, A. Schechter, M. Tubiana, W. Vainchenker, N. Young).

We are grateful to Drs David Badman and Alan Levine (National Institutes of Health) for guidance and support throughout the planning stages of this conference.

It is worth mentioning our gratitude
> – to Pr Jean Loygue, Dr Philippe Denormandie, Sophia Sébille and Henri Santarelli (Assistance Publique-Hôpitaux de Paris),
> – and to Yvette Coadou (Bureau des Colloques, INSERM) and Martine Krief (John Libbey Eurotext) for the publication of this book.

Remerciements

Nous sommes heureux de remercier pour leur aide financière
> l'Institut National de la Santé et de la Recherche Médicale (INSERM)
> le National Institutes of Health (National Institute of Diabetes and Digestive and Kidney Diseases, National Heart, Lung and Blood Institute, National Cancer Institute)
> la Ligue Nationale contre le Cancer
> l'Assistance Publique-Hôpitaux de Paris.

Nous remercions également les laboratoires
> Ipsen-Biotech, Roche et Amgen (France), Wellcome, Sarget, Smith Kline Beecham, Schering Plough, Immunotech, Coultronics France SA, Biosys, Immunex Research and Development Corporation, Amgen (USA), Hoechst, ERIA Diagnostics Pasteur, Cell Science Products-Becton Dickinson, Upjohn

qui ont contribué par leur aide au succès de cette conférence.

Nous sommes reconnaissants au Pr Jean-Claude Imbert, Doyen de la Faculté de Médecine Saint-Antoine et à M. Yves Thomas et son équipe qui nous ont aidés à organiser cette réunion à la Faculté de Médecine Saint-Antoine.

Nous remercions chaleureusement Claude Baillou et tous les membres du Laboratoire, Judith Brami-Najman qui a assuré le secrétariat de la conférence et Frédéric Alran pour leur aide efficace et sympathique.

Nous tenons à remercier le Pr Jacques Picard (Faculté de Médecine Saint-Antoine), M. Dominique Deroubaix (Assistance Publique-Hôpitaux de Paris), le Pr Jean Rosa (INSERM) et le Pr Henri Rochant (Société Française d'Hématologie) qui ont introduit la Conférence ainsi que les Présidents de séance qui ont eu à cœur de respecter un emploi du temps serré et d'organiser des discussions de haut niveau (C. Bréchot, H. Broxmeyer, J. Caen, Y. Cayre, N. Dainiak, M. Dexter, C. Eaves, C. Hesdorffer, E. Gluckman, N.C. Gorin, M. Guigon, F.M. Lemoine, H. Lodish, B. Lowenberg, A. Najman, I. Pragnell, J. Rosa, A. Schechter, M. Tubiana, W. Vainchenker, N. Young).

Nous voudrions exprimer notre gratitude
> – au Pr Jean Loygue, au Dr Philippe Denormandie, à Sophia Sébille et à Henri Santarelli (Assistance Publique-Hôpitaux de Paris),
> – ainsi qu'à Yvette Coadou (Bureau des Colloques et Conférences de l'INSERM) et à Martine Krief (John Libbey Eurotext) qui ont permis la publication de ce livre.

Organizing Committee
Comité d'organisation

Albert Najman
Martine Guigon
François M. Lemoine
Françoise Isnard
Jean-Philippe Laporte
Norbert-Claude Gorin

*Département d'Hématologie
Faculté de Médecine et Hôpital Saint-Antoine*

International Scientific Committee
Comité Scientifique International

Christian Bréchot, France
Hal Broxmeyer, USA
Jacques Caen, France
Nicholas Dainiak, USA
Connie Eaves, Canada
Bob Lowenberg, The Netherlands
Ian Pragnell, UK
Alan Schechter, USA
Pierre Tambourin, France
William Vainchenker, France
Neal Young, USA
Dov Zipori, Israel

List and address of contributors
Liste et adresse des intervenants

Ballerini P. Deparments of Physiology and Medicine, Columbia University, New York, NY 10032, USA

Banchereau J. Schering-Plough, Laboratory for Immunological Research, 27, chemin des Peupliers, 69571 Dardilly, France

Baudier J. INSERM U 309, GEN-G 85X, 38041 Grenoble Cedex, France

Bauvois B. INSERM U 365 «Interférons et Cytokines», Institut Curie, Section de Biologie, Pavillon Pasteur, 26, rue d'Ulm, 75231 Paris Cedex 05, France

Bi S. LRF Centre for Adult Leukaemia, Department of Haematology, Royal Postgraduate Medical School, Hammersmith Hospital, Du Cane Road, London W12 ONN, UK

Bonnet D. Laboratory of Hematology, CHU Saint-Antoine, Paris, France

Bouchard C. INSERM U 255, Institut Curie, 26, rue d'Ulm, 75231 Paris Cedex 05, France

Bréchot C. INSERM U 370, CHU Necker, 156, rue de Vaugirard, 75742 Paris Cedex 15 et Unité d'Hépatologie, Hôpital Laennec, rue de Sèvres, 75007 Paris, France

Brosh N. Department of Cell Biology, The Weizmann Institute of Science, Rehovot 76100, Israel

Broxmeyer H.E. Departments of Medicine (Hematology/Oncology), Microbiology/Immunology, and the Walter Oncology Center, Indiana University School of Medicine, 975 West Walnut Street, IB 501, Indianapolis, IN 46202-5121, USA

Brûlet P. Unité d'Embryologie Moléculaire, Institut Pasteur, 25, rue du Docteur-Roux, 75724 Paris Cedex 15, France

Calenda V. INSERM U 322, Unité des Rétrovirus et Maladies Associées, Campus Universitaire de Luminy, BP 33, 13273 Marseille Cedex 9, France

Carde P. Institut Gustave-Roussy, 39, rue Camille-Desmoulins, 94805 Villejuif, France

Carlo-Stella C. Department of Hematology, Bone Marrow Transplantation Unit, University of Parma, Parma, Italy

Chatelain C. Laboratory of Experimental Hematology and Oncology, Oncology Unit, Catholic University of Louvain, UCL Mont-Godinne, Av. G. Therasse 1, B-5530 Yvoir, Belgium

Coutton C. Centre de Recherche en Immunologie, Laboratoire d'Hématologie et d'Immunothérapie des Cancers, 1, place de l'Hôpital, 67091, Strasbourg Cedex, France

Craig S. British Bio-Technology, Cowley, Oxford, OX4 5LY, UK

Dainiak M. University of Connecticut School of Medicine, Departments of Medicine, Laboratory Medicine and Radiology, Biomolecular Structure Analysis Center, Connecticut Cancer Institute, Farmington, CT, USA

Davis B. California Pacific Medical Center Research Institute, 2330 Clay St., San Francisco, CA, USA

Dinarello C.A. Department of Medicine, Tufts University School of Medicine and New England Medical Center Hospital, Boston, MA 02111, USA

Dorée M. CNRS UPR 9008 and INSERM U 249, BP 5051, 34033 Montpellier Cedex, France

Douay L. Formation Associée Claude-Bernard, Unité de Recherche sur les greffes de cellules souches hématopoïétiques, Laboratoire de Cultures Cellulaires, CHU Saint-Antoine, 27, rue de Chaligny, 75012 Paris et Laboratoire d'Hématologie, Hôpital Armand-Trousseau, 26, avenue du Docteur-Arnold-Netter, 75571 Paris Cedex 12, France

Dumenil D. Institut Gustave-Roussy, U 362, 94800 Villejuif, France

Durand B. Hôpital de la Croix-Rousse, Lyon, France

Eaves C.J. Terry Fox Laboratory, British Columbia Cancer Agency and University of British Columbia, 601 West 10th Avenue, Vancouver, BC, Canada V5Z 1L3

Fardoun D. Department of Research on Hematopoietic Stem Cells and Extracellular Matrix Molecules, School of Medicine, BP 97, 76803 St Etienne-du-Rouvray Cedex, France

Fisher D.E. Center for Cancer Research, Massachusetts Institute of Technology, Cambridge, MA 02139, and Dana Farber Cancer Institute, Harvard Medical School, Boston, MA 02115, USA

Freedman M.H. Divisions of Immunology/Cancer Research, and Hematology/Oncology, The Hospital for Sick Children, 555 University Ave., Toronto, Ontario, Canada, M5G 1X8

Gallichio V.S. Hematology/Oncology Division, Departments of Internal Medicine and Clinical Sciences, University of Kentucky Medical Center and Department of Veterans Affairs, Lexington, KY, USA

Gewirtz A.M. Hematology/Oncology and Molecular Diagnosis Sections, Departments of Pathology, and Internal Medicine, University of Pennsylvania School of Medicine, Philadelphia, Pennsylvania 19104, USA

Gluckman E. Bone Marrow Transplant Unit and Laboratoy of Bone Marrow Biology; and Institut d'Hématologie, Hôpital Saint-Louis, Paris, France

Godden J. Beatson Institute for Cancer Research, Bearsden, Glasgow, UK

Gorodinsky M. Faculty of Health Sciences, Clinical Biochemistry Unit and Department of Hematology, Ben-Gurion University, Beer-Sheva, Israel

Gothelf Y. Department of Cell Biology, The Weizmann Institute of Science, Rehovot 76100, Israel

Grognet J.M. Service de Pharmacologie et d'Immunologie, CEA/Saclay, 91191 Gif-sur-Yvette Cedex, France

Guigon M. Department of Hematology, CHU Saint-Antoine, Paris, France

Hamood M. Laboratory of Experimental Hematology, Brugmann University Hospital and Hematology/Oncology Unit, Pediatric University Hospital, Free University of Brussels, Brussels, Belgium

Han Z.C. Institut des Vaisseaux et du Sang, Hôpital Lariboisière, 8, rue Guy-Patin, 75010 Paris, France

Hatzfeld J. CNRS Laboratory of Cell and Molecular Biology of Cytokines, GBGM, UPR 272, IRSC, 7, rue Guy-Môquet, BP 8, 94801 Villejuif Cedex, France

Hérodin F. Department of Radiobiology, Centre de Recherches du Service de Santé des Armées, 38702 La Tronche-Grenoble, France

Hesdorffer C. Department of Medicine, College of Physicians and Surgeons of Columbia University, New York, USA

Heyworth C.M. CRC Department of Experimental Haematology, Paterson Institute for Cancer Research, Manchester M20 9BX, UK

Holyoake T.L. CRC Beatson Laboratories, Glasgow, G61 1BD, UK

Khoury E. Laboratoire d'Hématologie, Faculté de Médecine Saint-Antoine, 27, rue de Chaligny, 75012 Paris, France

Kirby S. Departments of Pathology and Medicine, University of North Carolina, Chapel Hill, NC, USA

Koury M.J. Vanderbilt University and Veterans Administration, Medical Centers, Nashville, TN 37232, USA

Kravtsov V. Department of Histology and Cell Biology, Sackler School of Medicine, Tel Aviv University, Tel Aviv, Israel

Kretser de T. Peter MacCallum Cancer Institute, 481 Little Lonsdale Street, Melbourne, Victoria 3000, Australia

Kretzner L. Fred Hutchinson Cancer Research Center, 1124 Columbia Street, Seattle, WA 98104, USA

Lasfar A. INSERM U 365, Institut Curie, Section de Biologie, Pavillon Pasteur, 26, rue d'Ulm, 75231 Paris Cedex 05, France

Lebeurier I. Institut des Vaisseaux et du Sang, Hôpital Lariboisière, 8, rue Guy-Patin, 75010 Paris, France

Lenfant M. Institut de Chimie des Substances Naturelles, CNRS, 91198 Gif-sur-Yvette, France

Lodish H.F. Whitehead Institute for Biomedical Research, Nine Cambridge Center, Cambridge, MA 02142, USA

Mahon F.X. Laboratoire de Greffe de Moelle, URA CNRS 1456, Université de Bordeaux II, Bordeaux, France

Mao N. Department of Experimental Hematology-Oncology, Institute of Basic Medical Sciences, Beijing 100850, PO Box 130, PR China

Martyré M.C. INSERM U 365, Institut Curie, 26, rue d'Ulm, 75231 Paris Cedex 05, France

Milenković P. Institute for Medical Research, Faculty of Medicine, Beograd, Yugoslavia

Mire-Sluis A.R. Division of Immunobiology, NIBSC, Blanche Lane, South Mimms, Herts, EN6 3QG, UK

Moreb J. Department of Medicine, Division of Medical Oncology, University of Florida, Gainesville, FL 32610, USA

Mouchiroud G. Centre de Génétique Moléculaire et Cellulaire, UMR CNRS 106, Université Claude-Bernard Lyon I, 69622 Villeurbanne Cedex, France

Najman A. Department of Hematology, CHU Saint-Antoine, Paris, France

Neta R. Department of Experimental Hematology, Armed Forces Radiobiology Research Institute, Bethesda, MD, USA

Paukovits W.R. Laboratory of Growth Regulation, Institute for Tumor Biology-Cancer Research, University of Vienna, Vienna, Austria

Peled A. Department of Cell Biology, The Weizmann Institute of Science, Rehovot 76100, Israel

Pelus L.M. Departments of Anti-Infectives and Peptidomimetic Research, SmithKline Beecham Pharmaceuticals, 709 Swedeland Road, King of Prussia, PA 19406, USA

Pouysségur J. Centre de Biochimie-CNRS, Université de Nice, Parc Valrose, 06108 Nice, France

Quesenberry P. Department of Internal Medicine, University of Massachusetts Medical Center, Worcester, Massachusetts, USA

Riches A. School of Biological and Medical Sciences, University of St. Andrews, Scotland, UK

Robinson S. School of Biological and Medical Sciences, University of St. Andrews, Scotland, UK

Rogalsky V. D.H. Ruttenberg Cancer Center, Mount Sinai Medical Center, New York, NY 10029, USA

Roussel M.F. Department of Tumor Cell Biology, St. Jude Children's Research Hospital, Memphis, Tennessee 38105, USA

Ruscetti F.W. Laboratory of Leukocyte Biology, Biological Response Modifiers Program and Biological Carcinogenesis and Development Program, PRI/DynCorp, Inc., National Cancer Institute, Frederick Cancer Research and Development Center, Frederick, Maryland 21702, USA

Samarut J. Laboratoire de Biologie Moléculaire et Cellulaire, CNRS UMR49, INRA, Ecole Normale Supérieure de Lyon, 46, allée d'Italie, 69364 Lyon Cedex 7, France

Sanders M. University of Connecticult School of Medicine, 263 Farmington Avenue, Farmington, 06030 USA

Schwartz G.N. Transplantation Therapy Section, National Cancer Institute, Bethesda, MD, USA

Segovia J.C. Departemento de Biología Molecular y Celular, CIEMAT, Av. Complutense 22, 28040, Madrid, Spain

Sredni B. C.A.I.R. Institute, Department of Life Sciences, Bar Ilan University, Ramat Gan, 52900 Israel

Stevenson E.C. Department of Haematology, The Queen's University of Belfast, BT12 6BA, N. Ireland, UK

Stratford May W. The Johns Hopkins Oncology Center, 424 N. Bond Street, Baltimore, MD, USA

Stryckmans P. Institut Jules-Bordet, Brussels and Eurogenetics, Tessenderlo, Belgium

Sumereau-Dassin E. Department of Research on Hematopoietic Stem Cells and Extracellular Matrix Molecules, School of Medicine, BP 97, 76803 Saint-Etienne-du-Rouvray Cedex, France

Torok-Storb B. Fred Hutchinson Cancer Research Center and University of Washington School of Medicine, Seattle, Washington, USA

Vittet D. INSERM U 300, Faculté de Pharmacie, 34060 Montpellier, France

Wdzieczak-Bakala J. Institut de Chimie des Substances Naturelles, CNRS, 91198 Gif-sur-Yvette, France

Wierenga P.K. Department of Radiobiology, State University, Groningen, The Netherlands

Wietzerbin J. INSERM U 365 «Interférons et Cytokines», Institut Curie, Section de Biologie, 26, rue d'Ulm, 75231 Paris Cedex 05, France

Young N.S. Hematology Branch, National Heart, Lung and Blood Institute, 9000 Rockville Pike, Bethesda, MD 20892, USA

Zauli G. Institute of Human Anatomy, University of Ferrara, Via Fossato di Mortara 66, 44100 Ferrara, Italy

Contents
Sommaire

- **V** Preface
- **VI** Préface
- **VII** Acknowledgments
- **VIII** Remerciements
- **XI** List and address of contributors
  Liste et adresse des intervenants

I. RECEPTORS AND THEIR INHIBITORS/*RÉCEPTEURS ET LEURS INHIBITEURS*

3 **H.F. Lodish, S.S. Watowich, G.D. Longmore, D.I. Hilton**
The erythropoietin receptor: structure, function and disease
Le récepteur de l'érythropoïétine : structure, fonction et anomalies

15 **O. Gandrillon, M. Garcia, N. Ferrand, J. Samarut**
Negative and positive control of erythrocytic progenitor cells by the nuclear receptors for thyroid hormone T3 and retinoic acid: a target for v-erbA oncogene and other mutated nuclear receptors
Contrôle négatif et positif des progéniteurs érythrocytaires par les récepteurs nucléaires de l'hormone thyroïdienne T3 et l'acide rétinoïque : une cible pour l'oncogène v-erbA et d'autres récepteurs nucléaires mutés

23 **F.W. Ruscetti, S.E.W. Jacobsen, C.M. Dubois, K. Hestdal, J.R. Keller**
Hematopoietic cell growth is regulated by a balance of positive and negative regulators: potential role for receptor modulation
La croissance hématopoïétique est régulée par une balance entre facteurs positifs et négatifs : rôle potentiel de la modulation des récepteurs

31 **J. Wietzerbin**
TNF-interferon interaction: molecular and cellular aspects
Interactions entre le TNF et l'IFN : aspects moléculaires et cellulaires

37 **E. Khoury, S. Pontvert-Delucq, C. Baillou, M. Guigon, A. Najman, F.M. Lemoine**
TNFα downregulates the expression of c-kit on normal bone marrow CD34⁺ cells
Le TNFα diminue l'expression de c-kit sur les cellules CD34⁺ de moelle normale

41 **C.A. Dinarello**
Blocking interleukin-1
Le blocage de l'interleukine-1

49 **P. Milenković, S. Stošić-Grujičić, Z. Ivanović, M. Lukić**
The effect of IL-1 receptor blockade on proliferative activity of haemopoietic stem cells in regenerating bone marrow
Effet du blocage du récepteur de l'IL-1 sur la prolifération des cellules souches hématopoïétiques dans la moelle en régénération

51 **M.C. Martyré, J. Wietzerbin**
Characterization of specific receptors for HuIFN-α on a human megakaryocytic cell line (DAMI): regulation of their expression in PMA-differentiated cells
Caractérisation des récepteurs spécifiques de l'IFNα humain sur une lignée humaine mégacaryocytaire (DAMI) : régulation de leur expression dans les cellules différenciées par le PMA

II. CELL CYCLE INHIBITORS/*INHIBITEURS DU CYCLE CELLULAIRE*

55 **J. Pouysségur, A. Brunet, J.C. Chambard, G. L'Allemain, P. Lenormand, G. Pagès**
MAP kinase cascade: an essential signalling route that controls cell proliferation
La cascade des MAP kinases : un signal essentiel dans le contrôle de la prolifération cellulaire

65 T. Lorca, J.C. Cavadore, A. Devault, D. Fesquet, J.C. Labbé, M. Dorée
Mechanisms involved in activation and inactivation of cyclin-dependent protein kinases
Mécanismes impliqués dans l'activation et l'inactivation des protéines-kinases cycline-dépendantes

75 L. Kretzner, E.M. Blackwood, J. Mac, R.N. Eisenman
Transcriptional repression by Max proteins p21 and p22
Répression transcriptionnelle par les protéines Max p21 et p22

83 M.F. Roussel, H. Matsushime, J. Kato, C.J. Sherr
CSF-1 regulation of D-type cyclins during the G1 phase of the cell cycle
Régulation par le CSF-1 des cyclines de type D pendant la phase G1 du cycle cellulaire

91 F. Zindy, J. Wang, E. Lamas, X. Chenivessel, B. Henglein, C. Bréchot
Cyclins and carcinogenesis
Cyclines et cancérogenèse

103 J. Baudier
Regulations of the wild type p53 protein functions: a role for calcium and the calcium-binding S100 proteins
Régulation des fonctions de la protéine p53 de type sauvage : rôle du calcium et des protéines S100 liées au calcium

111 S. Bi, F. Lanza, J.M. Goldman
The tumour suppressor p53 is expressed in CML CD34[+] cells and inhibition of p53 expression can affect the pattern of CFU-GM colony formation
Le suppresseur de tumeur p53 est exprimé dans les CD34[+] de LMC et l'inhibition de l'expression de p53 peut affecter la formation des CFU-GM

115 V. Rogalsky, G. Todorov, D. Moran
Retinoblastoma protein translocation. A possible mechanism of negative regulation of cell growth
Translocation de la protéine du rétinoblastome, mécanisme possible de régulation négative de la croissance

119 A.R. Mire-Sluis, L. Randall, M. Wadhwa, R. Thorpe
Transforming growth factor-β_1 mediates its inhibition of interleukin-4 induced proliferation by interfering with interleukin-4 signal transduction
Le TGF-β_1 intervient dans l'inhibition de la prolifération induite par l'IL-4 en interférant avec les signaux de transduction de l'IL-4

III. *IN VITRO* AND *IN VIVO* EFFECTS OF NEGATIVE HEMATOPOIETIC REGULATORS/*EFFETS* IN VITRO *ET* IN VIVO *DES INHIBITEURS*

123 C.J. Eaves, J.D. Cashman, A.C. Eaves
Control mechanisms for primitive hemopoietic cells
Mécanismes de contrôle des cellules hématopoïétiques primitives

133 P. Quesenberry, D. Deacon, P. Lowry
Cytokine responsiveness of murine high-proliferative potential colony-forming cells
Réponse aux cytokines des colonies à haut potentiel prolifératif de la souris

141 H.E. Broxmeyer, L. Benninger, N. Hague, P. Hendrie, S. Cooper, C. Mantel, K. Cornetta, S. Vadhan Raj, A. Sarris, L. Lu
Suppressive effects of the chemokine (macrophage inflammatory protein) family of cytokines on proliferation of normal and leukemia myeloid cell proliferation
Effets suppresseurs des chémokines (MIP) sur la prolifération myéloïde normale et leucémique

155 B. Durand, B. Samal, A.R. Migliaccio, G. Migliaccio, J.W. Adamson
Macrophage inhibitory protein 1 (MIP-1α) and leukemia inhibitory factor (LIF) increase the number of colony-forming cells (CFC) in long-term cultures of human CD34$^+$ cord blood cells
Le MIP 1α et le LIF augmentent le nombre de CFC dans les cultures à long terme de CD34$^+$ de sang de cordon

159 T.L. Holyoake, M.G. Freshney, W.P. Steward, D.J. Dunlop, E. Fitzsimons, I.B. Pragnell
Contrasting effects of rh-MIP-1α and TGF-β_1 on chronic myeloid leukemia progenitors *in vitro*

Contraste entre l'effet in vitro *du rh-MIP1α et du TGFβ₁ sur les progéniteurs de leucémie myéloïde chronique*

161 **C.M. Heyworth, J. Owen-Lynch, M.A. Pearson, M.G. Hunter, S. Craig, B.I. Lord, A.D. Whetton, T.M. Dexter**
An assessment of the cell biological and biochemical effects of the stem cell inhibitor, MIP-1α, on the multipotent cell line, FDCP-Mix A4
Evaluation des effets biologiques et biochimiques de l'inhibiteur des cellules souches MIP 1α sur la lignée multipotente FDCP-Mix A4

163 **C. Caux, J. Banchereau**
TNFα enhances the CSF dependent growth of human hematopoietic progenitors and induces their differentiation into dendritic cells
Le TNFα augmente la croissance CSF-dépendante des progéniteurs humains hématopoïétiques et induit leur différenciation en cellules dendritiques

169 **D. Bonnet, F.M. Lemoine, L. Liu, S. Pontvert-Delucq, C. Baillou, A. Najman, M. Guigon**
Effects of the tetrapetide AcSDKP (Seraspenide) on normal and malignant hematopoietic cells
Effets du tétrapeptide AcSDKP (Séraspenide) sur les cellules hématopoïétiques normales et leucémiques

177 **D. Bonnet, F.M. Lemoine, A. Najman, M. Guigon**
Comparison of the inhibitory effect of Seraspenide (AcSDKP), Tumor Necrosis Factor α (TNFα), Thymosin β4, Transforming Growth Factor β1 (TGFβ1) and Macrophage Inflammatory Protein 1α (MIP 1α) on CD34⁺ cell growth
Comparaison de l'effet inhibiteur du Séraspenide (AcSDKP), du TNFα, de la Thymosine β4, du TGFβ1 et du MIP 1α sur la croissance des cellules CD34⁺

179 **S. Robinson, M. Lenfant, J. Wdzieczak-Bakala, A. Riches**
The specificity of action of the tetrapeptide Acetyl-N-Ser-Asp-Lys-Pro (AcSDKP) in the control of haematopoietic stem cell proliferation
La spécificité d'action du tétrapeptide Acétyl-N-Ser-Asp-Lys-Pro (AcSDKP) dans le contrôle de la prolifération des cellules souches hématopoïétiques

181 K.J. Rieger, N. Saez-Servent, J. Wdzieczak-Bakala, A. Rousseau, W. Voelter, M. Lenfant

Involvement of human plasma angiotensin converting enzyme in the degradation of the tetrapeptide N-Ac-Ser-Asp-Lys-Pro (AcSDKP), an inhibitor of hematopoietic stem cell (CFU-S) proliferation

Implication de l'enzyme de conversion de l'angiotensine du plasma humain dans la dégradation du tétrapeptide N-Ac-Ser-Asp-Lys-Pro (AcSDKP), un inhibiteur de la prolifération des cellules souches hématopoïétiques (CFU-S)

183 J. Wdzieczak-Bakala, C. Grillon, S. Robinson, A. Riches, P. Carde, M. Lenfant

Degradation rate of the tetrapeptide N-Ac-Ser-Asp-Lys-Pro (AcSDKP), an inhibitor of hematopoietic stem cell (CFU-S) proliferation, in tissues from normal and leukemic mice

Etude de la dégradation du tétrapeptide N-Ac-Ser-Asp-Lys-Pro (AcSDKP), un inhibiteur de la prolifération des cellules souches hématopoïétiques (CFU-S), dans les tissus de souris normales et leucémiques

185 J. Godden, A. Riches, G. Graham, I. Pragnell

The tetrapeptide AcSDKP-preventing stimulation of stem cells into cycle

Prévention de la stimulation des cellules souches en cycle par le tétrapeptide AcSDKP

187 R. Bourette, M. Guigon, J.P. Blanchet, G. Mouchiroud

Effects of interleukin-4 (IL-4) and the tetrapeptide Acetyl-Ser-Asp-Lys-Pro (AcSDKP) on the proliferation of the multipotent FDCP-Mix cell line

Effets de l'interleukine-4 et du tétrapeptide AcSDKP sur la prolifération de la lignée multipotente FDCP mixte

189 F. Hérodin, N. Grenier, J.C. Mestries, E. Deschamps de Paillette, F. Thomas

In vivo effects of the tetrapeptide N-Acetyl-Ser-Asp-Lys-Pro (Seraspenide) on hematopoiesis of normal primates and mice

Effets in vivo du tétrapeptide Acétyl-Ser-Asp-Lys-Pro (Séraspenide) sur l'hématopoïèse de souris et de primates normaux

191 J.M. Grognet, P. Carde, F. Isnard, X. Morge, F. Hérodin, E. Ezan, P. Pradelles, F. Thomas, E. Deschamps de Paillette
Seraspenide: a peptide with hematopoietic regulatory activities. Pharmacokinetics in animals and man
Le Séraspenide, un peptide possédant des activités régulatrices de l'hématopoïèse. Pharmacocinétique chez l'animal et chez l'homme

193 L.M. Pelus, P. DeMarsh, A. King, C. Frey, P. Bhatnagar
Novel hematoregulatory peptides: monomeric and dimeric forms determine opposite biological activity
Nouveaux peptides hémorégulateurs : les formes monomère et dimère entraînent des activités biologiques opposées

201 N. Dainiak, A. Guha, R. Preston Mason
Physical interaction of a negative regulator of erythropoiesis with the membrane lipid bilayer
Interaction physique d'un régulateur négatif de l'hématopoïèse avec la bicouche membranaire lipidique

213 M.J. Koury, L.L. Kelley, M.C. Bondurant
Apoptosis as a regulatory mechanism in the late stages of hematopoiesis
L'apoptose, mécanisme régulateur des stages tardifs de l'hématopoïèse

221 S. Craig, S. Patel, D. Brotherton, S. Evans, L. Czaplewski, N. Woods, L. Howard, R. Gilbert, J. Fisher, P. Morgan
Structural characterization of recombinant LD78 the human Stem Cell Inhibitor protein
Caractérisation de la structure du LD78 recombinant, protéine de l'inhibiteur humain des cellules souches

223 E.C. Stevenson, A.E. Irvine, T.C.M. Morris, G.J. Graham
Characterisation of granulopoietic inhibitory activity
Caractérisation d'une activité inhibitrice de la granulopoïèse

225 Z.C. Han, I. Lebeurier, J.P. Caen
Negative regulation of megakaryocytopoiesis
Régulation négative de la mégacaryopoïèse

235 I. Lebeurier, J. Amiral, J.P. Caen, Z.C. Han
Inhibition of megakaryocytopoiesis by molecules related to the platelet factor 4 (PF4)

Inhibition de la mégacaryopoïèse par des molécules en relation avec le facteur plaquettaire 4

237 D. Vittet, C. Duperray, C. Chevillard
Cyclic AMP inhibits cell growth and glycoprotein Ib expression on a human megakaryocytic cell line
L'AMP cyclique inhibe la croissance cellulaire et l'expression de la glycoprotéine Ib dans une lignée humaine mégacaryocytaire

239 C. Chatelain, S. Baatout, P. Staquet, B. Chatelain, C.B. Chatelain
Inhibition of actin polymerization induces endomitosis of human megakaryocyte cell lines
L'inhibition de la polymérisation de l'actine induit des endomitoses dans des lignées mégacaryocytaires humaines

241 T. de Kretser
Cellular architecture of the foetal liver with respect to the production of haemopoietic growth factors and inhibitors
Architecture cellulaire du foie fœtal et production de facteurs stimulants et inhibiteurs de l'hématopoïèse

243 Y. Gothelf, A. Peled, N. Brosh, J. Honigwachs-Shaanani, B.C. Lee, D. Sternberg, P. Carmi, A. Levy, J. Toledo, D. Zipori
The restrictive role of stromal factors in hemopoiesis
Le rôle restrictif des facteurs d'origine stromale dans l'hématopoïèse

251 A. Peled, D. Zipori
Restrictive microenvironments: the mechanism controlling stromal-dependent myelopoiesis
Micro-environnements restrictifs : mécanismes contrôlant la myélopoïèse dépendante du stroma

253 N. Brosh, J. Honigwachs-Shaanani, D. Sternberg, A. Levi, D. Zipori
A stroma-derived protein induces apoptotic cell death of plasmacytoma cell lines
Une protéine dérivée du stroma induit la mort par apoptose de lignées plasmocytaires

255 A. Lasfar, J. Wietzerbin, C. Billard
Effects of interferons on IL-6 receptor expression in multiple myeloma cells
Effets des interférons sur l'expression des récepteurs à l'IL-6 des cellules de myélome multiple

257 C. Bouchard, D. Choquet, W.H. Fridman, C. Sautès
A recombinant soluble receptor for IgG inhibits G_1 progression of B cells stimulated by LPS or anti IgM
Un récepteur soluble recombinant des IgG inhibe la progression en G_1 de cellules B stimulées par le LPS ou des anti-IgM

259 P. Ballerini, C. Labbaye, Y.E. Cayre
Regulation of myeloblastin expression by retinoic acid in acute promyelocytic leukemia
Régulation de l'expression de la myéloblastine par l'acide rétinoïque dans la leucémie aiguë promyélocytaire

263 A. Laouar, B. Bauvois
Divergent regulation of cell surface proteases on human myeloblastic (HL-60) cells differentiated into macrophages or neutrophils. Comparison to normal myeloid cells
Régulation divergente des protéases de la surface cellulaire des cellules HL-60 différenciées en macrophages ou neutrophiles. Comparaison avec les cellules myéloïdes normales

265 D. Fardoun, E. Sumereau-Dassin, G. Lebrun-Texeira, B. Lenormand, J.P. Vannier
In vitro sensitivity of hematopoietic progenitors to NKH-1 positive cells
Sensibilité in vitro *des progéniteurs hématopoïétiques aux cellules NKH-1 positives*

267 M. Hamood, F. Corazza, W. Bujan, E. Sariban, P. Fondu
The role of Natural Killer (NK) cells on human umbilical cord blood erythropoiesis
Rôle des cellules NK dans l'érythropoïèse du sang de cordon humain

269 N. Mao, C.Y. Jiang, M.W. Zhang, X.S. Li, F.Z. Jiang, D.L. Du, P.H. Tang
Effects of cultured medium derived from purified lymphokine-activated killer cells (pLAK-CM) on survival and growth of normal blood cells and leukemia cells

Effets du milieu de culture dérivé de cellules tueuses purifiées activées par les lymphokines (pLAK-CM) sur la survie et la croissance de cellules du sang normal et de leucémies

271 **M. Gorodinsky, A. Dvilansky, I. Nathan**
In vitro effect of non-steroidal anti-estrogenic drugs on B lymphocytes from chronic lymphocytic leukemic patients
Effet in vitro d'anti-oestrogènes non stéroïdiens sur les lymphocytes B de patients atteints de leucémie lymphoïde chronique

IV. MOLECULAR APPROACHES FOR THE MANIPULATION OF HEMATOPOIETIC PROGENITORS/*APPROCHES MOLÉCULAIRES POUR LA MANIPULATION DES PROGÉNITEURS HÉMATOPOÏÉTIQUES*

275 **A.M. Gewirtz**
The c-kit protooncogene in normal and malignant human hematopoiesis
Le proto-oncogène c-kit dans l'hématopoïèse humaine normale et maligne

283 **J. Hatzfeld, M.L. Li, A. Cardoso, P. Batard, J.P. Levesque, E. Brown, H. Sookdeo, S.C. Clark, A. Hatzfeld**
Antisense oligonucleotides for inhibitors and tumor suppressor genes reveal the hematopoietic potential of quiescent progenitors
Des oligonucléotides antisens pour des inhibiteurs et gènes suppresseurs de tumeur révèlent le potentiel hématopoïétique des progéniteurs quiescents

291 **W. Stratford May, M.P. Carroll**
Antisense approach to examining the role of RAF-1 in erythropoietin action
Utilisation des antisens pour étudier le rôle de RAF-1 dans l'action de l'érythropoïétine

301 **M. Sanders, H. Lu, F. Walker, N. Dainiak**
Developmental regulation of the RAF-1 protein: potential role in erythropoietin responsiveness
Régulation au cours du développement de la protéine RAF-1 : rôle potentiel dans la capacité de réponse à l'érythropoïétine

305 F.X. Mahon, F. Belloc, A. Rice, J.M. Boiron, P. Bernard, A. Broustet, J. Reiffers
Inhibition of S phase in chronic myeloid leukemia by antisense oligonucleotides
Inhibition de la phase S dans la leucémie myéloïde chronique par des oligonucléotides antisens

307 D.E. Fisher
Positive and negative transacting factors and their modification
Facteurs positifs et négatifs agissant en trans et leurs modifications

315 J. Perreau, D. Dumenil, J.L. Escary, L. Tiret, F. Conquet, Y. Lallemand, P. Brûlet
Genetic analysis of the LIF cytokine during mouse embryogenesis
Analyse génétique du LIF au cours de l'embryogenèse de la souris

325 D. Dumenil, S. Ezine, J. Perreau, J.L. Escary, P. Brûlet
Hematopoietic disorders in LIF-mutant mice
Désordres hématopoïétiques chez les souris mutées pour le LIF

335 D. Cook, S. Kirby, T. Coffman, M. Plumb, I. Pragnell, O. Smithies
Inactivation of the Stem Cell Inhibitor (SCI) gene by homologous recombination
Inactivation du gène de l'inhibiteur des cellules souches par recombinaison homologue

V. INHIBITION OF HEMATOPOIESIS BY LEUKEMIC CELLS/*INHIBITION DE L'HÉMATOPOÏÈSE PAR LES CELLULES LEUCÉMIQUES*

341 A. Cohen, T. Grunberger, W. Vanek, M.H. Freedman
Role of cytokines in growth control of acute lymphoblastic leukemia cell lines
Rôle des cytokines dans le contrôle de la croissance de lignées leucémiques lymphoblastiques aiguës

349 C. Carlo-Stella, L. Mangoni, V. Rizzoli
Effect of interferons on normal and leukemic hematopoiesis
Effet des interférons sur l'hématopoïèse normale et leucémique

357 **A. Riches, S. Robinson**
The status of haematopoietic stem cell proliferation regulators in murine myeloid leukemia
Etat des régulateurs de la prolifération des cellules souches hématopoïétiques dans la leucémie myéloïde murine

359 **E. Sumereau-Dassin, D. Fardoun, P. Peulvé, G. Teixeira-Lebrun, J.P. Vannier**
Negative regulation of human myelopoiesis by a Burkitt lymphoblastoid cell line. A new class of inhibitor(s)?
Régulation négative de la myélopoïèse humaine par une lignée lymphoblastoïde de Burkitt. Nouvelle classe d'inhibiteur(s) ?

361 **P. Stryckmans, L. Lagneaux, C. Dorval, D. Bron, E. Bosmans, A. Delforge**
Excessive production of TGF-β by bone marrow stromal cells in B-cell chronic lymphocytic leukemia (B-CLL) inhibits growth of hematopoietic precursors and IL-6 production by the stromal cells
La production excessive de TGFβ par les cellules du stroma médullaire dans la leucémie lymphoïde chronique B inhibe la croissance des progéniteurs hématopoïétiques et la production d'IL-6 par les cellules stromales

VI. HEMATOPOIESIS AND VIRUSES/*HÉMATOPOÏÈSE ET VIRUS*

365 **E. Gluckman, G. Socié**
Mechanisms of aplastic anemia in humans
Mécanismes de l'aplasie médullaire chez l'homme

375 **N.S. Young, K.E. Brown**
The B19 parvovirus receptor: implications for disease pathogenesis
Le récepteur du parvovirus B19 : implications en pathogénie

381 **J.C. Segovia, J.M. Almendral, J.A. Bueren**
In vivo haemopoietic effects of the parvovirus MVMi in newborn mice
Effet in vivo *du parvovirus MVMi sur l'hématopoïèse de souris nouveau-nées*

383 **B. Torok-Storb, B. Fries, D. Myerson**
Cytomegalovirus mediated cytopenias
Cytopénies induites par le cytomégalovirus

391 **B.R. Davis, G. Zauli, D. Cen, V.P. Antao, M.C. Re, G. Visani, M. La Placa**
Inhibition of hematopoietic stem/progenitor cells by human immunodeficiency virus type 1
Inhibition des cellules souches/progéniteurs par le virus de l'immunodéficience humaine de type 1

399 **V. Calenda, J.C. Chermann**
HIV exerts a dual effect on hematopoietic progenitor cells: direct infection and modulatory activities on their growth
Le VIH exerce un double effet sur les progéniteurs hématopoïétiques : infection directe et modulation de leur croissance

401 **G. Zauli, M. Vitale, B.R. Davis, M.C. Re, D. Gibellini, S. Capitani, M. La Placa**
Human immunodeficiency virus type-1 triggers apoptotic cell death of hematopoietic progenitor ($CD34^+$) cells
Le virus de l'immunodéficience humaine de type 1 induit la mort par apoptose des progéniteurs $CD34^+$

403 **G.N. Schwartz, S.W. Kessler, M.L. Francis, D.L. Hoover, S.W. Rothwell, L.M. Burrell, M.S. Meltzer**
Negative regulators may mediate hematopoietic suppression in HIV-infected marrow long term cultures
Des régulateurs négatifs peuvent médier la suppression de l'hématopoïèse dans des cultures à long terme de moelle infectées par le VIH

407 **V.S. Gallichio, N.K. Hughes, K.F. Tse**
Retrovirus induced inhibition of hematopoiesis: failure to establish long-term marrow cultures using LP-BM5 murine retrovirus infected marrow cultures
Inhibition de l'hématopoïèse induite par les rétrovirus : incapacité d'établir des cultures à long terme à partir de moelle infectée par le rétrovirus murin LP-BM5

VII. BONE MARROW PROTECTION/*PROTECTION DE LA MOELLE OSSEUSE*

411 C. Hesdorffer, M. Ward, S. Podda, C. Richardson, L. Michel, L. Smith, M. Gottesman, I. Pastan, A. Bank
MDR gene transfer into human hematopoietic cells
Transfert du gène MDR dans les cellules hématopoïétiques humaines

419 R. Neta, J.M. Wang, J.J. Oppenheim, N. Davis, C.M. Dubois
Cytokine interactions in protection from lethal irradiation : synergy of IL-1 and c-kit ligand
Interactions entre les cytokines dans la protection contre l'irradiation létale : synergie entre IL-1 et c-kit ligand

429 J.R. Zucali, J. Moreb
Protective effects of TNF and IL-1
Effets protecteurs du TNF et de l'IL-1

437 W.R. Paukovits, M.H. Moser, J. Paukovits
Protection of non-self-renewing repopulating stem cells by the pentapeptide pEEDCK
Protection des cellules souches qui ne s'autorenouvellent pas et sont capables de repeupler la moelle par le pentapeptide pEEDCK

445 M. Guigon, C. Hamilton, D. Bonnet, C. Jiang, F. Lemoine, F. Isnard, A. Najman
Protective effects of AcSDKP (Seraspenide) against the *in vitro* toxicity of Asta Z to hematopoietic and stromal cells
Effets protecteurs d'AcSDK (Séraspenide) contre la toxicité in vitro *de l'Asta Z pour les cellules hématopoïétiques et les cellules stromales*

455 P.K. Wierenga, A.W.T. Konings
Effect of the tetrapeptide AcSDKP on the hyperthermic sensitivity of the murine hematopoietic stem cell compartment
Effet du tétrapeptide AcSDKP sur la sensibilité à l'hyperthermie du compartiment des cellules souches de la souris

457 C. Coutton, A. Bohbot, M. Guigon, F. Oberling
Photoprotection of the hematopoietic progenitors by the tetrapeptide N-AcSDKP

Photoprotection des progéniteurs hématopoïétiques par le tétrapeptide N-AcSDKP

459 P. Carde, N. Mathieu-Tubiana, E. Vuillemin et al.
Phase I-II trial of Seraspenide: a supressor of myelopoiesis protects against chemotherapy myelotoxicity
Essai phase I-II du Séraspenide : un inhibiteur de la myélopoïèse protège de la myélotoxicité de la chimiothérapie

469 L. Douay, C. Hu, M.C. Giarratana, S. Bouchet, D. Bardinet, N.C. Gorin
WR 2721 (Ethiofos) protects normal progenitor/stem cells from cyclophosphamide derivatives toxicity, with preservation of their anti-leukemic effects: application to *in vitro* marrow purging
Le WR 2721 (Ethiofos) protège les progéniteurs/cellules souches normaux contre la toxicité de dérivés du cyclophosphamide, avec préservation de leurs effets antileucémiques : application à la purge de moelle in vitro

477 B. Sredni, M. Albeck, Y. Kalechman
Immunomodulation by AS101: mode of action and clinical applications in cancer therapy
Immunomodulation par l'AS101 : mode d'action et applications cliniques dans le traitement du cancer

485 V. Kravtsov, I. Fabian
A new macrophage-associated activity protects mice from radiation induced death
Une nouvelle activité associée aux macrophages protège les souris contre la mort par irradiation

487 A. Najman
Chemotherapy bone marrow toxicity: stimulation or/and protection of stem cells
Toxicité médullaire de la chimiothérapie : stimulation et/ou protection des cellules souches

493 Author index
Index des auteurs

I. Receptors and their inhibitors

I. Récepteurs et leurs inhibiteurs

The negative regulation of hematopoiesis. Ed. M. Guigon et al. Colloque INSERM/John Libbey Eurotext Ldt.
© 1993, Vol. 229, pp. 3-13.

The erythropoietin receptor: structure, function and disease

Harvey F. Lodish, Stephanie S. Watowich, Gregory D. Longmore*, and Douglas J. Hilton**

*Whitehead Institute for Biomedical Research, Nine Cambridge Center, Cambridge, MA 02142, USA; The Department of Biology, Massachusetts Institute of Technology, Cambridge, MA 02139, USA; *Washington University School of Medicine, Division of Hematology-Oncology, St. Louis, MO 63110, USA; **The Walter and Eliza Hall Institute for Medical Research, P.O. Royal Melbourne Hospital, Parkville 3050, Victoria, Australia*

Here we focus on several aspects of the erythropoietin receptor (EPO-R): on its role in erythroid differentiation, on the structure of the receptor, on the nature of the signals generated by its cytosolic domain and, most importantly, on two ways in which the receptor can be activated for proliferation independent of the presence of erythropoietin. At least in animal models, such hormone-independent activation of the EPO-R is the first stage in the development of erythroleukemia.

Introduction

Erythropoietin (EPO) is a serum glycoprotein hormone required for the survival, proliferation, and differentiation of committed erythroid progenitor cells (Krantz, 1991), and is the principal hormone that regulates the level of circulating red cells. Its synthesis by kidney glomerular cells is induced by low levels of blood oxygen, such as occurs in normal individuals at high elevations. The administration of recombinant human EPO to anemic patients suffering from chronic renal failure, AIDS, or bone marrow suppression due to chemotherapy has dramatically alleviated their need for blood transfusions (Fischl, 1990).

The erythropoietin receptor (Figure 1) is a member of the cytokine receptor superfamily (Cosman et al, 1990; Bazan, 1990). Conserved structural features of the cytokine receptor superfamily include four similarly spaced cysteine residues and a sequence motif, WSXWS, located in the exoplasmic domain. The EPO-R and other members of the cytokine receptor family do not contain kinase-related sequences in their cytoplasmic domains and the intracellular signalling pathways they initiate following ligand binding have yet to be defined. The ligand binding subunits of the granulocyte colony stimulating factor receptor, prolactin receptor, growth hormone receptor, and, as we shall see, the EPO-R, form homodimers (Fukunaga et al, 1990; De Vos et al, 1992; Fuh et al, 1992). In contrast to many other hematopoietic growth factors that affect many kinds of cells, EPO acts primarily on erythroid progenitors.

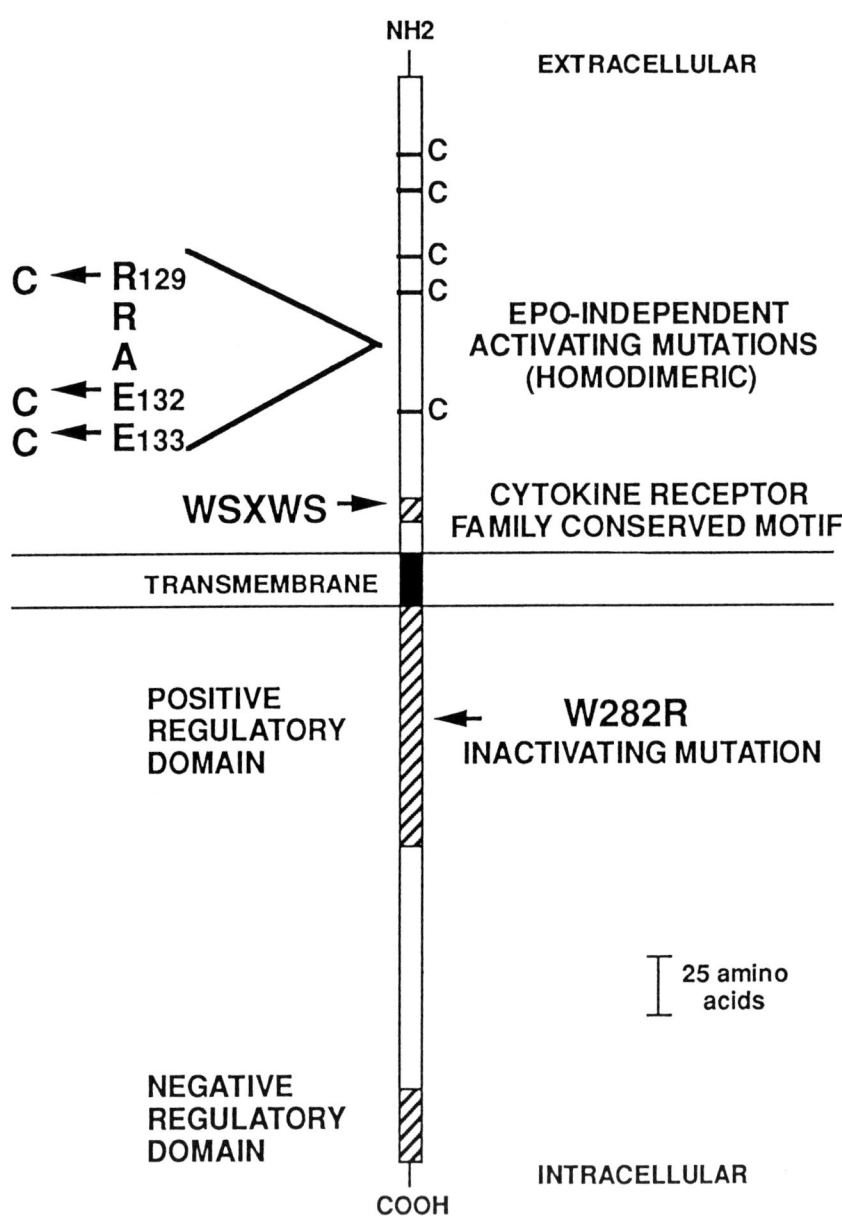

Figure 1. Functional organization of the EPO-R molecule. The schematic diagram shows the relative position of the four conserved Cys residues (black bars; C), the fifth Cys, closest to the membrane, which is not present in all members of the cytokine receptor superfamily, the activating mutations R129C, E132C, and E133C, the WSXWS motif, and the negative and positive regulatory domains of signal transduction.

Erythropoietin receptors and erythroid differentiation.

Within the fetal liver and adult bone marrow, pluripotential hematopoietic stem cells generate erythroid progenitors. The earliest committed erythroid progenitors identified by in vitro cell culture are the slowly proliferating burst-forming unit-erythroid (BFU-E) cells, which are not responsive to EPO and do not express an EPO receptor (EPO-R) (Sawada et al, 1990). After growth in the presence of additional growth factors (GM-CSF, IL-3 or SCF), "mature" BFU-E's appear which do express the EPO-R and are weakly responsive to EPO. After further proliferation in culture these give rise to colony-forming units-erythroid (CFU-E) which are highly responsive to EPO (Sawada et al, 1990) and which generate erythroblast colonies in 2 days in mouse, 7 days in humans. The sensitivity of these erythroid progenitors to EPO is transient. Beyond the late basophilic erythroblast stage, a nondividing cell, the level of the EPO-R drops and the cells are no longer dependent on EPO for continued maturation (Koury and Bondurant, 1988). Whether Epo directly affects erythroid differentiation, or is simply a proliferative/apoptotic factor for erythroid precursors, is not clear.

Other proteins associated with the cell-surface EPO-R

Initial binding studies on COS cells transiently transfected with the cloned receptor detected two affinities for EPO (K_d = 30pM and 210pM) (D'Andrea et al, 1989). However, subsequent studies have detected a single binding affinity for EPO in heterologous hematopoietic cells, or fibroblasts or COS cells transfected with the EPO-R cDNA (K_d = 300-800pM) (D'Andrea et al, 1990; D'Andrea et al, 1991; Miura et al, 1991; Miura and Ihle, 1993). Similarly, erythroleukemia cell lines or cells expressing a constitutively active form of the receptor (see below), bind EPO with a single affinity (Watowich et al, 1992). Binding studies on erythroid progenitor cells have detected either one (Mayeux et al, 1987; Broudy et al, 1991) or two (Sawada et al, 1988; Landschulz et al, 1989) affinities for radioiodinated EPO. Thus, whether there indeed are EPO-Rs of differing affinities is unresolved, although two affinities may suggest the presence of a second subunit of the EPO-R.

Proteins of ~100kD and ~85kD that cross-link to EPO are detected on the surface of a variety of cell types examined and limited peptide mapping suggests that these two proteins may be related (Sawyer, 1989). One recent study demonstrates that EPO-cross-linked proteins of ~110kD and ~75kD are immunologically related to the cloned EPO-R (Miura and Ihle, 1993) while a second suggests that the ~100kD and ~85kD proteins are unrelated to the isolated receptor (Mayeux et al, 1991). A third component of ~95kD, specific to hematopoietic cells, was also detected by EPO cross-linking (Miura and Ihle, 1993). Since this protein is not related to the cloned receptor it may be another cell surface molecule which interacts with the EPO-R. Clearly, further work remains to delineate the molecular structure of the cell-surface EPO-R complex.

Signaling by the erythropoietin receptor

Stable expression of the EPO-R in IL3-dependent murine hematopoietic cells - such as lymphoid Ba/F3 cells and myeloid/erythroid progenitor FDC-P1 cells and DA-3 cells - allows them to grow in the presence of either EPO or IL3, indicating that the cloned EPO-R is capable of generating a proliferative signal in these cells. Similarly, expression of the receptors for IL6, GM-CSF, and G-CSF in Ba/F3 cells also allows the corresponding hormones to replace IL3 in supporting growth, suggesting that a number of receptors can generate a similar growth-promoting signal.

Although the cytosolic domain of the EPO-R does not contain an obvious protein kinase, protein phosphorylation does appear to be important for EPO-induced proliferation (Miura et al, 1991; Miura et al, 1993). Several proteins are phosphorylated in response to EPO addition, including the EPO-R, and it is likely that intracellular tyrosine or serine/threonine kinases are activated by the receptor which in turn generate downstream signals to stimulate proliferation Figure 2). Both EPO and IL3 induce rapid phosphorylation of tyrosine residues of several common and several unique cellular proteins in EPO-dependent cell lines.

Several studies have implicated regions of the EPO-R cytoplasmic domain which are crucial for EPO-dependent proliferation (D'Andrea et al, 1991).In an effort to delineate the regions essential for signal transduction, deletions have been introduced in the cytoplasmic domain of the EPO-R. Two regions of opposing function were defined: a membrane-proximal region of about 100 amino acids that is sufficient for transduction of proliferative signals, and a distal region that down-modulates such signals (Fig.1). The proximal region is homologous to the corresponding segment of the IL2-R ß-chain, which is also sufficient for signal transduction. Mutant EPO-Rs lacking the entire cytoplasmic domain, with an internal deletion of 20 amino acids or, interestingly, with a single mutation of a strictly conserved Trp residue in the membrane-proxiamal part of the cytoplasmic domain (W282R), are non-functional in cell-proliferation assays. The W238R receptor is transported to the cell surface and is able to bind radio-iodinated EPO, but cannot signal EPO-dependent proliferation in IL-3 dependent DA-3 cells (Miura et al, 1993). It is essential to identify proteins interacting with the EPO-R and determine the role they play in EPO-induced proliferation; some of these proteins may be specific for the EPO-R while others may interact with several cytokine receptors. That the EPO-R stimulates proliferation of transfected IL-3 dependent cells but not of transfected fibroblasts suggests that hematopoietic-specific kinases may be part of the signal transduction pathway employed by several members of the cytokine receptor superfamily (Longmore and Lodish, 1991).

Erythroblastosis caused by a mutant human erythropoietin receptor

Deletion of the carboxy-terminal 40 or 90 amino acids of the EPO-R allows Ba/F3 cells to proliferate in as little as 1 pM EPO, one-tenth that required for proliferation of cells expressing the wild-type EPO-R. Since the deletion had no effect on the number or affinity of cell-surface EPO-Rs, occupancy of fewer cell-surface receptors by EPO then in cells expressing the wild-type EPO-R was sufficient to generate the growth-promoting intracellular signal.(D'Andrea et al, 1991; Yoshimura and Lodish, 1992).

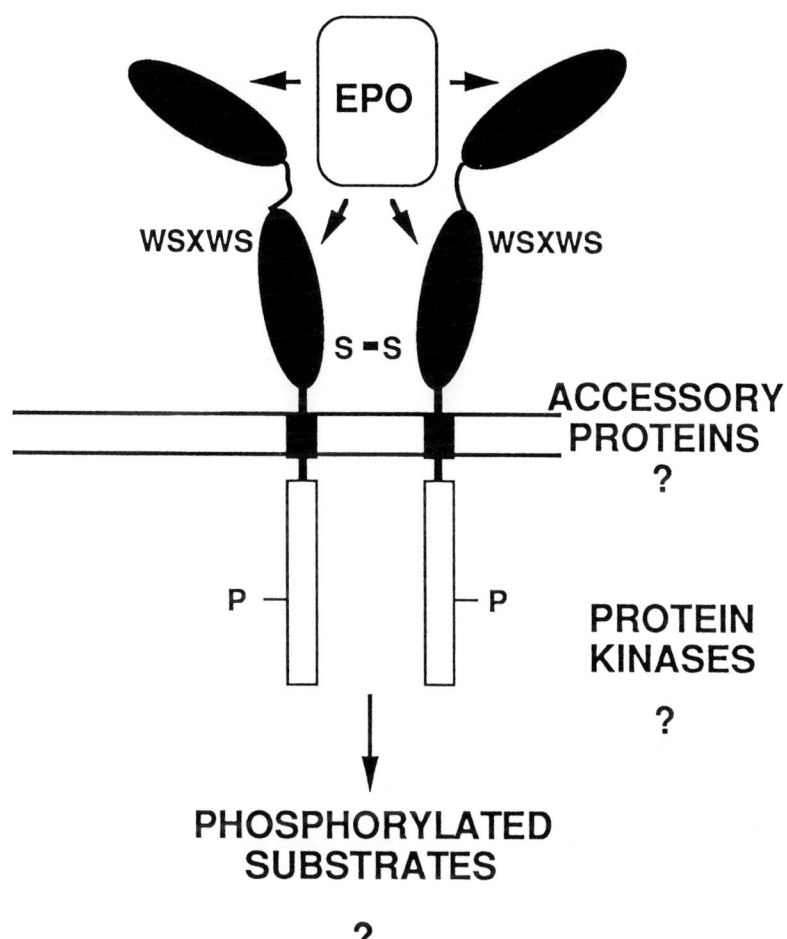

Figure 2. Schematic model of the EPO-R complex. By analogy to the GH-R complex, a single molecule of EPO is thought to bind to the extracellular domains of an EPO-R dimer. The EPO binding site is distinct from the WSXWS motif. Additional protein-protein interactions occur within the transmembrane domain of EPO-R (e.g. with gp55), and with putative protein kinases and the EPO-R cytoplasmic domain. Phosphorylation (P) of both the EPO-R and additional substrates may occur as a result of the mitogenic signal. By analogy with the structure of the growth-hormone receptor, the amino acids that generate the activating mutations R129C, E132C, and E133C, are thought to be at the dimer interface, schematically shown here as a disulfide bond connecting the two receptor subunits.

Affected members of a large Finnish family with autosomal dominant benign erythrocytosis have a mutation in one allele of the EPO-R, which introduces a premature stop codon and generates an EPO-R lacking the carboxy-terminal 70 amino acids (de la Chapelle et al, 1993). Cultured bone marrow erythroid progenitors from these individuals are hypersensitive to EPO (Juvonen et al, 1991). Thus, carboxy-terminal deletions have phenotypically the same effects on the murine and human EPO-Rs, and in humans contribute, in a dominant fashion, to a mild pathologic condition. Deletion of the carboxy-terminal 40 amino acids of the EPO-R reduces receptor tyrosine phosphorylation (Yoshimura and Lodish, 1992), suggesting that a phosphorylated tyrosine in this region may serve as a recognition site for a protein which negatively modulates EPO-R signaling. Alternatively, truncated EPO-Rs may preferentially interact with positive signaling molecules. Assuming that the functional EPO-R is a homodimer (see below), we do not know whether heterodimers of the wild-type and carboxy-terminal truncated EPO-Rs form and whether they are functionally distinct from wild-type receptor homodimers.

Activation of the erythropoietin receptor by a retroviral envelope protein - the first step in virus-induced erythroleukrmia

The role of growth factors and their receptors in the leukemic process has been the focus of a great deal of investigation. Constitutive (dysregulated) overexpression of hematopoietic growth factors, such as EPO, GM-CSF, and IL-3 in mice results in purely proliferative disorders, not leukemia (Chang et al, 1989; Semenza et al, 1989). On the other hand activating mutations in some hematopoietic growth factor receptors, such as the CSF-1 receptor and the EPO-R have, in experimental murine models, led to the development of leukemia in a multistep process (Heard et al, 1987; Longmore and Lodish, 1991). Specifically, the EPO-R has been implicated in two modes of leukemia induction in mice.

Friend virus, an acutely transforming murine erythroleukemia retrovirus, has served as a model for the study of the multi-step nature of leukemia (Ben-David and Bernstein, 1991). A replication-defective spleen focus-forming virus (SFFV) and a replication-competent Friend murine leukemia virus (F-MuLV) comprise the Friend virus complex. SFFV induces erythroleukemia in newborn and adult mice whereas F-MuLV induces leukemia only when injected into newborn mice, and only after a long latency period. One to two weeks following infection with SFFV mice develop polyclonal erythroblastosis and erythrocytosis. By five to six weeks the spleen contains growth factor- independent, erythroleukemic cells. The *env* gene of SFFV encodes a recombinant/deletion mutant membrane glycoprotein, gp55, that is directly involved in leukemogenesis (Li et al, 1986; Aizawa et al, 1990). Co-expression of the murine EPO-R and the gp55 envelope protein of SFFV in the IL-3 dependent pro-B cell line BA/F3 (Li et al, 1990), and in the EPO-dependent erythroleukemic cell line HCD57 (Ruscetti et al, 1990), confers hormone-independent cell proliferation. The EPO-R and gp55 interact, as judged by co-immunoprecipitation, in the endoplasmic reticulum (Yoshimura et al, 1990) as well as the cell surface (Casadevall et al, 1991): the vast majority of both gp55 and the EPO-R is retained in intracellular compartments (Yoshimura et al, 1990). The cell surface, disulfide-linked dimeric form of gp55 is the biologically active form, and can be cross-linked to radioiodinated EPO (Casadevall et al, 1991). Since gp55 itself cannot bind EPO, these results demonstrate that gp55 molecules are associated with cell surface EPO-Rs, and that the gp55 binding site on the EPO-R is distinct from the ligand-binding domain (Casadevall et al, 1991). Thus, induction of proliferation by the gp55 gene of SFFV is restricted to those cells that express a functional EPO-R. This association between the EPO-R and gp55 renders the receptor active in the absence of EPO and is likely to be the principle cause of the initial stage of EPO-independent erythroblastosis in Friend disease.

Factor-independent, disulfide-bonded homodimeric mutants of the EPO-R

Dimerization of the erythropoietin receptor (EPO-R) may be critical for signal transduction, as suggested by the properties of a mutant EPO-R (R129C) that does not require EPO for cell proliferation and that contains an arginine to cysteine mutation in the exoplasmic domain (Figure 2). Since R129E, R129P and R129S receptors are wild-type with respect to EPO-dependent proliferation, the presence of cysteine and not the loss of arginine at position, 129 is necessary for hormone-independent proliferation. All constitutively active EPO-Rs, containing the R129C mutation, form disulfide-linked homodimers found both in internal membranes and on the plasma membrane (Watowich et al, 1992). R129C receptors display a single affinity for EPO, similar to that of the wild-type receptor (K_D = 700 pM). We hypothesized that the disulfide-linked dimers might mimic the structure of the hormone-bound form of the EPO-R, and thus transmit a constitutive proliferative signal(Watowich et al, 1992).

In this respect, the growth hormone (GH) receptor (GH-R) may be similar to the EPO-R. The GH-R forms an unusual structure since two receptor molecules are complexed to a single, nonsymmetric ligand molecule (De Vos et al, 1992). The exoplasmic region of the GH-R receptor has two domains, each of which is built of seven ß-strands, and similar binding determinants in each receptor monomer interact with distinct sites (site I and site II) on GH (De Vos et al, 1992). Receptor dimerization is essential for signal transduction and occurs by sequential binding; one GH-R molecule binds to site I on GH and then a second binds to site II (Fuh et al, 1992). Residues in the membrane-proximal domains of the two GH-R monomers (termed here the dimer interface) interact with each other and stabilize the ligand-bound dimer (De Vos et al, 1992)

The extracellular domains of the GH and EPO receptors share a similar genomic organization, have approximately 20% amino acid identity, and are thought to form similar tertiary structures. The GH-R crystal structure demonstrates that many of the highly conserved residues maintain the three-dimensional structure of the molecule, making it likely that other members of the family adopt a similar conformation. By aligning the sequences of the GH-R and EPO-R exoplasmic domains exon by exon, we predicted that residue 129 of the EPO-R would be localized in the dimer interface region, and that the disulfide bond in R129C covalently links two subunits to generate a constitutively active receptor which mimics the ligand-bound form of the receptor (Figure 2). Several other residues- 130, 131, 132, and 133 - of the EPO-R were also predicted to be at the dimer interface. Mutants R130A, R130C, A131C, E132A, E133A functioned identically to the wild-type EPO-R. Importantly, 32-D cell clones expressing E132C and E133C were able to grow in the absence of added growth factor, and the mutant receptors also formed disulfide-linked dimers (manuscript submitted for publication), thus validating the model in Figure 2.

Transformation of hematopoietic progenitor cells induced by a factor-independent mutant of the EPO-R

If the *env* gene of SFFV is replaced by a cDNA encoding a constitutively active form of the erythropoietin receptor, EpoR(R129C), the resultant recombinant virus SFFVcEpoR also induces erythrocytosis and leukemia in adult mice (Longmore and Lodish, 1991). An analysis of mice infected by the SFFVcEpoR virus indicated that, in contrast to the gp55 of SFFV, EpoR(R129C) can signal proliferation of a number of types of hematopoietic progenitors and can induce transformation of cells other than erythroid progenitors. The increase in circulating erythroid cells following infection by SFFVcEpoR was one to two weeks later than seen after infection by SFFV, and, in contrast to infection by SFFV, there was a transient rise in platelet count. Thus the target cell of SFFVcEpoR may be a multipotent progenitor with predominate erythroid-megakaryocytic features, explaining the early thrombocytosis and delayed erythrocytosis (manuscript submitted for publication). Supporting this were results of clonogenic progenitor cell assays from the bone marrow of SFFVcEpoR infected mice. Within three weeks after infection there was an increase in the numbers of IL-3/IL-6- dependent granulocyte-macrophage, mixed myeloid-megakaryocytic, and megakaryocytic colonies. Marrow from mice five weeks after infection generated megakaryocytic, erythroid, granulocyte-macrophage, and mixed colonies in the absence of added growth factors, and several factor-independent leukemic cell lines, derived from the spleens of infected mice, exhibited properties of primitive erythroid, lymphoid, and monocytic cells. Some expressed proteins characteristic of more than one lineage. Thus, when aberrantly expressed in vivo, the EpoR is capable of transducing functional growth signals in nonerythroid as well as early erythroid progenitor cells (manuscript submitted for publication).

Given the importance of members of the cytokine receptor superfamily in controlling proliferation of all hematopoietic cells, it is striking that hormone-independent activation of receptors other than the EPO-R have not yet been described. Activated EPO-R's were detected by functional expression in IL-3- dependent hematopoietic cell lines, and similar analyses of cytokine receptors cloned from other types of human and animal tumors might reveal constitutively active mutants of other cytokine receptors.

References

Aizawa, S., Suda, Y., Furata, Y., Yagi, T., Takeda, N., Watanabe, N., Nagayoshi, M. and Ikawa, Y. (1990): Env-derived gp55 gene of Friend spleen focus-forming virus specifically induces neoplastic proliferation of erythroid progenitors. *EMBO J.* **9,** 2107-2116.

Bazan, J. F. (1990): Structural design and molecular evolution of a cytokine receptor superfamily. *Proc. Natl. Aced. Sci. USA* **87,** 6934-6938.

Ben-David, Y. and Bernstein, A. (1991): Friend virus-induced erythroleukemia and the multistage nature of cancer. *Cell* **66,** 831-833.

Broudy, V. C., Lin, N., Brice, M., Nakamoto, B. and Papyannopoulou, T. (1991): Erythropoietin receptor characteristics on primary human erythroid cells. *Blood* **77,** 2583-2590.

Casadevall, N., Lacombe, C., Muller, O., Gisselbrecht, S. and Mayeux, P. (1991): Multimeric structure of the membrane erythropoiten receptor of murine erythroleukemia cells (Friend cells). *J. Biol. Chem.* **266**, 16015-16020.

Chang, J. M., Metcalf, D., Lang, R. A., Gonda, T. J. and Johnson, G. R. (1989): Nonneoplastic hematopoietic myeloproliferative syndrome induced by dysregulated multi-CSF (IL-3) expression. *Blood* **73**, 1487-1497.

Cosman, D., Lyman, S. D., Idzerda, R. L., Beckman, M. P., Park, L. S., Goodwin, R. G. and March, C. J. (1990): A new cytokine receptor superfamily. *Trends Biochem. Sci.* **15**, 256-268.

D'Andrea, A. D., Fasman, G. D., Zon, L. I., Li, J. P. and Lodish, H. F. (1990): Structure of the erythropoietin receptor in stable fibroblast transfectants. *Prog Clin Biol Res* **352**, 153-159.

D'Andrea, A. D., Lodish, H. F. and Wong, G. G. (1989): Expression cloning of the murine erythropoietin receptor. *Cell* **57**, 277-285.

D'Andrea, A. D., Yoshimura, A., Youssoufian, H., Zon, L. I., Koo, J. W. and Lodish, H. F. (1991): The cytoplasmic region of the erythropoietin receptor contains nonoverlapping positive and negative growth-regulatory domains. *Mol Cell Biol* **11**, 1980-1987.

de la Chapelle, A., Traskelin, A.-L. and Juvonen, E. (1993): Truncated erythropoietin receptor causes dominantly inherited benign human erythrocytosis. *Proc. Natl. Acad. Sci. USA* **90**, 4495-4499.

De Vos, A. M., Ultsch, M. and Kossiakoff, A. A. (1992): Human growth hormone and extracellular domain of its receptor: crystal structure of the complex. *Science* **255**, 306-311.

Fischl, M. (1990): Recombinant human erythropoietin for patients with AIDS treated with Zidovudine. *N. Engl. J. Med.* **322**, 1488-1493.

Fuh, G., Cunningham, B. C., Fukunaga, R., Nagata, S., Goeddel, D. V. and Wells, J. A. (1992): Rational design of potent antagonists to the human growth hormone receptor. *Science* **256**, 1677-1680.

Fukunaga, R., Ishizaka-Ikeda, E. and Nagata, S. (1990): Purification and characterization of the receptor for murine granulocyte colony-stimulating factor. *J. Biol. Chem.* **265**, 14008-14115.

Heard, J. M., Roussel, M. F., Rettenmier, C. W. and Sherr, C. J. (1987): Multilineage hematopoietic disorders induced by transplantation of bone marrow cells expressing the v-fms oncogene. *Cell* **51**, 663-673.

Juvonen, E., Ikkala, E., Fyhrquist, F. and Ruutu, T. (1991): Autosomal Dominant Erythrocytosis caused by increased sensitivity to erythropoietin. *Blood* **78**, 3066-3069.

Koury, M. J. and Bondurant, M. C. (1988): Maintenance by erythropoietin of viability and maturation of murine erythroid precursor cells. *J. Cell. Physiol.* **137**, 65-73.

Krantz, S. (1991): Erythropoietin. *Blood* **77**, 419-429.

Landschulz, K. T., Noyes, A. N., Rogers, O. and Boyer, S. H. (1989): Erythropoietin receptors on murine erythroid colony-forming units: Natural history. *Blood* **73**, 1476-1485.

Li, J.-P., Bestwick, R. K., Machida, C. and Kabat, D. (1986): Role of membrane glycoprotein in Friend virus erythroleukemia: nucleotide sequences of nonleukemogenic mutant and spontaneous revertant viruses. *J. Virol.* **57**, 534-538.

Li, J. P., D'Andrea, A. D., Lodish, H. F. and Baltimore, D. (1990): Activation of cell growth by binding of Friend spleen focus-forming virus gp55 glycoprotein to the erythropoietin receptor. *Nature* **343**, 762-764.

Longmore, G. D. and Lodish, H. F. (1991): An activating mutation in the murine erythropoietin receptor induces erythroleukemia in mice: a cytokine receptor superfamily oncogene. *Cell* **67**, 1089-1102.

Mayeux, P., Billat, C. and Jacquot, R. (1987): The erythropoietin receptor of rat erythroid progenitor cells. *J. Biol. Chem.* **262**, 13985-13992.

Mayeux, P., Lacombe, C., Casadevall, N., Chretien, S., Dusanter, I. and Gisselbrecht, S. (1991): Structure of the murine erythropoietin receptor complex. *J. Biol. Chem.* **266**, 23380-23385.

Miura, O., Cleveland, J. L. and Ihle, J. N. (1993): Inactivation of the erythropoietin receptor function by a point mutation in a region having homology with other cytokine receptors. *Mol. Cell. Biol.* **13**, 1788-1795.

Miura, O., D'Andrea, A. D., Kabat, D. and Ihle, J. N. (1991): Induction of tyrosine phosphorylation by the erythropoietin receptor correlates with mitogenesis. *Mol. Cell. Biol.* **11**, 4895-4902.

Miura, O. and Ihle, J. N. (1993): Subunit structure of the erythropoietin receptor analyzed by 125I-Epo Cross-linking in cells expressing wild type or mutant receptors. *Blood* **81**, 1739-1744.

Ruscetti, S. K., Janesch, N. J., Chakraborti, A., Sawyer, S. T. and Hankins, W. D. (1990): Friend spleen focus-forming virus induces factor independence in an erythropoietin-dependent erythroleukemia cell line. *J. Virol.* **64**, 1057-1064.

Sawada, K., Krantz, S. B., Dai, C. H., Koury, S. T., Horn, S. T., Glick, A. D. and Civin, C. I. (1990): Purification of human blood burst forming units-erythroid and demonstration of the evolution of the erythropoietin receptor. *J. Cell. Physiol.* **142**, 219-230.

Sawada, K., Krantz, S. B., Sawyer, S. T. and Civin, C. I. (1988): Quantitation of specific binding of erythropoietin to human erythroid colony-forming cells. *J. Cell. Physiol.* **137**, 337-445.

Sawyer, S. T. (1989): The two proteins of the erythropoietin receptor are structurally similar. *J. Biol. Chem.* **264**, 13343-13347.

Semenza, G. L., Traystman, M. D., Gearhart, J. D. and Antonarakis, S. E. (1989): Polycythemia in transgenic mice expressing the human erythropoietin gene. *Proc. Natl. Acad. Sci. USA.* **86**, 2301-2305.

Watowich, S. S., Yoshimura, A., Longmore, G. D., Hilton, D. J., Yoshimura, Y. and Lodish, H. F. (1992): Homodimerization and constitutive activation of the erythropoietin receptor. *Proc Natl Acad Sci USA* **89**, 2140-2144.

Yoshimura, A., D'Andrea, A. D. and Lodish, H. F. (1990): Friend spleen focus-forming virus glycoprotein gp55 interacts with the erythropoietin receptor in the endoplasmic reticulum and affects receptor metabolism. *Proc Natl Acad Sci U S A* **87**, 4139-4143.

Yoshimura, A. and Lodish, H. F. (1992): In vitro phosphorylation of the erythropoietin receptor and an associated protein, pp130. *Mol Cell Biol* **12**, 706-715.

Résumé

Cet article concerne plusieurs aspects du récepteur de l'érythropoïétine: son rôle dans la différenciation érythroïde, la structure du récepteur, la nature des signaux générés par son domaine cytosolique et surtout, les deux voies par lesquelles le récepteur peut être activé en vue d'une prolifération indépendante de l'érythropoïétine. Au moins dans les modèles animaux, il apparait qu'une activation hormone-indépendante du récepteur de l'érythropoïétine est le premier stade de développement d'une érythroleucémie.

Negative and positive control of erythrocytic progenitor cells by the nuclear receptors for thyroid hormone T3 and retinoic acid: a target for v-erbA oncogene and other mutated nuclear receptors

Olivier Gandrillon, Madeleine Garcia, Nathalie Ferrand and Jacques Samarut

Laboratoire de Biologie Moléculaire et Cellulaire, CNRS UMR49, INRA, Ecole Normale Supérieure de Lyon, 46, allée d'Italie, 69364 Lyon Cedex 7, France

Summary

Retinoic acid (RA) and the thyroid hormone triiodothyronine (T3) exert a wide range of biological effects related to cell proliferation and differentiation. We have focused on their role in the control of erythrocytic differentiation in the chicken. We describe here the use of an original model system to study the differentiative influences of RA and T3. The results we obtained led us to propose a schematical 3-phase model of the erythrocytic differentiation pathway. We then discuss the respective contributions of the study of v-erbA, the oncogenic counterpart of T3 receptor and of the β receptor of RA to our understanding of the molecular processes controlling the self-renewal and the differentiation of erythrocytic progenitor cells.

I) The role of T3 and RA in the erythrocytic differentiation pathway.

Initial in vitro experiments showed that T3 is a positive modulator of erythrocytic differentiation in the mammals and in the chicken (Popovic et al., 1977; Dainiak et al., 1978; Schroeder et al., 1992). Interestingly, retinoic acid (RA) was also shown to modulate erythrocytic differentiation in vitro and in vivo (Correa and Axelrad, 1992; Visani et al., 1992) and to cooperate with T3 for its action (Schroeder et al., 1992). Two majors questions were nevertheless left unanswered; (i) by which mechamism do those hormones act? and (ii) are those hormones dispensable co-regulators of the differentiation process or do the activation of the corresponding receptors represent a critical step in that process?

1.) The model system: the TECs.

In order to investigate the precise function of T3 and RA in erythrocytic differentiation we used a recently described in vitro model in the chicken. This model consists of semi-solid cultures of embryonic bone marrow cells grown in the presence of Transforming Growth Factor α (TGF-α) (Pain et al., 1991). This growth factor stimulates the self-renewal of chicken progenitor cells at both the BFU-E and early CFU-E stages of differentiation. These cells are called TGF-α-induced erythrocytic cells (TECs). It has recently been shown that estradiol can cooperate with TGF-α for the self-renewal of TECs (Schroeder et al., 1993).

TECs can be grown in an undifferentiated state and can be induced to differentiate at any time into mature erythrocytes by the addition of Erythropoietin plus Insulin (Pain et al., 1991; Schroeder et al., 1992) thereby demonstrating that they represent a population of normal hormone-responsive cells. Hallmarks of this model are that it uses normal hematopoietic cells and that it allows to clearly distinguish between an early self-renewal phase in which proliferation is not coupled to differentiation and a late differentiation phase during which limited proliferation is strictly coupled to maturation into erythrocyte. The percentage of cells in the self-renewing compartment can be directly measured by estimating the subcloning ability of cells reseeded in semi-solid medium.

2.) Both T3 and RA inhibit the self-renewal of erythrocytic progenitor cells (Gandrillon et al., submitted).

The vast majority of T3- and RA-induced biological effects is believed to result from the recognition, binding and activation of the respective nuclear receptors. We thus first studied the expression of the different receptors in TECs. We found by RT-PCR analysis that TECs expressed the mRNAs encoding T3Rα, RARα and RARγ. The RARβ mRNA was detected only after RA treatment.

We then were able to show that both T3 and RA severely reduced the subcloning ability of TECs and therefore induced a strong inhibition of the self renewal potential of those erythrocytic progenitor cells. Different non-mutually exclusive mechanisms can account for such an effect. T3 and RA could either (i) induce the self-renewing cells to enter a quiescent state or (ii) induce the self-renewing cells to die or (iii) induce those cells to differentiate beyond a point where they would loose their self-renewing ability. The last hypothesis would be in line with the positive modulation of chicken erythropoiesis proposed by Schroeder et al. (1992) and to our own results (see below).

3.) A role for RA and T3 in the commitment phase of the differentiation process?

In order to clarify the role played by RA and T3 in the differentiation process, we induced the differentiation of TECs in a culture medium containing erythropoietin plus insulin that was specifically depleted in endogenous RA and T3 by the addition of specific antibodies. The results we obtained indicate that absolutely no differentiation occurs in such an hormone-depleted medium and that a normal differentiation rate can be recovered by the addition of all-trans RA, T3 or 9-cis-RA. This suggests that T3 and RA on one side and erythropoietin plus insulin on the other side are necessary but not sufficient to induce the differentiation of TECs. Preliminary results indicate that T3 and RA are only needed during the initial commitment phase.

4.) The working hypothesis.

Taken together all of our data led us to propose a model for the erythrocytic differentiation (figure 1). RARs, RXRs and T3Rα are depicted as factors which trigger TECs from their selfrenewing program into a commitment phase where the cells further proceed through final differentiation in the presence of insulin and erythropoietin. Cells in the commitment phase are induced to express the RARβ gene in response to RA and the carbonic anhydrase II gene in response to T3 which shows that they have engaged a new genetic program. Beyond the commitment phase erythropoietin and insulin drive the cells into the final differentiation program during which maturation into erythroblasts takes place through a limited number of cell divisions. This model can now be validated only by upravelling the genetical programs that are operating under each of those phases. One might imagine that each specific factor combinations control either one (or a few) master genes or an entire bank of genes, each of which plays out its own peculiar and discrete role in the combined phenomenon of erythrocytic differentiation. The identification of those genes

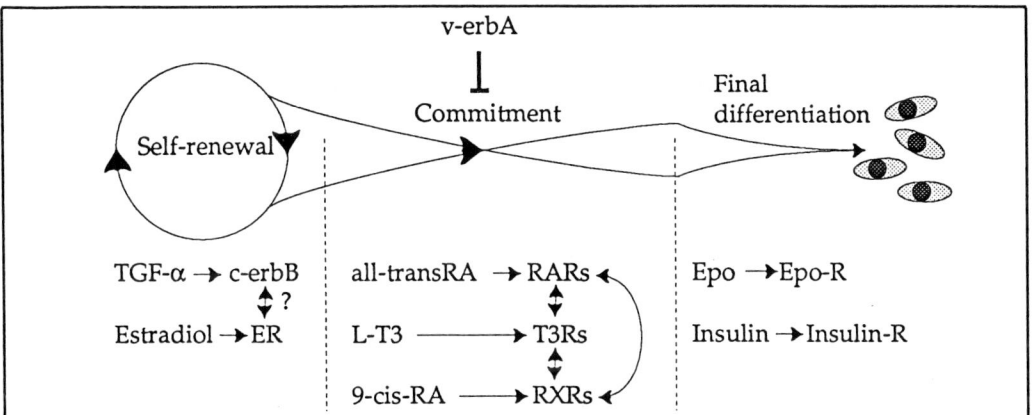

Figure 1: Schematical model of the erythrocytic differentiation pathway. The entire differentiation sequence can be subdivided in three steps, each of which is controlled by a specific combination of hormones acting through their respective receptors: (1) the self-renewal phase is under the control of both a membrane-bound receptor (c-erbB) and a nuclear hormonal receptor (Estrogen Receptor; ER). The vertical arrow suggests a possible cross-talk between the receptors deduced from the synergistic action of their respective ligands; (2) the commitment phase would be under the control of T3, all-trans-RA and 9-cis-RA. The vertical arrows indicate the existence of heterodimers between all those receptors suggesting a possible redundancy between the pathway controlled by those different receptors; (3) the final differentiation step is controlled by the archetypal erythrocytic hormone, erythropoietin, in synergy with insulin in the avian model. The v-erbA oncogenic activity probably results from its ability to block the commitment induced by T3R and RAR.

represent a crucial step toward a molecular understanding of the processes by which vertebrate cells control their differentiation and development and whose abnormal regulation results in neoplasia.

II) The v-erbA oncogene.

The v-erbA oncogene is one of the two oncogenes carried by the Avian Erythroblastosis Virus (AEV). The other oncogene is v-erbB which encodes a mutated form of the EGF/TGF-α receptor. AEV induces acute erythroleukemias in chicken and blocks the differentiation of erythrocytic progenitors in vitro at precisely the Colony-Forming Unit (CFU-E) stage (Samarut and Gazzolo, 1982). The v-erb A oncogene represents the most clearcut case of an oncogene specifically interfering with the differentiation process. As such, it is an invaluable tool to investigate molecular alterations in the differentiation control leading to leukemic development.

1.) Role of v-erbA in the erythrocytic transformation.

Initial experiments aimed at establishing the respective roles of v-erbA and v-erbB oncogenes in erythroleukemic transformation clearly showed that although the v-erbB oncogene alone was able to induce a sustained proliferation of immature cells it was not able to induce a strict differentiation block as the one observed with AEV (Frykberg et al., 1983; Graf and Beug, 1983).

More recent data using a selectable retroviral expression vector showed that the v-erbA oncogene was able by itself to induce a differentiation block at the CFU-E stage of chicken erythrocytic progenitor cells (Gandrillon et al., 1989). Those v-erbA-transformed cells were nevertheless dependent for their sustained growth upon a seric factor. The quest for that factor led us to identify a new erythrocytic self-renewal factor, the Transforming Growth Factor α (TGF-α) (Pain et al., 1991; see I. 1.). Beyond its ability to arrest differentiation, the v-erbA oncogene also cooperates with v-erbB and other tyrosine-kinase oncogenes to allow the transformed cells to grow in less highly supplemented medium and under wide extremes of pH (Damm et al., 1987). It nevertheless seems that those two properties of v-erb A are not linked to each other (Fuerstenberg et al., 1992).

Furthermore, we were able to show that the v-erbA oncogene expression rendered the TECs insensitive to RA- or T3-induced inhibition of self renewal. A non transforming mutant of v-erbA was unable to induce such a protective effect. Comparison between the molecular activities of the wild-type v-erbA oncogene and its non-transforming mutant suggests that self renewal inhibition is mediated through an indirect regulatory pathway similar to that of functional interference between the nuclear receptors and AP-1 transcription factors (Desbois et al., 1991; Gandrillon et al., submitted; see below).

2.) Origin of the v-erbA oncogene.

The first clues concerning the mode of action of the v-erbA oncogene arose when the proto-oncogene c-erbA from which v-erbA is derived was cloned, sequenced and shown to encode the nuclear receptor for the thyroid hormone T3 (Sap et al., 1986; Weinberger et al., 1986). The v-erbA protein is a mutated version of the T3 receptor α (T3Rα) that has lost its ability to bind T3 but that still binds to DNA (Munoz et al., 1988; Privalsky et al., 1988). It has been shown to function in animal cells as a constitutive T3 receptor antagonist by repressing transcription from promoters linked to idealized or natural T3 response element (T3RE) (Damm et al., 1989; Sap et al., 1989). Furthermore, T3 receptors α were shown to be expressed in erythrocytic cells (Hentzen et al., 1987; Bigler and Eisenmann, 1988). Taken together these observations suggest that v-erbA contributes to cellular transformation by interfering constitutively with the function of normal endogenous T3 receptors α. It was indeed found that v-erbA constitutively represses the carbonic anhydrase II (CA II) gene expression, an erythrocyte specific gene whose expression is strongly enhanced by T3 (Pain et al., 1990; Disela et al., 1991).

3.) Mechanisms of action of v-erbA

The simple picture of v-erbA blocking the differentiation sequence by acting like a dominant negative version of T3Rα and therefore blocking the transcriptional regulation of differentiation -linked genes normally regulated by T3-activated T3Rα has nevertheless been shattered by recent findings implicating numerous other transcription factors in both the T3 response and the v-erbA activity. v-erbA has indeed been shown to antagonize the transcription activation by retinoic acid receptors (RARs) which are structurally and functionally very close to T3R (Sharif and Privalsky, 1991; our own results: II.1.). Since a mutant of v-erbA, which does not antagonize RARs but still antagonizes T3R, does not block the erythrocytic differentiation, it has been proposed that RARs should play an essential role in normal erythrocytic differentiation (Sharif and Privalsky, 1991).

Furthermore it has recently been shown that the v-erbA activity requires the association between v-erbA and RXR, a new member of the nuclear receptor family that is activated by 9-cis RA, a stereoisoform of all-trans RA (Hermann et al., 1993). Both T3Rs and RARs have also been shown to be able to form heterodimers with RXRs (Green, 1993), besides their ability to form homodimers (T3R/T3R and RAR/RAR) and heterodimers (T3R/RAR). An other factor, triiodothyronine receptor auxilliary protein (TRAP), has also been shown to be able to interact with T3R to form T3R/TRAP heterodimers (Darling et al., 1991; Beebe et al., 1991). One of the main question now arising is the precise role of each of the combinations of transcription factors in eliciting the T3 or RA response and whether specific and/or common sets of target genes are controlled by each combination. A first hint came from recent studies showing that different complexes are not equivalent in their DNA binding stability in the presence of T3: T3R/T3R homodimers and T3R/RAR heterodimers are inhibited in their DNA binding of a T3RE whereas T3R/TRAP and T3R/RXR are unaffected by T3 addition (Yen et al., 1992a; 1992b). This suggest that T3R/TRAP and T3R/RXR heterodimers are more likely to be involved in T3-regulated transcription that T3R/T3R homodimers and T3R/RAR heterodimers.

Besides those associations, that can be described as a "direct" pathway of v-erbA action (i.e. acting through direct interference with nuclear receptor), an other, indirect, mechanism has recently been described. It consist in the v-erbA-induced maintenance of a fully active AP-1 complex in RA-treated fibroblasts. In normal cells, such a treatment results in a dramatic decrease of AP-1 activity (Desbois et al., 1991). The molecular nature of the interference between nuclear receptors and the AP-1 complex is presently unknown.

III) Involvement of other mutated forms of nuclear receptors in the transformation process.

Three different RAR subtypes (α, β and γ) have to date been isolated in humans and other animal species. They are believed to exert their functions by regulating the transcription of various genes at different times of development and in specific cells (Evans, 1988).

So far, two observations have raised the possibility that altered retinoid receptors may play an important role in oncogenesis :

1) in one human hepatocellular carcinoma, the RARβ gene is rearranged as a result of a hepatitis B virus (HBV) integration (Dejean et al., 1986). In the resulting chimeric protein, HβR, the amino terminus of RARβ was replaced by a coding sequence from the pre-S1 large envelope protein of HBV.

2) in acute promyelocytic leukaemias, the RARα subtype is truncated by 15:17 and 11:17 translocations (de Thé et al., 1990, Borrow et al., 1990, Alcalay et al., 1991, Chen et al., 1993).

Moreover, recent data support the hypothesis that RARβ functions as a tumor suppressor gene in epidermoid lung tumorigenesis (Houle et al., 1993).

1.) Transformation of erythrocytic cells by HβR, a natural mutant of RARβ.

We have investigated the oncogenic potential of the RARβ gene by expressing into hematopoietic cells the RARβ coding sequences derived either from the normal human cDNA or from the rearranged form (HβR) found in the original human hepatocellular carcinoma.

We have shown that, like v-erbA, the HβR mutant was able to block the differentiation of erythrocytic cells at the CFU-E stage, whereas the wild-type RARβ was unable to do so (Garcia et al., 1993). It is then likely that the HBV sequences present in the HβR fusion protein strongly enhances the oncogenic potential of the rearranged RARβ in erythroid cells. In contrast with v-erbA transformed cells, the HβR-transformed erythrocytic cells grew independently of TGFα.

2.) Molecular mechanisms governing the oncogenic effect of HβR.

In vitro and in vivo analysis of the transforming ability of HβR have suggested that the mechanisms of cellular transformation by RARβ might be different from that of v-erbA. Different molecular mechanisms could account for the ability of the HBV-RARβ chimera to transform erythrocytic cells. The mutant RARβ might either compete with normal RARβ for specific response element or be involved in the formation of inactive heterodimers with c-erbA or other related transcription factors. Both pathway would results in the silencing of genes required for the normal differentiation program of erythrocytic progenitor cells.

We have previously shown that, like v-erbA, the HβR chimera was able to repress the CAII gene expression. We are currently analysing more precisely the molecular interactions of the HBV-RARβ oncoprotein with the regulatory sequences of CAII and with other potentially involved nuclear proteins.

IV) Conclusions

Early steps of chicken erythrocytic differentiation appear to be controlled by several hormone-regulated pathways that work antagonisticaly or synergistically. Estrogens are involved in maintaining the progenitors in a self-renewing process. In contrast T3 and RA control the commitment of self-renewing progenitors into the final differentiation steps. Interstingly all these hormones or related mediators work through nuclear receptors that function as transcription factors and thereby induce specific gene expression programs.

REFERENCES

1. Alacalay, M., Zangrilli, D., Pandolfi, P. P., Lo,go, L., Mencarelli, A., Giacomucci, A., Rocchi, M., Biondi, A., Rambaldi, A., Lo Coco, F., Diverio, D., Donti, E., Grignani, F. and Pelicci, P. G. (1991). Translocation breakpoint of acute promyelocytic leukemia lies within the retinoic acid receptor α locus. Proc. Natl. Acad. Sci. USA 88, 1977-1981.
2. Beebe, J. S., Darling, D. S. and Chin, W. W. (1991). 3,5,3'-triodothyronine receptor auxiliary protein (TRAP) enhances receptor binding by interactions within the thyroid hormone response element. Mol. Endocrinol. 5, 85-93.
3. Bigler, J. and Eisenman, R. N. (1988). c-erbA encodes multiple proteins in chicken erythroid cells. Mol. Cell. Biol. 8, 4155-4161.
4. Borrow, J., Goddard, A., Sheer, D. and Salomon, E. (1990). Molecular analysis of acute promyelocytic leukemia breakpoint cluster region on chromosome 17. Science 249, 1577-1580.
5. Chen, Z., Brand, N. J., Chen, A., Chen, S.-J., Tong, J.-H., Wang, Z.-Y., Waxman, S. and Zelent, A. (1993). Fusion between a novel Krüppel-like zinc finger gene and the retinoic acid receptor α locus due to a variant t(11;17) translocation associated with acute prolmyelocytic leukemia. Embo J. 12, 1161-1167.
6. Correa, P. N. and Axelrad, A. A. (1992). Retinyl Acetate and All-Trans-Retinoic Acid Enhance Erythroid Colony Formation Invitro by Circulating Human Progenitors in an Improved Serum-Free Medium. Int J Cell Cloning 10, 286-291.
7. Dainiak, N., Hoffman, R., Maffei, L.A. and Forget, B.G. (1978). Potentiation of human erythropoiesis in vitro by thyroid hormone. Nature 272, 260-262.
8. Damm, K., Beug H., Graf T. and Vennström B. (1987). A single point mutation in erb A restores the erythroid transforming potential of a mutant avian erythroblastosis virus (AEV) defective in both erb A and erb B oncogenes. Embo, 6, 375-382.
9. Damm, K., Thompson, C. C., and Evans, R. M. (1989). Protein encoded by v-erbA functions as a thyroid-hormone receptor antagonist. Nature 339, 593-597.
10. Darling, D. S., Beebe, J. S., Burnside, J., Winslow, E. R. and Chin, W. W. (1991). 3,5,3'-triodothyronine receptor auxiliary protein (TRAP) binds DNA and forms heterodimers with the T3 receptor. Mol. Endocrinol. 5, 73-84.
11. Dejean, A., Bouguelerey, L., Grzeschik, K. H. and Tiollais, P. (1986). Hepatitis B virus DNA integration in a sequence homologous to v-erbA and steroid receptor genes in a hepatocellular carcinoma. Nature 322, 70-72.
12. Desbois, C., Aubert, D., Legrand, C., Pain, B. and Samarut, J. (1991). A novel mechanism of action for the v-erbA oncogene: abrogation of the inactivation of AP-1 transcription factor by retinoic acid receptor and thyroid hormone receptor. Cell 67, 731-740.
13. De Thé, H., Chomienne, C., Lanotte, M., Degos, L. and Dejean, A. (1990). The t(15;17) translocation of acute promyelocytic leukemia fuses the retinoic acid receptor α gene to a novel transcribed locus. Nature 347, 558-561.
14. Disela, C., Glineur, C., Bugge, T., Sap., J., Stengl, G., Dodgson, J., Stunnenberg, H., Beug, H. and Zenke, M. (1991). v-erbA overexpression is required to extinguish c-erbA function in erythroid cell differentiation and regulation of the erb A target CAII. Genes and Development 5, 2033-2047.
15. Evans, R. M. (1988). The steroid and thyroid hormone receptor superfamily. Science 240, 889-895.
16. Frykberg, L., Palmieri, S., Beug, H., Graf, T., Hayman, M. J. and Vennström, B. (1983). Transforming capacities of avian erythroblastosis virus mutants deleted in the erbA or erbB oncogenes. Cell 32, 227-238.
17. Fuerstenberg, S., Leitner, I., Schroeder, C., Schwarz, H., Vennström, B. and Beug, H. (1992). Transcriptional repression of band 3 and CA II in v-erb A transformed erythroblasts accounts for an important part of the leukaemic phenotype. Embo J. 11, 3355-3365.

18. Gandrillon, O., Jurdic, P., Pain, B., Desbois, C., Madjar, J.-J., Moscovici, M.G., Moscovici, C. and Samarut, J. (1989). Expression of the v-erbA product, an altered nuclear hormone receptor, is sufficient to transform erythrocytic cells in vitro but not to induce acute erythroleukemia in vivo. Cell 58, 115-121.
19. Garcia, M., de Thé, H., Tiollais, P., Samarut, J. and Dejean, A. (1993). A hepatitis B virus pre-S-retinoic acid receptor β chimera transforms erythrocytic progenitor cells in vitro. Proc. Natl. Acad. Sci. USA 90, 89-93.
20. Graf, T. and Beug, H. (1983). Role of the v-erbA and v-erbB oncogenes of avian erythroblastosis virus in erythroid cell transformation. Cell 34, 7-9.
21. Green, S. (1993). Promiscuous liaisons. Nature 361, 590-591.
22. Hentzen, D., Renucci, A., Le Guellec, D., Benchaibi, M., Jurdic, P., Gandrillon, O., and Samarut, J. (1987). The chicken c-erbA protooncogene is preferentially expressed in erythrocytic cells during late stages of differentiation. Mol. Cell. Biol. 7, 2416-2424.
23. Hermann, T., Hoffmann, B., Piedrafita, F. J., Zhang, X.-K. and Pfahl, M. (1993). v-erbA requires auxiliary proteins for dominant negative activity. Oncogene 8, 55-65.
24. Houle, B., Rochette-Egly, C. and Bradley, W. E. C. (1993). Tumor-suppressive effect of the retinoic acid receptor β in human epidermoid lung cancer cells. Proc. Natl. Acad. Sci. USA 90, 985-989.
25. Munoz, A., Zenke, M., Gehring, U., Sap, J., Beug, H. and Vennström, B. (1988). Characterization of the hormone binding domain of the chicken c-erbA/thyroid hormone receptor protein. Embo J., 7, 155-159.
26. Pain, B., Melet, F., Jurdic, P. and Samarut, J. (1990). The carbonic anhydrase II gene, a gene regulated by thyroid hormone and erythropoietin, is repressed by the v-erbA oncogene in erythrocytic cells. The New Biol. 2, 284-294.
27. Pain, B., Woods, C. M., Saez, J., Flickinger, T., Raines, M., Kung, H. J., Peyrol, S., Moscovici, C., Moscovici, G., Jurdic, P., Lazarides, E. and Samarut, J. (1991). EGF-R as a hemopoietic growth factor receptor: The c-erbB product is present in normal chicken erythrocytic progenitor cells and controls their self-renewal. Cell 65, 37-46.
28. Popovic, W.J., Brown, J.E. and Adamson, J.W. (1977). The influence of thyroid hormones on in vitro erythropoiesis. Mediation by a receptor with beta adrenergic properties. J. Clin. Invest. 60, 907-913.
29. Privalsky, M. L., Boucher, P., Koning, A. and Judelson, C. (1988). Genetic dissection of functional domains within the avian erythroblastosis virus erb A oncogene. Mol. Cell. Biol. 8, 4510-4517.
30. Samarut, J. and Gazzolo, L. (1982). Target cells infected by avian erythroblastosis virus differentiate and become transformed. Cell 28, 921-929.
31. Sap, J., Munoz, A., Damm, K., Goldberg, Y., Ghysdael, J., Leutz A,., Beug, H. and Vennström, B. (1986). The c-erbA protein is a high-affinity receptor for thyroid hormone. Nature 324, 635-640.
32. Sap, J., Munoz, A., Schmitt, J., Stunnenberg, H. and Vennström, B. (1989). Repression of transcription mediated at thyroid hormone response element by the v-erbA oncogene product. Nature 340, 242-244.
33. Schroeder, C., Gibson, L. Zenke, M. and Beug, H. (1992). Modulation of normal erythroid differentiation by the endogenous thyroid hormone and retinoic acid receptors: a possible target for v-erbA oncogene action. Oncogene 7, 217-227.
34. Schroeder, C., Gibson, L., Nordström, C and Beug, H. (1993). The estrogen receptor cooperates with the TGFa receptor (c-erbB) in regulation of chicken erythroid progenitor self-renewal. EMBO J. 12, 951-960.
35. Sharif, M. and Privalsky, M.L. (1991). v-erbA oncogene function in neoplasia correlates with its ability to repress retinoic acid receptor action.Cell 66, 885-893.
36. Visani, G., Cenacchi, A., Tosi, P., Finelli, C., Fogli, M., Gamberi, B., Martinelli, G. and Tura, S. (1992). All-trans retinoic acid improves erythropoiesis in myelodysplastic syndromes: a case report. Brit. J. Haemat. 81, 444-448.
37. Weinberger, C., Thompson, C. C., Ong, E. S., Lebo, R., Gruol, D. J. and Evans, R. M. (1986). The c-erb-A gene encodes a thyroid hormone receptor. Nature 324, 641-646.
38. Yen, P. M., Darling; D. S., Carter, R. L., Forgione, M., Umeda, P. and chin, W. W. (1992a). Triiodothyronine (T3) decreases binding to DNA by T3-receptor homodimers but not receptor auxiliary protein heterodimers. J. Biol. Chem. 267, 3565-3568.
39. Yen, P. M., Sugawara, A. and Chin, W. W. (1992b). Triiodothyronine (T3) differentially affects T3-receptor/retinoic acid receptor and T3-receptor/retinoid X receptor heterodimer binding to DNA. J. Biol. Chem. 267, 23248-23252.

Résumé

L'acide rétinoïque et l'hormone thyroïdienne triiodothyronine (T3) exercent des effets biologiques très variés et liés à leur capacité de moduler la prolifération et la différenciation cellulaire. Nous nous sommes intéressés au rôle joué par l'acide rétinoïque et la T3 dans la différenciation érythrocytique chez le poulet. Nous avons utilisé un modèle de culture original nous permettant d'étudier les influences différenciantes de ces molécules. Les résultats obtenus nous ont conduit à proposer un modèle en trois phases de la différenciation érythrocytique. Nous discutons ensuite les contributions respectives apportées par l'étude de l'oncogène v-erbA, une version onco-génique du récepteur de la T3, et du récepteur β de l'acide rétinoïque à la compréhension en termes moléculaires des processus régissant l'auto-renouvellement et la différenciation des précurseurs érythrocytiques.

Hematopoietic cell growth is regulated by a balance of positive and negative regulators: potential role for receptor modulation

Francis W. Ruscetti, Sten E.W. Jacobsen, Claire M. Dubois, Kjetil Hestdal and Jonathan R. Keller

Laboratory of Leukocyte Biology, Biological Response Modifiers Program and Biological Carcinogenesis and Development Program, PRI/DynCorp, Inc., National Cancer Institute, Frederick Cancer Research and Development Center, Frederick, Maryland 21702, USA

Summary

A number of well characterized cytokines have been implicated in the positive and negative regulation of hematopoietic progenitor cell growth. To initiate and maintain the growth and differentiation of primitive progenitor cells, multiple cytokine stimulation is required. This has led to the concept of "growth factor synergy". Here, we show that such cooperativity also occurs between negative regulators of hematopoietic cell growth, and that the ability of primitive progenitors to proliferate depends on the balance of positive and negative signals the cell receives. Interestingly, most inhibitory cytokines also have direct and/or indirect stimulatory effects on hematopoietic progenitors. The effects of these bifunctional cytokines depend on the differentiation state of the target cell and the other cytokines interacting with the cell. Studies on the surface expression of cytokine receptors on primitive progenitor cell populations have shown that they possess a heterogenous number of receptors (ranging from 100-200 per cell to undetectable levels). Also, TGF-β and TNF-α have bidirectional effects on the cell surface expression of many cytokine receptors that directly correlates with their effects on growth. Specifically, stem cell factor receptor (c-kit) expression is down-regulated by TGF-β in part by affecting c-kit mRNA stability. Thus, TGF-β may inhibit growth by modulating the expression of growth factor receptors.

Introduction

The growth and differentiation of hematopoietic cells can be stimulated both in vitro and in vivo by numerous cytokines (for review, see Lord and Dexter, 1992). Similarly, hematopoietic progenitor cell growth can also be negatively regulated by interferons (IFN), tumor necrosis factor-α (TNF-α), transforming growth factor-β (TGF-β), chemokines and several small peptides (for review, see Broxmeyer, 1992). Growth of highly purified progenitor cells requires the action of multiple cytokines, leading to the concept of "growth factor synergy" (Lowry et al, 1992). Thus, it has been proposed that hematopoiesis might be regulated, in part, by the balance of both positive and negative growth signals received by a cell (Ruscetti et al, 1992).

In this regard, we have investigated the interactions of TGF-β, a potent inhibitor of hematopoietic cell growth (for review, see Keller et al, 1992) with multiple stimulatory and inhibitory cytokines. In addition, we have compared the effects of two other hematopoietic cell growth inhibitors, MIP-1α and TNF-α, with TGF-β. Finally, we examined the role of receptor modulation in the control of hematopoietic progenitor cell growth.

Bidirectional Regulation of Hematopoiesis by Individual Cytokines

While neither TNF-α, MIP-1α or TGF-β alone have any effect on the growth of bone marrow cells in vitro, they inhibit cytokine-mediated growth of progenitor cells in vivo and in vitro. Also, TNF-α (Jacobsen, S. et al 1992a), MIP-1α (Broxmeyer et al, 1990 and TGF-β (Keller et al, 1992) have bifunctional effects on hematopoeitic cell growth since all enhance GM-CSF-induced growth. In vitro colony formation assays on purified progenitors do not necessarily eliminate indirect effects, and questions of indirect effects of biological modifiers on such cells can only be properly addressed by using single cell cloning experiments. Therefore, we compared the bidirectional effects of these cytokines on a range of hematopoietic progenitors and whether those interactions were directly mediated.

Using either murine Lin- bone marrow cells or FACS sorted Thy-1+ Lin- and Thy-1- Lin- populations, single cells were seeded in Torasaki plates with different combinations of cytokines. The results summarized in Table 1 show that MIP-1α can act to directly enhance IL-3 and GM-CSF-induced growth. This is in contrast to TGF-β1 which inhibited IL-3-induced but enhanced GM-CSF-induced growth of Lin-cells. In contrast, MIP-1α and TGF-β1 both inhibited the IL-3 and GM-CSF-induced growth of isolated Thy-1+ Lin- cells. Similar to their effects on Lin- cells, MIP-1α enhanced IL-3 and GM-CSF-induced growth of Thy-1- Lin- cells, while TGF-β1 inhibited IL-3-induced but enhanced GM-CSF-induced growth of Thy-1- Lin- cells. In contrast, the direct effects of TNF-α on murine bone marrow cells and enriched Lin- progenitors were only inhibitory. The different direct effects of TGF-β (and MIP-1α) on GM-CSF induced growth and Lin- cells and Lin- Thy-1+ cells point out the need of purification of subpopulations of progenitor cells to understand the multifunctional effects of cytokines in hematopoiesis. In this regard, it is important to determine whether the effects of GM-CSF and TNF-α on the differentiation of progenitor cells to dendritic cells are direct stimulatory effects (see J. Banchereau, this volume).

Since these results suggest that TNF-α, MIP-1α and TGF-β have overlapping but distinct effects on progenitor cells, we next compared their effects on the most primitive progenitors detectable in vitro including the high proliferative potential-colony forming cells (HPP-CFCs). HPP-CFC-1 and HPP-CFC-2 are progenitor cells detected in the bone marrow of mice given a single sublethal injection of 5-FU and require multiple cytokines for growth in vitro. While MIP-1α did not affect HPP-CFC-1 or HPP-CFC-2 colony formation, all HPP-CFC colonies regardless of the factors used for stimulation are inhibited by TGF-ß1 (Keller et al, 1990) and TNF-α (Jacobsen, F. et al, 1993). Thus, although MIP-1α is also a direct bidirectional regulator of hematopoietic cell growth, we find that its inhibitory effects are more modest and restricted to a subpopulation of primitive progenitors than the other inhibitory cytokines. This is supported by recent data suggesting that MIP-1α has no effect on the marrow repopulating cells in vivo (Quesniaux et al, 1993).

Table 1. Summary of Direct effects of TNF-α, MIP-1α and TGF-β on CSF-dependent growth of separated bone marrow progenitor cells.

Cells	CSF	TNF-α	MIP-1α	TGF-β	Growth Response in Terasaki plates
Lin-	-	-	-	-	None
Lin-	GM-CSF	-	-	-	+
Lin-	GM-CSF	+	-	-	↓
Lin-	GM-CSF	-	+	-	↑
Lin-	GM-CSF	-	-	+	↑
Lin-	IL-3	-	-	-	+
Lin-	IL-3	+	-	-	↓
Lin-	IL-3	-	+	-	↑
Lin-	IL-3	-	-	+	↓
Lin-Thy-1+	-	-	-	-	None
Lin-Thy-1+	GM-CSF	-	-	-	+
Lin-Thy-1+	GM-CSF	+	-	-	↓
Lin-Thy-1+	GM-CSF	-	+	-	↓
Lin-Thy-1+	GM-CSF	-	-	+	↓
Lin-Thy-1+	IL-3	-	-	-	+
Lin-Thy-1+	IL-3	+	-	-	↓
Lin-Thy-1+	IL-3	-	+	-	↓
Lin-Thy-1+	IL-3	-	-	+	↓

Murine Lin- and Lin-Thy-1+ bone marrow cells were separated and seeded as single cells as described (Keller et al, 1990). Results of four separate experiments with a minimum of 1200 wells scored per group are summarized.

TGF-β and TNF-α: Bidirectional regulators of CSF receptor expression

The ability of TGF-β1 and TNF-α to affect CSF-induced proliferation of bone marrow progenitor cells is correlated to modulation of CSF receptor expression (Table 2). Specifically, TGF-β1-mediated inhibition of IL-3-stimulated murine bone marrow proliferation was preceded by reduced IL-3 receptor expression, while upregulation of GM-CSF receptors preceded the synergistic effect of TGF-β1 on GM-CSF-stimulated proliferation (Jacobsen, S. et al, 1993). While, down-modulation of CSF receptor expression by homologous ligands is normally a rapid event occuring within minutes, TGF-β1-induced trans- modulation of IL-1 (Dubois et al,

1990), IL-3 and GM-CSF receptors (Jacobsen, S. et al, 1991) was slower with significant modulation seen after 6 and 24 hrs incubation but prolonged to 48 and 72 hrs. Also, the ED_{50}s for TGF-β1-induced CSF receptor modulation and cell growth regulation are similar. TGF-β is also a bidirectional modulator of CSF receptor expression in vivo (Hestdal et al, 1993). Furthermore, the inability of TGF-β1 to affect G-CSF-stimulated bone marrow progenitor cell growth (Jacobsen, S. et al, 1993) was correlated to unchanged expression of G-CSF receptors following TGF-β1 treatment. We had previously demonstrated that TGF-β1 inhibits the proliferation of murine hematopoietic progenitor cell lines regardless of the growth promoting CSF including G-CSF and GM-CSF (Keller et al, 1990). In addition, this inhibition was preceded by reversible prolonged down-modulation of the expression of all investigated CSF receptors, (Jacobsen, S. et al, 1991). Thus, the actions of G-CSF and GM-CSF in the presence of TGF-β vary on different myeloid progenitor cell populations, but are still directly correlated with receptor modulation by TGF-β (Table 2).

TNF-α rapidly reduced the expression of four CSF receptors on bone marrow cells and Lin- progenitor cells (Jacobsen, S. et al., 1992a). This correlated to reduced responsiveness to all four CSFs in short-term tritiated thymidine incorporation assays with the magnitude of inhibition correlated to the magnitude of receptor downregulation. The down-modulation of CSF receptors was specific and distinct from TGF-β since TNF-α did not affect the expression of a number of other cell surface proteins and down-regulated G-CSF receptor expression on Lin- cells while TGF-β did not (Table 2). Thus, TNF-α and TGF-β-induced receptor trans-downmodulation is only associated with inhibitory responses. In contrast to its direct inhibitory effects, the later appearing stimulatory effects of TNF-α on CSF-1-, GM-CSF-, and IL-3-induced growth and receptor expression on murine progenitors are indirect, and mediated through induction of, at least, G-CSF. In this regard, antibodies to TNF-α did not block the ability of TNF-α stimulated conditioned media to increase growth and receptor expression. In addition, this increase in GM-CSF and IL-3 receptor expression preceded the synergistic activity seen in proliferation assays.

Regulation of c-kit cell surface expression and mRNA stability by TGF-β

As summarized above, TGF-β is a potent inhibitor of stem cell factor (SCF) mediated synergistic growth of murine and human progenitor cell growth and the growth of myeloid cell lines in response to SCF. Treatment of both cell lines and primary purified Lin-Thy-1+ by TGF-β progenitor cells resulted in a marked reduction of cell surface expression of c-kit (less than 5% of control binding of radiolabelled SCF to progenitor cells was present after six hours). Using the FDC-P1 cell line, the reduction of cell surface c-kit expression by TGF-β is preceded by a marked reduction in c-kit mRNA levels which is observed two hours post TGF-β treatment and reaches a maximum by 6 hours. This inhibition of steady-state mRNA levels is explained, in part, by a decrease in the half-life of c-kit mRNA (3-4 hrs for control versus 30-90 min after TGF-β treatment. These findings suggest that one mechanism by which TGF-β regulates the responsiveness of hematopoietic progenitors to SCF is through decreasing c-kit mRNA stability leading to decreased cell surface expression of c-kit.

Table 2. Summary of the effects of TGF-β and TNF-α on bone marrow cell growth and CSF receptor expression

Growth Factor	Target Cell	Growth		Receptor Expression	
		TGF-β	TNF-α	TGF-β	TNF-α
G-CSF	Cell Lines	↓	ND	↓	ND
G-CSF	Lin- progenitors	NE	↓	NE	↓
GM-CSF	Cell Lines	↓	ND	↓	ND
GM-CSF	Lin- progenitors (SC)	↑	↓	ND	ND
GM-CSF	Lin- progenitors (Col)	↑	↑	↑	↑
IL-3	Cell Lines	↓	ND	↓	ND
IL-3	Lin- progenitors (SC)	↓	↓	ND	ND
IL-3	Lin- progenitors (Col)	↓	↑	↓	↑
SCF	Cell Lines	↓	ND	↓	ND
SCF	Lin- progenitors	↓	ND	↓	ND

(↑) increased proliferation or receptor expression; (↓) decreased proliferation or receptor expression; NE, no effect; ND, not determined; SC, single cell assay; Col, colony assay.

Cooperative Inhibition of Stem Cell Growth by Inhibitory Cytokines

As previously stated, multiple factors are needed to stimulate the growth and differentiation of purified progenitors. Preliminary data shows that combined action of IL-3 and SCF give approximately the same frequency of response of Thy-1+ Lin- cells in single cell cloning assays as does the combination of IL-3, SCF, GM-CSF, CSF-1, G-CSF, IL-1 and IL-6. However, the concentration of TGF-β1 (20 ng/ml) which inhibited >80% of the growth in response to the combination of IL-3 and SCF did not inhibit growth when stimulated by the combination of seven cytokines (Table 3). This suggests that the inhibitory effects of TGF-β can be reversed by the combined action of multiple cytokines and that a complex balance of positive and negative regulation of hematopoietic cell growth exists. Since TGF-β does not inhibit the growth of Lin-Thy-1+ progenitors stimulated by seven cytokines, we also studied whether the addition of other inhibitors of this population could act in combination with TGF-β to inhibit cell growth. Neither MIP-1α, TNF-α or γ-INF alone inhibited the growth of these cells when stimulated by seven cytokines. However, all four together inhibited 70% of cell growth (Table 3), indicating that negative cooperativity in inhibiting hematopoietic progenitor cell growth occurs. Surprisingly, MIP-1α did not contribute to this negative cooperativity in suppressing growth. However, it does contribute to negative cooperativity among chemokines in suppressing growth (see H. Broxmeyer, this volume).

Table 3. Balance between Multiple Positive and Negative Regulators in the Growth of Lin-Thy-1+ cells.

CSF	TGF-β	TNF-α	MIP-1α	γINF	Growth Inhibition
IL-3 + SCF	+	-	-	-	+
Seven	+	-	-	-	-
Seven	-	+	-	-	-
Seven	-	-	+	-	-
Seven	-	-	-	+	-
Seven	+	+	+	+	+
Seven	+	+	-	+	+

Lin-Thy-1+ were purified and assays performed as described (Keller et al, 1990). Cytokines (IL-1, IL-6, IL-3, SCF, GM-CSF, G-CSF and M-CSF) were used at predetermined optimal doses.

Role of Modulation of Receptor Expression in Regulation of Primitive Hematopoietic Progenitor Cell Growth

The importance of receptor modulation as a mechanism for regulation of growth factor responsiveness has been questioned due to the spare receptor concept which proposes that only a low percentage of receptors need a ligand bound to achieve a maximum biological response. However, a more recent study (Nicola et al, 1988) suggests that as many as up to 50% of the CSF receptors on hematopoietic progenitor cells need to be occupied to elicit a maximum response. Moreover, these results were obtained using cell lines which possess >1000 receptors/cells. In contrast, many studies suggest that primary bone marrow cells have 200 or less receptors for cytokines per cell. For example, our finding that bone marrow possess on the average of 80 IL-3 receptors per cell which can be reduced to 37 receptors per cell by TGF-β, and that this receptor modulation correlates with the biological response suggest that receptor modulation is mechanistically important (Jacobsen, S. et al, 1993). In support of this, autoradiography reveals profound heterogeneity with regard to CSF receptor expression on Lin- bone marrow progenitors, ranging from undetectable grains to >30. The majority of the blasts contain <3 grains per cell. After treatment with a synergizing cytokine only 14% of the blast cells still had 0-3 grains, suggesting that induction of CSF receptor expression on low- or non-responsive progenitors is one mechanism pathway for HGF synergy (Jacobsen, S. et al, 1992b) and that reduction of CSF receptor expression on responsive progenitors having low numbers of receptors is one mechanism for HGF mediated inhibition.

REFERENCES

Broxmeyer, H.E. (1992) Suppressor cytokines and regulation of myelopoiesis. Amer J Ped Hematol/Oncol 14:22-35.
Broxmeyer, H., Sherry, Lu, L., Cooper, S., Oh, K., P. Tekamp-Olson, Kwon, B. and Cerami, A. (1990): Enhancing and suppressing effects of recombinant murine macrophage inflammatory proteins on colony formation in vitro by bone marrow

myeloid progenitor cells. Blood 76:1110-1117.
Dubois, C., Ruscetti, F., Palazynski, E., Falk, L., Oppenheim, J. and Keller, J. (1990) Down modulation of IL 1 receptors by TGF-β: Proposed mechanism of inhibition of IL 1 action. J Exp Med 172: 737-744.
Hestdal, K., Jacobsen, S.E.,Ruscetti, F., Longo, D.L., Oppenheim, J. and Keller, J. (1993): Increased granulopoiesis after sequential administration of transforming growth factor $\beta 1$ and granulocyte-macrophage CSF. Exp Hematol in press.
Jacobsen, F., Rothe, M., Rusten, L., Goeddal, D., Smeland, E., Slordal, L., Warren, D. and Jacobsen, S.E. (1993): TNF-α, a potent inhibitor of primitive murine hematopoietic stem cells in vitro: Differential Role of Type I and Type II TNF receptors. J Exp Med in press.
Jacobsen, S.E., Ruscetti, F., Dubois, C. and Keller, J. (1991) Transforming growth factor β modulates the expression of GM-CSF and G-CSF receptors on murine hematopoietic progenitor cell lines. Blood 77: 1706-1716.
Jacobsen, S.E., Ruscetti F., Dubois, C. and Keller, J. (1992a) Tumor necrosis factor alpha is a bidirectional modulator of CSF receptor expression on progenitor cells: Proposed mechanisms of action for its multifunctional effects in hematopoiesis. J Exp Med 175: 1759-1772.
Jacobsen, S.E., Ruscetti, F., Dubois, C., Wine, J., and Keller, J. (1992b): Induction of CSF Receptor Expression on Hemato-poietic Progenitor Cells: Proposed Mechanism for Growth Factor Synergy. Blood 80: 678-687.
Jacobsen, S.E., Ruscetti F., Roberts, A. and Keller, J. (1993): Transforming growth factor-β is a bidirectional modulator of colony stimulating factor receptor expression on murine bone marrow progenitor cells. J Immunol in press.
Keller, J., Ellingsworth, L., McNiece, I., Quesenberry, P., Sing, G., and Ruscetti, F. (1990) Transforming growth factor-β directly regulates primitive murine hematopoietic cell proliferation. Blood 75: 596-602.
Keller, J., Jacobsen, S. E., Dubois, C., Hestdal, K. and Ruscetti, F. (1992) Transforming growth factor β and its role in hematopoiesis. Internat J Cell Cloning 10: 2-11.
Lord, B. and Dexter, T.M. (1992) Growth Factors In Hematopoiesis Bailliere's Clinical Haematology, London, W.B. Saunders Co.
Lowry, P., Deacon, D., Whitefield, P., McGrath, H. and Quesen-berry, P. (1992) Stem cell factor induction of in vitro murine hematopoietic colony formation by "subliminal' cyto-kines combinations: Role of anchor factors. Blood 80 663-669.
Nicola, N., Peterson, L., Hilton, D. and Metcalf, D. (1988): Cellular processing of murine colony stimulating factor (multi-CSF, GM-CSF, G-CSF) receptors by normal hematopoietic cells and cell lines. Growth Factors 1: 41-52.
Quesinaux, V., Graham, G., Pragnell, I., Donaldson, D., Wolpe, S., Iscove, N. and Fagg, B. (1993) Use of 5-fluorouracil to analyse the effect of MIP-1α on long-term reconstituting stem cells in vivo. Blood 81: 1497-1504
Ruscetti, F., Dubois, C., Jacobsen, S.E., and Keller, J.R. (1992) Transforming growth factor β and interleukin 1:A paradigm for opposing regulation of hematopoiesis. In Growth Factors in Haemopoiesis, eds Lord, B. Lord and T.M. Dexter. Bailliere's Clinical Haematology, London, W.B. Saunders Co., pp. 703-721.

Résumé

Un certain nombre de cytokines bien caractérisées sont impliquées dans la régulation positive et négative de la croissance des progéniteurs hématopoiétiques. Pour initier et maintenir la croissance des progéniteurs primitifs, une stimulation par de multiples cytokines est nécessaire. Ceci a conduit à la notion de "synergie de facteurs de croissance". Ce travail montre qu'une telle coopération survient aussi entre les régulateurs négatifs de la croissance hématopoïétique et que la capacité des progéniteurs primitifs à proliférer dépend de l'équilibre entre les signaux positifs et négatifs reçus par la cellule. Il est intéressant de noter que la plupart des cytokines négatives ont aussi un effet stimulant direct/ou indirect sur les progéniteurs hématopoiétiques. Les effets de ces cytokines bifonctionnelles dépendent de l'état de différenciation de la cellule cible et des autres cytokines interagissant sur la cellule. Des études de l'expression des récepteurs des cytokines sur la surface de populations de progéniteurs primitifs ont montré qu'elles possèdent un nombre hétérogène de récepteurs (de 100-200 /cellule jusqu'à un niveau indétectable). Le TGF β et le TNF α ont aussi des effets bidirectionnels sur l'expression à la surface des cellules des récepteurs de nombreuses cytokines qui sont corrélés directement avec leur effet sur la croissance. De façon spécifique, l'expression du récepteur du stem cell factor (c-kit) est diminué par le TGF β en partie par une modification de la stabilité du mRNA de c-kit. Ainsi, le TGF β peut inhiber la croissance en modulant l'expression des récepteurs des facteurs de croissance.

TNF-interferon interaction: molecular and cellular aspects

Juana Wietzerbin

INSERM U 365 «Interférons et Cytokines», Institut Curie, Section de Biologie, 26, rue d'Ulm, 75231 Paris Cedex 05, France

ABSTRACT

Tumor necrosis factor (TNF) often acts synergistically with or amplifies the biological effect of other cytokines including IFN-γ and IFN-α. The results reported here are an overview of recent work performed in our laboratory concerning the molecular mechanisms of functional interaction of IFN and TNF in monocytic/macrophage cells and hairy cell leukemia.

INTRODUCTION

TNF regulates the homeostasis of hematopoiesis by exerting both stimulatory and inhibitory effects (Beutler & Cerami, 1986). TNF acts synergistically with other cytokines like IFN-γ and IFN-α (Wietzerbin et al., 1990; Billard et al., 1990). On the other hand, IFN-γ is a potent activator of macrophage functions. As its effects on monocytes/macrophages include stimulation of TNF-α production (Philip & Epstein, 1986), TNF-α might thus act as an autocrine factor, together with IFN-γ, in inducing cell responses.

IFN-γ/TNF INTERACTION FOR IL-6 AND IL-6 RECEPTOR EXPRESSION

IL-6 is a multifunctional cytokine involved in the control of many cell functions, including antibody synthesis by B cells, T cell proliferation and differentiation, and stem cell differentiation. IL-6 and IL-3 act synergistically by enhancing the proliferation of multipotential progenitors (Kishimoto, 1989; Seghal, 1990). A large number of recent investigations showed that dysregulation of IL-6 expression is involved in lymphoid malignancies, such as multiple myeloma (Revel, 1989). We have explored the possibility that functional interactions between IFN-γ and TNF represent an important step in the regulation of IL-6 expression in monocytic cells. Although in other cell systems, such as fibroblasts, TNF-α stimulates IL-6 expression, we showed that IFN-γ is required for TNF to induce IL-6 expression in human monocytic THP-1 cells, chosed as a model since IL-6 gene is transcriptionally inactive (Sancéau et al., 1990).

IFN-γ or TNF-α alone failed to induce significant levels of IL-6 mRNA whereas combined treatment of cells with IFN-γ + TNF-α triggered the IL-6 gene expression. These data demonstrate that stimulation of IL-6 gene expression by IFN-γ and TNF-α requires complementary signals delivered by the two inducers. In fact, significant amounts of IL-6 activity were only detected in the supernatants of THP-1 cells stimulated by the combination of IFN-γ + TNF.

Stimulation of IL-6 secretion by monocytes also requires complementary signals delivered by IFN-γ and TNF. A marked synergistic effect was observed since more than 15,000 units/ml of IL-6 were produced by 10^6 cells, compared with 640 units/ml for either inducer alone (Sancéau & Wietzerbin, 1992).

Interestingly, IFN-γ alone, IFN-γ + TNF-α or LPS enhanced the level of IL-6 receptors mRNA already at 6 hours of stimulation of THP-1 cells. However, while IL-6 receptor level remains high at 18 hours in control and cytokine-treated cells, its expression appears to be down-regulated in LPS-treated cells (Sancéau et al., 1991).

IL-6 receptor mRNA level correlates with the expression of surface receptors. An increase of 40-50 % was already observed in the number of IL-6R after 6 hours of IFN-γ treatment ; the effect of IFN-γ on IL-6R seems to be specific because no significant modification was observed after TNF-α treatment. LPS or IFN-γ + TNF treatment decreased the binding capacity by 40 % and this finding may be due to a ligand-mediated down-regulation since both inducers led to IL-6 secretion (Sancéau et al., 1991).

Although, sequential exposure of THP-1 cells to IFN-γ followed by TNF-α did not allow IL-6 gene induction. Restimulation of IFN-γ pretreated cells with IFN-γ combined to TNF-α led to a superinduction of the IL-6 gene expression and of IL-6 secreted activity. The effect of IFN-γ was shown to be related to a marked increase in IL-6 mRNA stability.

Thus, these findings showed that IFN-γ and TNF-α act together to induce IL-6 synthesis in monocytic cells, and that IFN-γ is an essential signal in triggering IL-6 receptor expression (Sancéau & Wietzerbin, 1992).

MODULATION OF IFN-γ RECEPTOR EXPRESSION BY TNF AND IL-6

We next addressed the question of whether TNF-α and IL-6 play a role on the modulation of IFN-γ receptor expression in monocytic cells (Sancéau et al., 1992b)

Pretreatment of THP-1 cells with TNF or IL-6 markedly increased the number of IFN-γ cell surface receptors without any modification in the affinity. Interestingly, no modification of the binding was observed after LPS treatment, indicating that the effect observed is not simply linked to monocytic cell activation.

The increased IFN-γ surface receptor expression correlates with an increase in the mRNA receptor levels.

These results are the first evidence of cytokine-mediated up-regulation of IFN-γ receptors in human monocytic cells. This up-regulation by TNF-α and IL-6 -which are both produced by monocytic cells- occurs through different mechanisms. TNF, but not IL-6, increases IFN-γ receptor gene transcription. IL-6 increases IFN-γ mRNA receptor stability. Our results suggest that autocrine or paracrine TNF-α and IL-6-mediated up-regulation of IFN-γ receptors might be of physiological relevance. The *in vivo* effect of TNF on IFN-γ receptors will be investigated in cancer patients treated with TNF.

IFN-α / TNF-α INTERACTION

IFN-γ but also IFN-α influence the expression of TNF-α receptors. In this regard, during IFN-α treatment of hairy cell leukemia patients, an increase in the number of TNF-α receptors was observed in hairy cells. Furthermore, IFN-α treatment also appears to induce their expression when the TNF-α receptors were not detectable before treatment (Billard et al., 1990a).

In addition, not only the TNF-α receptors are up-regulated during IFN-α treatment but also biologically active TNF becomes detectable in the serum. IFN-α also induced hairy cells in vitro to produce TNF activity (Billard & Wietzerbin, 1990b; Billard & Lasfar, 1993).

These findings favour the hypothesis of the involvement of TNF-α in the therapeutic effects of IFN-α on hairy cell leukemia and illustrate that in addition to IFN-γ, IFN-α also interacts with the TNF system.

SURFACE PROTEASES AND TNF

The biological effect of cytokines is dependent on their interactions with their specific receptor, and also on the level of available intact molecule. In this context, we have been interested in studying the role of surface proteases on cytokine degradation (Bauvois et al., 1992).

We have shown that human cells from the monocytic lineage possess at least two distinct classes of membrane-bound proteases on their cell surface : a group of bestatin-sensitive N-aminopeptidases, and two DFP-inhibitable serine proteases : a DPP-IV-like protease and a phenylalanine endopeptidase.

Interestingly, enhancement of protease expression occurred during U937 cell differentiation induced by IFN-γ. Although both IFN-α and -γ inhibited cell proliferation, IFN-α had no effect on protease expression.

We have analyzed the contribution of these hydrolytic enzymes recovered in highly purified U937 cell membranes to the cleavage of biologically active radiolabelled cytokines, including IFN-γ and TNF-α (Bauvois, 1992).

Autoradiography of a gel electrophoresis analysis of cytokines incubated with cell membranes or intact cells showed that TNF-α is totally hydrolyzed to inactive fragments of less than 2 kDa, in contrast to IFN-γ, which appears to be resistant. We have shown that at least two serine proteases present in U937 plasma membranes, a DPP-IV-like enzyme (inhibited by the specific inhibitor, diprotin A) and an endophenyl peptidase (inhibited by the Ala-Ala-Phe-ketone) are involved in TNF digestion. This is a specific phenomenon since neither elastase nor cathepsin are able to degrade this cytokine (Laouar et al., 1993).

Our results support the idea that IFN-γ contributes, by enhancing the level of surface proteases, to limit the effects of TNF on the producing cell.

IFN-γ-TNF-α INTERACTION TO INDUCE NITRIC OXIDE (NO) SYNTHESIS

When activated for cytotoxicity, murine peritoneal macrophages develop inhibition of mitochondrial enzymes with iron sulfur prosthetic groups, like aconitase (Drapier & Hibbs, 1986). This inhibition depends on the synthesis of nitrogen oxides from L-arginine (Hibbs et al., 1987).

Also, NO is an intercellular messenger involved in immunomodulation, neurotransmission and vascular tone and is the effector molecule of macrophage cytotoxicity. Nitric oxide is produced by oxidation of the guanido-group of L-arginine to citrulline, by a family of recently discovered enzymes called NO synthase (Drapier, 1991a). The biosynthesis of NO from L-arginine is induced in macrophages by IFN-γ and requires the contribution of TNF. Neutralization of autocrine produced TNF by specific antibodies inhibits NO synthase induction by IFN-γ (Drapier et al., 1988). When target cells were cocultured with NO producing macrophages, nitrosylation of mitochondrial [Fe-S] containing enzymes takes place and this process is responsible for macrophage cytotoxicity (Pellat et., 1990; Drapier et al., 1991b; Henry et al., 1991). In addition, it was shown that nitrosylation also affects the activity of other iron-dependent enzymes, such as ribonucleotide reductase, a limiting enzyme in DNA synthesis (Lepoivre et al., 1991).

Recent findings suggest a potential role of NO in the hematopoietic system. NO gas or NO donors were shown to modulate the differentiation of human leukemia HL-60 cells with the concomitant increase in TNF gene expression and down-regulation of mRNA for c-myc and c-myb (Madrinat et al., 1992). NO was also shown to activate the transcription factor NFκB (Lander et al., 1993). In addition, nitric oxide biosynthesis modulates IRF (iron regulatory factor), a cytoplasmic bifunctional transactivator which modulates both ferritin RNA translation and transferrin receptor mRNA stability (Drapier et al., 1992).

These findings suggest that endogenous produced NO, whose synthesis is regulated by cytokines (IFN-γ and TNF), can modulate cellular differentiation and gene expression and suggest that NO produced in the bone marrow may have a role in normal and malignant hematopoietic cell differentiation.

Résumé

Le facteur nécrosant de tumeur (TNF) agit fréquemment en synergie avec d'autres cytokines, y compris les interférons. Les résultats rapportés ici résument les travaux récents de notre laboratoire concernant les mécanismes moléculaires responsables des interactions fonctionnelles des interférons avec le TNF dans les cellules monocytaires/macrophages et la tricholeucémie.

ACKNOWLEDGEMENTS

This work was supported by grants from Institut National de la Santé et de la Recherche Médicale (I.N.S.E.R.M.) and Association pour la Recherche sur le Cancer (A.R.C.).

REFERENCES

Bauvois, B., Sancéau, J. & Wietzerbin, J. (1992): Human U937 cell surface peptidase activities : characterization and degradative effect on tumor necrosis factor-α. *Eur. J. Immunol.* 22, 923-930.

Beutler, B. & Cerami, A. (1986): Cachectin and tumor necrosis factor as two sides of the same biological coin. *Nature* 320, 584.

Billard, C. & Lasfar A. (1993): Production of tumor necrosis factor in response to interferon-α in hairy cell leukemia. *Leukemia* 7, 331-332.

Billard, C., Sigaux, F. & Wietzerbin, J. (1990a): Interferon-α *in vivo* up-regulates tumor necrosis factor receptors on tumor cells from hairy cell leukemia. *J. Immunol.* 145, 1713-1718.

Billard, C. & Wietzerbin, J. (1990b): On the mechanism of action of IFN-α on hairy cell leukemia. *Eur. J. Cancer* 26, 67-69.

Drapier, J.C. (1991): L-arginine-derived nitric oxide and the cell-mediated immune response. Forum in Immunology. *Res. Immunol.*, 142, 553-602.

Drapier, J.C., Pellat, C. & Henry, Y. (1991): Generation of EPR-detectable nitrosyl-iron complexes in tumor target cells cocultured with activated macrophages. *J. Biol. Chem.* 266, 10162-10167.

Drapier, J.C., Hirling, H., Wietzerbin, J., Kaldy, P. & Kühn, L.C. (1993): Biosynthesis of nitric oxide activates iron regulatory factor in macrophages. *EMBO J.* In press.

Drapier, J.C. & Hibbs Jr., J.B. (1986): Murine cytotoxic activated macrophages inhibit aconitase in tumor cells. Inhibition involves the iron-sulfur prosthetic group and is reversible. *J. Clin. Invest.* 78, 790-797.

Drapier, J.C., Wietzerbin, J. & Hibbs Jr., J.B. (1988): Interferon-gamma and tumor necrosis factor induce the L-arginine dependent cytotoxic effector mechanism in murine macrophages. *Eur. J. Immunol.* 18, 1587-1592.

Henry, Y., Ducrocq, C., Drapier, J.C., Servent, D., Pellat, C. & Guissani, A. (1991): Nitric oxide, a biological effector. Electron paramagnetic resonance detection of nitrosyl-iron protein complexes in whole cells. *Eur. J. Biophys.* 20, 1-15.

Hibbs Jr., J.B., Taintor, R.R. & Vavrin, Z. (1987): Macrophages cytotoxicity : role for L-arginine deiminase activity and imino nitrogen oxidation to nitrite. *Science* 235, 473-476.

Kishimoto, T. (1989): The biology of interleukin-6. *Blood* 34, 1-10.

Lander, H.M., Sehajpal, P., Levine, D.M. & Novogrodsky, A. (1993): Activation of human peripheral blood mononuclear cells by nitric oxide-generating compounds. *J. Immunol.* 150, 1509-1516.

Laouar, A., Villiers, C., Sancéau, J., Maison, C., Colomb, M., Wietzerbin, J. & Bauvois, B. (1993): Inactivation of interleukin-6 *in vitro* by monoblastic U937 cell plasma membranes involves both protease and peptidyl-transferase activities. *Eur. J. Biochem.* In press.

Lepoivre, M., Chenais, B., Yapo, A., Lemaire, G., Thelander, L. & Tenu, J.P. (1990): Alterations of ribonucleotide reductase activity following induction of the nitrite-generating pathway in adenocarcinoma cells. *J. Biol. Chem.* 265, 14143-14149.

Magrinat, G., Mason, S.N., Shami, P.J. & Weinberg, B. (1992): Nitric oxide modulation of human leukemia cell differentiation and gene expression. *Blood* 80, 1880-1884.

Pellat, C., Henry, Y. & Drapier, J.C. (1990): IFN-γ -activated macrophages : detection by electron paramagnetic resonance of complexes between L-arginine-derived nitric oxide and non-heme iron proteins. *Biochem. Biophys. Res. Commun.* 166, 119-125.

Philip, R. & Epstein, L.B. (1986): Tumor necrosis factor as immuno-modulator and mediator of monocyte cytotoxicity induced by itself, γ-interferon and interleukin-1. *Nature* 323, 86-89.

Revel, M. (1989): Host defence against infection and inflammation IL-6/IFN-β2. *Experiencia* 45, 549.

Sancéau, J., Béranger, F., Gaudelet, C. & Wietzerbin, J. (1990): Interferon-γ is an essential cosignal for triggering IFN-β2/BSF-2/IL-6 gene expression in human monocytic cell lines. *Ann. New York Acad. Sci.* 557, 130-143.

Sancéau, J., Merlin, G. & Wietzerbin, J. (1992): TNF-α and IL-6 up-regulate IFN-γ receptor gene expression in human monocytic THP-1 cells by transcriptional and post-transcriptional mechanisms. *J. Immunol.* 149, 1671-1675.

Sancéau, J., Wijdenes, J., Revel, M. & Wietzerbin, J. (1991): IL-6 and IL-6 receptor modulation by IFN-γ and tumor necrosis factor-α in human monocytic cell line (THP-1). ming effect of IFN-γ. *J. Immunol.* 147, 2630-2637.

Sancéau, J. & Wietzerbin, J. (1992): Priming effect of IFN-γ on IL-6 and IL-6 receptor expression in human monocytic cells. In *IL-6, Physiopathology and Clinical potentials*, ed. M. Revel, pp. 74-84.

Seghal, P.B. (1990): Interleukin in infection and cancer. *Proc. Soc. Exp. Biol. Med.* 195, 183-191.

Wietzerbin, J., Gaudelet, C., Catinot, L., Chebath, J. & Falcoff, R. (1990): Antiviral enhancing activity and (2'-5')oligo(A) synthetase induction mediated by rTNF-α in myelomonocytic cells primed with IFN-γ. *J. Leuk. Biol.* 48, 149-155.

TNFα downregulates the expression of c-kit on normal bone marrow CD34+ cells

E. Khoury, S. Pontvert-Delucq, C. Baillou, M. Guigon, A. Najman and F.M. Lemoine

Laboratoire d'Hématologie, Faculté de Médecine Saint-Antoine, 27, rue de Chaligny, 75012 Paris, France

TNFα seems to play a key role in the regulation of hematopoiesis. In a previous report, we studied the effects of TNFα in long term bone marrow culture (LTBMC) and we showed that the addition of TNFα to LTBMC was accompagnied by a faster exhaution of hematopoiesis(1). This might be due in part to the interaction with hematopoietic growth factors or their receptors. Indeed, TNFα can upregulate the expression of IL3 and GM-CSF receptors(2) and downregulate G-CSF receptors (3) on the surface of hematopoietic cells. In addition, it synergizes the hematopoietic proliferation induced by IL3 and/or GM-CSF(4). Recently, a stromal derived factor, called Stem Cell Factor (SCF) or Kit ligand, active both as soluble and transmembrane forms, has been described. It has a strong synergistic activity for various growth factors. Therefore, we studied the interaction between TNFα and c-Kit on normal bone marrow progenitor cells.

In a first set of experiments, normal bone marrow mononuclear cells were incubated in IMDM with 20% FCS, in the presence or the absence of 200U/ml TNFα for 20hours and then stained with anti-CD34 and anti-c-Kit monoclonal antibodies and analysed by flow-cytometry. Table I shows a decrease in the c-Kit expression on CD34+ gated cells. Similar results were obtained using purified CD34+ cells.

	% of CD34+ coexpressing c-Kit		% of inhibition
	Control	TNF	
Exp 1	35,4	20,1	43,2
Exp 2	23	16	30,4
Exp 3	30,1	15,5	48,5
Exp 4	53,1	42,8	19,3

Table I

In further experiments, we studied whether the decrease of c-Kit expression was accompagnied with a diminution of the proliferative response of CD34+ cells to SCF. Thus, 1×10^4 cells/well of CD34+ sorted cells were plated in 96 wells microtiter plates, in IMDM 20% FCS in the presence or the absence of 200U/ml TNFα. 20 hours later, the SCF was added at a concentration of 100 ng/ml. After a culture period of 48 hours at 37°C, the cells were exposed to a 16 hours pulse of 1μCi ^{3}H thymidine. Finally, the cells were harvested on glass fiber strips and the incorporated radioactivity measured in a scintillation counter. Our results show (table II) that TNFα diminishes the proliferative response of CD34+ purified cells to SCF.

	no GFs cpm/well	Proliferative Index		
		TNF	SCF	TNF+SCF
Exp 1	7319±623	0,88	2,46	1,23
Exp 2	4378±442	0,83	1,29	0,94
Exp 3	5285±505	0,89	1,25	0,96
Exp 4	13931±1962	1,23	3,36	1,78

Table II

Proliferative Index is calculated as follows: cpm in test well (i.e cells+ TNFα or SCF or SCF+TNFα)
cpm in control well (cells alone)

The overall data from 4 separate experiments indicate that TNFα decreases the expression of c-Kit on the surface of CD34+ cells by 35,5±13% and diminishes their response to SCF suggesting that TNFα downregulates c-Kit. This downregulation may account of the inhibitory effects of TNFα on normal hematopoiesis.

References:

1. Khoury E, Lemoine FM, Baillou C, Kobari L, Deloux J, Guigon M, Najman A (1992) Tumor necrosis factor alpha in human long-term bone marrow cultures: distinct effects on nonadherent and adherent progenitors. Exp Hematol 20: 991.
2. Elbaz O, Budel LM, Hoogerbrugge H, Touw PI, Delwel R, Mahmoud AL, Löwenberg B (1991) Tumor necrosis factor regulates the expression of granulocyte-macrophage colony-stimulating factor and interleukin-3 receptors on human acute myeloid leukemia cells. Blood 77: 989.

3. Elbaz O, Budel LM, Hoogerbrugge H, Touw PI, Delwel R, Mahmoud AL, Löwenberg B (1991) Tumor necrosis factor downregulates granulocyte-colony-stimulating factor receptor expression on human acute myeloid leukemia cells and granulocytes. J Clin Invest 87: 838.

4. Caux C, Sealand S, Favre C, Duvert V, Mannoni P, Banchereau J (1990) Tumor necrosis factor-alpha strongly potentiates interleukin 3 and granulocyte-macrophage colony-stimulating factor-induced proliferation of human CD34+ hematopoietic progenitor cells. Blood 75: 2292.

Blocking interleukin-1

Charles A. Dinarello

Department of Medicine, Tufts University School of Medicine and New England Medical Center Hospital, Boston, MA 02111, USA

Summary   Interleukin-1 (IL-1) affects nearly every cell type by increasing the expression of a variety of genes associated with the promotion of inflammatory processes. In animal models of infectious, inflammatory or metastatic disease, the role of IL-1 has been defined by specifically preventing the ability of IL-1 to trigger its type I receptor (IL-1R). The type I IL-1R transmits the signal whereas the type II IL-1R lacking a cytoplasmic segment does not. Antibodies to the type I receptor, the IL-1 receptor antagonist (IL-1Ra) and soluble (extracellular) type I receptors have been used to reduce the severity of disease in these animal models. IL-1Ra binds to both IL-1 receptors but does not transmit a biological signal. IL-1Ra has been given to humans in pharmacologic levels without toxicity, accomplishing a reduction in endotoxin-induced neutrophilia and mortality rate in the septic shock syndrome. Soluble IL-1 receptors are also in clinical trials. The IL-1Ra and the soluble form (extracellular) of IL-1RII are produced naturally and are elevated in the circulation in a variety of diseases. The ratios of IL-1Ra to IL-1β in the circulation favor IL-1Ra by a 100-fold molar excess. However, in acute myelogenous leukemia, this ratio is reversed.

Introduction.   The interleukin-1 (IL-1) family consists of three structurally related polypeptides; the first two are IL-1α and IL-1β, each of which has a broad spectrum of both beneficial and harmful biological activities. The third member is IL-1 receptor antagonist which inhibits the activities of IL-1. Among the properties of IL-1 (α and β) are the ability to induce fever, sleep, anorexia, and hypotension. IL-1 stimulates the release of pituitary hormones, increases the synthesis of collagenases, resulting in cartilage destruction, and stimulates prostaglandin production, leading to a decrease in pain threshold. IL-1 also has been implicated in destruction of beta-cells of the islets of Langerhans, growth of acute and chronic myelogenous leukemia cells, inflammation associated with arthritis and colitis, and development of atherosclerotic plaques. In addition, IL-1 has some host defense properties. For example, IL-1 stimulates T and B lymphocytes and, in animals, protects bone marrow stem cells from radiation-induced death; in Phase I trials, IL-1 increased the numbers of bone marrow precursor cells and circulating platelets and neutrophils (Smith *et al.*, 1990; Tewari *et al.*, 1990). However, increasing doses caused fever, gastrointestinal disturbances, myalgia, arthralgia and hypotension.

There is a third member of the interleukin-1 family, interleukin-1 receptor antagonist (IL-1Ra) which provides some protection against the disease-provoking effects of IL-1. IL-1Ra is a specific inhibitor of IL-1 activity that acts by blocking the binding of IL-1 to its cell surface receptors (Arend, 1991; Seckinger *et al.*, 1987). The preliminary results of clinical studies with IL-1Ra suggest that it may be beneficial in some patients with sepsis syndrome and some forms of chronic myelogenous leukemia and that it is safe.

<u>Structure, Synthesis, Processing and Secretion</u>. There are two distinct genes for IL-1: these have been named IL-1α and IL-1β. Each gene codes for the IL-1α and β proteins respectively. IL-1α and IL-1β have different amino acid sequences, but are structurally related at the three-dimensional level. They act via the same cell surface receptors and share similar biological activities (Dinarello, 1991). IL-1 is rapidly synthesized by mononuclear cells, primarily monocytic phagocytes, when stimulated by microbial products or inflammation. IL-1 is first synthesized as a larger precursor molecule. The precursors for IL-1α and β have molecular weights of 31,000 Daltons, but, unlike most proteins, both forms of IL-1 lack the classic chain of signal amino acids that enable cells to cleave precursor proteins into smaller "mature" sizes and tranport them out of the cell. The mature size for both forms of IL-1, is 17,500 Daltons.

Because of the lack of signal amino acids, most IL-1α remains in the cytosol of cells where it may function as an autocrine messenger. There is also evidence that the precursor of IL-1α is transported to the surface of the cell where it is expressed as "membrane bound" IL-1. Membrane-bound IL-1α precursor is biologically active perhaps serving as a paracrine messenger for adjacent cells. On the other hand, a considerable amount of IL-1β is transported out of the cell where it enters the extracellular space and also the circulation. Various mechanisms for transport include exocytosis from vesicles, active transport via multiple drug resistance proteins, "leakiness" or following cell death.

Unlike the IL-1α precursor, the IL-1β precursor requires cleavage for optimal biological activity. Several common enzymes will cut the IL-1β precursor into smaller and more active forms. However, one particular protease appears highly specific for cutting the IL-1β precursor from 31,000 to 17,500, its most active form (Cerretti *et al.*, 1992; Thornberry *et al.*, 1992). This enzyme is known as the IL-1β converting enzyme (ICE). ICE is an intracellular protease and member of the cysteine protease family. ICE does not cleave the IL-1α precursor but non-specific proteases for cutting IL-1α have been identified. ICE-specific substrate inhibitors reduce the amount of mature IL-1β produced by activated monocytes (Thornberry, Bull et al., 1992).

<u>IL-1 as a Mediator of Disease</u>. IL-1 has actions both directly on specific target tissues and as an effector molecule. An example of a direct effect of IL-1 is the septic shock syndrome in which infection leads to the synthesis and release of large amounts of IL-1 that directly induce hypotension. An example of IL-1 as an effector molecule is autoimmune disease in which IL-1 contributes to tissue destruction because of its ability to stimulate collagenases and prostaglandins which induce bone resorption.

In most autoimmune diseases, T lymphocytes and not IL-1-producing phagocytic cells are the critical determinants of disease. The various treatments such as cyclosporine affect T-lymphocytes rather than mononuclear phagocytes. Any reduction in IL-1 production associated with immunosuppressive therapies is likely secondary to reduced T-lymphocyte stimulation of IL-1-producing cells. Reducing the production of IL-1 in autoimmune diseases reduces the inflammatory effects of IL-1 on target tissues such as the synovium, pancreatic islet cells or intestinal lining cells.

<u>Inhibition of Synthesis or Processing of IL-1.</u> A number of drugs, cytokines and other substances inhibit IL-1 production. With the exception of anti-sense IL-1 DNA, their action is not specific, since the production of other cytokines is usually inhibited as well. Inhibition of the proteolytic cleavage of the precursor of IL-1β is a potential mechanism for reducing IL-1 activity but only amino acid substrate inhibitors have been identified (Thornberry, Bull et al., 1992).

<u>Neutralization of IL-1 Activity</u>. Neutralizing antibodies to IL-1β decrease the proliferation of leukemia cells in vitro (Cozzolino *et al.*, 1989). There are two IL-1 cell surface receptors (IL-1R). The type I IL-1R (IL-1RI) is found on most cells and appears to be important for transducing IL-1 activity. The type II IL-1 receptor (IL-1RII), is found primarily on neutrophils, monocytes, bone marrow cells and B-lymphocytes; its function is appears to be one of binding IL-1 and preventing it from interacting with the signal transducing type I receptor.

Soluble IL-1R receptors produced by recombinant techniques act like antibodies in that they specifically bind IL-1 and thereby prevent IL-1 binding to its receptors on cells. A naturally occurring soluble form of an IL-1 receptor has been identified in body fluids (Symons *et al.*, 1991). This soluble IL-1 receptor is generated by proteolytic cleavage of the extracellular portion of the type II receptor. It specifically binds IL-1 and therefore prevents IL-1 binding to IL-1 receptors on cell surfaces. Low levels of this soluble receptor are found in the circulation of healthy subjects.

<u>Blocking IL-1 Receptors</u>. Considerable information has accumulated pertaining to the blockade of IL-1 receptors using the naturally occurring IL-1Ra (Arend, 1991; Dinarello and Thompson, 1991). This IL-1 antagonist belongs to the IL-1 gene family, it is produced by the same cells, and it has the same molecular size of mature IL-1 and is structurally related to it. Similar to the findings on IL-1, IL-1Ra is not present in peripheral blood monocytes or plasma from normal subjects but is found in the keratinocytes of the skin and in neuronal cells. IL-1Ra binds to cellular IL-1 receptors but that binding does not initiate any biological response. Lacking biological activity, it can be given in doses sufficient to occupy IL-1 receptors and therefore prevent IL-1 from binding to these same receptors. Since both IL-1Ra and soluble IL-1R are specific inhibitors of IL-1, any modification of disease with the use of these agents would indicate a role for IL-1 in that particular disease.

IL-1Ra appears to be a pure receptor antagonist. In a Phase I trial in normal subjects, raising the plasma IL-1Ra concentrations to 25 to 30 µg/ml caused no symptoms or changes in vital signs and did not alter white blood cell counts or routine biochemical and endocrinologic tests (Granowitz et al., 1992). These results are consistent with the concept that IL-1 does not play an important role in normal hematopoiesis

<u>Acute and Chronic Myelogenous Leukemia.</u> Increasing evidence supports the concept that IL-1 is a growth factor for acute and chronic myelogenous leukemia cells. Although IL-1β mRNA cannot be detected in peripheral blood cells or bone marrow aspirates from normal subjects, it can be detected in cells from patients with acute myelogenous leukemia (Rambaldi et al., 1991). In addition, these leukemia cells produce IL-1 activity (Griffin et al., 1987) as do chronic granulocytic leukemia cells of the juvenile type (Bagby et al., 1988). These results do not necessarily implicate IL-1 as a cause of these leukemias. However, antibodies to IL-1β significantly reduce the spontaneous proliferation and colony formation of these cells *in vitro* (Cozzolino, Rubartelli et al., 1989), and IL-1Ra and an IL-1 inhibitor from human M20 leukemia cells similarly decrease the proliferation and colony formation of leukemic cells (Estrov et al., 1991; Peled et al., 1992; Rambaldi, Torcia et al., 1991). Anti-IL-1-based therapy may be useful to inhibit or synchronize the growth of the leukemia cells so that they become more vulnerable to chemotherapeutic agents.

The postulated mechanism for the role of IL-1 as a stimulator of the growth of myelogenous leukemia cells is consistent with the known ability of IL-1 to stimulate the synthesis of colony stimulating growth factors and to act as a co-factor in the proliferation of stem cells. In tumors such as multiple myeloma, IL-1 may also play a role by inducing IL-6, a cytokine growth factor for B-lymphocytes. In patients with this tumor, blocking IL-1 production or action may be beneficial. The IL-1 specific precursor-converting enzyme is found in large amounts in myelogenous leukemia cells (Cerretti, Kozlosky et al., 1992). One approach to reduce the action of IL-1 in myelogenous leukemia therefore would be to prevent the proteolytic cleavage of the IL-1 precursor to an active form.

Differential Regulation of Production of IL-1 and IL-1Ra. The production of IL-1 and IL-1Ra is differentially regulated even in the same cell. For example, in cells stimulated with endotoxin, the production of IL-1β preceds that of IL-1Ra. In addition, activation of immunoglobulin receptors on the monocytes stimulates the production of IL-1Ra but not that of IL-1. Monocytes from patients on chronic hemodialysis produce more IL-1Ra than IL-1β. In humans injected with a small amount of endotoxin, the plasma concentrations of IL-1Ra are 100-fold greater than those of IL-1β (Granowitz *et al.*, 1991). In contrast, acute myelogenous leukemia cells produce IL-1β but little IL-1Ra (Rambaldi, Torcia et al., 1991).

Other cytokines may also contribute to the balance between IL-1 and IL-1Ra production in disease. For example, IL-4, transforming growth factor-β and IL-10 increase the production of IL-1Ra but at the same time decrease the production of IL-1 (Vannier *et al.*, 1992). The actions of these cytokines in various inflammatory diseases may include regulation of the balance between IL-1 and IL-1Ra. A special example of this balance exists in the skin. Skin keratinocytes contain large amounts of IL-1α and 10 to 100-fold higher concentrations of IL-1Ra.

Does Blocking IL-1 Action Impair Host Defense Mechanisms? Patients with sepsis have plasma concentrations of IL-1β in the range of 250 to 500 pg/ml and occasionally over 1,000 pg/ml. Raising the plasma concentrations of IL-1Ra to 20 to 30 µg/ml (representing a 20,000 fold molar excess) by intravenous infusion of IL-1Ra for three days in patients with septic shock is associated with improved survival and in normal subjects, the same plasma IL-1Ra concentrations had no deleterious effects on immune responses (Granowitz, Porat et al., 1992). These results are consistent with in vitro experiments in which blocking IL-1 action did not diminish the response of human T-cells to antigens (Nicod *et al.*, 1992). Thus, blocking IL-1 action is likely to be safe, at least in the short term. Since IL-1Ra treatment of some diseases thought to be mediated by IL-1 may require prolonged use, the question arises whether sustained blockade of IL-1 action will weaken host defense mechanisms.

There is no dearth of experimental data illustrating a role for IL-1 in boosting natural host defense mechanisms. For example, IL-1 protects the early stem cells against radiation damage (Neta, Oppenheim et al., 1988). How can both IL-1 administration and blocking IL-1 action have the same effect? An example of this type of duality in biology is well documented for trace elements - low levels are essential for several metabolic pathways whereas high levels are toxic(Mertz, 1981). Thus, small amounts of IL-1 may be necessary for maintenance of host defenses, but large amounts of IL-1 are injurious.

Conclusions The IL-1 family consists of two agonists (IL-1α and IL-1β) as well as its own antagonist (IL-1Ra). IL-1 plays a pathogenic role in several diseases and reducing IL-1 production or blocking its actions are appropriate strategies for treating patients with both acute and chronic diseases. Although reducing or blocking IL-1 action for short periods of time is not likely to impair host defenses, there may be a risk to complete and prolonged anti-IL-1 treatment.

Acknowledgements These studies were supported by NIH Grant AI 15614 from the National Institutes of Health.

References

Arend, W.P. (1991): Interleukin-1 receptor antagonist. J Clin Invest. 88: 1445-1451.

Bagby, G.C.J., Dinarello, C.A., Neerhout, R.C. et al. (1988): Interleukin 1-dependent paracrine granulopoiesis in chronic granulocytic leukemia of the juvenile type. J Clin Invest. 82: 1430-1436.

Cerretti, D.P., Kozlosky, C.J., Mosley, B. et al. (1992): Molecular cloning of the IL-1β processing enzyme. Science. 256: 97-100.

Cozzolino, F., Rubartelli, A., Aldinucci, D. et al. (1989): Interleukin 1 as an autocrine growth factor for acute myeloid leukemia cells. Proc Natl Acad Sci (USA). 86: 2369-2373.

Dinarello, C.A. (1991): Interleukin-1 and interleukin-1 antagonism. Blood. 77: 1-26.

Dinarello, C.A. and Thompson, R.C. (1991): Blocking IL-1: effects of IL-1 receptor antagonist in vitro and in vivo. Immunol Today. 12: 404-410.

Estrov, Z., Kurzrock, R., Wetzler, M. et al. (1991): Suppression of chronic myelogenous leukemia colony growth by IL-1 receptor antagonist and soluble IL-1 receptors: a novel application for inhibitors of IL-1 activity. Blood. 78: 1476-1484.

Granowitz, E.V., Porat, R., Mier, J.W. et al. (1992): Effects of intravenous interleukin-1 receptor antagonist in healthy humans subjects. Cytokine. 4: in press.

Granowitz, E.V., Santos, A.A., Poutsiaka, D.D. et al. (1991): Production of interleukin-1-receptor antagonist during experimental endotoxaemia. Lancet. 338: 1423-4.

Griffin, J.D., Rambaldi, A., Vellenga, E. et al. (1987): Secretion of interleukin-1 by acute myeloblastic leukemia cells in vitro induces endothelial cells to secrete colony stimulating factors. Blood. 70: 1218-1221.

Mertz, W. (1981): The essential trace elements. Science. 213: 1332-1338.

Neta, R., Oppenheim, J.J. and Douches, S.D. (1988): Interdependence of the radioprotective effects of human recombinant interleukin 1 alpha, tumor necrosis factor alpha, granulocyte colony-stimulating factor, and murine recombinant granulocyte-macrophage colony-stimulating factor. J Immunol. 140: 108-11.

Nicod, L.P., El Habre, F. and Dayer, J.-M. (1992): Natural and recombinant interleukin 1 receptor antagonist does not inhibit human T-cell proliferation induced by mitogens, soluble antigens or allogeneic determinants. Cytokine. 4: 29-35.

Peled, T., Rigel, M., Peritt, D. et al. (1992): Effect of M20 interleukin-1 inhibitor on normal and leukemia human myeloid progenitors. Blood. 79: 1172-1177.

Rambaldi, A., Torcia, M., Bettoni, S. et al. (1991): Modulation of cell proliferation and cytokine production in acute myeloblastic leukemia by interleukin-1 receptor antagonist and lack of its expression by leukemic cells. Blood. 78: 3248-53.

Seckinger, P., Lowenthal, J.W., Williamson, K. et al. (1987): A urine inhibitor of interleukin-1 activity that blocks ligand binding. J Immunol. 139: 1546-1549.

Smith, J., Urba, W., Steis, R. et al. (1990): Interleukin-1 alpha: Results of a phase I toxicity and immunomodulatory trial. Am Soc Clin Oncol. 9: 717.

Symons, J.A., Eastgate, J.A. and Duff, G.W. (1991): Purification and characterization of a novel soluble receptor for interleukin-1. J Exp Med. 174: 1251-1254.

Tewari, A., Buhles, W.C., Jr. and Starnes, H.F., Jr. (1990): Preliminary report: effects of interleukin-1 on platelet counts. Lancet. 336: 712-714.

Thornberry, N.A., Bull, H.G., Calaycay, J.R. et al. (1992): A novel heterodimeric cysteine protease is required for interleukin-1 beta processing in monocytes. Nature. 356: 768-74.

Vannier, E., Miller, L.C. and Dinarello, C.A. (1992): Coordinated antiinflammatory effects of interleukin 4: interleukin 4 suppresses interleukin 1 production but up-regulates gene expression and synthesis of interleukin 1 receptor antagonist. Proc Nat'l Acad Sci. 89: 4076-4080.

Résumé

L'interleukine-1 (IL-1) affecte presque chaque type cellulaire en augmentant l'expression d'une variété de gènes associés avec le développment des processus inflammatoires. Dans des modèles animaux de maladies infectieuses, inflammatoires ou métastatiques, le rôle de l'IL-1 a été défini en empêchant spécifiquement sa capacité à stimuler son récepteur de type I (IL-1R). L'IL-1R de type I transmet le signal, ce que ne peut faire l'IL-1R de type II dépourvu d'un segment cytoplasmique. Des anticorps contre le récepteur de type I, l'antagoniste du récepteur à l'IL-1 (IL1-Ra) et des récepteurs solubles (extracellulaires) de type I ont été utilisés pour réduire la sévérité de la maladie dans ces modèles animaux. L'IL1-Ra se lie aux 2 récepteurs de l'IL-1 mais ne transmet pas de signal biologique. L'IL1-Ra a été administré chez l'homme à doses pharmacologiques sans signes de toxicité, mais en entrainant une diminution de la neutrophilie induite par l'endotoxine et le taux de mortalité par choc septique. Les récepteurs solubles font aussi l'objet d'essais cliniques. L'IL1-Ra et la forme soluble (extracellulaire) de l'IL1-RII sont produits de façon naturelle et ont des taux circulants élevés dans certaines maladies. Le rapport IL1-Ra à l'IL-1β dans la circulation est en faveur de l'IL1-Ra (molarité 100 fois plus élevée). Cependant, dans la leucémie aigue myéloblastique, ce rapport est inversé.

The effect of IL-1 receptor blockade on proliferative activity of haemopoietic stem cells in regenerating bone marrow

Pavle Milenković, Stanislava Stošić-Grujičić*, Zoran Ivanović and Miodrag Lukić**

*Institute for Medical Research, *Institute for Biological Research, **Institute for Microbiology and Immunology, Faculty of Medicine, Beograd, Yugoslavia*

Interleukin-1 (IL-1) influences haemopoietic stem cell proliferation in synergistic pattern with other colony-stimulating factors which involves up-regulation of their receptors on bone marrow cells (Dinarello, 1991). It is possible that haemopoietic effects of IL-1 occur through both direct and indirect mechanisms. In this work the effect of IL-1 receptor blockade with IL-1 receptor antagonist (IL-1ra) on proliferation of day 12 or day 8 spleen colony-forming units (CFU-Sd12, CFU-Sd8) was assessed during the recovery of haemopoiesis after sublethal damage by irradiation. Regenerating bone marrow cells of CBA mice 3 days after 2 Gy irradiation were incubated for 5h in the presence of increasing concentrations of recombinant human IL-1ra* (rhIL-1ra-blocking both type I and type II IL-1 receptors) or an IL-1ra activity (50 - 100 kDa molecular weight range) derived from glucocorticoid treated rat keratinocytes (kIL-1ra), which we have described as type I IL-1ra activity (Stošić-Grujičić & Lukić, 1992). To determine the proliferative activity of CFU-Sd12 or CFU-Sd8 Ara-C was added for the last 1h of incubation (Milenković et al., 1991).

Table 1. Influence of rhIL-1ra and kIL-1ra on proliferative activity of CFU-S in regenerating bone marrow

rhIL-ra (ng/ml)	CFU-Sd8 (% in S phase)	CFU-Sd12 (% in S phase)	kIL-1ra (dilutions)	CFU-Sd12 (% in S phase)
0	34.1 ± 9.0	37.0 ± 3.8	0	41.5 ± 6.7
5	-	30.7 ± 8.3	1 : 20	51.2 ± 3.2
10	35.1 ± 8.0	2.2 ± 5.7 ***	1 : 10	69.6 ± 4.6
50	26.1 ± 10.0	9.3 ± 2.7 ***	1 : 5	82.6 ± 4.3
100	19.9 ± 6.9	3.3 ± 4.0 ***		

(mean ± SE; *** = $p < 0.001$)

The results revealed (Table 1) inhibition of CFU-Sd12 but not of CFU-Sd8 cycling after blocking IL-1 receptors with recombinant human IL-1ra. However, glucocorticoid induced IL-1ra activity did not suppress cycling of CFU-Sd12. It is possible that down regulation of stimulatory IL-1 activity in stem cell proliferation influences primarily primitive populations and does not involve equally type I and type II receptors known to be differentially expressed on various populations of haemopoietic cells and cells belonging to haemopoietic microenvironment. This hypothesis is presently being tested.
* kindly provided by Dr. C.A. Dinarello

REFERENCES

Dinarello, C.A. (1991): Interleukin-1 and interleukin-1 anagonism. *Blood.* 77: 1627 - 1652.
Milenković, P., Ivanović, Z., Kataranovski, M., and Lukić, M.L. (1991): Stimulator of proliferation of spleen colony-forming cells in T-cell deprived mice treated with cyclophosphamide or irradiation. *Cell Prolif.* 24: 507 - 515.
Stošić-Grujičić, S. & Lukić, M.L. (1992): Glucocorticoid-induced keratinocyte-derived interleukin-1 receptor antagonist(s). *Immunology.* 75: 293 - 298.

Characterization of specific receptors for HuIFN-α on a human megakaryocytic cell line (Dami): regulation of their expression in PMA-differentiated cells

Marie-Claire Martyré and Juana Wietzerbin

INSERM U 365, Institut Curie, 26, rue d'Ulm, 75231 Paris Cedex 05, France

Interferon-alpha (IFN-α) treatment has been shown to be highly effective in controlling thrombocytosis in patients with myeloproliferative disorders. These observations suggest that IFN-α might play some role on the biologic feature of the megakaryocytic lineage and led us to investigate for the presence of specific receptors for IFN-α on a human megakaryocytic cell line, Dami, and to study the regulation of their expression. Our study demonstrates that [^{125}I]-recombinant human IFN-α ([^{125}I]-rHu-IFN-α) binds to high affinity specific receptor on these cells. Scatchard analysis of binding data indicates the presence of homogeneous binding sites estimated in the range of 3,000-5,000, with an apparent equilibrium dissociation constant, Kd, of $1-2 \times 10^{-9}$ M. Also, [^{125}I]-rHuIFN-α binding capacity decreased in Dami cells incubated with unlabeled rHuIFN-α. This down-regulation which was dose-dependent appeared to result from a reduction of IFN-α cell surface receptors and was observed at doses that elicited antiproliferative effects in Dami cells. Cross-linking of [^{125}I]-rHuIFN-α to Dami membrane proteins using a bifunctional reagent yielded to a radioactive complex of ≈ 150,000 kDa on SDS-PAGE. Furthermore, in response to PMA, which induces the differentiation/maturation of the Dami cells as evaluated by surface marker and ploidy analysis, a three-fold increase of the number of specific membrane receptors for IFN-α was observed, without any modification of either the affinity or the Mr value of the cross-linked complex. Such an increase appeared to be restricted to IFN-α receptors, actually it was not observed in [^{125}I]-IFN-γ binding experiments. Transcript analysis indicated that down-regulation and increased expression of the IFN-α receptor after PMA treatment are post-transcriptional events.

II. Cell cycle inhibitors

II. Inhibiteurs du cycle cellulaire

MAP kinase cascade: an essential signalling route that controls cell proliferation

J. Pouysségur, A. Brunet, J.-C. Chambard, G. L'Allemain, P. Lenormand and G. Pagès

Centre de Biochimie-CNRS, Université de Nice, Parc Valrose, 06108 Nice, France

ABSTRACT

Mitogen-Activated Protein kinases ($p42^{mapk}$ and $p44^{mapk}$) are serine/threonine kinases rapidly activated in cells stimulated with various extracellular signals. Activation is mediated *via* MAP kinase kinase ($p45^{mapkk}$), a dual specificity kinase that phosphorylate two key regulatory threonine and tyrosine residues of MAP kinases. Here we demonstrate that: (i) synergistic mitogens synergistically activate MAP kinases, (ii) the persistent phase of MAP kinase activation correlates with G0 to S-phase progression, (iii) MAP kinases rapidly translocate to the nucleus in response to growth factors and (iv) suppression of MAP kinase activation (antisense or dominant-negative allele expression) inhibits AP1 activity and growth factor-stimulated cell cycle re-entry. We conclude that MAP kinases are absolutely required to transmit the mitogenic response, relaying extracellular signals to the nucleus.

INTRODUCTION

Mitogen Activated Protein Kinases (MAP kinases) also described as extracellular signal regulatory kinases (ERKs), belong to a group of protein serine/threonine kinases that are activated in response to extracellular stimuli in virtually all cell types (see Sturgill and Wu, 1991; Cobb *et al.*, 1991; Thomas, 1992; Pelech and Sanghera, 1992 for reviews). Two highly related mammalian MAP kinases, $p44^{mapk}$ (ERK1), and $p42^{mapk}$ (ERK2) have been cloned and found ubiquitously expressed in cultured cells (Boulton *et al.*, 1991; Boulton and Cobb, 1991; Her *et al.*, 1991; Meloche *et al.*, 1992) and highly homologous to the yeast kinases SLT2 (Torres et al., 1991), KSS1 (Courchesnes *et al.* 1989) and FUS3 (Elion *et al.* 1990). Dual phosphorylation of these MAP kinases on both tyrosine and threonine residues has been shown to be required for full activity of the enzyme (Anderson et al 1990). The sites of phosphorylation, identified in $p42^{mapk}$ (Payne *et al.* 1991) and conserved in all members of the family, were found to reside on a single phosphopeptide separated by only one residue T*EY*. Recently a MAP kinase activator was identified, purified and cloned (Crews *et al.*, 1992; Ashworth *et al.* 1992; Kosako *et al.*, 1992; Pagès, G., Brunet, A. and Pouysségur, J., in preparation), and shown to phosphorylate *in vitro* a kinase-deficient mutant of $p44^{mapk}$ and re-activate a dephosphorylated wild type MAP kinase in the presence of ATP. This MAP kinase activator referred now to as MAP kinase kinase could be phosphorylated and activated *in vitro* by v-Raf or constitutively active forms of c-Raf (Dent *et al.* 1992; Kyriakis *et al.* 1992;

Howe et al., 1992). Thus, c-Raf appears to be, as we previously proposed (Pouysségur et al., 1991), one transducer integrating mitogenic and differentiating signals leading to activation of the MAP kinase cascade.

Here we bring direct evidence that this kinase cascade, that integrate various extracellular signals, plays an essential role in transmitting the proliferative response in fibroblasts.

SYNERGISTIC ACTIVATION OF MAP KINASES

Reinitiation of DNA synthesis in resting cells is the result of cooperation between multiple signalling pathways. For example in CCL39 cells — a fibroblastic cell line that we have analyzed in details over the last ten years — the strong mitogen α-thrombin stimulates phospholipases {PIP2-PLC (L'Allemain et al. 1986), PC-PLD, PLA2 (McKenzie et al., 1992)} and inhibits adenylyl cyclase (Magnaldo et al., 1988), implicating activation of pertussis toxin-sensitive and -insensitive G proteins (Pouysségur et al., 1991). On the other hand, serotonin which is not a mitogen on its own although it activates G protein-coupled effectors systems, synergizes with FGF to induce mitogenesis (Seuwen et al., 1988). We therefore analyzed the magnitude and time course of MAP kinase activation in response to individual growth factor or to the synergistic combination serotonin/FGF. We showed that α-thrombin induces a biphasic activation of $p44^{mapk}$ in CCL39 cells: a rapid and sharp phase appearing at 5-10 min was followed by a late and sustained phase lasting at least 4 hours (Meloche et al., 1992).

A set of experiments strongly suggest that the second and persistent phase of activation is required for G0-arrested cells to enter the S-phase. The non-mitogenic agents for CCL39 cells: TPA, serotonin, thrombin-receptor 7-mer peptide (SFFLRNP) activate only the first peak (Vouret-Craviari et al., 1993). Early interruption of the α-thrombin signal with hirudin completely abolished the late phase of $p44^{mapk}$ activation as well as DNA synthesis reinitiation (Meloche et al., 1992). Furthermore, pretreatment of the cells with pertussis toxin, which inhibits more than 95% α-thrombin-induced mitogenicity (Chambard et al., 1987), resulted in the complete loss of late $p44^{mapk}$ activation phase, while the early peak was partially attenuated. Finally, serotonin and the thrombin-receptor peptide which can induce neither mitogenicity nor the sustained $p44^{mapk}$ activation phase on their own, do so when they are associated with a suboptimal concentration of FGF (Meloche et al., 1992; Vouret-Craviari et al., 1993). Taken together, these results point to a very close correlation existing between the ability of a growth factor to induce late and sustained $p44^{mapk}$ activation and its mitogenic potential. Similar results have also been obtained for the activation of $p42^{mapk}$ in mesangial cells (Wang et al. 1992; Wang, Y., Pouysségur, J. and Dunn, M., in preparation).

MAP KINASES TRANSLOCATE TO THE NUCLEUS

At the present time the respective roles of the two isoforms of MAP kinases ($p44^{mapk}$ and $p42^{mapk}$) have not been identified. Both isoforms are phosphorylated on conserved threonine and tyrosine residues *via* the action of the MAP kinase kinase and their kinetics of activation are identical reflecting the action of a common activator. In addition, both isoforms seem to possess the same substrate specificity, at least *in vitro*. In an attempt to determine whether each isoform exerts a specific function, we examined their subcellular distribution by indirect immuno-fluorescence microscopy in G0-arrested and serum-stimulated CCL39 cells. Antisera directed against individual MAP kinases were raised with 16 and 14 amino-acid synthetic peptides corresponding respectively to the predicted carboxyl terminus of rat p44^{mapk} and p42^{mapk}. The 837 antisera (Boulton and Cobb, 1991) was made p44^{mapk} specific by absorption with the p42 carboxy peptide. With this antisera we detected the endogenous p44^{mapk} mainly in the cytoplasm of resting CCL39 cells. Interestingly a clear nuclear translocation appeared within 15 min following serum stimulation. The signal became predominantly nuclear after one hour and reached its maximum three hours following stimulation, then six hours later, the p44^{mapk} specific signal returned to the cytoplasm. The nuclear accumulation of MAP kinase was not permanent and required the continuous presence of serum to be maintained. Removal of serum induced a rapid return of MAP kinases to the cytoplasm (almost complete within 30 min). This rapid serum-dependent MAP kinase translocation to the nucleus also occurred in response to potent mitogens such as α-thrombin or FGF but not in response to non-mitogenic agents such as α-thrombin receptor peptides or phorbol esters. α-Thrombin and FGF are less potent than 10% serum in their ability to trigger MAP kinase translocation when they are added individually, yet almost equivalent to serum when added together. Interestingly, the intensity and percentage of nuclei that were labelled correlated well with the relative potency of mitogens to reinitiate DNA synthesis. It is worthy to note that the nuclear translocation of p44^{mapk} coincides with the second phase of MAP kinase activation. Thus the MAP kinase molecules that translocate to the nucleus are potentially active. It has been shown *in vitro* that MAP kinases were able to phosphorylate nuclear proteins such as c-Jun, c-Myc (Alvarez *et al.*, 1991; Baker *et al.*, 1992; Pulverer *et al.*, 1991) and p62^{TCF} (Gille *et al.*, 1992). In the latter case, the phosphorylation of p62^{TCF} by MAP kinase has been correlated with increased binding of a multiprotein complex on the transcriptional response element SRE (Serum Responsive Element) of the *c-fos* gene.

As far as the p42^{mapk} localization is concerned, unfortunately, all the antisera that we raised recognized both isoforms *in vivo* as judged by immunofluorescence of ectopic expression of each MAP kinase isoforms (see next section). We circumvented this problem by introducing at the extreme carboxylic end of each cloned MAP kinase isoforms a specific epitope recognized by a monoclonal antibody. We added 17 residues corresponding to the Vesiculo Stomatitis Virus Glycoprotein (VSVG) (Kreis, 1986) and a steric spacer between the body of the kinase and the epitope. We showed that both p44^{mapk} and p42^{mapk} VSVG-tagged isoforms were able to translocate to the nucleus upon growth factor stimulation. This signal was specific as the VSVG monoclonal antibody did not elicit any significant immuno-fluorescence in non-transfected cells. This result confirms similar work performed with HeLa cells showing nuclear localization of both isoforms of MAP kinases (Chen *et al.*, 1992).

In order to shed light on the mechanism of MAP kinase translocation, we transfected several point mutants of the cloned p44^{mapk}. We transfected both kinase-deficient (T192A, Y194F) and mutants for which the phosphorylation sites were mutated to glutamic acid in order to mimic the negative charges brought by the phosphorylations: T192E, Y194E and TY-EE. All of these kinase-deficient mutants showed nuclear localization, and for those studied in detail (T192A, Y194F and TY-EE) the nuclear translocation in response to serum was not affected. We thus conclude that MAP kinase translocation is independent of its intrinsic activity (Lenormand, P., Sardet, C., Pagès, G. and Pouysségur J. in preparation).

CLONING AND ECTOPIC EXPRESSION OF MAP KINASES

We cloned a near full length cDNA of the p44^{mapk} from Chinese hamster lung fibroblasts (CCL39) model system (Meloche *et al.*, 1992). The hamster p44^{mapk} revealed

98.6% homology with the rat protein. To analyze the expression and the biological function of the cloned p44^{mapk}, an artificial initiation codon and 9 aminoacid corresponding to an epitope of the HA1 influenza hemagglutinin were added at the amino terminal end of the MAP kinase (Meloche et al., 1992). The chimeric kinase under transcriptional control of the cytomegalovirus promoter, was stably expressed in CCL39 cells in a functional form. We showed that its basal activity, measured by phosphorylation of the substrate myelin basic protein, is activated several-fold (up to 25) by the mitogens α-thrombin, platelet-derived growth factor, fibroblast growth factor and fetal calf serum. In response to α-thrombin, this ectopically expressed p44^{mapk} shows two peaks of activation that parallels increased phosphorylation of tyrosine and threonine residues. From all criteria (hormonal activation, phosphorylation, time-course of activation, nuclear localization) this expressed form of p44^{mapk} has apparently retained all the regulatory features of the endogenous form. We therefore decided to use this cloned isoform to see whether expression of the wild type or mutated forms of the kinase could alter growth control.

First, we isolated different clones of CCL39 cells stably expressing the recombinant p44^{mapk} to levels representing up to 5 times the endogenous one. All these clones remained growth factor-dependent with no sign of increased sensitivity to mitogens. This result suggested that MAP kinase basal level is thighly regulated. Indeed, in spite of an increased level of p44^{mapk}, MAP kinase basal level was not elevated by overexpression of the enzyme. We then expressed p44^{mapk} mutants in which the two key phosphorylation sites, T192 and Y194, were replaced by glutamic acid, hoping to partially mimic the negative charges of the phosphate groups. The simple or double mutant T192E/Y194E did not show any 'constitutive' activity when expressed in CCL39 cells. Therefore, to interfere with the endogenous MAP kinases, our second approach was to overexpress p44^{mapk} antisense RNA or kinase-deficient mutants of p44^{mapk}.

INHIBITION OF MAP KINASES STOPS CELL PROLIFERATION

In order to assess the biological roles of MAP kinases, we decided to specifically suppress MAP kinase activation in fibroblasts by transiently expressing either the entire p44^{mapk} antisense RNA, or p44^{mapk} kinase-deficient mutants (T192A or Y194F) (Pagès et al., 1993). We set up a transient transfection assay to markedly overexpress either the active and inactive forms of p44^{mapk} and to investigate the incidence of the transgene on the activation of endogenous MAP kinases. Thus, we co-transfected PS120, a Na$^+$/H$^+$ antiporter-deficient derivative of CCL39 (Pouysségur et al., 1984) cells, with either of the p44^{mapk} constructs along with a plasmid bearing the antiporter gene (Sardet et al., 1989). Two days after transfection, application of an acid-load selection specifically enriched the population with cells which had been efficiently transfected. Under these conditions, the transfected mutant or wild type p44^{mapk} were highly overexpressed and the activation of endogenous p42^{mapk} activity by serum was analyzed. We found that expression of the mutant forms specifically inhibited by up to 70% growth factor-stimulated endogenous MAP kinases (Pagès et al., 1993). In an alternative approach to reduce endogenous MAP kinase activation, but using the same transient transfection assay, we expressed p44^{mapk} antisense RNA. Several antisense constructs specifically reduced, in a dose-dependent manner, the expression of both endogenous isoforms of MAP kinases. As a consequence the antisense constructs suppressed growth factor activation of MAP kinases, whereas the sense construct did not. Interestingly, the inhibition of MAP kinase activation upon growth factor stimulation had striking effects on cell proliferation: it reduced the number of colonies obtained one week after transfection of either the antisense or the kinase-deficient forms of p44mapk. The growth inhibitory effects are specific since for example, p44mapk sense RNA expression reverted the effect of the p44mapk antisense RNA expression. Similar effects were observed on the transcription of a reporter gene under the control of the collagenase minimal promoter (Angel et al., 1987). Growth factor-induced transcriptional activity of the reporter gene was inhibited by expression of the p44mapk kinase-deficient mutants.

CLONING AND EXPRESSION OF THE MAP KINASE KINASE OF HAMSTER FIBROBLASTS

Nishida's group has identified and purified a *Xenopus* MAP kinase activator (Kosako et al., 1992). It is a 45 kDa phosphoprotein capable to reactivate an inactive recombinant MAP kinase. As expected, this reactivation is coincidental with a dual phosphorylation of the tyrosine and threonine residues of the T*EY* conserved cluster. A key finding in the identification of this activator was the demonstration that the dual phosphorylation also took place with a kinase-dead MAPK (Kosako et al., 1992); this finding left no doubt that this activator is a dual specificity protein kinase and not a kind of 'allosteric' activator inducing an auto MAP kinase activity as originally thought. Indeed the partial sequence of this activator revealed the existence of two subdomains of protein kinases (Kosako et al., 1992). From this peptide partial sequence, we designed two degenerate oligonucleotides corresponding to the N-terminal and the conserved subdomain VI respectively and isolated a PCR fragment from CCL39 cells. With this PCR fragment, used as a probe, we screened a CCL39-derived cDNA library and isolated a ~ 2 kb full length cDNA. The complete sequence revealed a protein kinase of 393 residues (Pagès, G., Brunet, A. and Pouysségur, J., in preparation) that is highly homologous (~50%) to byr1 and STE7, two yeast protein kinases involved in pheromone signalling (Sprague et al., 1992) and >98% identical with the mouse MAP kinase kinase recently reported (Crews et al., 1992). We epitope-tagged the N-terminal of this MAP kinase kinase cDNA and expressed the kinase into PS120 cells. With two antibodies, one directed against the *influenza* hemagglutinin epitope (HA1), the second against the amino terminus of MAP kinase kinase, we demonstrated a strong expression of the transfected MAP kinase kinase and a change in mobility upon growth factor stimulation (Figure 5). This mobility change that occurs rapidly reflects activation of MAP kinase kinase via phosphorylation (Kosako et al., 1992). These biological tools will now facilitate the investigation of the upstream partners involved in this activation and the key phosphorylation sites that participate in the activating mechanism.

CONCLUSIONS

The results presented herein indicate that the HA1-tagged p44mapk transfected into CCL39 cells behaves similarly to the endogenous MAP kinase. For example stimulation of quiescent cells by mitogenic growth factors induced a biphasic stimulation of MAP kinase activity, the first peak occurring 5 min after stimulation and the second peak between 1 and 3 hours. Non mitogenic signals induced the first peak of activation but not the second one. This correlation indicated for the first time a potential involvement of MAP kinase in the transmission of mitogenic signalling.
We then established that both p42mapk and p44mapk translocate to the nucleus upon growth factor activation. This nuclear localization was rapid, being detectable 15 min after agonist addition and maximal after 3 hours. This nuclear translocation was transient since MAP kinase returned to the cytoplasm 6 hours after stimulation, and the

nuclear accumulation was not permanent since removal of serum induced a rapid efflux of MAP kinase to the cytoplasm. Furthermore, the nuclear translocation occurred only in response to strong mitogenic agonists. Finally, we were able to specifically suppress MAP kinase activation in fibroblasts by transiently overexpressing either the antisense MAP kinase RNA or kinase-deficient mutants of MAP kinase. We found that either of these two means of markedly reducing growth factor-stimulated MAP kinase activation, inhibited cell growth and the expression of a reporter gene under the control of the collagenase minimal promoter. *Therefore, we conclude that growth factor activation of p44mapk and p42mapk is an absolute requirement for triggering the proliferative response.*

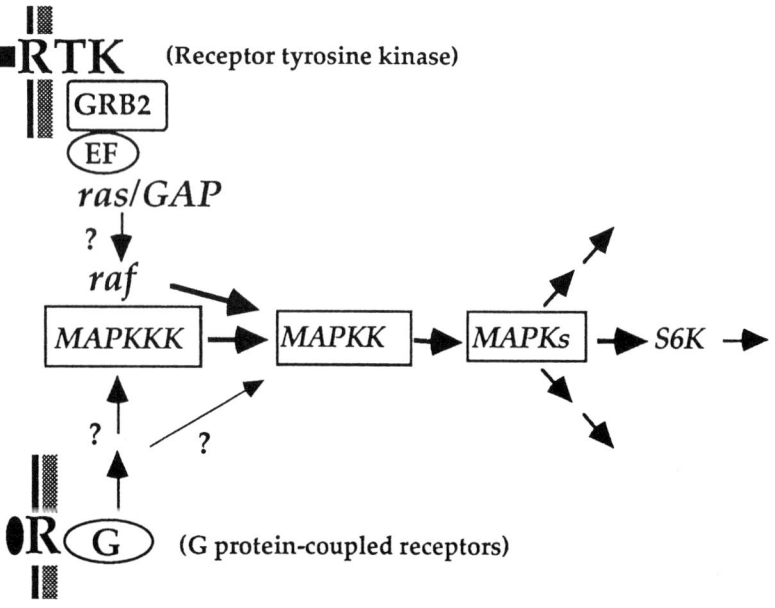

Figure 1: *MAP kinase cascade outlining the hypothetical or unknown (?) activating steps. GRB2: SH2/SH3 adaptor, EF: GDP/GTP exchange factor.*

We predict that these MAP kinases are also essential for establishing differentiating programs. For example, we expect to suppress NGF-induced differentiation of PC12 cells if we specifically block MAP kinase activation. If this happen to be the case, then this kinase cascade, conserved in yeast (Sprague, 1992), should be considered as a *master signalling route* activating multiple cellular targets, in particular, transcription factors essential for growth and differentiation.

We have outlined in Figure 1 the first known elements of the MAP kinase cascade together with the question marks that remain to be solved.

ACKNOWLEDGEMENTS

We thank our colleagues Drs. C. Sardet and S. Meloche for discussion and helpful collaboration at some stage of this work, Drs. F. McKenzie and Y. Wang for raising specific MAP kinase antisera and M. Cobb for kindly providing the 837 antiserum. This work was supported by the Centre National de la Recherche Scientifique (UMR 134), the Institut National de la Santé et de la Recherche Médicale, and the Association pour la Recherche contre le Cancer.

REFERENCES

Alvarez, E., Northwood, I. C., Gonzalez, F. A., Latour, D. A., Seth, A., Abate, C., Curran, T. and Davis, R. J. (1991): Pro-Leu-Ser/Thr-Pro is a consensus primary sequence for substrate protein phosphorylation. J. Biol. Chem. 266, 15277-15285.

Anderson, N. G., Maller, J. L., Tonks, N. K. and Sturgill, T. W. (1990): Requirement for integration of signals from two distinct phosphorylation pathways for activation of MAP kinase. Nature. 343, 651-653.

Angel, P., Baumann, I., Stein, B., Delius, H. and Rahmsdorf, H. J. (1987): 12-0-tetradecanoy-phorbol-13-acetate induction of the human collagenase gene is mediated by an inducible enhancer element located in the 5'-flanking region. Mol. Cell. Biol. 7, 2256-2266.

Ashworth, A., Nakielny, S., Cohen, P. and Marshall, C. (1992): The amino acid sequence of a mammalian MAP kinase kinase. Oncogene. 7, 2555-2556.

Baker, S.J., Kerppola, T.K., Luk, D., Vandenberg, M.T., Marshak, D.R., Curran T., Abate, C. (1992): Jun is phosphorylated by several protein kinases at the same sites that are modified in serum-stimulated fibroblasts. Mol. Cell. Biol. 12, 4694-4705

Boulton, T. G. and Cobb, M. H. (1991): Identification of multiple extracellular signal-regulated kinases (ERKs) with antipeptide antibodies. Cell Regul. 2, 357-371.

Boulton, T., Yancopoulos, G., Gregory, J., Slaughter, C., Moomaw, C., Hsu, J., Cobb, M. (1990): An insulin-stimulated protein kinase similar to yeast kinases involved in cell cycle control. Science 249, 64-67.

Chambard, J. C., Paris, S., L'Allemain, G. and Pouysségur, J. (1987): Two growth factor signalling pathways in fibroblasts distinguished by pertussis toxin. Nature. 326, 800-803.

Chen, R. H., Sarnecki, C. and Blenis, J. (1992): Nuclear localization and regulation of erk- and rsk-encoded protein kinases. Mol. Cell. Biol. 12, 915-927.

Cobb, M.H., Boulton, T.G. and Robbins, D.J. (1991): Extracellular signal-regulated kinases: ERKs in progress. Cell. Regul. 2, 965-978.

Courchesnes, W.E., Kunisawa, R. and Thorner, J. (1989): A putative protein kinase overcomes phromone-induced arrest of cell cyling in S. cerevisiae. Cell 58, 1107-1119.

Crews, C. M., Alessandrini, A. and Erikson, R. L. (1992): The primary structure of MEK, a protein kinase that phosphorylates the ERK gene product. Science. 258, 478-480.

Dent, P., Haser, W., Haystead, T., Vincent, L. A., Roberts, T. M. and Sturgill, T. W. (1992): Activation of Mitogen-Activated Protein kinase kinase by v-raf in NIH 3T3 cells and in vitro. Science. 257, 1404-1407.

Elion, E.A., Grisafi, P.L.and Fink, G.R. (1990): FUS3 encodes a cdc^{2+}/CDC28-related kinase required for the transition from mitosis into conjugation. Cell, 64, 649-664.

Gille, H., Sharrocks, A.D. and Shaw, P.E. (1992): Phosphorylation of transcription factor p62TCF by MAP kinase stimulate ternary complex formation at c-fos promoter. Nature 358, 414-421.

Her, J.-H., Wu, J., Rall, T. B., Sturgill, T. and Weber, M. J. (1991): Sequence of pp42/MAP kinase, a serine/threonine kinase regulated by tyrosine phosphory-lation. Nucl. Ac. Res. 19, 3743.

Kosako, H., Gotoh, Y. and Nishida, E. (1992): cDNA cloning of MAP kinase kinase reveals kinase cascade pathways in yeasts to vertebrates. EMBO J. (in press).

Howe, L.R. (1992): Activation of the MAP kinase pathway by the protein kinase raf. Cell 71, 335-342.

Kreis, T. (1986): Microinjected antibodies against the cytoplasmic domain of vesicular stomatitis virus glycoprotein block it transport to the cell surface. EMBO. J. 5,931-941.

Kyriakis, J.M., App, H., Zhang, X-F, Banerjee, P., Brautigan, D.L., Rapp, U.R. and Avruch, J. (1992): *Raf*-1 activates MAP kinase-kinase. Nature 358, 417-421.

L'Allemain, G., Paris, S., Magnaldo, I. and Pouysségur, J. (1986): α-thrombin-induced inositol phosphate formation in G0-arrested and cycling hamster lung fibroblasts: evidence for a protein kianse C-mediated desensitization response. J. Cell. Physiol. 129, 167-174.

Magnaldo, I., Pouysségur, J. and Paris, S. (1988): Thrombin exerts a dual effect on stimulated adenylate cyclase in hamster fibroblasts, an inhibition *via* a GTP-binding protein and a potentiation *via* activation of protein kinase C. Biochem.J. 253, 711-719.

McKenzie, F. R., Seuwen, K. and Pouysségur, J. (1992): Stimulation of phosphatidyl-choline breakdown by thrombin and carbachol but not by tyrosine kinase receptor ligands in cells transfected with M1 muscarinic receptors. J. Biol. Chem. 267, 22759-22769.

Meloche, S., Pagès, G. and Pouysségur, J. (1992): Functional expression and growth factor activation of an epitope-tagged p44 mitogen-activated protein kinase, p44mapk. Mol. Biol. Cell. 3, 63-71.

Meloche, S., Seuwen, K., Pagès, G. and Pouysségur, J. (1992): Biphasic and synergistic activation of p44mapk (ERK1) by growth factors : correlation between late phase activation and mitogenicity. Mol. Endocrinol. 638, 845-854.

Pagès, G., Lenormand, P., L'Allemain, G., Chambard, J. C., Meloche, S. and Pouysségur, J. (1993): The mitogen-activated protein kinases p42mapk and p44mapk are required for fibroblast cell proliferation. Proc. Natl. Acad. Sci. (in press).

Payne, D. M., Rossomando, A. J., Martino, P., Erickson, A. K., Her, J. H., Shabanowitz, J., Hunt, D. F., Weber, M. J. and Sturgill, T. W. (1991): EMBO J. 10, 885-892.

Pelech, S.L. and Sanghera, J.S. (1992): Mitogen-activated protein kinases : versatile transducers for cell signaling. TIBS 17, 233-238.

Pouysségur, J., Kahan, C., Vouret, V., L'Allemain, G., Van Obberghen-Schilling, E. and Seuwen, K. (1991): Transmembrane signaling pathways that control cell-cycle reentry in fibroblasts. In "Origins of Human Cancer", Cold Spring Harbor Laboratory Press. 285-295.

Pouysségur, J., Sardet, C., Franchi, A., L'Allemain, G. and Paris, S. (1984): A specific mutation abolishing Na^+/H^+ antiport activity in hamster fibroblasts precludes growth at neutral and acidic pH. Proc.Natl.Acad.Sci.USA. 81, 4833-4837.

Pulverer, B., Kyriakis, J., Avruch, J., Nikolakaki, E., and Woodgett, J. (1991): Phosphory-lation of c-*Jun* mediated by MAP kinases. Nature 353, 670-673.

Sardet, C., Franchi, A. and Pouysségur, J. (1989): Molecular cloning, primary structure, and expression of the human growth factor-activatable Na^+/H^+ antiporter. Cell. 56, 271-280.

Seuwen, K., Magnaldo, I. and Pouysségur, J. (1988): Serotonin stimulates DNA synthesis in fibroblasts acting through $5-HT_{1B}$ receptors coupled to a G_i-protein. Nature. 335, 254-256.

Sprague, G. (1992): Kinase cascade conserved. Current Biol. 2, 587-589.

Sturgill, T. W. and Wu, J. (1991): Recent characterization of protein kinase cascade for phosphorylation of ribosomal protein S6. Biochim.Biophys.Acta. 1092, 350-357.

Thomas, G. (1992): MAP kinase by any other name smells just as sweet. Cell. 68, 3-6.

Torres, L., Martin, H., Garcia-Saez, I., Arroyo, J., Molina, M., Sanchez, M. and Nombela, C. (1991): A protein kinase gene complements the lytic phenotype of *Saccharomyces cerevisiae* lyt2 mutants. Mol. Microbiol. 5, 2845-2854.

Vouret-Craviari, V., Van Obberghen-Schilling, E., Scimeca, J-C., Van Obberghen, E. and Pouysségur, J. (1993): Differential activation of p44mapk (ERK1) by α-thrombin and thrombin-receptor peptide agonist. Biochem. J. (in press)

Wang, Y., Simonson, M.S., Pouysségur, J. and Dunn, M.J. (1992): Endothelin rapidly stimulates mitogen-activated protein kinase activity in rat mesangial cells. Biochem. J. 287, 589-594

Résumé

Les 'Mitogen-Activated Protein' kinases ($p42^{mapk}$ et $p44^{mapk}$) sont des kinases à spécificité sérine/thréonine qui ont la remarquable propriété d'être activées par de nombreux stimuli externes y compris tous les facteurs de croissance. Leur activation est médiée par phosphorylation de deux sites clef sur tyrosine et thréonine par la MAP kinase kinase ($p45^{mapkk}$) qui possède une double spécificité. Ici nous démontrons que: (i) la synergie entre mitogènes se révèle au niveau de l'activation des MAP kinases, (ii) la phase persistante d'activation des MAP kinases corrèle remarquablement bien avec la progression G0/phase S, (iii) les MAP kinases se transloquent très rapidement dans le noyau en réponse aux facteurs de croissance, (iv) l'inhibition de l'activation endogène des MAP kinases (expression d'antisens ou d'allèles dominant-négatifs) inhibe l'activation de AP1 et l'entrée des cellules dans le cycle cellulaire. Nous concluons que l'activation des MAP kinases est une étape requise dans la transmission du signal mitogénique de la surface au noyau.

Mechanisms involved in activation and inactivation of cyclin-dependent protein kinases

Thierry Lorca, Jean-Claude Cavadore, Alain Devault, Didier Fesquet, Jean-Claude Labbé and Marcel Dorée

CNRS UPR 9008 and INSERM U 249, BP 5051, 34033 Montpellier Cedex, France

Abstract

Changes in the kinase activity of cdc2/cdks along cell cycle depend both on their association with a variety of cyclins (including G1/S and G2/M cyclins), but also on post-translational modifications by phosphorylation-dephosphorylation reactions. Phosphorylation of Thr161 in cdc2 (or the equivalent position in cdks) is required for cyclin-cdc2 complexes to express their protein kinase activity. This phosphorylation is catalyzed by MO15, another kinase conserved through evolution. MO15 has a companion protein that may regulate its activity. Even when phosphorylated on Thr161, cdc2 kinase may be maintained in an inactive form, due to its phosphorylation on Tyr 15. The cdc25 tyrosine phosphatase removes this inhibitory phosphorylation and induces cells to enter mitosis. Exit from mitosis requires inactivation of MPF (cyclin B-cdc2 kinase), that itself depends on proteolytic degradation of the cyclin subunit. In vertebrate, unfertilized eggs are prevented from exiting metaphase of meiosis II, due to a cytostatic factor (CSF), a component of which is the proto-oncogene *c-mos*. Activation of Ca^{2+}-calmodulin dependent protein kinase II at fertilization is both necessary and sufficient for the cyclin degradation pathway to be released from its CSF-inhibited state and eggs to escape metaphase arrest.

Transitions of the cell cycle are controlled in all eukaryotes by oscillations in the activity of kinase complexes between cdc2 or cdc2-like proteins and cyclins. The first of these complexes to be identified was MPF (M-phase promoting factor), that universally controls entry into and exit from M-phase (for reviews see Nurse, 1990 ; Dorée, 1990). Biochemical purification of MPF to near homogeneity in several species has firmly established that it consists of an heterodimeric complex between 1 molecule of cdc2 and 1 molecule of cyclin B.
Cdc2 (CDC 28 in budding yeast) is the catalytic subunit of a protein kinase first demonstrated by genetic analysis in fission yeast to control the G2 to M-phase transition. Mitotic cyclins are a family of proteins, first discovered in embryos of marine invertebrates, whose abundance follows a sawtooth pattern throughout each cell cycle, due to their accumulation then sudden degradation at the metaphase to anaphase transition (reviewed in Hunt, 1989). This behavior actually characterizes two sub-families, cyclins A and B, that can be distinguished by specific motifs that are conserved throughout evolution (Nugent et al, 1991).While in both fission and budding yeast another major control point of the cell cycle, the G1/S transition, is controlled through association of a unique catalytic subunit, cdc2/CDC28, with different types of cyclins, including G1 cyclins (CLN-type), a family of cdc2-related proteins (called cdks for cyclin-dependant kinases) mediate this function in vertebrate, through association with a variety of cyclins, classified as C, D and E (reviewed in Reed, 1992).
One of the most characteristical features of many cdks is the periodic and specific oscillation of their activity along cell cycle. For example, cyclin A- and cyclin B-cdc2 kinases activities are low at interphase and high at mitosis, whilst cyclin A- and cyclin E-cdk2 activities are already high at S-phase. Cyclin E-cdk2 activity also drops after S-phase, whilst cyclin A-cdk2 activity keeps high (Pagano et al., 1992 ; Dulic et al., 1992). These periodical changes of cdks activities may depend on transcriptional mechanisms that control their expression or that of the associated cyclin, on changes of their respective localization within the cell, or on post-translational modifications occurring at the level of the catalytic or the regulatory subunit. The present paper focuses on these post-translational modifications, with special emphasis to entry into and exit from mitosis.

PHOSPHORYLATION OF CDC2 ON THREONINE 161 (OR THE EQUIVALENT POSITION IN OTHER CDKS) AND ACTIVATION OF CYCLIN-DEPENDENT PROTEIN KINASES.

Changes in the kinase activity of cdc2/cdks along cell cycle depend not only on their association with a variety of cyclins, but also on post-translational modifications by phosphorylation-dephosphorylation reactions (see Fig. 1). Phosphorylation of Thr 161 (Thr 167 in fission yeast) is associated with activity of cdc2 along cell cycle (Krek and Nigg, 1991 ; Gould et al., 1991). In cell-free systems derived from Xenopus eggs, formation of cyclin B-cdc2 complexes is also correlated with Thr 161 phosphorylation (Solomon et al., 1992). This has been extended to complexes of cyclin A with either cdc2 or cdk2 (Devault et al., 1992 ; Gu et al., 1992), and most likely holds true for other cyclin-cdks complexes.
Phosphorylation of Thr 161 in cdc2 (or the equivalent position in other cdks) was first believed to be required for either formation or stabilization of cyclin-cdc2 complexes (Ducommun et al., 1991). Next, it was realized that such complexes could form in the presence of EDTA and without any phosphorylation when cyclin was added in excess (Solomon et al., 1992). Unphosphorylated complexes may either demonstrate reduced protein kinase activity or no activity at all, depending on the catalytic subunit. No activity was detected in cdc2-containing complexes and ectopic expression of cdc2 mutated at Thr 167 failed to rescue thermosensitive mutants of cdc2 (Gould et al., 1991 ; Ducommun et al., 1991). In contrast a small but significant H1 histone kinase activity was demonstrated in cyclin A-cdk2 complexes (Connel-Crowley et al, 1993 ; Fesquet et al., 1993). This small activity was considerably increased upon further phosphorylation of Thr 160 in cdk2. Whilst cyclin A dramatically increases phosphorylation of Thr 161 in cdc2 in vitro, it remains unclear whether a cyclin subunit is an absolute requirement for cdc2/cdks to undergo phosphorylation at this position.

Although Thr 161 is localized in the so-called "autophosphorylation" domain of serine/threonine kinases (Hanks et al., 1988), it is not possible to transfer ^{32}P from ^{32}P-γ-ATP to Thr 161 in highly purified cyclin B-cdc2 kinase (Labbé et al., 1991), even after phosphatase treatment. Moreover a kinase distinct of cdc2 and able to phosphorylate Thr 161 in cyclin B-cdc2 complexes was characterized in *Xenopus* (Solomon et al., 1992) and starfish oocytes (Fesquet et al., 1993). In starfish this kinase was purified to near homogeneity. The final preparation contained two polypeptides. One was shown to be the starfish homolog of the MO15 gene product, a kinase previously cloned by homology probing from a *Xenopus* cDNA library (Shuttleworth et al., 1990). Moreover, immunodepletion of the MO15 protein also depleted *Xenopus* egg extracts from their ability to phosphorylate Thr 161 *in vitro*. This demonstrated that MO15 is a gene conserved throughout evolution which encodes the catalytic subunit of a protein kinase that activates cdc2 through phosphorylation of Thr 161. The purified kinase was found to phosphorylate and activate not only cdc2, but also cdk2 and other cdc2-like proteins in chimeric complexes including a variety of cyclins. By analogy with MAPKK (Mitogen Activated Protein Kinase Kinase) that exerts a similar function through phosphorylation of an homologous residue in domain VIII of MAP kinases, this kinase was called CDKK (for Cyclin-Dependent-Kinase Kinase). Although MO15 has a companion subunit and undergoes post-translational modifications by phosphorylation that appear to depend on its cytological localization in both *Xenopus* and starfish oocytes, nothing is as yet known on regulation of CDKK activity. Possibly, future investigations will reveal that phosphorylation of Thr 161 and its homologs in cdc2/CDKs complexes is regulated in a sophisticated manner, as previously reported for phosphorylation of MAP kinases.

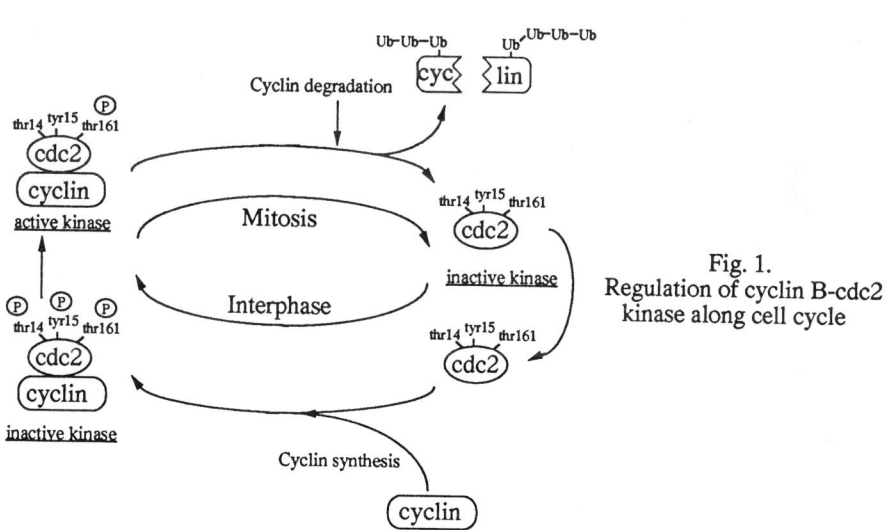

Fig. 1.
Regulation of cyclin B-cdc2 kinase along cell cycle

THE PRE-MPF TO MPF TRANSITION

In many animals, including starfish and *Xenopus*, oocytes arrested before first meiotic metaphase contain a pre-formed cyclin B-cdc2 complex. Although cdc2 is already phosphorylated on Thr 161 in this complex, it has no protein kinase activity, and for this reason it is called pre-MPF. Pre-MPF is maintained in an inactive form due to phosphorylation of cdc2 on inhibitory residues, Thr 14 and Tyr 15, localized in the ATP-binding domain (Fig. 1). Pre-MPF has been purified to near homogeneity from starfish oocytes and shown to undergo activation when treated *in vitro* with the bacterially-produced cdc25 gene product, that specifically catalyzes dephosphorylation of pre-MPF on Thr 14 and Tyr 15 (Strausfeld *et al.*, 1991). Thus, cdc25, first recognized as a mitotic inducer in fission yeast (Russell and Nurse 1986), encodes a phosphatase with potentially dual specificity (Gautier *et al.*, 1991 ; Dunphy and Kumagai, 1991 ; Millar *et al.*, 1991), even if it is still unclear whether it plays actually this role during meiotic maturation.

The pre-MPF to MPF transition is not only observed during meiotic maturation, but also at late G2 in somatic cells. Indeed cyclin B-cdc2 complexes are maintained in an inactive form during interphase due to phosphorylation at Thr 14 and Tyr 15 (Norbury *et al.*, 1991 ; Krek and Nigg, 1991). Several observations support the view that inhibition of the pre-MPF to MPF conversion by unreplicated DNA prevents cells to enter mitosis prematurely before DNA replication has been completed. For example, if inhibitors of DNA synthesis are added to dividing cells, conversion of pre-MPF to MPF is suppressed, and cells do not enter mitosis until the inhibitors are removed and replication is completed. It seems that cdc25 is part of the linkage between DNA replication and mitosis, since fission yeast mutants in cdc2 that are insensitive to cdc25 divide in the presence of hydroxyurea, an inhibitor of DNA replication (Enoch and Nurse, 1990). However, H1 histone kinase activity of CDC28, the homolog of cdc2 in budding yeast, at least partially due to its association with CLB2, an homolog of cyclin B, is already high during S-phase, as assessed in mutants that arrest in S-phase or in wild type cells blocked in S-phase by hydroxyurea treatment (Sorger *et al.*, 1992 ; Amon *et al.*, 1992). This indicates that MPF activity may not by itself inhibit DNA replication and that mechanisms unrelated to the pre-MPF to MPF conversion may be involved in the feed back mechanism that prevents mitosis to occur before DNA replication has been completed.

Enzymes that regulate the equilibrium between pre-MPF and MPF may themselves be controlled by phosphorylation-dephosphorylation reactions. Indeed, inhibition of type 2A phosphatase by okadaic acid strongly reduces tyrosine phosphorylation of cyclin B-associated cdc2. Since this is observed even in the presence of vanadate, an inhibitor of cdc25, but not in the presence of 6-dimethyl-aminopurine, a reported inhibitor of serine-threonine kinases, it seems likely that the kinase(s) which phosphorylate(s) cdc2 on inhibitory sites can be inactivated by a phosphorylation event, itself antagonized by type 2A phosphatase (Devault *et al.*, 1992). Periodic changes in phosphorylation of cdc25 have also been proposed to regulate its activiy (Izumi *et al.*, 1992 ; Kumagai and Dunphy, 1992). Cyclin B-cdc2 kinase may itself positively control activity of cdc25 (Hoffmann *et al.*, 1993). This may be part of an amplification process that would allow a small amount of active MPF to accelerate conversion of pre-MPF to MPF in oocytes during meiotic maturation or in somatic cells at the end of G2 phase.

At variance with cyclin B, cyclin A appears to form an active complex without lag phase with cdc2 in many systems. As a result, protein kinase activity of cyclin A-cdc2 peaks earlier than that of cyclin B-cdc2 at late G2. Moreover, *in vitro* experiments demonstrated that cyclin A-cdc2 kinase accelerates conversion of pre-MPF to MPF. This led to propose that the active cyclin A-cdc2 kinase generated without a lag phase from neo-synthesized cyclin A and cdc2 may play a key role in the rapid switch in the equilibrium of cyclin B-cdc2 complexes to the tyrosine-dephosphorylated and active form of cdc2 at the G2 to mitosis transition, perhaps owing to strong inhibition of the tyrosine kinase that phosphorylates cdc2 (Devault *et al.*, 1992).

INACTIVATION OF MITOTIC CDC2 KINASES AND EXIT FROM MITOSIS

Exit from mitosis requires inactivation of MPF, and this inactivation itself depends on proteolytic degradation of the cyclin subunit (Murray *et al.*, 1989 ; Ghiara *et al.*, 1991). Proteolytic degradation of cyclin A in cyclin A-cdc2 complexes is also required for cells to exit

M-phase of the cell cycle (Luca et al., 1991). Cell cycle-regulated proteolysis of cyclins A and B is thought to be mediated by a ubiquitin-dependent process, since polyubiquitin chain formation was shown to be required for cyclin destruction (Glotzer et al., 1991 ; Hershko et al., 1991). The N-terminal domain of cyclins A and B contains a conserved stretch of aminoacids (R X A L X X I). Mutations within this domain that inhibit ubiquitin conjugation also inhibit degradation (Glotzer et al., 1991 ; Lorca et al., 1991a and 1992a). Whilst the cyclin degradation pathway is normally switched off after cdc2 kinase inactivation, it remains on if type 2A phosphatase is blocked by okadaic acid (Lorca et al., 1991b)

Degradation of cyclins A or B does not occur spontaneously in extracts prepared at interphase from *Xenopus* or *Spisula* eggs. Yet, it can be triggered in such extracts by adding either purified cyclin B-cdc2 kinase (Felix et al., 1990) or cyclin B, which forms a complex in extracts with cdc2 (Glotzer et al., 1991 ; Luca et al., 1991). Therefore, cyclin B-cdc2 kinase was proposed to promote degradation of mitotic cyclins by initiating a cascade of reactions including cyclin ubiquitinylation and ending with proteolysis. In contrast, addition of purified cyclin A-cdc2 kinase, or recombinant cyclin A, which forms an active cyclin A-cdc2 kinase complex in cell extracts, failed to turn on the cyclin degradation pathway (Luca et al., 1991 ; Lorca et al., 1992b). Rather, cyclin A-cdc2 kinase was found to delay cyclin degradation induced by cyclin B-cdc2 kinase (Lorca et al., 1992b). Free cyclin A has no such effect. Since cyclins are believed not to require association with cdc2 for ubiquitin-dependent degradation (Glotzer et al., 1991), it seems unlikely that cyclin A could merely saturate the ubiquitin-dependent pathway in such experiments. The finding that cyclin A-cdc2 kinase cannot turn on, and even delays, cyclin degradation may be essential to prevent premature MPF inactivation before the genetic material has been correctly packaged into highly condensed chromosomes on the metaphasic plate. Indeed cyclin A-cdc2 kinase peaks earlier than cyclin B-cdc2 kinase during mitosis and already reaches its highest value at prophase.

Although cyclin proteolysis is required to inactivate cyclin A-or cyclin B-cdc2 kinases, it may not be sufficient. Indeed dephosphorylation of cdc2 on Thr 161 accompanies cyclin degradation (Lorca et al., 1992c). Moreover, okadaic acid prevents cdc2 kinase inactivation when it is used at a concentration sufficient to completely block type 1 phosphatase. Genetic evidence actually supports the view that activity of type 1 phosphatase may be required to inactivate cdc2 kinase and for cells to exit from mitosis (reviewed in Yanagida et al., 1992).

INACTIVATION OF MPF AT FERTILIZATION : ROLE OF CA2+-CALMODULIN-DEPENDENT PROTEIN KINASE II

In vertebrate, unfertilized eggs are prevented from exiting metaphase of meiosis II, due to the presence of a cytostatic factor (CSF). Although CSF has never been purified, the product of the proto-oncogene *c-mos* has been shown to be intimately associated with this activity. Indeed, microinjection of synthetic *mos* mRNA into two-cell embryos induces cleavage arrest at metaphase. By contrast, cytosol extracts of unfertilized eggs, when immunodepleted of endogenous p39^{mos}, lose their cleavage-arresting activity in injected embryos (Sagata et al., 1989). In CSF-arrested cells, cyclin is stable though cyclin B-cdc2 kinase in high, suggesting either that a p39^{mos}-catalyzed phosphorylation event makes cyclin resistant to proteolysis or inhibits some enzyme required for cyclin degradation. The disappearance of CSF activity at fertilization is believed to be caused by a transient increase in cytoplasmic free Ca^{2+} upon fertilization. In the intact *Xenopus* egg, free Ca^{2+} rises from a resting value of 0.4 µM to a peak of ~ 1.2 µM over the course of 2 min after fertilization (Busa and Nuccitelli, 1985). The calcium transient propagates with a velocity of ~ 8 µm/s as a 45s "wide" ring of elevated free Ca^{2+} from the sperm entry point to its antipodes (Kubota et al., 1987). The 45 s width of the Ca^{2+} wave fits the finding that increasing free Ca^{2+} to the micromolar level for only 30s is sufficient to release the cyclin degradation pathway from its inhibited state in extracts prepared from metaphase-arrested eggs (Lorca et al., 1991a).
What may be the nature of the Ca^{2+}-dependent event required to free the cyclin degradation pathway ? A synthetic peptide corresponding to the auto-inhibitory domain of myosin light chain kinase (MLCK [781-814]), which binds Ca^{2+}-calmodulin with high affinity, was first found to prevent Ca^{2+} to specifically trigger cyclin degradation. The peptide had no effect when added either simultaneously with EGTA to terminate the transient rise in Ca^{2+}, or together with

calmodulin. This provided strong evidence that formation of a Ca^{2+}-calmodulin complex is required at fertilization as an early event (the first 30s) following elevation of intracellular free Ca^{2+}, to release the cyclin degradation pathway from its inhibited state, resulting in exit from M-phase (Lorca et al., 1991a).
Next it was found that a constitutively active mutant of calmodulin-dependent protein kinase II readily released the cyclin degradation machinery from its MLCK [781-814]-inhibited state, even in the absence of Ca^{2+}. The ability of egg extracts to arrest the cell cycle at metaphase when microinjected into a blastomere of a two-cell embryo (CSF activity) also disappeared upon addition of the constitutively active mutant. Moreover, an activity that phosphorylates syntide 2, a specific substrate for calmodulin-dependent protein kinase II, was found to increase dramaticaly, earlier than the drop of H1 kinase activity associated with cyclin degradation, when Ca^{2+} was added to CSF extracts. Finally, a conserved peptide corresponding to the autoinhibitory domain of Ca^{2+}-calmodulin-dependent protein kinase II that specifically inhibits this enzyme was found to suppres both the burst of syntide phosphorylation that follows Ca^{2+} addition, and cyclin degradation (Lorca et al., 1993). Taken together, these results show that activation of Ca^{2+}-calmodulin-dependent protein kinase II is both necessary and sufficient for the cyclin degradation machinery to be released from its CSF-inhibited state upon fertilization in vertebrate eggs. The molecular mechanisms by which it governs CSF and MPF inactivation remain to be elucidated.

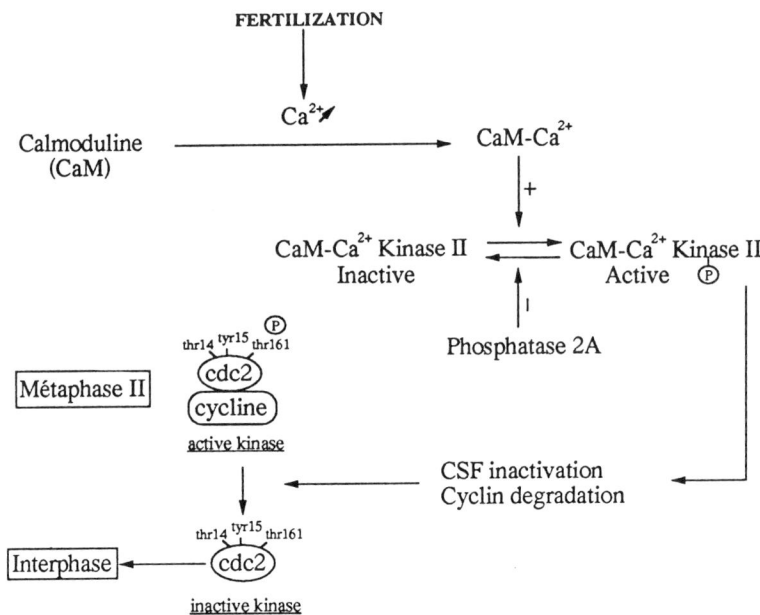

Fig. 2. Inactivation of MPF and CSF at fertilization of vertebrate eggs

REFERENCES

Amon, A., Surana, U., Muroff, I., & Nasmyth, K. (1992) : regulation of p34CDC28 tyrosine phosphorylation is not required for entry into mitosis in *S. cerevisiae*. Nature, 355:368-371.

Busa, W.B., and Nuccitelli, R. (1985) : an elevated free cytosolic Ca2+ wave follows fertilization in eggs of the frog *Xenopus laevis*. J. Cell Biol, 100:1325-1329.

Connell-Crowley, L., Salomon, M.J., Wei, N. and Harper, J.W. (1993) : phosphorylation-independent activation of human cyclin-dependent protein kinase by cyclin A *in vitro*. Mol. Biol. Cell., 4:79-92.

Devault, A., Fesquet, D., Cavadore, J. C., Garrigues, A. M., Labbe, J. C., Lorca, T., Picard, A., Philippe, M., & Doree, M. (1992) : cyclin A potentiates maturation-promoting factor activation in the early *Xenopus* embryo *via* inhibition of the tyrosine kinase that phosphorylates cdc2. J Cell Biol. 118:1109-1120.

Doree, M. (1990) : control of M-phase by maturation-promoting-factor. Curr. Opin. Cell Biol. 2:269-273.

Ducommun, B., Brambilla, P., Felix, M. A., Franza, B. J., Karsenti, E., & Draetta, G. (1991) : cdc2 phosphorylation is required for its interaction with cyclin. EMBO J. 10:3311-3319.

Dulic, V., Lees, E., & Reed, S. I. (1992) : association of human cyclin E with the periodic G1-S phase protein kinase. Science. 257:1958-1961.

Dunphy, W. G., & Kumagai, A. (1991) : the cdc25 protein contains an intrinsic phosphatase activity. Cell. 67:189-196.

Enoch, T. and Nurse, P. (1990) : mutations of fission yeast cell cycle control genes abolish dependence of mitosis on DNA replication. Cell. 60:665-673.

Felix, M. A., Labbé, J.C., Dorée, M. Hunt, T. and Karsenti; E. (1990) : triggering of cyclin degradation in interphase extracts of amphibian eggs by cdc2 kinase. Nature. 346:379-382.

Fesquet, D., Labbé, J.C., Derancourt, J., Capony, J.P., Galas, S., Girard, F., Lorca, T., Shuttleworth, J., Dorée, M. and Cavadore, J.C.(1993) : the MO15 gene encodes the catalytic subunit of a protein-kinase that activates cdc2 and other cyclin-dependent kinases (CDKs) through phosphorylation of threonine 161 and its homologs. EMBO J. submitted.

Gautier, J., Solomon, M. J., Booher, R. N., Bazan, J. F., & Kirschner, M. W. (1991) : cdc25 is a specific tyrosine phosphatase that directly activates p34^{cdc2}. Cell. 67:197-211.

Ghiara, J.B. Richardson, H.E., Sugimoto, K., Heuze, M., Lew, D.J., Wittenberg, C. and Reed, S.I. (1991) : a cyclin B homolog in *Saccharomyces cerevisiae* : chronic activation of the CDC 28 protein kinase by cyclin prevents exit from mitosis. Cell. 65:163-174.

Girard, F., Strausfeld, U., Fernandez, A., & Lamb, N. J. (1991) : cyclin A is required for the onset of DNA replication in mammalian fibroblasts. Cell. 67:1169-1179.

Glotzer, M., Murray, A.W. and Kirschner, M. W. (1991) : cyclin is degraded by the ubiquitin pathway. Nature, 349:132-138.

Gould, K. L., Moreno, S., Owen, D. J., Sazer, S., & Nurse, P. (1991) : phosphorylation of Thr 167 is required for *Schizosaccharomyces pombe* p34^{cdc2} function. EMBO J., 10:3297-3309.

Hershko, A., Ganoth, D., Pehrson, J., Palazzo, R.E. and Cohen, L.H. (1991) : methylated ubiquitin inhibits cyclin degradation in clam embryo extracts. J. Biol. Chem., 266:16376-16379.

Hunt, T. (1989) : maturation-promoting factor, cyclin and the control of M-phase. Curr. Opin. Cell Biol., 1:268-274.

Hoffmann, I., Clarke, P.R., Jesùs Marcote, M., Karsenti, E., and Draetta, G. (1993) : phosphorylation and activation of human cdc25-C by cdc2-cyclin B and its involvement in the self-amplification of MPF at mitosis. EMBO J., 12:53-63.

Izumi, T., Walker, D.H., and Maller, J.L. (1992) : periodic changes in phosphorylation of the *Xenopus* cdc25 phosphatase regulate its activity. Mol. Biol. Cell.,3:927-939.

Krek, W. and Nigg, E. A. (1991) : differential phosphorylation of vertebrate p34cdc2 kinase at the G1/S and G2/M transitions of the cell cycle : identification of major phosphorylation sites. EMBO J., 10:305-316.

Krek, W., & Nigg, E. A. (1992) : mutations of p34cdc2 phosphorylation sites induce premature mitotic events in Hela cells : evidence for a double block to p34cdc2 kinase activation in vertebrates. EMBO J., 10:3331-3341.

Kubota, H.Y., Yoshimoto, Y., Yoneda, M. and Hiramoto, Y. (1987) : free calcium wave upon activation in *Xenopus* eggs. Dev. Biol, 119:129-136.

Kumagai, A. and Dunphy, W.G. (1992) : regulation of the cdc25 protein during the cell cycle in *Xenopus* extracts. Cell, 70:139-151.

Labbe, J. C., Cavadore, J. C., & Doree, M. (1991) : Purification of M-phase specific protein kinase from starfish and amphibian oocytes. Methods Enzymol, 200:291-301.

Lorca, T., Galas, S., Fesquet, D., Devault, A., Cavadore, J.C. and Dorée, M. (1991, a) : degradation of the proto-oncogene product p39mos is not necessary for cyclin proteolysis and exit from meiotic metaphase : requirement for a Ca2+-calmodulin dependent event. EMBO J., 10:2087-2093.

Lorca, T., Fesquet, D., Zindy, F., Le Bouffant, F., Cerruti, M., Brechot, C., Devauchelle, G. and Dorée, M. (1991, b) : an okadaic acid-sensitive phosphatase negatively controls the cyclin-degradation pathway in amphibian eggs. Mol. Cell Biol., 11:1171-1175.

Lorca, T., Devault, A., Colas, P., Van Loon, A., Fesquet, D., Lazaro, J.B. and Dorée, M. (1992, a) : cyclin A-Cys 41 does not undergo cell cycle-dependent degradation in *Xenopus* extracts. FEBS Lett., 1:90-93.

Lorca, T., Labbe, J. C., Devault, A., Fesquet, D., Strausfeld, U., Nilsson, J., Nygren, P. A., Uhlen, M., Cavadore, J. C., & Doree, M. (1992b) : cyclin A-cdc2 kinase does not trigger but delays cyclin degradation in interphase extracts of amphibian eggs. J. Cell Sci., 102, 55-62.

Lorca, T., Labbe, J. C., Devault, A., Fesquet, D., Capony, J. P., Cavadore, J. C., Le Bouffant,.F., & Doree, M. (1992, c) : dephosphorylation of cdc2 on threonine-161 is required for cdc2 kinase inactivation and normal anaphase.EMBO J., 11, 2381-2390.

Lorca, T., Cruzalegui, F.H., Fesquet, D., Cavadore, J.C., Mery, J., Means, A. and Dorée, M.(1993) : calmodulin-dependent protein kinase II mediates Ca^{2+}-dependent inactivation of MPF and CSF activities upon fertilization of *Xenopus* eggs. Nature, submitted.

Luca, F.C., Shibuya, E.K., Dohrmann, C.E. and Ruderman, J.V. (1991) : both cyclin A-delta 60 and B-delata 97 are stable and arrest cells in M-phase but only cyclin B-delta 97 turns on cyclin destruction. EMBO J., 10:4311-4320.

Millar, J.B., Mc Gowan, C.H., Lenaers, G., Jones, R., and Russell, P. (1991) : p80cdc25 mitotic inducer is the tyrosine phosphatase that activates p34cdc2 kinase in fission yeast. EMBO J., 10:4301-4309.

Murray, A. W., Solomon, M. J., & Kirschner, M. W. (1989) : the role of cyclin synthesis and degradation in the control of maturation-promoting factor activity. *Nature*, 339:280-286.

Norbury, C., Blow, J., & Nurse, P. (1991) : regulatory phosphorylation of the p34cdc2 protein kinase in vertebrate. EMBO J., 10, 3321-3329.

Nugent, J. H., Alfa, C. E., Young, T., & Hyams, J. S. (1991) : conserved structural motifs in cyclins identified by sequence analysis. J Cell Sci ., 99:669-674.

Nurse, P. (1990) : universal control mechanism regulating onset of M-phase.Nature, 344:503-508.

Pagano, M., Pepperkok, R., Verde, F., Ansorge, W., & Draetta, G. (1992) : cyclin A is required at two points in the human cell cycle. EMBO J., 11:961-971.

Reed, S.I. (1992) : the role of p34 kinases in the G1 to S-phase transition. Ann. Rev. Cell Biol. , 8:529-561.

Russell, P., & Nurse, P. (1986) : cdc25+ functions as an inducer in the mitotic control of fission yeast. Cell, 45:145-53.

Sagata, N., Watanabe, N., Vande Woude, G.W. and Ikawa, Y. (1989) : the c-mos proto-oncogene product is a cytostatic factor responsible for meiotic arrest in vertebrate eggs. Nature, 342:512-518.

Shuttleworth, J., Godfrey, R., & Colman, A. (1990) : p40MO15, a cdc2-related protein kinase involved in negative regulation of meiotic maturation of *Xenopus* oocytes. EMBO J., 9:3233-40.

Solomon, M. J., Lee, T., & Kirschner, M. W. .(1992) : role of phosphorylation in p34cdc2 activation : identification of an activating kinase. Mol Biol Cell, 3:13-17.

Sorger, P. K., & Murray, A. W. (1992) : S-phase feedback control in budding yeast independent of tyrosine phosphorylation of $p34^{cdc28}$.Nature, 355, 365-368.

Strausfeld, U., Labbe, J. C., Fesquet, D., Cavadore, J. C., Picard, A., Sadhu, K., Russell, P., & Doree, M. (1991) : dephosphorylation and activation of a p34cdc2-cyclin B complex *in vitro* by human cdc25 protein. Nature, 351:242-245.

Yanagida, M., Kinoshita, N., Stone, E.M. and Yamano, H. (1992) : protein phosphatases and cell division cycle control. In Regulation of the Eukaryotic Cell Cycle, Ciba Foundation Symp. 170, pp 130-146. Chichester : John Wiley and sons.

Zindy, F., Lamas, E., Chenivesse, X., Sobczak, J., Wang, J., Fesquet, D., Henglein, B., & Brechot, C. (1992) : cyclin A is required in S-phase in normal epithelial cells. Biochem. Biophys. Res. Commun., 182:1144-54.

Résumé

Les changements d'activité des kinases cdc2/cdks au cours du cycle cellulaire dépendent non seulement de leur association avec une variété de cyclines (cyclines G1/S et cyclines G2/M), mais aussi de réactions de phosphorylation-déphosphorylation. La phosphorylation de Thr161 (ou la position équivalente des cdks) est requise pour que les complexes cyclin-cdc2 acquièrent leur activité proteine kinase. Cette phosphorylation est catalysée par une autre kinase conservée au cours de l'évolution, M015. MO15 est associée à une autre protéine qui pourrait contrôler son activité. Même phosphorylée sur Thr161, la kinase cdc2 peut être maintenue inactive si elle est phosphorylée sur la Tyr15. La phosphatase cdc25 élimine cette phosphorylation inhibitrice et induit l'entrée des cellules en mitose. La sortie de mitose requiert l'inactivation de MPF (la kinase cycline B-cdc2), qui dépend elle-même de la dégradation protéolytique de la sous-unité cycline. Chez les vertébrés, les oeufs non fécondés sont maintenus en métaphase du 2ème division de meïose par un facteur cytostatique (CSF), dont le proto-oncogène c-mos est un constituant. L'activation de la kinase de type II dépendante du Ca^{2+} et de la calmoduline est nécessaire et suffisante pour libérer la voie de dégradation des cyclines de l'inhibition par CSF et permettre aux oeufs de sortir de l'arrêt métaphasique.

Transcriptional repression by Max proteins p21 and p22

Leo Kretzner, Elizabeth M. Blackwood*, Jaclynn Mac and Robert N. Eisenman

*Fred Hutchinson Cancer Research Center, 1124 Columbia Street, Seattle, WA 98104, USA; *Department of Biology, University of California, San Diego, La Jolla, CA 92093, USA*

INTRODUCTION

The c-*myc* protooncogene has been the subject of a great deal of research in the past decade, due to its association with a wide variety of animal neoplasias, including in humans (Lüscher and Eisenman, 1990, and references therein). c-*myc* activity is tightly regulated at virtually every level of control of gene expression, and while the forms of *myc* activation associated with oncogenesis vary, they have in common the constitutive overexpression of Myc proteins. Further evidence for Myc involvement in cellular growth control came from observations that its expression levels normally increase upon growth factor stimulation as cells progress from the G_0 to G_1 phase of the cell cycle. Conversely, expression levels are reduced in some cell types upon differentiation (Spencer and Groudine, 1991).

Thus, the involvement of Myc in transformation and cell growth has been well established phenomenologically. However the molecular basis of *myc* action has been harder to ascertain. It has been variously speculated that Myc played a role in transcription, DNA replication, RNA processing, and/or nuclear architecture (Lüscher and Eisenman, 1990). Key breakthroughs in defining Myc's molecular actions came with the identifications of a specific DNA sequence (CACGTG) to which Myc can bind (Blackwell et al, 1990), and of a heterodimeric Myc binding partner, Max (Blackwood and Eisenman, 1991). Myc and Max are both nuclear phosphoproteins of the Basic/Helix-Loop-Helix/Leucine-Zipper class, but differ in several key respects. While Myc proteins have a very short half-life (20 - 30 minutes), Max proteins are quite long-lived ($t_{1/2}$ = 12+ hours). Further, Max protein levels remain constant even during the G_0 phase of the cell cycle, when Myc is absent (Blackwood et al, 1992). Max can also form homodimers, while full-length Myc fails to do so. Thus it was hypothesized that Myc acts by binding DNA as a heterodimer with Max. Indeed, Myc-Max heterodimers bind to the same DNA sequence, and with greater affinity, as Max homodimers (Blackwood and Eisenman, 1991).

With the specific Myc DNA binding sequence and heterodimeric partner both in hand, we set out to directly test the notion that Myc and Max are transcription factors (Kretzner et al, 1992). We used a reporter construct with a four-fold reiteration of CACGTG linked to a minimal promoter driving the CAT gene (M4MinCAT). A low level of activation was seen with this reporter gene alone, due not surprisingly to endogenous Myc and other CACGTG-binding factors. However, Myc overexpression clearly enhances reporter activity while Max expression has an opposite, repressive effect. Max repression can be relieved in a concentration-dependant manner by cotransfection with Myc expression vector. Exogenous Myc apparently acts through endogenous Max, as its activation effects can be abrogated by cotransfection of dominant negative forms of Max (Ayer et al, 1993). All these responses were shown to be sequence-specific responses, as they do not occur with promoters linked to CACCTG sites, targets of MyoD activation (Weintraub et al, 1990). Further, Myc activation and relief of repression depend on its C-terminal dimerization and N-terminal activation domains. Similarly, Max repression requires its DNA-binding basic region (Kretzner et al, 1992). These experiments suggested a model in which Max homodimers act as transcriptional repressors, while Myc-Max heterodimers, tightly regulated by Myc abundance, are activators of transcription. This model has been further elaborated with the discovery of another Max dimerization partner, Mad, which augments the repressive activity of Max when the two are coexpressed (Ayer et al, 1993). In either form of the model, genetic targets of Myc activation are presumed to be important positive regulators of cell growth, while their downregulation by Max or Mad+Max is hypothesized to be concomittant with cellular programs of stasis and differentiation. Thus, the opposing transcriptional actions of Myc and Max may be important in the positive and negative regulation of cell growth. In this report, we extend our observations of Myc and Max transcriptional activity, and show new data on Max's behavior as a transcriptional repressor.

RESULTS AND DISCUSSION

Generality of the responses

Our original observations of the opposing transcriptional behaviors of Myc and Max were derived from transient transfection assays in two cell lines. While the appropriate control experiments indicated that the effects seen were real and biologically significant, we wish to summarize here the generality of these effects as seen in a variety of cell types and with a number of variations in the expression and reporter constructs (summarized in Table 1). To begin with, similar conclusions about Myc and Max transcriptional activity were reached by others working with a yeast system (Amati et al, 1992), though actual transcriptional repression by Max was not assayable in that system. We have since obtained qualitatively similar results in a wide variety of cell types, including primary and transformed cells, and others have seen these responses in lymphoblastoid cells (Gu et al, 1993). As regards activation by Myc, we have found that another *myc* family member known to dimerize with Max, N-Myc, activates reporter gene transcription as well or better than c-Myc (Fig. 1). Also like c-Myc, N-Myc is capable of relieving Max repression, though we note that the relief of repression seen in the particular experiment shown is rather modest for both forms of Myc.

TABLE ONE

GENERALITY OF MYC ACTIVATION AND MAD/MAX REPRESSION

CELL TYPES:
fibroblasts (NIH3T3), epithelial (CV-1), primary (REF),
transformed (HeLa; C33A), lymphoid (Gu et al, 1993),
Xenopus (XTC) and Drosophila (SL2) lines, Yeast (*S. cer.*) (Amati et al, 1992)

VARIATIONS:
N-Myc works as well as c-Myc. (Fig. 1)
c-Myc 1 & 2 (p67 & p64) activate M4MinCAT equally. (Blackwood et al, in prep.)
Basic responses not affected by:
  Co-transfection with T24 *ras*, *bcl-2*, or *Rb*.
  Transfection in Rb⁻ cells (C33A).

REPORTERS:
TK-CAT:
1x, 2x, 4x copies of CACGTG; CATGTG also works
M4 works at 3' end of reporter gene.
M4 works when stably integrated in cells (NIH3T3).
ODC: needs two CACGTGs in first intron (Fernandez et al, 1993) (LK, unpub.)

The basic responses to Myc and Max (with or without Mad) are also seen using a number of variations in the reporter construct (Table 1; data not shown). For example, reporters containing different numbers of the binding site give qualitatively similar results, though the strength of the response diminishes with decreasing site number, suggesting that as for many transcription factors cooperativety of binding may be involved (Weintraub et al, 1990). M4MinCAT responds similarly to Myc and Max whether transiently transfected or stably integrated in cells, indicating these transcription factors can interact productively with promoters in a natural chromosomal context. We also note that CACGTG is necessary and sufficient for Myc activation and Max repression when located at the 3' end of the reporter gene, consistent with the notion that this is a transcriptional enhancer element whose location is flexible with respect to the obligate proximal elements. Finally and importantly, similar responses to Myc and Max are seen by ourselves and others (Fernandez et al, 1993) from the promoter/control region of a putative Myc responsive gene, ornithine decarboxylase (ODC), the rate limiting enzyme in the polyamine biosynthetic pathway necessary for DNA synthesis. These responses depend on two CACGTG sites within the first intron of ODC, and suggest that more complex gene promoters respond to Myc and Max similarly to our simpler artificial constructs.

Differential effects of Max p21 and p22

Max proteins exist in several alternative forms believed to reflect alternative mRNA splicing. Consistent with this idea, the two most common isoforms seen in vivo, p21 and p22, were isolated as separate cDNA clones, (Blackwood and Eisenman, 1991; Blackwood et al, 1992). The *max9* cDNA encodes for the protein p22, containing nine additional amino acids N-terminal to the basic region relative to p21, which is encoded by the *max*. cDNA. We have shown previously that p22/Max9 represses reporter gene transcription to a greater degree than p21/Max, and similarly that Myc can better relieve repression by p21 than p22 (Kretzner et al, 1992). We wish to address here the molecular basis for this.

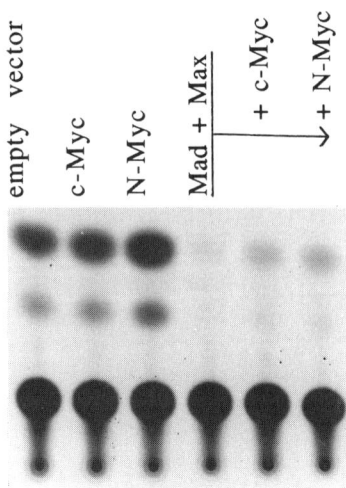

Fig. 1: N-Myc behaves similarly to c-Myc in transactivation assays. Shown here are CAT assays using NIH3T3 fibroblasts, transiently transfectd with the M4MinCAT reporter and expression vectors for the proteins indicated. Methods for this and Fig. 2 are as in Kretzner et al, 1992.

Max Titrations
NIH-3T3 cells, with pMinCAT reporter +/- M4

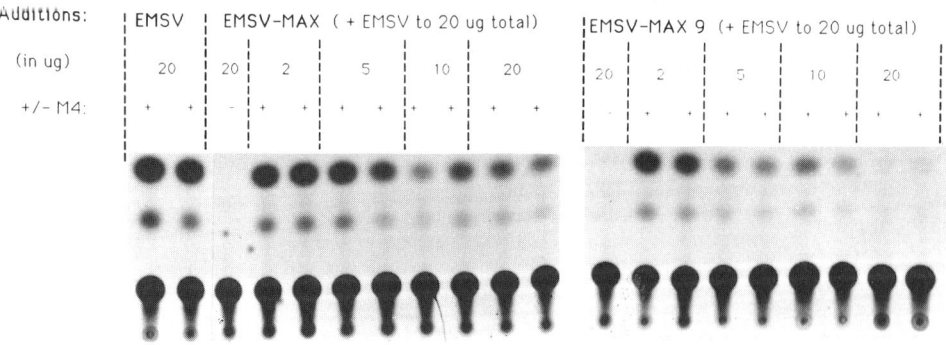

Fig. 2: Max9 is a stronger transcriptional repressor than Max. Parallel titrations with the two Max isoforms are shown. Total amount of expression vector was held constant by appropriate addition of empty vector (pEMSV). Background activity is seen with EMSV alone, and the background dependence on the binding sequence is evident by comparison of the + and - lanes.

The differential repression of p22 versus p21 is depicted in Fig. 2. Parallel titrations of each cDNA show that at any concentration of expression vector added repression by Max9 is greater than that by Max. This effect is not attributable to non-specific "squelching" effects, as Max mutants lacking the basic region are expressed at similar levels yet are inert at even very high concentrations of expression vector (Kretzner et al, 1992). Immune precipitations of metabolically labelled cell lysates show that similar amounts of protein are produced by each Max vector (Fig. 3). Furthermore, both Max isoforms have similarly long half-lives (Blackwood et al, 1992). Taken together, these observations rule out gross differences in protein level as an explanation for the differential effect of p21 and p22 Max.

The molecular basis for differential repression by Max isoforms was revealed when off-rate experiments were performed, shown in Fig. 4. As in other electrophoretic mobility shift assays, radiolabelled oligonucleotide probe was preincubated with bacterially derived Max (or Max + Myc) proteins. To study off-rates, an excess of unlabelled probe is added at time zero, and aliquots of the reaction mixtures were loaded onto nondenaturing gels at the times indicated. The rate of decrease in shifted complexes seen in the autoradiogram reflects the extent of competition between labelled and unlabelled probe. Protein complexes with a slower off-rate (ie, longer half-life on the labelled probe) are concluded to have a greater stabilty on DNA than those with faster off-rates. These analyses show that p22/Max9 has greater stability on specific DNA than p21/Max, both as a homodimer and as a heterodimer with Myc (Fig. 4). We believe this difference in binding affinity accounts for at least part of the differential repression by Max isoforms.

The biological significance of differential Max repression is not yet clear, as both forms of Max protein (i.e., +/- nine residues) have been found in all cell types thus far examined. However, we note that the additional amino acids of Max9 move two mapped N-terminal CKII phosporylation sites (serines 2 and 11; E.M.B and B. Lüscher, unpublished data) further away from the basic region. Thus, there may be differences in the degree to which phosphorylation negatively regulates DNA binding of p21 and p22 (Berberich and Cole, 1992). It also remains possible that differences in the expression pattern of p21 and p22 will emerge in certain cell types or particular cases of cellular differentiation. That Max regulation is likely to be complex is suggested by the discovery of cDNAs coding for truncated forms of the protein, termed ΔMax (Mäkela et al, 1992). ΔMax cDNAs were also found with and without the nine residue insert. Though ΔMax proteins have not yet been shown to occur naturally in cells, they do possess biological activity in cotransformation assays.

Further studies will be required to understand the nature of Max repression more fully, including in the context of Mad coexpression. As poorly as transcriptional activation is understood, transcriptional repression is understood even less well. In particular, it remains to be shown whether Max and Mad+Max repress transcription in some active way, or if they are simply nonactivators competing for binding sites. Furthermore, we need to know what the actual genetic targets of these molecules are. Experiments to address these questions are actively under way.

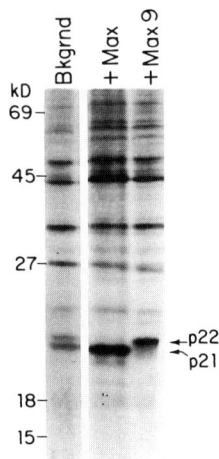

Fig 3: Max isoforms are expressed at similar levels when transiently transfected. Immunoprecipitations with lysates from ^{35}S-methionine labelled NIH3T3 cells are shown. Lane 1: endogenous levels; Lanes 2 and 3: Max and Max9 overexpression, respectively. We note that ectopic Max expression appears to affect endogenous levels of the proteins.

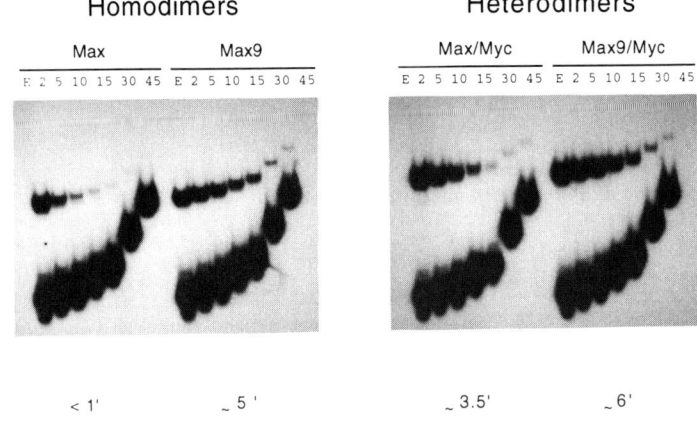

Fig. 4: Off rates indicate Max 9 has greater stability on DNA than Max. Shown here are electrophorectic mobility shift assays comparing Max and Max9 homodimers, as well as the corresponding heterodimers with Myc. Cold competing oligonucleotide was added at time zero, and aliquots of the reactions were loaded onto a running native gel at the times indicated. E = equilibrium, ie, with no competing oligo. Half-lives, calculated from densitometric analyses, are shown below the data.

REFERENCES

Amati, B., S. Dalton, M.W. Brooks, T.D. Littlewood, G.I.. Evans, & H. Land (1992) Transcriptional activation by the human cMyc oncoprotein in yeast requires interaction with Max. *Nature* 359, 423-426.

Ayer, D.E., L. Kretzner, & R.N. Eisenman (1993) Mad: a heterodimeric partner for Max that antagonizes Myc transcriptional activity. *Cell* 72, 211-222.

Berberich, S.J., & M.D. Cole (1992) Casein kinase II inhibits the DNA-binding activities of Max homodimers but not Myc/Max heterodimers. *GenesDev* 6, 166-176.

Blackwell, T.K., L. Kretzner, E.M. Blackwood, R.N. Eisenman, & H. Weintraub (1990) Sequence-specific DNA-binding by the c-Myc protein. *Science* 250, 1149-1151.

Blackwood, E.M.. & R.N. Eisenman (1991) Max: A helix-loop-helix zipper protein that forms a sequence-specific DNA-binding complex with Myc. *Science* 251, 1211-1217.

Blackwood, E.M., B. Lüscher, & R.N. Eisenman (1992) Myc and Max associate in vivo. *GenesDev* 6, 71-80.

Fernandez, C.B., G. Packham, & J.L. Cleveland (1993) Ornithine decarboxylase is a transcriptional target of c-Myc. *Proc. Natl.Acad. Sci. USA* , in press.

Gu, W., K. Cechova, V. Tassi, & Riccardo Dalla-Favera (1993) Opposite regulation of gene transcription and cell proliferation by c-Myc and Max. *Proc. Natl.Acad. Sci. USA* 90. 2935-2939.

Kretzner, L., E.M. Blackwood, & R.N. Eisenman (1992) The Myc and Max proteins possess distinct transcriptional activities. *Nature* 359, 426-429.

Lüscher, B., & R.N. Eisenman (1990) New light on Myc and Myb. Part 1, Myc. *Genes Dev* 4, 2025-2035.

Mäkela, T.P., P.J. Köskinen, I. Västrik, & K. Alitalo (1992) Alternate forms of Max as enhancers or suppressors of Myc-Ras cotransformation. *Science* 256, 373-377.

Spencer, C.A., and M. Groudine (1991) Control of *c-myc* regulation in normal and neoplastic cells. *Adv Cancer Res* 56, 1-48.

Weintraub, H., R. Davis, D. Lockshon, & A. Lassar (1990) MyoD binds cooperatively to two sites in a target enhancer sequence: occupancy of two sites is required for activation. *Proc. Natl.Acad. Sci. USA* 87, 5623-5627.

Résumé

Nous avons montré précédemment que les protéines Myc et Max peuvent agir comme facteurs de transcription, la surexpression de Max réprimant l'activité du gène reporter et la surexpression de Myc l'activant et supprimant la répression de Max (Kretzner et al. 1992). Ces expériences ont été réalisées avec un promoteur artificiel dans 2 types cellulaires. Ici, les observations ont été étendues à un certain nombre de variations de promoteurs dans de nombreux types cellulaires. De plus, nous avons comparé particulièrement la répression transcriptionnelle par les isoformes p21 et p22 de la protéine Max, la seconde étant un répresseur plus fort que la première. Nous présentons de nouveaux résultats montrant que cette différence est due au moins en partie à une plus grande stabilite de l'ADN de p22. Ce fait est vérifié à la fois pour les homodimères de p22, les hétérodimères de p22 et Myc.

CSF-1 regulation of D-type cyclins during the G1 phase of the cell cycle

Martine F. Roussel*, Hitoshi Matsushime*,**, Jun-ya Kato*
and Charles J. Sherr*,**

*Department of Tumor Cell Biology and **Howard Hughes Medical Institute, St. Jude Children's Research Hospital, Memphis, Tennessee 38105, USA

The macrophage colony-stimulating factor, CSF-1, is required throughout the G1 phase of the cell cycle to regulate both immediate and delayed early responses necessary for cell proliferation. Binding of the growth factor to its receptor, a transmembrane glycoprotein encoded by the FMS proto-oncogene, activates its intrinsic tyrosine kinase activity, thereby triggering a cascade of signals that induce gene expression. The immediate early response to CSF-1 involves the transcriptional activation of a series of genes independently of new protein synthesis, but progression of cells through the G1 phase requires the subsequent expression of delayed early genes, whose induction is protein synthesis dependent. Among the latter are the D-type G1 cyclins, which play important roles in enforcing G1 progression and determining the cell's commitment to enter S phase. Three distinct D-type cyclins (D1, D2 and D3) are differentially expressed in response to growth factor stimulation in various cell types, and each acts as a regulatory subunit for novel cyclin-dependent kinases (cdks) related to $p34^{cdc2}$/cdk1. The targets of cyclin D-cdk complexes are likely to include a series of critical substrates whose phosphorylation facilitates G1 exit and the onset of cellular DNA synthesis. Therefore, deregulation of cyclin D expression might perturb the normal cell cycle and contribute to oncogenesis.

CSF-1 regulates expression of G1 cyclins.

CSF-1 supports the proliferation, differentiation and survival of cells of the mononuclear phagocyte series (Sherr & Stanley, 1990). When bone marrow-derived macrophages are deprived of CSF-1, they arrest early in G1 and eventually die. However, if transiently starved macrophages are restimulated with the growth factor before an apoptotic program is initiated, the cells synchronously progress through G1 and enter S phase 10 to 12 hours later. Removal of CSF-1 before the cells transit a late G1 restriction point prevents them from entering S-phase, but once the commitment to replicate their DNA is made, the cells will enter S phase and complete the cell cycle in the absence of CSF-1 (Tushinski and Stanley, 1985). In the continuous presence of CSF-1, the second and subsequent G1 intervals are shorter than the first, reducing the overall generation time.

To identify delayed early response genes that might play a role in G1 progression and S phase entry, a CSF-1 dependent mouse macrophage cell line was arrested by CSF-1 starvation and then restimulated, after which total RNA was harvested from cells in mid G1 and used to prepare a cDNA library. A radiolabeled single-stranded cDNA synthesized from the RNA of stimulated cells was negatively selected with RNA from starved macrophages to generate a subtracted probe that would pinpoint transcripts specifically expressed in CSF-1 stimulated cells. We isolated 23 cDNA clones, one of which showed limited homology to known cyclins (Matsushime et al., 1991b). This clone, originally called CYL-1 for cyclin-like gene 1, is now designated cyclin D1. Importantly, its human cognate was independently identified at a chromosomal breakpoint in parathyroid adenomas (designated PRAD1 by Motokura et al., 1991) and through its ability to complement G1 cyclin-deficient yeast (Xiong et al., 1991; Lew et al., 1991; Koff et al., 1991). Using the mouse cyclin D1 cDNA, we also isolated two related genes, cyclins D2 and D3, from T cell and NIH-3T3 fibroblast cDNA libraries. The cyclin D2 polypeptide is 63% identical to both cyclins D1 and D3, whereas cyclins D1 and D3 share 52% identity (Matsushime et al., 1991a). Although the sequences of the three mouse D-type cyclin genes exhibit significant differences, each is highly related (>95% identity) to its human cognate (Inaba et al., 1992; Xiong et al., 1992a). These data, together with their differential and lineage-specific patterns of expression (Matsushime et al., 1991b; Won et al., 1992; Ajchenbaum et al., 1993), suggest that the individual D-type cyclins do not have strictly redundant functions.

Temporal expression of D-type cyclins.

In macrophages induced to progress synchronously through G1, two cyclin D1 mRNAs of 4.5 and 3.8 kilobases (kb) were detected within 1-2 hours after CSF-1 stimulation (Matsushime et al., 1991a,b). Cyclin D1 mRNA levels become maximal by mid-G1, after which their expression is maintained as long as CSF-1 stimulation of the cells is continued. A 6.2 kb cyclin D2 mRNA was first induced 4 hours after CSF-1 stimulation, peaked in abundance at the G1/S boundary (10 to 12 hours post stimulation), and then decreased during the S and G2 phases of the cell cycle (Matsushime et al., 1991a). Detectable levels of cyclin D3 mRNA were not found in macrophages. The induction of cyclin D1 and D2 mRNAs was prevented by addition of the protein synthesis inhibitor, cycloheximide, suggesting that their transcription required new protein synthesis.

A polyclonal rabbit antiserum prepared to the bacterially synthesized cyclin D1 protein detected two polypeptides of about 36 kilodaltons (kd) in CSF-1 stimulated macrophages whose appearance and abundance paralleled that of the cyclin D1 mRNAs. Both forms are phosphoproteins, and the differences between them are still unknown. Only small amounts of cyclin D2 are synthesized by these cells, corresponding to the lower levels of transcription of the cyclin D2 gene later in G1. The cyclin D mRNAs and proteins are unstable, so that withdrawal of CSF-1 results in their rapid disappearance. Their degradation during G1 in response to growth factor starvation correlates with the failure of macrophages to enter S phase (Matsushime et al., 1991b).

Different combinations of D-type cyclins are expressed in other cell types. For example, in peripheral blood T lymphocytes, only cyclins D2 and D3 are induced during G1, whereas cyclin D1 is not expressed (Ajchenbaum et al., 1993). In these cells, cyclin D2 mRNA and protein appear before cyclin D3 is expressed. Rodent fibroblasts express each of the D-type cyclin mRNAs, but cyclin D1 is the predominant species, and its induction is highly dependent upon serum growth factors (Won et al., 1992; Baldin et al., 1993). Here, the gene can be transiently induced in the presence of cycloheximide . The D-type cyclins localize to the nucleus, and in fibroblasts, their abundance progressively diminishes once cells enter S phase (Baldin et al., 1993).

Cyclin D-dependent protein kinases.

A 34 kd polypeptide which coprecipitated with antiserum to cyclin D1 reacted with antibodies raised to yeast p34^{cdc2}, but not with antiserum directed to a C-terminal peptide of human p34^{cdc2}, suggesting that cyclin D forms complexes with a protein related to, but distinct from, the authentic cdc2 gene product. To identify this putative catalytic partner, the cDNA library prepared from mouse macrophages synchronized in mid-G1 was screened under low stringency conditions using cdk probes, and candidate cDNAs were sequenced. Several related genes were isolated, including cdk2 and several "PCTAIRE" kinases (Meyerson et al., 1992; Okuda et al., 1992). Another clone, first isolated from human HeLa cells and designated PSK-J3, had predicted hallmarks of a 34 kd protein kinase but had previously been found to be devoid of detectable enzymatic activity (Hanks, 1987). When transcribed and translated in vitro, $p34^{PSK-J3}$ formed complexes with D-type cyclins, but cdc2, cdk2 and PCTAIRE-1 did not (Matsushime et al., 1992). When lysates of proliferating macrophages were immunoprecipitated with antibodies to cyclin D1, the coprecipitating 34 kd polypeptide reacted with antiserum to $p34^{PSK-J3}$ and yielded a V8 protease fingerprint identical to that of $p34^{PSK-J3}$ transcribed and translated in vitro. Most importantly, the kinase activity of $p34^{PSK-J3}$ could be activated by D-type cyclins (Matsushime et al., 1992). The PSK-J3 polypeptide has now been designated cdk4. Unlike the previously isolated cdks (cdc2/cdk1, cdk2, cdk3) which can complement $p34^{cdc2}$-deficient yeast strains (Meyerson et al., 1992), cdk4 cannot. Although cdk4 is the major partner of D-type cyclins in macrophages and fibroblasts, the D-type cyclins can also form complexes with cdk2 and with yet another potential catalytic subunit, cdk5 (Xiong et al., 1992b).

The RB connection.

When insect Sf9 cells are coinfected with baculoviruses encoding cdk4 and any of the D-type cyclins, but not with regulatory or catalytic subunits alone, high levels of serine/threonine protein kinase activity are generated (Matsushime et al., 1992; Kato et al., 1993). These complexes are poorly active on histone H1, a substrate commonly used to assay cdk activity, but they phosphorylate the retinoblastoma protein, pRb, to high stoichiometry and at physiologically relevant sites. Using either pRb or the pRb-related protein, p107, as the substrate, cyclins D2 and D3 can also form active pRb kinase complexes with cdk2, but cyclin D1 does not, suggesting that the D-type cyclins are not identical regulators (Ewen et al., 1993).

The RB gene functions as a tumor suppressor, and elimination of both functional alleles predisposes to retinoblastoma in humans (reviewed in Weinberg, 1991). DNA tumor virus oncoproteins, including adenovirus E1A, SV40 T antigen, and the human papillomavirus E7 protein, bind directly to pRb, and their interactions depend upon domains within pRb that are necessary for its growth suppressive activity, as well as upon regions in the oncoproteins that are required for cell transformation (DeCaprio et al., 1988; Whyte et al., 1988; Dyson et al., 1989). pRb is normally dephosphorylated as cells exit mitosis and is rephosphorylated at cdk sites late in G1. Because these oncoproteins bind only to the underphosphorylated, G1-specific forms of pRb, it was reasoned that pRb normally acts to prevent G1 exit, and that either its phosphorylation by cyclin-dependent kinase(s) or its binding to oncoproteins can inactivate this growth suppressive function (Buchkovich et al., 1989; DeCaprio et al., 1989; Chen et al., 1989; Mihara et al., 1989; Ludlow et al., 1990).

The fact that D-type cyclin/cdk complexes preferentially phosphorylate pRb versus histone H1 appears to depend on their ability to bind directly to pRb and to pRb-related proteins, thereby targeting catalytic subunits to these substrates (Kato et al., 1993). D-type cyclins contain a Leu-X-Cys-X-Leu motif near their aminotermini, which they share with T antigen, E1A, and E7, and like the oncoproteins, the ability of D-type cyclins to bind to pRb depends in part upon the integrity of this motif (Ewen et al., 1993; Dowdy et al., 1993). In insect Sf9 cells, cyclins D2 and D3 form stable, high stoichiometry complexes with pRb, but coexpression of cdk4 induces pRb phosphorylation and prevents cyclin D/pRb complex formation (Kato et al., 1993). However, ternary complexes between pRb, cyclins D2 or D3, and a catalytically inactive cdk4 mutant can be assembled, suggesting that, in the presence of a wild-type cdk, the complexes would be destabilized as pRb undergoes phosphorylation. Cyclin D1 can also interact directly with pRb, but these complexes are less stable than those formed with cyclins D2 or D3 (Kato et al., 1993; Ewen et al., 1993).

When the RB protein or gene is introduced into RB-negative, human Saos-2 osteosarcoma cells, the cells undergo arrest in G1 (Goodrich et al., 1991; Qin et al., 1992). Cointroduction of RB together with expression vectors encoding cyclins A, E, or D2 induces pRb hyperphosphorylation and facilitates G1 exit (Hinds et al., 1992; Ewen et al., 1993). It seems likely that these cyclins can interact with either cdk2 or cdk4 to phosphorylate and inactivate pRb's growth suppressive function. Cyclin D1 is less active in this assay and does not trigger pRb phosphorylation. It has been suggested that cyclin D1 is sequestered by pRb and released to interact with cdks when pRb is phosphorylated (Dowdy et al., 1993). In these models, then, cyclin D2 is viewed to act upstream of pRb in regulating its phosphorylation (Ewen et al., 1993), whereas cyclin D1 is thought to act downstream (Dowdy et al., 1993). Whatever the exact mechanisms, cyclin D1 and D2 clearly differ from one another in this assay.

D-type cyclins are rate limiting for G1 progression.

Data that D-type cyclins regulate G1 have been largely inferential. However, microinjection of antibodies to cyclin D1 into fibroblasts has now been shown to prevent their entry into S phase (Baldin et al., 1993). When cells were stimulated with serum to enter the cell cycle and antibodies to D1 were injected at various times thereafter, those injected in mid-G1 failed to enter S phase, whereas injections near the G1/S boundary were without effect. Thus, cyclin D1 exerts its growth promoting function during the mid to late G1 interval. Rodent fibroblasts engineered to ectopically overexpress mouse cyclin D1 exhibit accelerated G1 transit times and a shortened generation time, have a decreased serum dependency, and become smaller in size. The enforced expression of cyclin D2 also accelerated G1 progression and reduced the cells' serum requirement but had a less significant effect on cell size (Quelle et al., 1993). In spite of these effects, cells overexpressing D-type cyclins remained serum- and anchorage-dependent, were contact-inhibited, and remained nontumorigenic in nude mice. Therefore, the D-type cyclins did not act as transforming genes.

Cyclin D and oncogenesis.

Cyclins D1, D2 and D3 (official genetic designations CCND1, CCND2 and CCND3) have been mapped to chromosomes 11q13, 12p13 and 6p21, respectively (Inaba et al., 1992; Xiong et al., 1992b). Interestingly, the human cyclin D1 gene was independantly isolated at the breakpoint of a chromosomal inversion in parathyroid adenoma (Motokura et al., 1991). In these tumors, cyclin D1 overexpression is driven by the parathyroid hormone promoter, which is translocated immediately 5' to intact cyclin D1 coding sequences. Chromosome 11q13 contains the t(11;14) breakpoint cluster region (*BCL1*) identified in centrocytic B-cell lymphomas, in which cyclin D1 is joined to the immunoglobulin heavy chain locus and represents the first transcriptional unit telomeric to the breakpoint (Rosenberg et al., 1991; Withers et al., 1991). Again, cyclin D1 expression is deregulated in these tumors, suggesting that it is the relevant genetic target of these chromosomal rearrangements. The cyclin D1 locus is amplified in about 20% of breast carcinomas and squamous cell tumors of the head and neck (Lammie et al., 1991), and in an even higher percentage of esophageal carcinomas (Jiang et al., 1992). By analogy, cyclins D2 and D3 might also contribute to malignancy, and indeed, deregulation of cyclin D2 expression consequent to retroviral insertions has been detected in rodent thymomas (Hanna et al., 1993). Although cyclin D expression alone is probably insufficient to account for tumor formation, the involvement of the cyclin D1 and D2 loci in various malignancies suggests that perturbations of normal cell cycle controls can ultimately contribute to cancer.

ACKNOWLEDGMENTS

Our special thanks go to Drs. Mark E. Ewen and David M. Livingston (Dana Farber Cancer Center, Boston) who were responsible for many of the experiments involving the retinoblastoma gene product. We also thank colleagues at our own institution who made important contributions to various aspects of this this work, including Drs. Richard A. Ashmun, Toshia Inaba, A. Thomas Look, Dawn E. Quelle, Sheila Shurtleff, David Strom, and Susan Vallance. Our studies were supported in part by the American Lebanese Syrian Associated Charities of St. Jude Children's Research Hospital.

REFERENCES

Ajchenbaum, F., Ando K., DeCaprio J.A., & Griffin J.D. (1993): Independent regulation of human D-type cyclin gene expression during G1 phase in primary human T lymphocytes. *J. Biol. Chem.* 268, 4113-4119.

Baldin, V., Likas, J., Marcote, M.J., Pagano, M., Bartek, J., & Draetta, G. (1993): Cyclin D1 is a nuclear protein required for cell cycle progression in G1. *Genes & Devel.* 7, 812-821.

Buchkovich, K., Duffy L.A., & Harlow E. (1989): The retinoblastoma protein is phosphorylated during specific phases of the cell cycle. *Cell* 58, 1097-1105.

Chen, P-L., Scully P., Shew J-Y., Wang J.Y.J., & Lee W-H. (1989): Phosphorylation of the retinoblastoma gene product is modulated during the cell cycle and cellular differentiation. *Cell* 58, 1193-1198.

DeCaprio, J.A., Ludlow J.W., Figge J., Shew J-Y., Huang C-M., Lee W-H., Marsilio E., Paucha E., & Livingston D.M. (1988): SV40 large tumor antigen forms a specific complex with the product of the retinoblastoma susceptibility gene. *Cell* 54, 275-283.

DeCaprio, J.A., Ludlow J.W., Lynch D., Furukawa Y., Griffin J., Piwnica-Worms H., Huang C-M., & Livingston D.M. (1989): The product of the retinoblastoma susceptibility gene has properties of a cell cycle regulatory element. *Cell* 58, 1085-1095.

Dowdy, S.F., Hinds, P.W., Louis, K., Reed, S.I., Arnold, A., & Weinberg, R.A. (1993): Physical interactions of the retinoblastoma protein with human cyclins. *Cell* 73, 499-511.

Dyson, N., Howley P.M., Munger K., & Harlow E. (1989): The human papilloma virus-16 E7 oncoprotein is able to bind to the retinoblastoma gene product. *Science* 243, 934-937.

Ewen, M.E., Sluss H.K., Sherr C.J., Matsushime H., Kato J-Y., & Livingston D.M. (1993): Functional interactions of the retinoblastoma protein with mammalian D-type cyclins. *Cell* 73, 487-497.

Goodrich, D.W., Wang, N.P., Qian, Y-W., Lee, Y-H.P., & Lee, W-H (1991): The retinoblastoma gene product regulates progression through the G1 phase of the cell cycle. *Cell* 67, 293-302.

Hanks, S.K. (1987): Homology probing: Identification of cDNA clones encoding members of the protein-serine kinase family. *Proc. Natl. Acad. Sci. USA* 84, 388-392.

Hanna, Z., Jankowski M., Tremblay P., Xiaoyan J., Milatovich A., Francke U., & Jolicoeur P. (1993): The *vin-1* gene, identified by provirus insertional mutagenesis, corresponds to the G1-phase cyclin D2. *Oncogene* 8, in press.

Hinds, P.W., Mittnacht, S., Dulic, V., Arnold, A., Reed, S.I., & Weinberg, R.A. (1992): Regulation of retinoblastoma protein functions by ectopic expression of human cyclins. *Cell* 70, 993-1006.

Inaba, T., Matsushime H., Valentine M., Roussel M.F., Sherr C.J., & Look A.T. (1992): Genomic organization, chromosomal localization, and independent expression of human CYL (cyclin D) genes. *Genomics* 13, 565-574.

Jiang, W., Kahn, S.M., Tomita, N., Zhang, Y-j., Lu, S-H., & Weinstein, I.B. (1992): Amplification and expression of the human cyclin D gene in esophageal cancer. *Cancer Res.* 52, 2980-2983.

Kato, J-Y., Matsushime H., Hiebert S.W., Ewen M.E., & Sherr C.J. (1993): Direct binding of cyclin D to the retinoblastoma gene product (pRb) and pRb phosphorylation by the cyclin D-dependent kinase, CDK4. *Genes & Devel.* 7, 331-342.

Koff, A., Cross, F., Fisher, A., Schumacher, J., Leguellec, K., Phillippe, M. & Roberts, J.M. (1991): Human cyclin E, a new cyclin that interacts wity two members of the CDC2 gene family. *Cell* 66, 1217-1228.

Lammie, G.A., Fantl V., Smith R., Schuuring E., Brookes S., Michalides R., Dickson C., Arnold A., & Peters G. (1991): D11S287, a putative oncogene on chromosome 11q13, is amplified and expressed in squamous cell and mammary carcinomas and is linked to BCL-1. *Oncogene* 6, 439-444.

Lew, D.J., Dulic, V., & Reed, S.I. (1991): Isolation of three novel human cyclins by rescue of G1 cyclin (cln) function in yeast. *Cell* 66, 1197-1206.

Ludlow, J.W., Shon J., Pipas J.M., Livingston D.M., & DeCaprio J.A. (1990): The retinoblastoma susceptibility gene product undergoes cell cycle-dependent dephosphorylation and binding to and release from SV40 large T. *Cell* 60, 387-396.

Matsushime, H., Ewen M.E., Strom D.K., Kato J-Y., Hanks S.K., Roussel M.F., & Sherr C.J. (1992): Identification and properties of an atypical catalytic subunit (p34^{PSKJ3}/CDK4) for mammalian D-type G1 cyclins. *Cell* 71, 323-334.

Matsushime, H., Roussel, M.F., & Sherr, C.J. (1991a): Novel mammalian cyclin (CYL) genes expressed during G1. *Cold Spring Hbr. Symp. Quant. Biol.* 56, 69-74.

Matsushime, H., Roussel M.F., Ashmun R.A., & Sherr C.J. (1991b): Colony-stimulating factor 1 regulates novel cyclins during the G1 phase of the cell cycle. *Cell* 65, 701-713.

Meyerson, M., Enders G.H., Wu C-L., Su L-K., Gorka C., Nelson C., Harlow E., & Tsai L-H. (1992): The human cdc2 kinase family. *EMBO J.* 11, 2909-2917.

Mihara, K., Cao X.R., Yen A., Chandler S., Driscoll B., Murphree A.L., T'Ang A., & Fung Y.K. (1989): Cell cycle-dependent regulation of phosphorylation of the human retinoblastoma gene product. *Science* 246, 1300-1303.

Motokura, T., Bloom T., Kim H.G., Juppner H., Ruderman J.V., Kronenberg H.M., & Arnold A. (1991): A novel cyclin encoded by a *bcl1*-linked candidate oncogene. *Nature (London)* 350, 512-515.

Okuda, T., Cleveland J.L., & Downing J.R. (1992): PCTAIRE-1 and PCTAIRE-3, two members of a novel cdc2/CDC28-related protein kinase gene family. *Oncogene*, 7, 2249-2258.

Qin, X-Q., Chittenden, T., Livingston, D.M., Kaelin, W.G. Jr. (1992): Identification of a growth suppression domain within the retinoblastoma gene product. *Genes & Devel.* 6, 953-964.

Quelle, D.E., Ashmun, R.A., Shurtleff, S.A., Kato, J-Y., Bar-Sagi, D., Roussel, M.F. & Sherr, C.J. (1993): Overexpression of mouse D-type cyclins accelerates G1 phase in rodent fibroblasts. *Genes & Devel.*, in press.

Rosenberg, C.L., Wong, E., Petty, E.M., Bale, A.E., Tsujimoto, Y., Harris, N.L., & Arnold, A. (1991): PRAD1, a candidate BCL1 oncogene: mapping and expression in centrocytic lymphoma. *Proc. Natl. Acad. Sci. USA* 88, 9638-9642.

Sherr, C.J. & Stanley, E.R. (1990): Colony-stimulating factor-1. In *Peptide Growth Factors and their Receptors*, ed., M.B. Sporn & A.B. Roberts, pp 667-698. Heidelberg: Springer-Verlag.

Tushinski, R.J., & Stanley E.R. (1985): The regulation of mononuclear phagocyte entry into S phase by the colony stimulating factor CSF-1. *J. Cell. Physiol.* 122, 221-228.

Weinberg, R.A. (1991): Tumor suppressor genes. *Science* 254, 1138-1146.

Whyte, P., Buchkovich K.J., Horowitz J.M., Friend S.H., Raybuck M., Weinberg R.A., & Harlow E. (1988): Association between an oncogene and an anti-oncogene: The adenovirus E1A proteins bind to the retinoblastoma gene product. *Nature (London)* 334, 124-129.

Withers, D.A., Harvey R.C., Faust J.B., Melnyk O., Carey K., & Meeker T.C. (1991): Characterization of a candidate *bcl-1* gene. *Mol. Cell. Biol.* 11, 4846-4853.

Won, K-A., Xiong, Y., Beach, D., & Gilman, M. (1992): Growth-regulated expression of D-type cyclin genes in human diploid fibroblasts. *Proc. Natl. Acad. Sci. USA* 89, 9910-9914.

Xiong, Y., Connolly, T., Futcher, B., & Beach, D. (1991): Human D-type cyclin. *Cell* 65, 691-699.

Xiong, Y., Menninger, J., Beach, D., & Ward, D.C. (1992a): Molecular cloning and chromosomal mapping of CCND genes encoding human D-type cyclins. *Genomics* 13, 575-584.

Xiong, Y., Zhang, H., & Beach, D. (1992b): D-type cyclins associate with multiple protein kinases and the DNA replication and repair factor PCNA. *Cell* 71, 505-514.

Résumé

Le facteur de croissance macrophagique CSF-1 est nécessaire pendant toute la durée de la première phase (G1) du cycle cellulaire à la régulation des réponses immédiates et différées essentielles à la prolifération cellulaire. La liaison du facteur de croissance à son récepteur, une glycoprotéine transmembranaire codée par le proto-oncogène c-*fm*s, déclenche une cascade de signaux qui induit la transcription d'une série de gènes faisant partie de la réponse immédiate. Cette réponse immédiate, qui a lieu en l'absence de nouvelle synthèse de protéine, permet le passage du stade de quiescence (G_0) à la phase G1. Cependant, la progression à travers la phase G1 et l'engagement dans la synthèse de DNA (phase S) exigent l'expression des gènes faisant partie de la réponse dite différée qui, elle, nécessite la synthèse de nouvelles protéines. Parmi les gènes faisant partie de la réponse différée, nous avons récemment cloné ceux des cyclines de type D, qui jouent un rôle important dans la capacité des cellules à traverser G1 et à répliquer leur ADN. L'expression des trois cyclines D (D1, D2, et D3) est réalisée de façon variable en fonction du type de cellules et du facteur de croissance nécessaire à leur prolifération. De plus les cyclines D exercent leur fonction en s'associant à une sous-unité catalytique, une nouvelle kinase, cycline-dépendante, cdk-4, homologue à la kinase $p34^{cdc2}$/cdk1. Les substrats cibles des complexes cycline D-cdk4 sont probablement des protéines dont la phosphorylation est importante pour la progression en phase G1 et l'entrée en phase S. Par conséquent, la dérégulation de l'expression des cyclines de type D pourrait perturber le cycle cellulaire et être responsable de la formation de tumeurs.

Cyclins and carcinogenesis

Frédérique Zindy*, Jian Wang*, Eugénia Lamas*, Xavier Chenivessel, Berthold Henglein* and Christian Bréchot*,**

*INSERM U 370, CHU Necker, 156, rue de Vaugirard, 75742 Paris Cedex 15; **Unité d'Hépatologie, Hôpital Laennec, rue de Sèvres, 75007 Paris, France

We have previously reported the identification of hepatitiis B virus (HBV) integration in an intron of cyclin A gene in an early hepatocellular carcinoma and hence the isolation of human cyclin A cDNA.
We have constructed a cDNA library of the original tumor (tumor HEN) and isolated several hybrid HBV-cyclin A cDNAs. These cDNAs have the coding capacity for a HBV-cyclin A fusion protein. In the chimeric protein, the N-terminal of cyclin A, including the signals for signals for cyclin degradation, was deleted and replaced by viral preS2/S sequences while the rest of cyclin A remained intact. HBV integration in the cyclin A gene resulted in the overexpression of hybrid HBV-cyclin A transcripts that code for a stabilized cyclin A. In addition, we have investigated cyclin A expression in a primary culture of normal rat hepatocytes and during rat liver regeneration after partial hepatectomy. In both cases, cyclin A mRNA and protein accumulate as the cells enter S phase. Moreover we microinjected anti-sense DNA constructs for cyclin A, resulting in effective inhibition of S phase entry.
   In conclusion, we showed in this paper an analysis of the expression pattern of cyclin A gene in the original tumor which supports the hypothesis of insertional mutagenesis of HBV, and a study of the role of cyclin A in a normal cell cycle which indicates its involvement in G1/S transition. That cyclin A is involved in S phase may provide new clues as to its potential role in carcinogenesis.

INTRODUCTION :
Cyclins play a major role in the cell cycle regulation. Two classes of cyclins, G1 and M cyclins, have been identified in fission (Booher and Beach, 1988 ; Forsburg and Nurse, 1991) and budding yeast (Hadwiger et al., 1989 ; Nash et al., 1988 ; Ghiara et al., 1991). They cooporate with the gene products of cdc2/cdc28 kinases in driving the cell through G1/S and G2/M boundaries. In higher eukaryotes, several cyclins (cyclins A, B, C, D and E) have been isolated. The cyclin B is a mitotic cyclin, which associates to the $p34^{cdc2}$ protein kinase to initiate mitosis and meiosis (Draetta et

al., 1989 ; Labbé et al., 1989 ; Meijer et al., 1989 ; Gautier et al., 1990). The C, D and E type cyclins seem to act in G1/S boundary since they can rescue G1 cyclin deficient mutants in S. Cerevisiae (Xiong et al., 1991 ; Matsushime et al., 1991 ; Motokura et al., 1991 ; Koff et al., 1991).

Cyclin A can also associate with $p34^{cdc2}$, but this complex is formed and active in advance of the $p34^{cdc2}$/cyclin B complex (Swenson et al., 1986 ; Draetta et al., 1989 ; Minshull et al., 1990 ; Pines and Hunter, 1990). However, several lines of evidence suggest that cyclin A may also have a role in the S phase. First, addition of cyclin A to a G1 phase extract was sufficient to initiate SV40 DNA replication in vitro (D'urso et al., 1990) ; second, microinjection of anti-sense cyclin A DNA into cultured cell blocked the initiation of DNA synthesis (see section 2 of this paper) and third, cyclin A associates to a $p34^{cdc2}$ related protein ($p33^{cdc2}$ or CDK2) that functions in S phase (Pines and Hunter, 1990 ; Fang & Newport, 1991 ; Tsai et al., 1991).

We have previously reported the identification of hepatitis B virus (HBV) DNA integration in human cyclin A gene in an early hepatocellular carcinoma (HCC) (Wang et al., 1990). Chronic HBV infection has been associated with the development of HCC by extensive epidemiological studies (Beasley, 1988). There are several mechanisms which account for the role played by HBV in hepatocarcinogenesis. HBV induces cirrhosis, a premalignant state (Chisari et al, 1990). In addition, HBV likely exerts a direct effect on cell transformation by transactivation of the x and truncated PréS2/S viral proteins (Kim et al, 1991) as well by insertional mutagenesis. This latter mechanism has been well illustrated in woodchucks infected with the woodchuck hepatitis virus (WHV) where WHV DNA was found integrated in the C-myc or N-myc gene in 30% of the tumor studied, most of them with an elevated expression of these oncogenes (Fourel et al., 1990). In human, however, there is only one report in which HBV DNA was found to integrate in the gene coding for a retinoic acid receptor β in HCC (Dejean et al., 1986 ; de The et al., 1987).

In the present manuscript we will focus on two implications of our results :
1) a detailed analysis of the expression pattern of cyclin A gene in the original tumor (tumor HEN) which supports the hypothesis of insertional mutagenesis.
2) An analysis of the role of cyclin A in a normal cell cycle which indicates its involvement in G1/S transition.

RESULTS :
1) HBV and cyclin A expression in tumor HEN
In Northern blots of the original tumor HEN, both cyclin A and HBV probes detected the same two polyadenylated transcripts of 2.7 and 1.7 kb, which were nearly the same size as the normal cyclin A transcripts but quite different from that of HBV. These bands were undetectable in the non-tumorous liver of the same patient. A cDNA library was constructed in lambda gt10 with mRNAs from the tumor and hybrid HBV cyclin A transcripts were caracterised.

The genomic structure of the human cyclin A gene has been established in our laboratory (Henglein et al., in preparation). A comparaison with the HBV integration site indicated that the HBV sequences integrated in the first intron of cyclin A gene (Fig. 1). Two representative cDNAs, 1.7 and 2.7 kb respectively, have been completely sequenced. The results indicated that hybrid transcripts

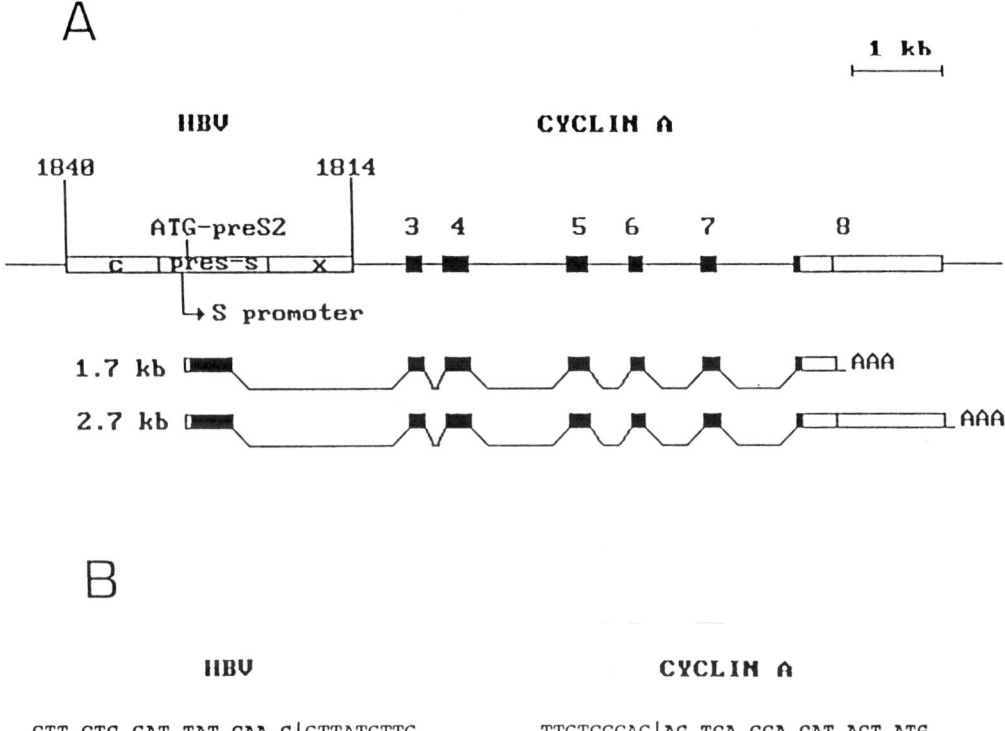

Fig. 1 A : Genomic structure of human cyclin A gene in the HBV integration site (upper) and structure of the hybrid cDNAs (lower). The cyclin A exons are numbered, the black boxs represent the coding sequences and the blanked boxs represent non coding exons, arrow indicate the initiation site of transcription from viral S promoter. The poly (A) tails in the two cDNAs mark the two commonly used polyadenylation sites for normal cyclin A transcription. B : **Sequence of HBV-cyclin A junction in the hybrid cDNAs. The splicing manner is indicated.**

were produced by splicing between HBV and cyclin A sequences, using a cryptic splice donnor site in the middle of viral S gene and the normal splice acceptor site of the third exon of cyclin A gene. A primer extension assay, confirmed that the hybrid transcripts were initiated from the viral Pre S2/S promoter (data not shown).
Sequence analysis showed that the HBV open reading frame was fused to that of the cyclin A in the hybrid cDNAs which code for an HBV-cyclin A fusion protein of 430 amino acids. In the chimeric protein, the N-terminal 152 amino acids of cyclin A were reimplaced by 150 amino acids from the viral PreS2 and a part of S regions whereas the

C-terminal two third of cyclin A, including the cyclin box, remained intact (Fig. 1). In vitro translation of the hybrid cDNA produced a 54 KD protein (slightly smaller than normal cyclin A). The N-terminal domain of cyclin A contains the signals for its degradation by the ubiquitin pathway, (Glotzer et al., 1991). An in vitro degradation assay recently indicated that the deletion of the N-terminal of cyclin A has indeed stabilized cyclin A (data not shown).

2) Cyclin A is required in S phase in normal epithelial cells.

Until now, the investigations on cyclins have been performed either in invertebrates or in transformed mammalian cells. Therefore, we chose to investigate cyclin A expression in normal epithelial cells. With this purpose, we have studied cyclin A mRNA and protein in primary culture of rat hepatocytes and in rat liver regeneration. This approach was associated to experiments based on microinjecting an anti-sense cyclin A cDNA.

In vitro cultured hepatocytes can be maintained for 8 days in serum-free medium supplemented with dexamethasone and insulin (referred to as untreated hepatocytes). Alternatively, culturing cells in serum-free medium supplemented with insulin, pyruvate and epidermal growth factor (EGF) (referred to as treated hepatocytes), stimulates DNA synthesis in most hepatocytes (around 80%), with a maximum at day 3 of culture (Mc Gowan, 1986). Using these experimental conditions, we established primary rat hepatocytes culture confirming that maximum DNA synthesis, shown by ^{3}H-thymidine incorporation, occured on day 3 in treated hepatocytes (fig. 2A). When the hepatocytes were untreated, DNA synthesis was barely detectable. Initial experiments used Northern analysis of cyclin A mRNAs. Cyclin A mRNA was not detected in untreated hepatocytes (fig. 2B). In contrast, in treated hepatocytes, the accumulation of cyclin A mRNA increased to a maximum at day 2-4 and then decreased (fig. 2B). This accumulation of transcripts coincided with maximal ^{3}H-thymidine incorporation. To determine if these transcripts were effectively related to protein expression, we also analyzed total cellular protein from cultured hepatocytes by means of immunoblotting with affinity purified polyclonal antibodies raised against human cyclin A. In primary culture, cyclin A protein was detectable from day 2 to day 5 with a maximum at day 3 in treated hepatocytes (i.e. at the time of DNA synthesis) (fig. 2C). Furthermore, we analyzed the localization of cyclin A in cultured hepatocytes. The staining of cyclin A was preferentially localized into the nucleus. At day 3 of culture, high amount of cells were labelled for cyclin A whereas at day 1 of culture, only few cells were stained (data not shown).

To determine if this observation also true in-vivo, the expression of cyclin A was analysed in regenerating liver at various times after partial hepatectomy (PH). Indeed, PH partially synchronizes the hepatocytes proliferative phase (30-40%) during the first mitosis. The growth process includes a distinct prereplicative phase of hypertrophy which lasts for 12-16 hrs after PH and a replicative phase in which hepatocytes undergo DNA replication (peak et 24 hrs) and then division (peak at 30 hrs) (Bucher and Malt, 1971 ; Grisham, 1962).

Therefore, we analyzed liver samples obtained at 16 to 32 hrs after PH. After 32 hrs, further points were not analyzed due to the loss of cellular synchronisation. ^{3}H-thymidine incorporation was detected with a maximum at 24 hrs after PH (fig. 3A). Northern blot analysis revealed cyclin A mRNA at a very low level before and 16 hrs after

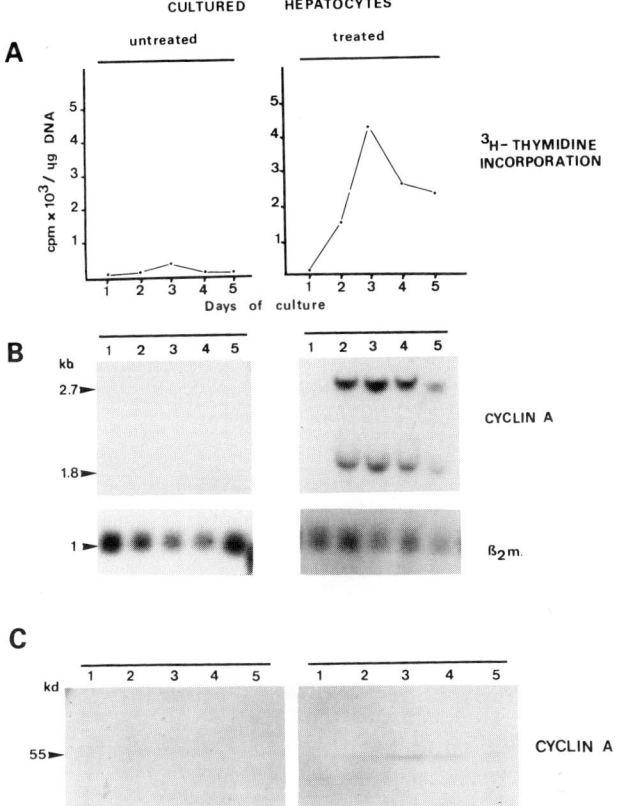

Fig. 2 : Cyclin A expression and ^{3}H-Thymidine incorporation in cultured hepatocytes. Hepatocytes were isolated from 2-month-old male Wistar rats and cultured in serum-free medium supplemented with dexamethasone (1µM) and insulin (200nUI/ml) (untreated hepatocytes) or in serum-free medium supplemented with insulin (20 m UI/ml), pyruvate (20 mM) and epidermal growth factor (50ng/ml) (treated hepatocytes). Hepatocytes untreated or treated by growth factors were harversted at the indicated times. ^{3}H-thymidine incorporation was measured (A), cyclin A mRNAs were analyzed by Northern blot normalized relative to B2 microglobulin mRNAs (B2m.) (B), and cyclin A protein was analyzed by Western blot (C).

PH (fig. 3B). Cyclin A mRNA accumulation increased at 20 hrs with a maximum level from 24 hrs to 32 hrs after PH corresponding respectively to the period in which maximal DNA synthesis and mitosis take place. Cyclin A protein was not detected before or 16 hrs after PH (fig. 3C) ; its level was maintained from 24-26 hrs (time of DNA synthesis) to 32 hrs (time of mitosis).

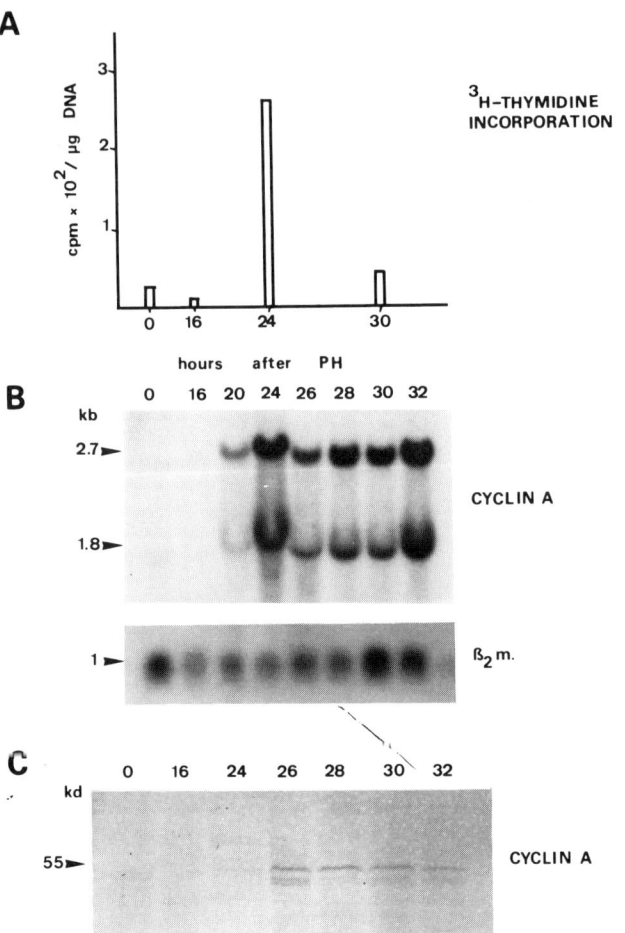

Fig. 3 : cyclin A expression and ^{3}H-thymidine incorporation in regenerating liver after partial hepatectomy. Male Wistar rats were subjected to 70% partial hepatectomy and liver was removed at the indicated times after hepatectomy. ^{3}H-thymidine incorporation was measured (A), cyclin A mRNAs were analyzed by Northern blot normalized relative to B2 microglobulin (B2m.) (B), and cyclin A protein was analyzed by Western blot.

To examine further the effects of changes in cyclin A mRNA and protein levels, we investigated the consequences on S phase transit of artificially inhibiting cyclin A synthesis. We inhibited cyclin A synthesis through microinjection of plasmid constructs encoding anti-sense human cyclin A cDNA under the control of SV40 promoter-enhancer element. At various times after plating, cultured rat hepatocytes stimulated by growth factors were microinjected with the anti-sense cyclin A construct, a sense cyclin A construct or an anti-sense human cyclin B cDNA under the control of an SV40 promotor-enhancer element. The effects of injection were assessed by following the incorporation of 5-bromo-deoxyuridine (5-Br-DU) (i.e. S-phase transit). Injected cells were relocated by inclusion of a non-specific antibody in the injection solutions which was subsequently stained after immunofluorescence with anti-5-Br-DU. Cells were injected 2-3 days after plating and labelled from injection until the end of day 4. Each microinjection experiment was performed three times, involving the injection of 30 to 40 cells every time. Under normal conditions, treated hepatocytes synthetized DNA between the end of day 2 and day 5 (as controlled by ^{3}H-thymidine and 5-Br-DU incorporation). The pattern of 5-Br-DU staining was similar in cells injected with the sense cyclin A construct (data not shown) and in cells injected with the anti-sense cyclin B construct (fig. 4 pannels A and B) which showed between 50 to 85% of DNA synthesis, a level similar to that achieved in the surrounding non injected cells. In contrast, cells injected with the anti-sense cyclin A construct (fig. 4 pannels C-F, and table 1), showed no evidence of DNA synthesis if injected during day 2 (C and D) or day 3 (E and F) ; we detected no evidence of nuclear staining for cyclin A in injected cells (data not shown). Surrounding uninjected cells proceeded normally to transit S-phase. This effect was observed in all the cells injected with the anti-sense constructs from the end of day 2 until day 3.

DISCUSSION :
In this paper, we have both analyzed the HBV and cyclin A expression in the tumor HEN and further caracterized the role of cyclin A in a normal cell cycle.
Concerning the tumor HEN, we have provided strong evidence for a role of HBV in a step of liver cell transformation by insertional mutagenesis. Indeed we showed an increased level of hybrid transcripts HBV-cyclin A, potentially coding for a stabilized chimeric protein. There are several possibilities to account for its effect on the cell phenotype. Loss of the degradation signals in the N-terminal part of the cyclin A, together with initiation from the Pre S2/S viral promoter, may lead to an increased and constitutive synthesis of cyclin A. In addition it is also plausible that the membranous PreS2/S protein may markedly change the localization in the cell of the protein. Finally the viral sequences at the N-terminal part include a truncated form of PreS/S which may have retained a transactivating effect on cellular oncogenes (Kekule et al., 1990 ; Caselman et al., 1990). These different possibilities are being explored through in vitro transfection of cell cultures and in vivo experiments on transgenic mice.
In order to clarify the involvement of cyclin A in carcinogenesis, we have also adressed the issue of its potential role in the G1/S phase of the cell cycle. The present study provides two lines of evidence for the requirement of cyclin A for the cell to proceed to S phase. First, cyclin A mRNA and protein are accumulated as the cells enter S phase in hepatocytes in-vitro and in-vivo. Secondly, the microinjection of an anti-sense cyclin A cDNA inhibited DNA synthesis in cultured normal epithelial cells. The involvememt of cyclin A in the S phase was further supported by the specific inhibition of DNA synthesis following the microinjection of an anti-sense cyclin A while an anti-sense cyclin B cDNA did not inhibit DNA synthesis. Therefore, our observation, based on normal epithelial cells analyzed in-vitro and in-vivo, does imply that cyclin A is not only a mitotic cyclin but also acts at S phase. This finding is also

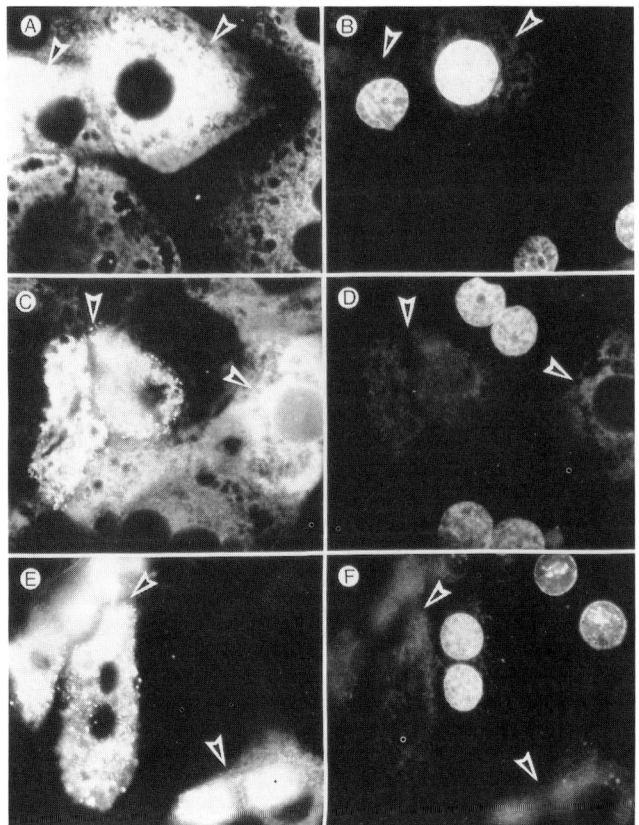

Fig. 4 : **effect on DNA synthesis in cultured hepatocytes of overexpression of anti-sense cyclin constructs.** To examine the involvement of cyclin A in S phase transit in cultured hepatocytes, cells stimulated by growth factors (day 0) were microinjected with either anti-sense human cyclin A or B cDNA under transcriptional regulation of a SV40 enhancer element. Immediately afterwards, cells were incubated in the presence of 5-Br-DU.
Cells were stained for the distribution of 5-Br-DU (panels B, D, F) and subsequently for the non-specific anti-serum co-injected with the DNA (panels A, C, E). Shown are cells injected on the second day of culture (panels A-D) and cells injected on the third day of culture (panels E and F). Cells were microinjected with either an anti-sense cyclin B (panels A-B), or anti-sense cyclin A (panels C-F) cDNA constructs. Arrowed are the injected cells.

consistent with recent in-vitro reports based on cell-free replication of simian virus 40 DNA (D'Urso et al, 1990). In view of our present observation on the role of cyclin A in S phase, it is interesting to note that cyclin A has been recently shown to associate to E_2F transcription factor in S phase (Mudryj et al,

1991), and to be included in a complex containing the retinoblastinoma protein and the DRTF1 transcription factor (related to E2F) (Bandara et al, 1991). It is therefore possible that cyclin A plays a role in regulation of transcription. Our result also raises the question as to the nature of cdc2 protein kinase which associates to cyclin A. Cyclin A has been shown to complex both to $p34^{cdc2}$ as well as to a $p33^{cdc2}$ (Pines and Hunter, 1990), recently referred to as CDCK2 (Tsai et al., 1991).
Therefore, this study might be important with regard to the potential involvement of cyclin A in cell transformation. Indeed, cyclin A has been shown to associate with the E1A protein of adenovirus in infected cells (Giordano et al 1983). In addition, hepatitis B virus DNA has integrated into the cyclin A gene in a human primary liver cancer. That cyclin A is involved in S phase may provide new clues as to its potential role in carcinogenesis.

REFERENCES

Bandara, L.R. & La Thangue, N.B. (1991) : Cyclin A and the retinoblastoma gene product complex with a common transcription factor. Nature 351, 494-497.
Beasley, R.P. (1988) : Hepatitis B virus : the major etiology of hepatocellular carcinoma. Cancer Res. 61, 1942-1956.
Booher, R., and Beach, D. (1988) : Involvement of cdc13+ in mitotic control in Schizosacchromyces pombe : possible interaction of the gene product with microtubules. EMBO J. 7, 2321-2327.
Bucher, N.L.R. & Malt, R.A. (1971) : Regeneration of liver and kidney. Little, Brown & Co, New-York, 1-278.
Caselman, W.H., Meyer, M., Kekule, A.S., Lauer, U., Hofscheneider, P.H. & Koshy, R. (1990) : A trans-activator function is generated by integration of hepatitis B virus preS/S sequences in human hepatocellular carcinoma DNA. Proc. Natl. Acad. Sci. USA 87, 2970-2974.
Chisari, F.V., Klopchin, K., Moriyama, T., Pasquinelli, C., Dunsford, H.A., Sell, S. Pinkert, C.A., Brinster, R.L. & Palmer, R.D. 1990 : Molecular pathogenesis of hepatocellular carcinoma in hepatitis B virus transgenic mice. Cell 59, 1145-1156.
Dejean, A., Bougueleret, L., Grzeschik, K.H., & Tiollais, P. (1986) : Hepatitis B virus DNA integration in a sequence homologous to v-erb-A and steroid receptor gene in a hepatocellular carcinoma. Nature 322, 70-72.
Draetta, G. Luca, F., Westendorf, J., Brizuela, L., Ruderman, J. & Beach, D. (1989) : cdc2 protein kinase is complexed with both cyclin A and B : evidence for proteolytic inactivation of MPF. Cell 56, 829-838.
Fang, F. & Newport, J.W. (1991) : Evidence that the G1-S and G2-M transitions are controlled by differnt cdc2 proteins in higher eukaryotes. Cell, 66, 731-742.
Forsburg, S.L., and Nurse, P. (1991) : Identification of a G1-type cyclin puc1+ in the fission yeast Schizosaccharomyces pombe. Nature 351, 245-248.
Fourel, G., Trepo, C., Bougueleret, L., Henglein, B., Ponzetto, A., Tiollais, P. & Buendia, M.A. (1990) : Frequent activation of N-myc genes by hepadnavirus insertion in

woodchuck liver tumours. <u>Nature</u> 347, 294-298.

Gautier, J., Minshull, J. Lohka, M., Glotzer, M., Hunt, & Maller, J.L. (1990) : Cyclin is a compoment of maturing-promoting-factor from Xenopus. <u>Cell</u> 60, 487-494.

Ghiara, J.B., Richardson, H.E., Sugimoto, K., Henze, M., Lew, D.J., Wittenberg, C. & Reed, S.I. (1991) : A cyclin B homolog in S. cerevisiae : chronic activation of the cdc28 protein kinase prevents exit from mitosis. <u>Cell</u> 65, 163-174.

Giordanno, A., Whyte, P., Harlow, Ed., Franza, B.R., Beach, D., Draetta, G. (1989) : A 60 kd cdc2-associated polypeptide complexes with the E1A proteins in adenovirus-infected cells. <u>Cell</u> 58, 981-990.

Glotzer, M., Murray, A.W. & Kirschner, M.W. (1991) : Cyclin is degraded by the ubiquitin pathway. <u>Nature</u> 349, 132-137.

Grisham, J.W., (1962) : A morphologic study of deoxyribonucleic acid synthesis and cell proliferation in regenerating rat liver ; autoradiography with Thymidine-^{3}H. <u>Cancer Res.</u> 22, 842-849.

Hadwiger, J.A., Wittenberg, C., Richardson, H.E., de Barros Lopes, M. & Reed, S.I. (1989) :A family of cyclin homologs that control the G1 phase in yeast. <u>Proc. Natl. Acad. Sci. USA</u> 86, 6255-6259.

Kekulé, A.S., Lauer, U., Meyer, M., Caselman, W.H., Hofschneider, P.H. & Koshy, R. (1990) : The preS2/S region of integrated hepatitis B virus encodes a transcriptionnal transactivator. <u>Nature</u> 343, 457-460.

Kim, C.M., Koike, K., Saito, I., Miyamura, T. & Jay, G. (1991) : HBx gene of hepatitis B virus induces liver cancer in transgenic mice. <u>Nature</u> 351, 317-320.

Koff, A., Cross, A., Fisher, A., Schuma, J., Le Guellec, K., Philippe, M. & Roberts, J.M. (1991) : Human cyclin E, a new cyclin that interacts with two members of the cdc2 gene family. <u>Cell</u> 66, 1217-1228.

Labbé, J.C., Capony, J.P., Caput, D., Cavadore, J.C., Derancourt, J., Kaghad, M., Lelias, J.M., Picard, A. & Dorée, M. (1989) : MPF from starfish oocytes at first meiotic metaphase is a heterodimer containing one molecule of cdc2 and one molecule of cyclin B. <u>EMBO J.</u> 8, 3053-3058.

Matsushime, H., Roussel, M.F., Ashum, R.A. & Sherr, C.J. (1991) : Colony-stimulating factor 1 regulates novel cyclins during the G1 phase of the cell cycle. <u>Cell</u> 65, 701-713.

Meijer, L., Arion, D., Golsteyn, R., Pines, J., Brizuela, L., Hunt, T. & Beach, D. (1989) :Cyclin is a compoment of the sea urchin egg M-phase specific histone H1 kinase. <u>EMBO J.</u> 8, 2275-2282.

Mc Gowan, J.A. (1986) : prolifération des hépatocytes en culture <u>Research in... Isolated and cultured hepatocytes.</u> INSERM/John Libbey Eurotext, 13-40.

Minshull, J., Golsteyn, R., Hill, C.S. & Hunt T. (1990) :The A- and B-type cyclin associated cdc2 kinase in Xenopus turn on and off at different times in the cell cycle. <u>Embo J.</u> 9, 2865-2875.

Motokura, T., Bloom, T., Kim, H.G., Juppner, H., Ruderman, J.V., Kronenberg, H.M. & Arnold, A. (1991) : A novel cyclin encoded by a bcl1-linked candidate oncogene. <u>Nature</u> 350, 512-515.

Nash, R., Tokiwa, G., Anand, S., Erickson, K. & Futcher, A.B. (1988) : The WHI1+ gene of Saccharomyces cerevisiae tethers cell division to cell size and is a cyclin homolog. EMBO J. 7, 4335-4346.

Pines, J., & Hunter, T. (1990) : Human cyclin A is adenovirus E1A-associated protein p60 and behaves differently from cyclin B. Nature 346, 760-763.

Swenson, K.I., Farell, K.M. & Ruderman, J.V. (1986) : The clam embryo protein cyclin A induces entry into M phase and resumption of meiosis in Xenopus oocytes. Cell 47, 861-870.

de Thé, H., Marchio, A., Tiollais, P. & Dejean, A. (1987) : A novel steroid thyroid hormone receptor-related gene inappropriately expressed in human hepatocellular carcinoma. Nature 330, 667-670.

Tsai, L.H., Harlow, E. & Meyerson, M. (1991) : Isolation of the cdk2 gene that encodes the cyclin A- and adenovirus E1A-associated p33 kinase. Nature 353, 174-177.

D'Urso, G., Marraciano, R.L., Marchak, D.K., & Roberts, J.M. (1990) : Cell cycle control of DNA replication by a homologue from human cells of the $p34^{cdc2}$ protein kinase. Science 250, 786-791.

Wang, J., Chenivesse, X., Henglein, B., Bréchot, C. (1990) : Hepatitis B virus integration in a cyclin A gene in a hepatocellular carcinoma. Nature 343, 555-557.

Xiong, Y., Connolly, T., Futcher, B. & Beach, D. (1991) : Human D-type cyclin. Cell 65, 691-699.

Résumé

Nous avons décrit précédemment l'intégration de l'ADN du virus de l'hépatite B (VHB) dans un intron du gène de la cycline A dans un cancer du foie. Ce travail nous a permis d'isoler l'ADNc de la cycline A humaine.

Nous avons construit une banque d'ADNc à partir de la tumeur originale (tumeur HEN) et isolé plusieurs ADNc hybrides VHB-cycline A. Ces ADNc hybrides présentent une phase ouverte de lecture codant pour une protéine de fusion qui comprend la partie N-terminale de la protéine préS2/S suivie de la cycline A. Cependant la partie N-terminale de la cycline A, impliquée dans sa dégradation, est délétée. L'intégration du VHB dans le gène de la cycline A rend compte de la surexpression des transcripts hybrides VHB-cycline A codant pour une protéine stabilisée.

De plus, nous avons étudié l'expression du gène de la cycline A dans les hépatocytes de rat en culture primaire et dans le foie de rat en régénération après hépatectomie partielle. Dans les deux cas, les ARNm et la protéine cycline A s'accumulent au moment où les cellules rentrent en phase S. La microinjection de plasmides contenant l'ADNc antisens de la cycline A inhibe la synthèse de l'ADN.

En conclusion, nous avons analysé le profil d'expression de la cycline A dans la tumeur HEN. Les résultats obtenus renforcent l'hypothèse d'une mutagénèse insertionnelle du VHB. Par ailleurs, nous avons montré, dans les hépatocytes, que la cycline A est impliquée dans la transition G1/S du cycle cellulaire. L'ensemble de ces résultats apportent de nouvelles données quant au rôle de la cycline A dans la carcinogénèse.

Regulations of the wild type p53 protein functions: a role for calcium and the calcium-binding S100 proteins

Jacques Baudier

INSERM U 309, CEN-G 85X, 38041 Grenoble Cedex, France

Summary

p53 is a key protein in protecting cells from neoplasia. Most human tumors have an abnormal or inactivated p53 protein. P53 is a multipartners protein subjected to multiple modes of regulation. Mutations within the p53 molecule result in conformational changes that can modify the interaction of p53 with partner-proteins and/or cause a defect in the regulatory processes that normally control w.t. p53 functions. We have characterized p53 as substrate for the calcium/phospholipid-dependent protein kinase C that also interacts with the cell cycle-associated calcium binding S100b protein. We here propose that these two calcium-dependent cell regulatory pathways, and thus calcium homeostasis, may play a role in the regulation of the p53 functions.

The p53 protein.

The p53 protein is a negative regulator of cell growth.

The p53 protein is considered as a product of an anti-oncogene that may negatively regulate the cell cycle. The concept that p53 functions as a tumor suppressor is sustained by several experimental approaches that show that wild type (wt.) p53 contributes to growth arrest of cells (Levine et al. 1991).
The p53 protein also plays a role in terminal cell differentiation processes (Rotter et al. 1993; Kastan et al. 1991). Other biological processes in which p53 may play a role are apoptosis (Yonish-Rouach et al.1991) and DNA repair (Lane, 1992).

Acknowledgement. I whish to thanks all the members of the laboratory who have contributed to this work and for stimulating discussions; particularly Chrisian Delphin, Saadi Khochbin, Didier Grunwald, Caroline Hoemann ,and Jean Jacques Lawrence. This work was supported by a grant (6777) from the Association pour la Recherche sur le Cancer.

To specify more precisely the role of the p53 protein in development and tumorigenesis, a p53 deficient mouse strain was generated (Donehower et al. 1992). These studies demonstrated that p53 is not essential for normal development during embryogenesis but that the p53 deficient animals are more succeptible to develop tumors than control animals. This observations, thus, confirmed the tumor suppressor role of p53 but they also suggest that other proteins or factors can replace p53 in cell differentiation and apoptosis, two key events during embryogenesis and cell organization into functional tissues.

P53 is a multipartners protein subjected to multiple modes of regulation.

Wt. p53 is a protein which is spatially regulated during the cell cycle. p53 is synthesized and sequestered within the cytoplasm until the protein translocates within the cell nucleus at the G1-S phase transition of the cell cycle. The cytoplasmic storage of p53 can be regulated by the self oligomerization of p53 (Levine et al. 1991) or the interaction of p53 with other cytoplasmic proteins (Gannon and Lane 1991). Nuclear localization of the p53 protein is dictated by nuclear localization signals (NLS) in the C-terminus of the protein (Shaulsky et al. 1991).
The tumor supressor function of wt. p53 is linked to its nuclear localization (Shaulsky et al. 1991) suggesting that its activity will be sustained by the interaction with nuclear targets (Fig.1). Several nuclear targets for wt. p53 have been identified, including the TATA binding protein (Seto et al. 1992) and the product of the oncogene mdm2 (Momand et al. 1992). P53 also binds to specific DNA target sequences (El-Deiry et al. 1992). In binding to DNA w.t. p53 may function as a transcription factor with the C-terminus acting as a DNA binding domain and the N-terminus as a transcriptional activator domain (Fields and Jang 1990).

The relatively low half-life of wt. p53 in the cell and the synchronized evolution of the protein during the cell cycle suggest that multiple, punctual, and probably interdependent control processes will regulate the p53 protein at the level of its synthesis, cytoplasmic storage, nuclear translocation, interactions with nuclear targets and degradation. Little is known about these control processes. The stability of p53, its cytoplasmic storage and DNA binding activity can be regulated by interactions with cellular protein factors . Posttranslational modifications of p53 by phosphorylation are likely also candidates for regulation of p53 activities. Five different modes of phosphorylation for p53 have already been characterized (Fig.1). The transcriptional activator N-terminal domain of p53 is a substrate for phosphorylation by a casein kinase I-like enzyme (Milne et al. 1992) and DNA-dependent protein kinase (Lees-Miller et al. 1990). The multifunctional C-terminal domain harbouring the nuclear targeting signal motifs, the oligomerization promoting domain and the DNA binding domain is phosphorylated by the cdc2 kinase (Bischoff et al. 1990), casein kinase II (Meek et al. 1990), and protein kinase C (Baudier et al. 1992). The phosphorylation of p53 by CKII positively regulates p53 DNA binding (Hupp et al.1992). A ribosomal RNA has also been found to bind to p53 through a covalent bound involving the C-terminal phosphorylated serine residue by CKII (Samad et al.1986).The functional consequences of the other modes of phosphorylation have not been yet characterized.

The p53 gene is frequently mutated in human tumors (Levine et al. 1991). These mutations are multiple and are mostly single missense point mutations within conserved coding sequences. The p53 mutants lose their tumor suppressor activity. Mutations of the p53 molecule result in conformational changes within the protein revealed by differences in immunoreactivity toward monoclonal antibodies. The mutations in p53 associated with tumorogenesis compromise p53 nuclear translocation and interactions with nuclear targets, proteins and DNA (Levine et al. 1991). It is likely that a defect in one of the multiple regulation processes acting on wt. p53 could contribute to the loss-of-function characteristic of mutant p53 proteins.

P53 is a substrate for the calcium/phospholipid-dependent protein kinase C.

Analysis of the p53 primary structure reveals that the protein has an extended basic C-terminal region that is predicted to form an amphipatic helix-like structure and is responsible for its binding to DNA and the formation of stable p53 oligomers (Sturzbecher et al. 1992). This domain on p53 shows a striking similarity with the protein kinase C (PKC) and calcium binding protein (CaBP)-binding domain on the MARCKS proteins. MARCKS is a major PKC substrate rapidly phosphorylated in a wide variety of cell types in response to growth factor stimulation (Aderem 1992). It was therefore predicted that p53 could be a PKC substrate that interacts with the CaBPs. The phosphorylation of p53 by PKC has been studied *in vitro* and *in situ* (Baudier et al. 1992). In vitro recombinant p53 protein incorporated up to 0.6 mol phosphate per mol protein suggesting a single phosphorylation site. Serine residue is the only phosphorylated amino acid on the human and mouse recombinant p53 (unpublished data). Although we have not yet been able to determine the PKC phosphorylation site on p53, preliminary studies have shown that V8 proteolysis of the phosphorylated p53 generated a single phosphopeptide whose sequence began at residue 350, thus including the predicted PKC phosphorylation site.

We also tested the effect of PKC phosphorylation on the DNA binding activity of wt. recombinant p53 in a gel mobility shift assay. PKC in the presence of ATP stimulated p53 binding to DNA (manuscript in preparation). This observation is consistent with an effect of the phosphorylation by PKC on the conformation of the C-terminal domain where the predicted PKC phosphorylation site domain and the DNA binding site are located.

p53 interacts with the calcium binding S100b protein.

To confirm the functional homology between the C-terminal domain of p53 with the PKC phosphorylation and CaBP binding domain on MARCKS, we also studied the interaction of cellular and recombinant p53 with the calcium binding S100b protein.

The choice of the S100b protein was motivated by the fact that MARCKS phosphorylation is strongly inhibited by the S100b protein (Albert et al.1984).

S100b belongs to the S100 protein family that plays a role in cell proliferation and differentiation (For reviews on S100 proteins see Hilt and Kligman 1991). S100b has long been considered as a protein specific to the nervous system, but is now recognized to be also present in most peripheral tissues although at a much lower concentration than in brain. The S100b is also overexpressed in several human cancers, and it is used as a prognostic indicator of the transformation state of the cells (Egan et al. 1987; Cochran et al. 1982). In human blood cells S100b has been found to be present in particular cases of T cell leukemia (Takahashi et al. 1988; Hanson et al. 1991). S100b expression is maximal in the G1 phase of the cell cycle (Marks et al. 1990) suggesting that it may contribute to regulation regulation, in a calcium-dependent manner, of cell cycle associated protein factors.

Calcium-dependent interaction of recombinant and cellular p53 with S100b was demonstrated by affinity chromatography and crosslinking experiments (Baudier et al. 1992). The interaction of S100b with p53 also inhibited p53 phosphorylation by PKC and p53 oligomerization (Baudier et al. 1992). These two results strongly argue that the S100b binding domain on p53 is likely located in the predicted C-terminus. We also have obtained evidence that *in vitro*, the interaction of S100b with p53 stimulates p53 binding to target DNA in a gel retardation assay. This latter observation is probably related to the S100b-induced inhibition of p53 oligomerization or induction of a conformational change in the C-terminus of p53 that unmasks the DNA binding domain. Both proteolysis of the C-terminal domain of p53 and binding of the monoclonal antibody PAb 421 to the C-terminus, prevent p53 oligomerization and also stimulate p53 DNA binding (Hupp et al. 1992). The interaction of p53 with S100b is specific in the sense that calmodulin do not inhibit PKC-dependent phosphorylation of p53 and p53 oligomerization. It should be noted that calmodulin does not stimulate p53 DNA binding either.

Conclusions: The roles of calcium, PKC and CaBPs in the regulation of p53 activities.

The characterization of p53 as a PKC substrate that also bind to the S100b protein supports the idea that p53 functions could be regulated by calcium homeostasis.

W.t. p53 functions as an antiproliferative protein. In non-transformed cells, its activity should then be turned off when cells are stimulated to proliferate by mitogenic growth factors. Because cytoplasmic calcium concentration generally increases in response to extracellular mitogenic stimulation, we propose that the inhibition of the p53 activity could be acomplished by the interaction of p53 with the calcium binding S100b or other S100b-related protein of the S100 familly that are mainly cytoplasmic proteins. The formation of a cytoplasmic S100b-p53 complex would inhibit the translocation of p53 to the nucleus. In favour of this hypothesis it is significant to note that the putative S100b-binding domain on p53 overlaps with two nuclear localization signals that mediate the migration of p53 to the cell nucleus (Shaulsky et al. 1991).

Altered calcium homeostasis is observed in tumor cells (Mac Manus et al. 1982; Tsuruo et al. 1984; Banyard and Tellam 1985). Cancer cells have elevated levels of cytosolic calcium which may in part be responsible for a permanent activation of DNA synthesis and/or increased motility/invasivness independent of extracellular stimulation (For review see Whitfield 1990). In tumoral cells, where the S100b concentration is elevated, a permanent inhibition of the w.t. p53 can thus be envisaged through the same mechanisms to that proposed above: sequestration of p53 by the S100 proteins. We are actually testing this hypothesis by studies on the effect of

S100b on the cytoplasmic sequestration of p53 in co-transfected cells with the p53 and S100b genes.

An other interesting aspect of the interaction of S100b with p53 is that it inhibits p53 phosphorylation by PKC. *In situ* phosphorylation studies of p53 using rat embryo fibroblasts transformed by a temperature sensitive p53 val135 mutant revealed a drastic stimulation of p53 phosphorylation by the PKC activator PMA at 32°C when the cells are growth arrested and p53 is predominantly located in the nucleus (unpublished data). This suggests that the PKC-mediated phosphorylation of p53 likely occurs within the cell nucleus. The S100b protein is mainly a cytoplasmic protein. One can therefore suppose that S100b can prevent p53 phosphorylation by the cytoplasmic PKC when the protein has not yet translocated to the nucleus. Once p53 has translocated within the nucleus, p53 can be phosphorylated by its specific nuclear enzyme. In this scenario, the calcium-dependent interaction of p53 with S100b could serve to protect excessive p53 phosphorylation by calcium activated PKC in a cell compartment, the cytoplasm, where p53 has no activity. Those are attractive hypotheses that we are currently testing.

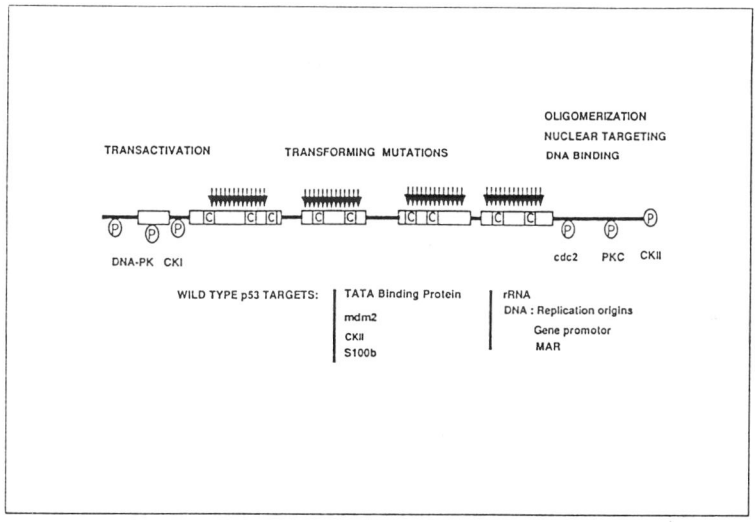

Figure 1. p53 is a multiple targets protein subjected to multiple regulations. p53 is a 48 kDa protein whose primary structure can be divided into three domains. The N-terminus is rich in di-carboxilic amino acids and is involved in gene transcription. The C-terminus is a multifunctional domain rich in basic amino acids. It has been implicated in p53 oligomerization, DNA binding and nuclear translocation. The third domain on p53 is the protein core characterized by amino acid sequences highly conserved among species (here represented as boxes). It is within these sequences that occur transforming mutations on p53 associated with cancers (represented as arrows). Within the protein core are found 9 cysteine residues (C) that may participate in the tertiary p53 protein structure and the regulation of its functions. This assumption is based on the fact that the state of oxydation of sulfhydryl groups in the w.t. p53 regulates both p53 binding to DNA and p53 phosphorylation (unpublished data). Numerous p53 targets have been identified, including proteins, rRNA and DNA.

REFERENCES.

Aderem, A. (1992): The MARCKS brothers: A family of protein kinase C substrates. Cell 71, 713-716.

Albert, K.A, Wu, W.C., Nairn, A.C., Greengard, P. (1984): Inhibition by calmodulin of calcium/phospholipid-dependent protein phosphorylation. P.N.A.S. USA 81, 3622-3625.

Banyard, M.R.C. and Tellam, R.L. (1985): The free cytoplasmic calcium concentration of tumorigenic and non tumorigenic human somatic cell hybrids. Br. J. Cancer 51, 761-766.

Baudier, J., Delphin, C., Grunwald, D., Khochbin, S., and Lawrence, J.J. (1992): Characterization of the tumor supressor p53 as a protein kinase C substrate and a S100b-binding protein. P.N.A.S. USA 89, 11627-11631.

Bischoff, J.R., Friedman, P.N., Marshak, D.R., Prives, C. and Beach, D. (1990): Human p53 is phosphorylated by p60-cdc2 and cyclin B-cdc2. P.N.A.S. USA 87, 4766-4770.

Cochran, A.J., Wen, D.R., Herschman, H.R., and Gaynor (1982): Detection of S100 protein as an aid to the identification of melanocytic tumors. Int. J. Cancer 30, 295-297.

Donehower, L.A., Harvey, M., Slagle, B., McArthur, M., Montgomery, C.A., Butel, J. and Bradley, A. (1992): Mice deficient for p53 are developmentally normal but susceptible to spontaneous tumors. Nature 356, 215-221.

Egan, M.J., Newman, J., Crocker, J. and Collard, M. (1987): Immunohistochemical localization of S100 protein in benign and malignant conditions of the breast. Arch. Pathol. Lab. Med. 111, 28-31.

El-Deiry, W.S., Kern, S.E., Pietenpol, S.E., Kinzler, K.W. and Vogelstein, B. (1992): Definition of a consensus binding site for p53. Nature Genet. 1, 45-49.

Fields, S., and Jang, S.J. (1990): Presence of a potent transcription activating sequence in the p53 protein. Science 249, 1046-1049.

Gannon, J.V. and Lane, D.P. (1991): Protein synthesis required to anchor a mutant p53 protein which is temperature sensitive for nuclear transport. Nature, 349, 802-806.

Hanson, C.A., Bockenstedt, P.L., Schnitzer, B., Fox, D.A., Kueck, B. and Braun, D.K. (1991): S100-positive, T-cell chronic lymphoproliferative disease: An aggessive disorder of an uncommon T-cell subset. Blood 78, 1803-1813.

Hilt, D.C. and Kligman, D. (1991): The S100 protein family: A biochemical and functional overview. In Novel Calcium Binding Proteins, Fundamental and Clinical Implication, C.W Heizmann ed., Berlin Springer Verlag, pp 65-103.

Hupp, T.R., Meek, D.W., Midgley, C.A. and Lane, D.P. (1992): Regulation of the specific DNA binding function of p53. Cell 71, 875-886.

Kastan, M.B., Radin, A.I., Kuerbitz, S.J., Onyekwere, O., Wolkow, C.A., Civin, C.I., Stone, K.D., Woo, T., Ravindranath, Y., and Craig, R.W. (1991): Level of p53 protein increase with maturation in human hematopoietic cells. Cancer Res. 51, 4279-4286.

Lane, D.P. (1992): p53 guardian of the genome. Nature 358, 15-16.

Lees-Miller, S.P., Chen, Y., and Anderson, C.W. (1990): Human cells contain a DNA-activated protein kinase that phosphorylates simian virus 40 antigen, mouse p53, and the human Ku autoantigen. Mol. cell. Biol. 10, 6472-6481.

Levine, A.J., Momand, J., and Finlay, C.A. (1991): The p53 tumour suppressor gene. Nature 351, 453-456.

Mac Manus, J.P., Boynton, A.L., Whitfield, J.F. (1982): The role of calcium in the control of cell reproduction. In The role of Calcium in Biological Systems, Vol. III, Anghileri, L.J., Tuffet-Anghileri, A.M. eds., CRC Press, Boca Raton, Florida, pp147-164.

Marks,A., Petsche, D., O Hanlon, D., Kwong, P.C., Stead, R., Dunn, R., Baumal, R. and Liao, S.K. (1990): S100 protein expression in human melanoma cells: Comparison of level of expression among different cell lines and individual cells in different phases of the cell cycle. Exp. Cell Res. 187, 59-64.

Meek, D.W., Simon, S., Kikkawa, U. and Eckhart, W. (1990): The p53 tumour suppressor protein is phosphorylated at serine 389 by casein kinase II. EMBO J. 9, 3253-3260.

Milne, D.M., Palmer, R.H., Campell, D.G. and Meek, D.W. (1992): Phosphorylation of the p53 tumor-suppressor protein at three N terminal sites by a novel casein kinase 1-like enzyme. Oncogene, 7, 1361-1370.

Momand, J., Zambetti, G.P., Olson, D.C., George, D., Levine, A.J. (1992): The mdm-2 oncogene product forms a complex with the p53 protein and inhibits p53-mediated transactivation. Cell 69, 1237-1245.

Rotter, V., Foord, O., and Navot,N. (1993): In search of the functions of normal p53 protein. Trends in Cell Biology 3, 46-49.

Samad, A., Anderson, C.W., and Caroll, R.B. (1986): Mapping of phosphomonoester and apparent phosphodiester bonds of the oncogene product p53 from simian virus 40-transformed 3T3 cells. P.N.A.S. USA 83, 897-901.

Seto,E., Usheva, A., Zambetti, G., Momand, J., Horikoshi, N., Weinmann, R., Levine, A.J., and Shenk, T. (1992): Wild-type p53 binds to the TATA binding protein and represses transcription. P.N.A.S. USA 89, 12028-12032.

Shaulsky, N, Goldfinger, N., Tosky, M.S., Levine, A.J. and Rotter, V. (1991): Nuclear localization is essential for the activity of p53 protein. Oncogene 6, 2055-2065.

Sturzbecher, H.W., Brain, R., Addison, C., Rudge, K., Remm, M., Grimaldi, M., Keenan, E. and Jenkins, J.R. (1992): A C-terminal a-helix plus basic region motif is the major structural determinant of p53 tetramerization. Oncogene 7, 1513-1523.

Takahashi, K., Ohtsuki, Y., Sonobe, H, Hayashi, K., Nakamura, S.I., Kotani, S., Kubonishi, I., Miyoshi, I., Isobe, T., Kita, K. and Akagi, T. (1988): S100b positive T cell leukemia. Blood 71, 1299-1303.

Tsuruo, T., Ida, H., Kawabata, H., Tsukagoshi, S. and Sakurai, Y. (1984): High calcium concentration of pleiotropic drug-resistant p388 and K562 leukemia and chines hamster ovary cells. Cancer Res. 44, 5095-5099.

Whitfield, J.F. (1990) Calcium, Cell Cycle and Cancer. CRC Press Inc., Boca Raton, Florida.

Yonish-Rouach, E., Resnitzky, D., Lotem, J., Sachs, L., Kimchi, A., Oren, M. (1991): Wild-type p53 induces apoptosis of myeloid leukaemia cells that is inhibited by interleukin-6. Nature 352, 345-347.

Résumé

La p53 est une protéine-clé pour la protection des cellules contre le cancer. La plupart des tumeurs humaines ont un protéine p53 anormale ou inactivée. P53 est une protéine interagissant à plusieurs niveaux et soumise à plusieurs modes de régulation. Des mutations de la molécule p53 entrainent des changements de conformation qui peuvent modifier les interactions de p53 avec les protéines-partenaires et/ou entrainer un défaut dans les processus de régulation qui contrôlent normalement les fonctions de p53 de type sauvage. Nous avons caractérisé la p53 comme substrat de la protéine-kinase C calcium/phospholipide-dépendante qui interagit également avec la protéine liée au calcium S100b associée au cycle cellulaire. Nous proposons ici que ces 2 voies de régulation dépendantes du calcium et donc l'homéostasie du calcium joue un rôle dans la régulation des fonctions de p53.

The tumour suppressor p53 is expressed in CML CD34+ cells and inhibition of p53 expression can affect the pattern of CFU-GM colony formation

Sucai Bi, Francesco Lanza and John M. Goldman

LRF Centre for Adult Leukaemia, Department of Haematology, Royal Postgraduate Medical School, Hammersmith Hospital, Du Cane Road, London W12 ONN, UK

Expression of the tumour suppressor p53 promotes haematopoietic cell differentiation, maturation and apoptosis (Shaulsky et al., 1991; Kastan et al., 1991; Yonish-Rouach et al., 1991). These different functions of p53 probably involve different mechanisms, with p53 acting as a regulator of the cell cycle or as a transcription factor. Alterations in the p53 gene have been found in various forms of leukaemia and lymphoma including blastic transformation of chronic myeloid leukaemia (CML) (Ahuja et al., 1989; Bi et al., 1992). Thus the p53 protein may play an important role in regulating the proliferation of haematopoietic cells. In order to characterize the involvement of p53 in haematopoiesis, we studied the expression of p53 in various types of haematopoietic cells. The effect of inhibiting p53 expression on proliferation of haematopoietic cells was also investigated in the granulocyte-macrophage colony forming unit (CFU-GM) assay.

Using two p53 monoclonal antibodies, PAb1801 and the mutant conformation-associated PAb240 (Gannon et al., 1990) and flow cytometry, we studied the expression of p53 in 13 samples of CML light density cells (<1.067 g/ml). An average of 14% of the cells expressed detectable p53. 56% of the CD34+ cells co-expressed p53. In 5 normal bone marrow samples, comparable results were obtained. The expression of p53 protein reacting with PAb240 was closely associated with CD34+/HLA-DR+ cells and with cells in active cell cycle, while majority of the p53 proteins recognised by PAb1801 were found in CD34+/HLA-DR- cells and in cells in G_0/G_1 phases of the cell cycle. Interestingly, the CD34- myeloid precursor cells had no detectable p53 protein in either normal or CML samples. These data indicate that expression of p53 is closely regulated during haematopoiesis. We conclude that expression of p53 protein reactive with PAb240 is not the result of mutations in the p53 gene. The conformation changes of p53 protein in different groups of cells and in cells in different stages of the cells cycle may also reflect different aspects of p53 function (Milner, 1991).

To assess possible involvement of p53 protein in the regulation of haematopoiesis, an anti-sense oligonucleotide approach was used. The effect of inhibiting p53 expression on proliferation of CML haematopoietic cells was studied. Light density CML bone marrow cells were pre-incubated with 5 µM of a 18-mer p53 anti-sense oligonucleotide (5'-CGG CTC CTC CAT GGC AGT-3') in serum-free medium (SFM) for 6 hours before being plated with CFU-GM medium supplemented with 5 µM of anti-sense oligonucleotides. A 18-mer sense oligonucleotide (5'-ACT GCC ATG GAG GAG CCG-3') was used in parallel in addition to controls which received no oligonucleotide treatment. Among the 18 CML samples studied, increased numbers of day 7 CFU-GM colonies in anti-sense oligonucleotides treated dishes were observed in 12 samples (66.7% of total cases, type 1 response). The increase was significant in 9 of the 12 cases. In another 6 samples, reduced day 7 CFU-GM colony numbers were observed in anti-sense oligonucleotide treated dishes (33.3% of total samples, type 2 response). The decrease was significant in 4 cases. In both groups, differences in CFU-GM colony numbers between anti-sense and sense oligonucleotide treated cells were highly significant (Table 1). The overall difference between anti-sense and sense oligonucleotides treated dishes was also significant (p=0.014). However, the cells treated with sense oligonucleotides generally formed fewer CFU-GM colonies compared with cells that had received no oligonucleotides (p=0.02), suggesting that the oligonucleotides have non-specific toxicity on haematopoietic cells.

Table 1. CFU-GM colony formation by anti-sense oligonucleotides treated CML cells according to their response to treatment (per 1 x 10^5 cells)

Patient nos.	Anti-sense	Sense	Nil	% change and p value*
12 (type 1)	471 ± 321	349 ± 282	394 ± 331	+35 p=0.002
6 (type 2)	174 ± 246	199 ± 237	210 ± 242	-12.6 p=0.02

Mean ± standard deviation. *: anti-sense compared with sense oligonucleotides treated cells.

In addition to differences in CFU-GM colony numbers, we observed that the size of the colonies formed by anti-sense oligonucleotide treated cells was generally larger than that of colonies formed by either sense oligonucleotide treated cells or cells that had received no oligonucleotides.

A similar approach was used to study 10 normal bone marrow samples. In contrast to CML cells, treatment of normal bone marrow cells with p53 anti-sense oligonucleotides produced either little change or reduced numbers of day 7 CFU-GM colonies in most samples. 60% of the samples showed no significant change in colony numbers between anti-sense and sense oligonucleotides treated cells. Only one case showed a significant increase in day 7 CFU-GM colony formation and another 3 cases showed significant reductions of day 7 CFU-GM colony numbers in anti-sense oligonucleotide treated dishes. However, the overall difference between anti-sense and sense oligonucleotide treated cells was not significant (p=0.27).

In summary, by using flow cytometry and p53 monoclonal antibodies, we found p53 protein was expressed by the immature haematopoietic CD34+ cells in both normal and CML samples while the CD34- myeloid precursors expressed no detectable p53 protein. Expression of different conformational p53 proteins was closely associated with different groups of CD34+ cells and cells in different stages of the cell cycle. Inhibition of p53 expression showed regulatory effects on proliferation of haematopoietic cells. Our data suggest that expression of p53 is precisely regulated in haematopoiesis and the p53 protein plays an important role in regulating proliferation of haematopoietic stem and progenitor cells, especially in CML.

References:

Ahuja, H., Bar-Eli, M., Advani, S.H., Benchimol, S. & Cline, M.J. (1989) Alterations in the p53 gene and the clonal evolution of the blast crisis of chronic myelocytic leukemia. Proc. Natl. Acad. Sci. USA. 86:6783-6787.

Bi, S., Hughes, T., Bungey, J., Chase, A., de Fabritiis, P. & Goldman, J.M. (1992) p53 in chronic myeloid leukemia cell lines. Leukemia. 6:839-842.

Gannon, J.V., Greaves, R., Iggo, R. & Lane, D.P. (1990) Activating mutations in p53 produce a common conformational effect. A monoclonal antibody specific for the mutant form. EBMO J. 9:1595-1602.

Kastan, M.B., Radin, A.I., Kuerbitz, S.J., Onyekwere, O., Wolkow, C.A., Civin, C.I., Stone, K.D., Woo, T., Ravindranath, Y. & Craig, R.W. (1991) Levels of p53 protein increase with maturation in human hematopoietic cells. Cancer Res. 51:4279-4286.

Milner, J. (1991) A conformation hypothesis for the suppressor and promoter functions of p53 in cell growth control and in cancer. Proc. R. Soc. Lond. (Biol). 245:139-145.

Shaulsky, G., Goldfinger, N., Peled, A. & Rotter, V. (1991) Involvement of wild-type p53 in pre-B-cell differentiation in vitro. Proc. Natl. Acad. Sci. USA. 88:8982-8986.

Yonish-Rouach, E., Resnitzky, D., Lotem, J., Sachs, L., Kimchi, A. & Oren, M. (1991) Wild-type p53 induces apoptosis of myeloid leukaemic cells that is inhibited by interleukin-6. Nature, 352:345-347.

Retinoblastoma protein translocation. A possible mechanism of negative regulation of cell growth

Vitaly Rogalsky, German Todorov and Donald Moran

D.H. Ruttenberg Cancer Center, Mount Sinai Medical Center, New York, NY 10029, USA

In our previous research we studied the subcellular localization of carcinoembryonic antigen (CEA)[1-3]. We observed that CEA localization in malignant and benign colonic tumors was different from that in normal goblet cells. We termed this phenomenon an "antigenic translocation" [2]. Antigenic translocation also occurs in the process of organ development[3]. Thus, in light of our previous findings, while studying the role of retinoblastoma protein (pRB) in the growth inhibition and differentiation of tumor cells, we explored the possibility of its intracellular translocation. We observed, that the growth inhibition of U-2 osteosarcoma cells and differentiation of U937 leukemic cells induced by specific agents were associated with translocation of pRB from the nucleoplasm to nucleoli. The present report describes this phenomenon of "regulatory translocation".

We have described elsewhere a low MW (7 KD) cytokine discovered in our laboratory and named NCPI (natural cell proliferation inhibitor)[4]. This factor is secreted by human fetal fibroblasts and is capable of inhibiting U-2 osteosarcoma cell growth, inducing adipose conversion and triggering leukemic cell differentiation. Inhibition of fibroblast growth by NCPI was associated with the stimulation of PKCα [5]. Medium conditioned by fibroblasts inhibits U-2 cell proliferation in dilutions up to 1:16. U-2 cells arrested in their growth by NCPI are able to survive for up to 6 months without change of medium.

Immunohistochemical study of subcellular localization of RB protein was performed. Within 15 - 20 days all cells in the control flasks detached and died. U-2 cells under NCPI propagated slowly for 3 - 5 days and then stopped growing. After a 6 month period about 10% of these cells were still alive. When fresh RPMI medium with 20% FBS was added to some of the flasks, cells resumed proliferation in about three days. The other flasks were cut, and bottoms with the attached cells were used for immunohistostaining with RB-specific antibody.

In control U-2 cells (not treated with NCPI) diffuse staining was observed in the nuclei. U-2 cells grown under NCPI revealed a different staining pattern. Nucleoli stained intensively, whereas the rest of the nuclei stained less than in control cells. This phenomenon was observed in all osteosarcoma cells.

The results obtained in U-2 cells indicated that the translocation of RB protein to nucleli could be associated with cell growth inhibition in some cell types. To verify this hypothesis another model for the study of cell growth inhibition was chosen. TPA is known to induce a rapid growth inhibition and differentiation in certain leukemic cells [6]. Moreover, significant amount of cells can be obtained in a relatively short time, which makes it possible to obtain purified subcellular fractions and perform Western blot analysis in addition to immunohistochemistry.

In control U937 cells diffuse staining of nuclei with RB specific antibodies was observed. Treatment with TPA for 36 hours resulted in the accumulation of RB protein in the nucleoli, similarly to that observed in U-2 cells grown under NCPI. The phenomenon of pRB translocation was clearly seen in about 30% of the cells. To monitor the process of growth inhibition of U937 cells induced by TPA, DNA synthesis was assayed. After 12 h DNA synthesis [H^3]thymidine incorporation was 60 ± 4% of control, 21 ± 9% after 36 h, and 19 ± 6% after 48 h.

Western blot analysis of purified nucleoli was chosen as an alternative method to confirm the observed translocation of RB protein. To obtain sufficient amount of cells for the isolation of the nucleoli U937 cells were grown to a relatively high density (10^6/ml before treatment).

Total nuclear fractions and purified nucleoli were isolated from both TPA treated and control cells. Equal amounts of protein from total nuclei and purified nucleoli were used in the Western blots. Polyclonal affinity purified antibodies raised against synthetic C-terminal RB peptide (15 C-terminal amino acids [7] were used. To confirm the antibody specificity, synthetic peptide was used in control blots to block the antibody (20-30 µg of peptide per 10 µg of antibody). The results of the Western blot are presented in Figure 1.

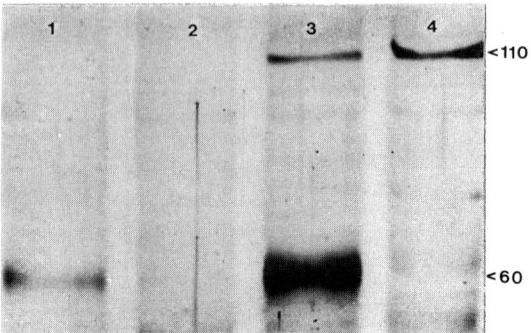

Fig. 1. Translocation of retinoblastoma protein (pRB) in U937 cells treated with 10^{-7} M TPA for 36 hours (Western blot analysis).
1-Extract of purified nucleoli of TPA treated cells
2-Extract of purified nucleoli of control cells
3- Total nuclear extract of TPA treated cells
4-Total nuclear extract of control cells.

The extract from total nuclei of control cells gave a major band of 110 KD (lane 4), while the extract from total nuclei of the treated cells had shown an intense 60 KD band and a weaker 110 KD band (lane 3). Purified nucleolar fraction of control cells yielded no bands (lane 2), whereas nucleoli of TPA treated cells produced a 60 KD band (lane 1). No bands were observed on the control blots where antibodies were preincubated with RB peptide.

To assess the phosphorylation state of RB protein Western Blot analysis of U937 cells growing at densities of 10^6/ml (exponential growth) and 2×10^6/ml (decline in the growth rate) was performed. In the nuclei of U937 cells growing at 10^6/ml most of RB protein was present in a highly phosphorylated form (115 KD), whereas in the cells growing at 2×10^6/ml most of RB protein was hypophosphorylated (major band at 110 KD, minor band at 115 KD) (data not shown).

DISCUSSION

The translocation of pRB was shown immunocytochemicaly in two cell lines inhibited in their growth with different agents. Another method, Western blot analysis has confirmed these findings.In the experiments inhibition was induced by different agents - TPA and NCPI. These results suggest that observed pRB transiocation to the nucleoli may be associated with negative regulation of cell growth and hence may be termed "regulatory translocation".

An important finding obtained from Western blot analysis was the appearance of major 60 KD band in nuclei and nucleoli of TPA treated cells. It can be supposed that 60 KD pRB specific band is a result of specific pRB proteolysis.

The results suggest that a special 60 kD form of pRB (RB (p60))representing its C-terminal fragment may occur in association with tumor cell growth inhibition. The truncated form of pRB form has been found in human fibroblasts [8], the cells with normal growth and unimpared RB function. It has also been elucidated that the 60 kD C-terminal fragment is a minimal portion of pRB needed for overt growth suppression [9,109,10]. These data indicate that appearance of the truncated form of pRB is likely to be a part of a physiologic mechanism of pRB action.

It is likely that during purification of nucleoli not only "free" pRB fragment, but also some of the nucleoli-bound one, is removed. This may account for the lesser intensity of 60 kD band for the purified nucleoli of TPA treated cells (compared to that for total nuclei). We believe that the truncated pRB may have higher capacity to bind DNA, RNA or certain proteins, than intact one. If this is the case, the hypothetical pRB cleaving protease may be of regulatory significance.

We did not observe a direct link between the phosphorylation of pRB and its translocation. In both control and TPA treated U937 cells pRB was largely hypophosphorylated at the end of the experiment. In control cells it may be due to a density-dependent decline in growth rate. However, since growth inhibition is believed to be associated with hypophosphorylated form of pRB [11,12], it is possible that only the latter can undergo proteolytic cleavage and translocation.

It seems quite possible that pRB translocation to the nucleolus is involved in the regulation of ribosome synthesis, which is a major nucleolar function. It has been reported that active pRB arrests cell growth at or prior to the restriction point of cell cycle, not allowing it to proceed past G1 phase [12]. This may be a result of pRB induced suppression of ribosome synthesis, which leads to a decline in overall protein synthesis. As a result the cell never reaches the protein/DNA ratio required to proceed past the restriction point of cell cycle and enter S phase.

It is well known that differentiation, quiescence or senescence is associated with suppression of ribosome synthesis[15,16]. It is, therefore, possible that antimitotic action of pRB is accomplished by ribosome synthesis downregulation.

Several conceivable mechanisms of pRB induced downregulation of ribosome synthesis can be proposed. rRNA production appears to be a limiting step in the regulation of ribosome synthesis. Several stages of rRNA transcription and maturation may be affected by pRB. RB protein may interfere with efficient initiation of rRNA transcription. Two promoter regions are known to be important for this process: proximal (or core) promoter (-40 to +15) and distal upstream promoter (-150 -110) (as reviewed in [15]). Core promoter is believed to be required for minimal and efficient initiation of transcription, whereas distal upstream promoter modulates the level of synthesis. It is attractive to suppose that pRB is involved in blocking the distal upstream promoter, so that ribosome production remains sufficient for cell maintenance, but not for mitosis. If pRB indeed prevents the initiation of rRNA transcription or reduces the initiation rate, one would suggest that pRB may interfere with the binding or propulsion of RNA-polymerase I itself, or block the binding of additional proteins required for the transcription initiation. Two DNA-binding proteins, SL1 [16] and hUBF [17], which cooperate with RNA-polymerase I to form initiation complex, have been described. hUBF binds to the downstream sequences of distal promoter (-75 -115) and directs the binding of SL1 to both distal and core promoters (presumably via protein-protein interactions)[15,17]. Either of this proteins may be a target for pRB action.

Another possible target is a protein named factor C, which does not bind directly to DNA, but is believed to modulate transcriptional activity of RNA-polymerase I [15,18,19].

Processing of 45S rRNA transcript into 18S, 28S and 5.8S rRNAs is also a potential regulatory step and may be affected by pRB. snRNAs, U3, U8 and U13 have been found associated with preribosomal RNP complexes and are believed to be essential for 45S rRNA processing (reviewed in [15,20,21]). The hypotheses presented above are testable . The investigation of pRB translocation mechanism and pRB role in the suppression of ribosome syntesis is in a progress.

Acknowledgement: This work was supported by the T. Martell Memorial Foundation for Cancer and Leukemia Reserch and by the grant #T35DK07420-11 from the National Institutes of Health.

REFERENCES
1. Rogalsky, V. (1974) The Lancet, 2, 7882, 729.
2. Rogalsky, V. (1975) J. Natl. Cancer Inst., 54, 5, 1061-1071.
3. Rogalsky, V. (1976) in: Onco-developmental Gene Expression, edit. Fishman, W. H. Academic Press, 593-598.
5. Rogalsky, V., Zinzar, S., Golub, E., Den, T. (1991) Ann. NY. Acad. Sci. 628, 323-325.
6. Rogalsky, V., Todorov, G., Den, T., Ohnuma, T. (1992) FEBS Letters, 304, 153-156. 7. 7. Whyte, P., Buckovich, K. J., Horowitz, J. M., Friend, S. H., Raybuck, M., Weinberg, R. A., Harlow, E. (1988) Nature 334, 124-129.
8. Xu H.-J., Hu S.-X, Hashimoto, T., Takahashi, R., Benedict, W. F. (1989) Oncogene, 22, 58(6), 1193-1198.
9. Edwards, G. M., Huber, H.E., DeFeo-Jones, D., Vuocolo, G., Goodhart, P. J., Maigetter, R. Z., Sanyal, G., Oliff, A., Heimbrook, D. C. (1992) J. Biol. Chem. 267(12) 7971-7974
10. Qin, X. Q., Chittenden. T., Livingston, D. M., Kselin, W. G. J. Gen.Dev. (1992) 6,953-964
11. Chen, P., Scully, P., Shew, J., Wang J. Y. J., Lee, W. H. (1989) Cell, 58, 1193-1198.
12. Goodrich, D. W., Wang, N. P., Qian, Y.W., Lee, E. Y., Lee, W. H. (1991) Cell , 67, 293-302.
13. Tushinsky, R. J., and Warner, J. R. (1982) J. Cell Physiol. 112, 128-135.
14. Bowman, L. H. (1987) Dev. Biol. 119, 152-163.
15. Larson, D. E., Zahradka, P., Sells, B. H. (1991) Biochem. Cell Biol. 69, 5-22.
16. Bell, J., Nielson, L., Pellegrini, M. (1988) Mol. Cell. Biol. 8, 91-95.
17. Jantzen, H.-M., Admon, A., Bell., S. P., Tjian, R. (1990) Nature 344, 860-836.
18. Mahajan, P. B., and Thompson, E. A., Jr. (1987) J. Biol. Chem. 262, 16150-16156.
19 Rubinstein, S.J., and Dasgupta, A. (1989) J. Virol. 63, 4689-4696.
20. Maser, R. L., and Calvet, J. P. (1989) Proc. Natl. Acad. Sci USA, 86, 6523-6527.
21. Tyc, K., and Steitz, J.A. (1989) EMBO J. 8, 3113-3119.

Transforming growth factor-β_1 mediates its inhibition of interleukin-4 induced proliferation by interfering with interleukin-4 signal transduction

A.R. Mire-Sluis, L. Randall, M. Wadhwa and R. Thorpe

Division of Immunobiology, NIBSC, Blanche Lane, South Mimms, Herts, EN6 3QG, UK

Interleukin-4 (IL-4) is a T-lymphocyte derived 20kDa glycoprotein exhibiting a broad spectrum of biological activity towards many of the cellular components of the haemapoietic system. IL-4 can induce the proliferation of several haemapoietic cell types including the leukaemic cell lines TF-1 and MO-7. We have used whole cell labelling procedures and a permeabilised system designed to detect rapid changes in dynamic phosphorylation events and show that IL-4 induces the activation of a phosphotyrosine specific phosphatase which leads to the rapid dephosphorylation of an 80kDa protein (p80) from tyrosine residues. TGF-β_1 is a complex molecule possessing potent anti-proliferative effects on many cell types and addition of TGF-β to both TF-1 and MO-7 cells simulaneously with IL-4 strongly inhibits IL-4 induced proliferation. The inhibitory effects of TGF-β_1 on IL-4 induced proliferation are time dependent, as preincubation with TGF-β_1 markedly increases the inhibitory capacity of the cytokine towards either TF-1 or MO-7 cells.

Data shows that cells preincubated with TGF-β_1 in this way no longer incorporate radiolabelled phosphate onto p80 suggesting that TGF-β_1 has downregulated either production of the 80kDa protein or the phosphatase activity required for IL-4 induced proliferation. If TGF-β_1 is added directly to cells it also inhibits the incorporation of radiolabel onto p80. Kinetic studies show that p80 undergoes continual and rapid exchange of phosphate on tyrosine residues. The use of Genistein (a TPK inhibitor) and sodium orthovanadate (a PTPase inhibitor) reveals that at any one time point, p80 tyrosine residues are fully phosphorylated. Therefore, phosphatase activity is required before new phosphate can be incorporated.

Time course and pulse-chase experiments show that TGF-β_1 inhibits the endogenous phosphatase activity, thus preventing the addition of any new phosphate onto p80. If cells are prelabelled with radioactive phosphate (thus labelling p80), it can be shown that TGF-β_1 can prevent IL-4 induced activation of phosphatase activity that normally leads to the removal of phosphate from p80.

This data suggests that TGF-β_1 inhibits the PTPase activated during IL-4 mediated signal transduction and that this may be the biochemical basis of the TGF-β_1 mediated inhibition of IL-4 induced proliferation.

III. *In vitro* and *in vivo* effects of negative hematopoietic regulators

III. *Effets* in vitro *et* in vivo *des inhibiteurs*

Control mechanisms for primitive hemopoietic cells

C.J. Eaves, J.D. Cashman and A.C. Eaves

Terry Fox Laboratory, British Columbia Cancer Agency and University of British Columbia, 601 West 10th Avenue, Vancouver, BC, Canada, V5Z 1L3

ABSTRACT

It has been appreciated for many years that blood cell production is regulated in part by changes in the cell cycle status of early progenitors. In adult life, many of these cells normally exist in a growth-arrested, albeit viable and reversible, G_0 state; however, a variety of treatments can transiently increase their rate of recruitment into S-phase and, in at least some instances, this is attributed to mechanisms local to the microenvironment of the bone marrow. The long-term culture system offers a unique experimental model to analyze the molecular mechanisms that may be involved in these intercellular regulatory processes. Continued production of the relevant primitive hemopoietic target cells occurs in this system for extensive periods and during this time the cycling status of these cells can be selectively and reversibly manipulated. In previous studies, we have identified a number of cytokines that are produced by adherent cells present in these cultures. Some of these factors, when added exogenously to long-term cultures can mimic the stimulatory activity of the stromal cells in the adherent layer. Similarly, other factors can mimic their inhibitory function. Primitive neoplastic progenitors from patients with chronic myeloid leukemia (CML) show deregulation of their cycling control both in vivo and in the long-term culture system. This deregulation can be explained by an absent or reduced responsiveness to some, but not all, inhibitors of primitive normal hemopoietic cell cycling. These findings provide support for a model of stromal cell regulation of primitive hemopoietic cell cycling both in the long-term culture system and in vivo that involves basal levels of production by stromal cells of a multiplicity of stimulators and inhibitors such that these factors are effective only as co-operating agonists or antagonists of cell cycle progression. This model explains how most normal stem cells would be maintained in a quiescent state under conditions where a mutated stem cell that acquired unresponsiveness to only a part of the control mechanism would proliferate continuously. The expected result would be an inexorable clonal expansion as is observed in CML. The strong parallels between the different cycling behaviour exhibited by normal and neoplastic hemopoietic cells in the long-term marrow culture system and in vivo underscores the importance of this experimental model for analysis of disease mechanisms in CML and for the future testing of new strategies for therapeutic intervention.

INTRODUCTION

Hemopoiesis is the process whereby multiple types of mature blood cells are generated from more primitive precursors. This involves a series of gene programming events which are, at least ultimately, irreversible. During hemopoietic cell differentiation, this process is particularly intriguing because of its complexity. For example, the number of distinct specialized blood cell types that appear to be produced from a single common precursor pool throughout adult life is amazingly large (>10), as is the number of cell divisions over which the various gene programming events occur in an apparently ordered sequence. In addition, it is now evident that there are multiple overlapping as well as non-overlapping mechanisms controlling the number of cells present at every level of the hemopoietic developmental hierarchy, at the same time providing remarkable scope for altering mature blood cell output as required.

The first two of these features of hemopoiesis predict that the most primitive cell types will be rare relative to the numbers of their differentiating progeny, even under conditions where some progenitors are lost from the system at intermediate stages of differentiation. Many findings support this prediction, which also explains in part why it has been difficult to develop simple direct methods for quantifying and characterizing primitive hemopoietic cells, forcing reliance on more cumbersome, retrospective functional assays. It is therefore not surprising that progress in understanding events that alter the behaviour of very early hemopoietic cells has been similarly slow since this requires not only the availability of quantitative endpoints that discriminate between sequential stages of early hemopoietic cell development, but also protocols that allow specific responses of defined cell types to be detected and quantified.

In this paper, we have restricted our attention to a discussion of extrinsic (extracellular) mechanisms that can participate in controlling the size of early hemopoietic progenitor populations by regulating the proportion of their members found in S-phase. The information that we have generated on this subject has been derived from studies of specific subpopulations of early (high proliferative potential) clonogenic erythroid and granulopoietic progenitors. In normal adult human marrow, these cell types are largely quiescent, as determined by the combined use of conventional thymidine suicide assays and reproducible and selective colony scoring and sizing criteria (Table I). Production of primitive erythroid and granulopoietic progenitors can be maintained for several weeks in long-term marrow cultures (LTC), i.e., cultures initiated with cell suspensions containing the most primitive hemopoietic cell populations known as well as certain types of fibroblasts, to provide a semi-confluent supportive adherent layer (Eaves et al, 1991a). The LTC system was thus an obvious experimental model for investigating exogenous regulatory mechanisms that may affect primitive progenitor cell cycling in vivo.

THE LTC SYSTEM AS A MODEL FOR ANALYZING MECHANISMS OF STROMAL CELL CONTROL OF PRIMITIVE HEMOPOIETIC CELLS

The first and key observation emanating from an analysis of primitive progenitor cycling behaviour in LTC of normal cells was a repeated oscillation in the proportion of these cells in the adherent layer that were in S-phase that accompanied each weekly change of the medium (Cashman et al, 1985; Eaves et al, 1986). Moreover, this oscillatory cycling

Table I. Differences in the cycling activity of various myeloid clonogenic progenitor populations in normal adults and in patients with CML.

Tissue Origin	Progenitor Type	% of Progenitors in S-phase Normal	CML
Marrow	CFU-E	43 ± 2	33 ± 2
	BFU-E (total)	27 ± 2	33 ± 4
	CFU-GM (total)	37 ± 3	31 ± 4
	Primitive BFU-E	4 ± 2	40 ± 4
	Primitive CFU-GM	4 ± 2	44 ± 7
Blood	CFU-E	1 ± 3	33 ± 3
	BFU-E (total)	0 ± 1	34 ± 3
	CFU-GM (total)	0 ± 2	30 ± 4
	CFU-GEMM	0 ± 20	48 ± 4

Data from Reference (Eaves & Eaves, 1987). % of progenitors in S-phase represents the proportion of progenitors that were unable to generate colonies after a 20 minute exposure to high specific activity ^{3}H-thymidine. Primitive BFU-E and CFU-GM represent minor subpopulations within the total clonogenic erythropoietic and granulopoietic progenitor populations, respectively, which are characterized by the greatest proliferative potential. Assay and colony scoring criteria have been described in detail previously (Gregory & Eaves, 1977; Cashman et al, 1985).

behaviour in the LTC system was exhibited only by the types of primitive progenitors that were normally quiescent in the marrow in vivo, regardless of their ultimate origin from marrow or blood cell precursors (Figure 1). This oscillatory behaviour was also restricted to primitive progenitors located within the adherent layer of the LTC system. Such findings provided initial confidence in this system for investigating mechanisms operative in the marrow in vivo, a view that has been further strengthened by the finding that early neoplastic progenitor populations from patients with CML, whose cycling control is deregulated in vivo (Table I), also proliferate continuously in the LTC system (Figure 1).

Our initial studies of primitive progenitor control mechanisms in LTC focussed on defining the possible role(s) of the stromal cells (non-hemopoietic cells of the fibroblast-adipocyte-endothelial lineage) present in the adherent layer of LTC established with normal hemopoietic cells. Experiments in which cultures were initiated with cell suspensions that did not contain any stromal cells or their precursors, even though they did contain large numbers of primitive hemopoietic cells, revealed two functions of the stromal cells: an essential supportive function (for both clonogenic cell production from LTC-IC as well as for LTC-IC maintenance (Eaves et al, 1986; Sutherland et al, 1991; Sutherland et al, 1993), and a reversible, growth inhibitory function (Cashman et al, 1985; Eaves et al, 1986). Thus, in the absence of a competent fibroblast feeder layer, all classes of clonogenic cells continue to be produced and in fact show a heightened turnover (no quiescent cells detected at any time) although, overall, the number of clonogenic cells present also declines rapidly over time. From these findings it was inferred that the stromal fibroblasts of the adherent layer would be found to produce both positive and negative factors with selective actions on very primitive hemopoietic cells, and furthermore, that the production of these factors might, itself, be regulated by exposure to certain cytokines. Subsequent studies have provided ample support for this model and have led to the identification of various molecules that may participate in the regulatory network it implies (Figure 2). One interesting outcome of these latter studies was the finding that, in this system, two factors combined could synergize to stimulate primitive progenitor cycling at concentrations and with a delivery schedule that was ineffective for either alone (Hogge et al, 1991; Otsuka et al, 1991b & c). In addition, analysis of the ability of a panel of murine fibroblast cell lines to stimulate clonogenic cell output from LTC-IC (which requires their proliferation) showed that this process can be supported by several different factors and factor combinations, including at least one that has not yet been identified (Sutherland et al, 1991; Sutherland et al, 1993).

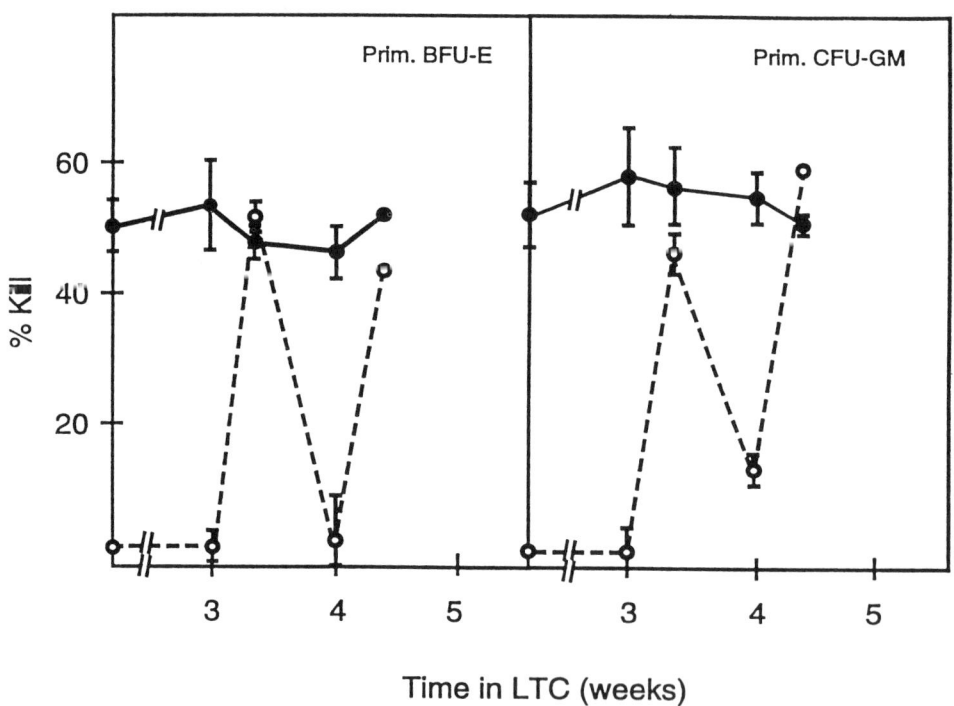

Figure 1. Cycling behaviour of primitive normal (open circles, dashed line) and CML (closed circles, solid line) progenitors in the adherent layer of LTC initiated by seeding light density normal or CML peripheral blood cells onto irradiated feeder layers subcultured from the adherent layer of LTC previously established from normal marrow. Values shown are from ^{3}H-thymidine suicide assays using a 20 minute exposure as in Table I. Data have been redrawn from that presented in Eaves et al (1986) to which the reader is referred for more details.

These findings provided a framework for the subsequent design of a similar series of experiments to investigate the basis of the uncontrolled cycling activity exhibited by primitive CML progenitors. These revealed no evidence of any change in the dependence of CML progenitors in LTC on the supportive function of marrow fibroblasts (Eaves et al, 1986), consistent with previous studies of the factor dependence of CML progenitors in semi-solid culture assays (Metcalf, 1977). Moreover, analysis of factor expression by normal stromal feeder layers that were co-cultured with exclusively leukemic primitive hemopoietic cells from CML patients failed to reveal any evidence of an activated paracrine stimulatory loop (Otsuka et al, 1991a). It thus seemed likely that the defect in cycling control exhibited by primitive CML cells would involve a default in the inhibitory arm of the mechanism that regulates primitive normal cells. Nevertheless, in spite of the fact that the low levels of TGF-β produced by the cells of the adherent layer had been shown to be responsible for maintaining primitive normal progenitors in this layer in a quiescent state in unperturbed cultures (Eaves et al, 1991b), CML cells were not found to show any alteration in responsiveness to TGF-β (Sing et al, 1988; Cashman et al, 1992a). Moreover, they also did not detectably alter the level of TGF-β produced in LTC containing stromal cells derived from either normal or CML marrow (Otsuka et al, 1991a).

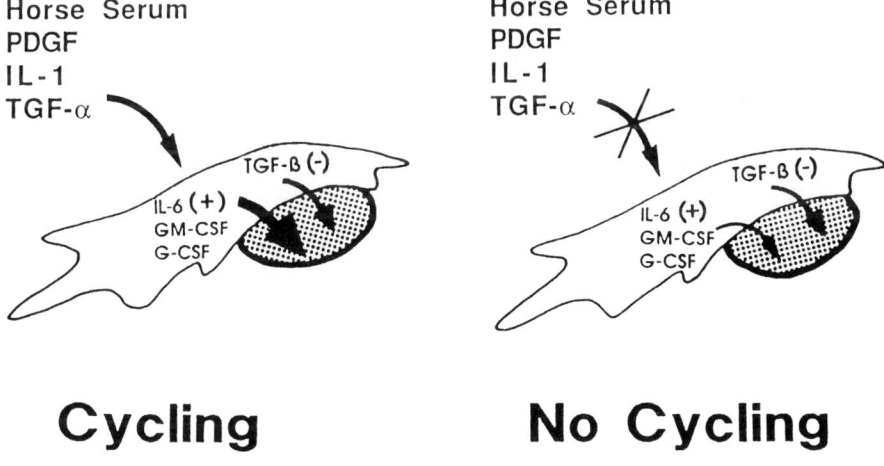

Figure 2. Diagrammatic representation of cytokines able to act on and produced by the cells of the adherent layer of LTC which may regulate the cycling activity of adjacent primitive hemopoietic progenitors. Factors presumed to act indirectly can stimulate the cycling of primitive hemopoietic progenitors in the adherent layer of LTC but not in methylcellulose assays (Cashman et al, 1990). They also lead to increased production of IL-6, GM-CSF and G-CSF by LTC adherent layer cells (Eaves et al, 1991b) which alone and/or in combination can stimulate increased progenitor turnover (Hogge et al, 1991; Otsuka et al, 1991c).

THE CONCEPT OF CO-OPERATIVE INHIBITION AND ITS RELEVANCE TO CML

At first, these results appeared paradoxical. However, given the precedent for cooperating stimulators of progenitor cycling, which at low levels might even appear as synergistic, we were led to consider the possibility of a similar mechanism involving two (or more) co-operating inhibitors, the blockade of either being sufficient to prevent the effective inhibitory action of the combination. Very recent experiments have identified MIP-1α to be a second, endogenously produced direct-acting inhibitor with selective and reversible ability to arrest the cycling of primitive progenitors in the adherent layer of normal LTC (Eaves et al, submitted). Moreover, neutralization of endogenous MIP-1α (by addition of excess MIP-1β), like the neutralization of endogenous TGF-β (by addition of specific antibody), prevents the return of normal primitive progenitors in the adherent layer to a noncycling state (Eaves et al, submitted). Thus it appears that these two inhibitors do, in fact, act in a co-operative faction in the LTC system. In addition, we have found that primitive CML cells in LTC exposed to exogenously added MIP-1α under conditions where this blocks the activation of primitive normal cells, continue to proliferate (Eaves et al, submitted), suggesting that primitive CML cells are unresponsive to, or are able to bypass, the inhibitory effect that MIP-1α can have on primitive normal hemopoietic cells. Since responsiveness to MIP-1α appears essential (albeit not sufficient) for controlling primitive normal progenitors in the adherent layer of the LTC system, the inability of CML cells to respond to this part of the inhibitory mechanism would be sufficient to explain their deregulated behaviour.

We have also now found that CML cells are unresponsive to the tetrapeptide AcSDKP, another molecule with a similarly selective and reversible inhibitory action on primitive normal progenitors (Cashman et al, 1992b). However, in this case, evidence for an indirect mode of action has been obtained (Cashman et al, 1992c). Thus, the regulation of primitive hemopoietic cells by stromal cells appears to be even more complex that previously anticipated. The fact that this complexity appears to be reproduced in the LTC system both in terms of normal cell regulation and a breakdown of this process in the regulation of primitive CML cells underscores its continuing utility as a model of the marrow microenvironment.

ACKNOWLEDGEMENTS

The studies reviewed here were supported in part by the National Cancer Institute of Canada and C.J. Eaves is a Terry Fox Cancer Research Scientist of the National Cancer Institute of Canada. The expert secretarial assistance of H. Calladine is also gratefully acknowledged.

REFERENCES

Cashman, J., Eaves, A.C. & Eaves, C.J. (1985): Regulated proliferation of primitive hematopoietic progenitor cells in long-term human marrow cultures. *Blood* 66, 1002-1005.

Cashman, J.D., Eaves, A.C. & Eaves, C.J. (1992a): Granulocyte-macrophage colony-stimulating factor modulation of the inhibitory effect of transforming growth factor-β on normal and leukemic human hematopoietic progenitor cells. *Leukemia* 6, 886-892.

Cashman, J.D., Eaves, A.C. & Eaves, C.J. (1992b): Primitive neoplastic hematopoietic cells from patients with chronic myeloid leukemia can ignore the reversible indirect inhibitory action of the tetrapeptide AcSDKP (Abstract). *Exp. Hematol.* 20, 782.

Cashman, J.D., Eaves, A.C. & Eaves, C.J. (1992c): Evidence for an indirect mechanism mediating the inhibitory effect of the tetrapeptide AcSDKP on primitive human hematopoietic cell proliferation (Abstract). *J. Cell. Biochem.* (Suppl. 16c), 95.

Cashman, J.D., Eaves, A.C., Raines, E.W., Ross, R. & Eaves, C.J. (1990): Mechanisms that regulate the cell cycle status of very primitive hematopoietic cells in long-term human marrow cultures. I. Stimulatory role of a variety of mesenchymal cell activators and inhibitory role of TGF-β. *Blood* 75, 96-101.

Eaves, A.C., Cashman, J.D., Gaboury, L.A., Kalousek, D.K. & Eaves, C.J. (1986): Unregulated proliferation of primitive chronic myeloid leukemia progenitors in the presence of normal marrow adherent cells. *Proc. Natl. Acad. Sci. USA* 83, 5306-5310.

Eaves, C.J. & Eaves, A.C. (1987): Cell culture studies in CML In Bailliere's Clinical Haematology. Vol. 1, #4. Chronic Myeloid Leukaemia. ed J.M. Goldman, pp. 931-961. London: Bailliere Tindall.

Eaves, C.J., Cashman, J.D. & Eaves, A.C. (1991a): Methodology of long-term culture of human hemopoietic cells. *J. Tissue Culture Methods* 13, 55-62.

Eaves, C.J., Cashman, J.D., Kay, R.J., Dougherty, G.J., Otsuka, T., Gaboury, L.A., Hogge, D.E., Lansdorp, P.M., Eaves, A.C. & Humphries, R.K. (1991b): Mechanisms that regulate the cell cycle status of very primitive hematopoietic cells in long-term human marrow cultures. II. Analysis of positive and negative regulators produced by stromal cells within the adherent layer. *Blood* 78, 110-117.

Eaves, C.J., Cashman, J.D., Wolpe, S.D. & Eaves, A.C. Unresponsiveness of primitive chronic myeloid leukemia cells to macrophage inflammatory protein-1α (MIP-1α) an inhibitor of primtive normal hematopoietic cells. *Manuscript submitted.*

Gregory, C.J. & Eaves, A.C. (1977): Human marrow cells capable of erythropoietic differentiation in vitro: Definition of three erythroid colony responses. *Blood* 49, 855-864.

Hogge, D.E., Cashman, J.D., Humphries, R.K. & Eaves, C.J. (1991): Differential and synergistic effects of human granulocyte-macrophage colony-stimulating factor and human granulocyte colony-stimulating factor on hematopoiesis in human long-term marrow cultures. *Blood* 77, 493-499.

Metcalf, D. (1977): Hemopoietic Colonies. In vitro cloning of normal and leukemic cells. Berlin Heidelberg: Springer-Verlag.

Otsuka, T., Eaves, C.J., Humphries, R.K., Hogge, D.E. & Eaves, A.C. (1991a): Lack of evidence for abnormal autocrine or paracrine mechanisms underlying the uncontrolled proliferation of primitive chronic myeloid leukemia progenitor cells. *Leukemia* 5, 861-868.

Otsuka, T., Thacker, J.D., Eaves, C.J. & Hogge, D.E. (1991b): Differential effects of microenvironmentally presented interleukin 3 versus soluble growth factor on primitive human hematopoietic cells. *J. Clin. Invest.* 88, 417-422.

Otsuka, T., Thacker, J.D. & Hogge, D.E. (1991c): The effects of interleukin 6 and interleukin 3 on early hematopoietic events in long-term cultures of human marrow. *Exp. Hematol.* 19, 1042-1048.

Sing, G.K., Keller, J.R., Ellingsworth, L.R. & Ruscetti, F.W. (1988): Transforming growth factor β selectively inhibits normal and leukemic human bone marrow cell growth in vitro. *Blood* 72, 1504-1511.

Sutherland, H.J., Eaves, C.J., Lansdorp, P.M., Thacker, J.D. & Hogge, D.E. (1991): Differential regulation of primitive human hematopoietic cells in long-term cultures maintained on genetically engineered murine stromal cells. *Blood* 78, 666-672.

Sutherland, H.J., Hogge, D.E., Cook, D. & Eaves, C.J. (1993): Alternative mechanisms with and without Steel factor support primitive human hematopoiesis. *Blood* 81, 1465-1470.

Résumé

Depuis plusieurs années, il a été montré que la production des cellules sanguines étaient en partie régulée par des changements de l'état cyclique des progéniteurs primitifs. Au cours de la vie adulte, beaucoup d'entre eux, bien que parfaitement viables, sont dans un état quiescent et réversible G_0; cependant différents types de traitement peuvent augmenter transitoirement leur niveau de recrutement en phase S, ce phénomène pouvant en partie être du à des mécanismes locaux provenant du microenvironnement médullaire. Le système de culture à long terme offre un modèle expérimental unique pour analyser les mécanismes moléculaires impliqués dans ces processus régulateurs intercellulaires. Dans ce système, il existe une production continue de progéniteurs primitifs pendant une période au cours de laquelle il est possible de manipuler de manière sélective et réversible l'état cyclique des cellules. Dans des études précédentes, nous avons identifiés plusieurs cytokines qui sont produites par les cellules adhérentes présentes dans les cultures à long-terme. L'ajout exogène aux cultures à long-terme de certains de ces facteurs peut mimer les effets stimulants des cellules stromales de la couche adhérente tandis que d'autres facteurs au contraire miment leurs effets inhibiteurs. Au cours des leucémies myéloïdes chroniques, il a été montré, aussi bien in vivo qu'en culture à long-terme, que l'état cyclique des progéniteurs néoplasiques primitifs était dérégulé. Cette dérégulation peut être expliquée par une réponse diminuée voire absente à certains inhibiteurs, (mais pas tous), connus pour agir normalement sur la mise en cycle des progéniteurs primitifs. A partir de ces données, on peut envisager pour les progéniteurs primitifs un modèle de régulation du cycle cellulaire où les cellules stromales produisent à un taux basal aussi bien in vivo qu'en culture à long-terme une multitude de stimulateurs et d'inhibiteurs qui agissent comme agonistes ou antagonistes de la progression du cycle cellulaire. Ce modèle permet d'expliquer comment dans ces conditions la plupart des cellules souches normales peuvent être maintenues à l'état quiescent alors qu'une cellule souche mutante et ne répondant partiellement plus aux mécanismes de contrôle pourrait continuellement proliférer. Le résultat attendu en serait inexorablement une expansion clonale comme cela est observé au cours des LMC. Les différences de comportement du cycle cellulaire des cellules hématopoïètiques normales et néoplasiques observées de façon parallèle aussi bien dans le système de culture à long-terme qu'in vivo montrent l'importance de ce modèle expérimental pour l'analyse des mécanismes pathologiques dans les LMC et pour tester dans le futur de nouvelles stratégies thérapeutiques.

Cytokine responsiveness of murine high-proliferative potential colony-forming cells

Peter Quesenberry, Donna Deacon and Phil Lowry

Departments of Internal Medicine, University of Massachusetts Medical Center, Worcester, Massachusetts, USA

ABSTRACT

The high proliferative potential colony-forming cell (HPP-CFC) is a primitive progenitor stem cell which may be an in-vitro model of the marrow repopulating cell. We have described a number of 2 factor cytokine combinations which stimulate HPP-CFC from unseparated marrow in a serum supplemented culture system, these include CSF-1 + GM-CSF, G-CSF + GM-CSF, GM-CSF + IL-3, G-CSF + CSF-1, and IL-3 + G-CSF. Increasing the number of cytokines in a combination and using plateau doses of IL-3, IL-1 alpha, GM-CSF, G-CSF, CSF-1 and Steel factor with unseparated marrow in serum replete culture leads to little additional growth of HPP-CFC, but progressive increases in growth were seen with increases in the number of cytokines in a combination, if marrow was purified (Lin- Sca+) or grown under serum-free conditions. If purified Lin- Sca+ marrow was grown in serum free conditions the inclusion of Steel factor was necessary for virtually any HPP-CFC formation; without Steel factor even 5 factor combinations stimulated little growth.

PMA was found to augment growth of HPP-CFC with most single or multi-factor combinations but with CSF-1 or GM-CSF included in a combination, PMA frequently inhibited HPP-CFC formation.

Finally, TGF-B, when added at the initiation of culture, inhibited 5-6 factor stimulated HPP-CFC but stimulated routine colony formation. There data indicate that HPP-CFC are regulated by multiple discreet stimulatory and inhibitory pathways.

Early murine hematopoietic stem cells respond to multiple growth factors. The high proliferative potential colony-forming cell (HPP-CFC) is a relatively early or primitive progenitor which appears to be related to the marrow repopulating cell (1-5). This cell was first described as a two factor responsive cell (1,3) the factor combinations including colony-stimulating factor-1 (CSF-1) and either interleukin-1 (IL-1) or interleukin-3 (IL-3). In further investigations, we found that a number of two factor combinations were capable of stimulating murine or human HPP-CFC (Table I) (6-8). The murine HPP-CFC contained only macrophages, while the post-5-FU human HPP-CFC consisted of mixed colonies with macrophages, granulocytes and probably megakaryocytes. We evaluated whether increasing the number of growth factors from 2 to 6 would further enhance HPP-CFC formation using unseparated murine (C57 BlK/KA-Thy1-1) marrow in serum supplemented cultures. We studied interleukin-1 alpha, interleukin-3, interleukin-6 (IL-6), CSF-1, granulocyte-macrophage colony-stimulating factor (GM-CSF) and granulocyte colony-stimulating factor (G-CSF). We tested all possible combinations of 3,4,5 and 6 factors. There was some slight augmentation of HPP-CFC number seen with some 3 factor combinations, but increasing the number of cytokines further did not augment HPP-CFC or total colony formation.

However, a different picture emerged when murine marrow was depleted of lineage positive cells (Lin-) and positively selected for the Sca antigen (Sca+). These Lin- Sca+ positive cells, separated by the techniques of Spangrude, Heimfeld and Weissman (9) as modified by Spangrude and Sollary (10), were cultured under serum replete conditions. Testing there purified Lin- Sca+ cells with increasing number of cytokines in a combination gave a progressive increase in the number of HPP-CFC observed. In this setting IL-6 gave varied results, but a combination of CSF-1, G-CSF, GM, CSF, IL-1 alpha and IL-3 was effective in stimulating HPP-CFC and total colonies with virtually all combinations (11). The combinations of CSF-1, GM-CSF and IL-3 and CSF-1, GM-CSF and IL-1 alpha were comparable with more growth seen with GM-CSF, IL-1 alpha and IL-3. The 4 factor combination of CSF-1, GM-CSF, IL-1 alpha and IL-3 induced more total colonies and HPP-CFC and the 5 factor combination of CSF-1, G-CSF, GM-CSF, IL-1 alpha and IL-3 stimulated significantly more total colonies and HPP-CFC as compared to the 4 factor combination. The addition of Steel factor further augmented HPP-CFC size (but not number) in the 5 factor combination.

In a similar manner if unseparated marrow is cultured under serum free conditions the growth of total colonies and HPP-CFC was similar to that seen with the purified lin- Sca+ cells under serum free conditions. Increasing the number of factors progressively from 2 to 5 lead to a progressive increase in total colonies and HPP-CFC number. The inclusion of Steel factor significantly augmented HPP-CFC and total colony formation with the different combinations (Table II).

Perhaps most striking were the results obtained when Lin- Sca+ cells were cultured in a serum-free culture system. Here combinations of up to 4-5 growth factors gave virtually no colony formation but the addition of Steel factor to a 3-5 factor base combination gave HPP-CFC and total colony formation equivalent to that seen with serum supplemented whole marrow or purified marrow cells. At 5-6 factors, with inclusion of Steel, the cell in the colonies were no longer only macrophage in phenotype; these colonies consisted of a mixture of granulocytes, macrophages, megakaryocytes and other less well defined cells.

In other studies we have determined that marrow stromal cells produce very low levels of IL-3 which are biologically active (12). These observations suggested that marrow cells might respond to very low levels of some cytokines when they are present in a synergistic combination. Initially we attempted to lower the concentration of all factors in a 5-6 factor combination, but HPP-CFC and total colony formation were rapidly lost. However, if Steel factor was included and kept at optimal levels, the concentrations of the other factors could be reduced 1-2 logs and HPP-CFC and total colony formation from either Lin- Sca+ cells or murine marrow harvested 2 days after 150mg/Kg 5-FU was maintained, and approached that seen with cytokine all at their "optimum" concentration (13). Thus Steel factor acted to anchor a synergistic combination. All of the individual factors were analyzed in a similar fashion (Table III). IL-1 alpha and to a lesser extent GM-CSF and IL-3 had an intermediate capacity to restore HPP-CFC and colony formation in the presence of reduced concentrations of the other factors.

Studies with serum-free and serum supplomented cultures and increasing numbers of cytokines (2-6) gave a progressive increase of growth with the serum-free system, but no or little increase with the serum replete conditions and with some combinations apparent inhibition of both total colony and HPP-CFC formation. These data indicate that serum provides both multiple stimulatory and inhibitory factors of HPP-CFC formation, and that HPP-CFC is a multi-factor responsive stem cell. Delayed addition experiments indicate that Steel factor can support the survival of HPP-CFC for up to 4 days in culture indicating that it may act as a survival factor.

Further studies have indicated that phorbol myristate acetate, a protein kinase-c stimulator, (PMA) augments single cytokine (G-CSF, GM-CSF, IL-1, IL-3 and c-Kit ligand) stimulation of CFU-C and HPP-CFC from normal bone marrow while effects with CSF-1 varied dependent upon the CSF-1 dose. PMA variably augments normal bone marrow CFU-C and HPP-CFC derived from 2-5 factor stimulation, dependent on the particular factor employed. Synergy was notably lacking in CSF-1 based combinations. In fact, marked to mild inhibition of HPP-CFC formation was seen in some cases. When purified SCA+ LIN-marrow cells were tested, PMA augmented CFU-C and HPP-CFC growth with many combinations. In general, inhibition of growth was seen when PMA was added to cytokine combinations including CSF-1 or GM-CSF.

TGF-beta was also assessed for its capacity to inhibit multiple-cytokine stimulated HPP-CFC formation. When added at the initiation of serum-free cultures to 5 or 6 factors and tested against murine (BALB/C) marrow harvested 2 days after 150 mg/kg 5-FU, TGF-beta inhibited HPP-CFC formation, but "increased" regular colony formation. When added at day 8 little effect was noted. similarly, little effect was noted when TGF-beta was added to multi-cytokine stimulated normal or day 2 post-5-FU marrow in a serum replete system. When TGF-beta was added to normal marrow in a serum-free system, total, but not HPP colony formation was enhanced.

These data indicate that multi-factor responsive HPP-CFC are regulated by discreet separate stimulatory and inhibitory pathways involving at a minimum the hemopoietic and tyrosine-kinase growth factor receptor families and protein-kinase C. Effects throughout the hematopoietic differentiaion pathway may vary dependent upon the stage or level of differentiation.

Table I

Two factor stimulation of HPP-CFC

Species	Factor I	Factor II	
Mouse	GM-CSF	CSF-1	
	G-CSF	CSF-1	
	G-CSF	IL-3	
	GM-CSF	IL-3	
	G-CSF	GM-CSF	
	IL-3	CSF-1	(ref.1)
	IL-1	CSF-1	(ref.1)
Human	GM-CSF	IL-3	
	GM-CSF	G-CSF	

Table II

Number of cytokines* giving maxiumu number of murine HPP-CFC and total colonies

unseparated cells serum replete	Lin- Sca+ cells	serum free
2-3	5-6	5-6

*Factors utilized in these studies include CSF-1, GM-CSF, G-CSF, IL-1 alpha, IL-3 and Steel Factor

Table III

Capacity of different cytokines to anchor a subliminal synergistic combination

Cytokine	Ability to "anchor" or restore growth with reduced concentrations of the other 5 factors*
Steel	++++
IL-1 alpha	++
GM-CSF	+
IL-3	+
G-CSF	0
CSF-1	0

*Factors evaluated included CSF-1, GM-CSF, G-CSF, IL-3, IL-1 alpha and Steel factor

Résumé

La "high-proliferative potential colony-forming cell" (HPP-CFC) est un progéniteur primitif qui peut représenter un modèle in vitro de cellule capable de repeupler la moelle osseuse. Nous avons décrit un certain nombre de combinaisons de 2 cytokines qui stimulent la croissance de HPP-CFCs de moelle non fractionnée dans un sytème de culture avec sérum, incluant CSF-1+ GM-CSF, G-CSF + GM-CSF, GM-CSF + IL-3, G-CSF + CSF-1 et IL-3 + G-CSF. L'augmentation du nombre de cytokines dans la combinaison et l'utilisation des doses au plateau d'IL-3, d'IL-1α, GM-CSF, G-CSF, CSF-1 and Steel factor avec de la moelle non fractionnée en cultures avec sérum, n'entrainent qu'une faible augmentation de la croissance des HPP-CFCs, mais une augmentation progressive est obtenue avec une augmentation du nombre de cytokines dans la combinaison si la moelle est purifiée (Lin-Sca$^+$) ou cultivée en l'absence de sérum. Si des cellules purifiées Lin-Sca$^+$ sont cultivées sans sérum, l'addition de Steel factor est nécessaire à la formation de HPP-CFCs; en l'absence de Steel factor, même la combinaison de 5 facteurs ne stimule que faiblement la croissance. Le PMA est capable d'augmenter la croissance des HPP-CFCs en présence d'un ou de plusieurs facteurs, mais si le CSF-1 et ou le GM-CSF sont inclus dans une combinaison, le PMA inhibe souvent la croissance des HPP-CFC. Finalement, le TGFβ ajouté au début de la culture inhibe la croissance des HPP-CFC stimulée par 5 ou 6 facteurs, mais stimule la formation de colonies dans les cultures de routine. Ces données indiquent que la croissance des HPP-CFCs est régulée par de multiples voies stimulantes et inhibitrices.

The negative regulation of hematopoiesis. Ed. M. Guigon *et al.* Colloque INSERM/John Libbey Eurotext Ldt.
© 1993, Vol. 229, pp. 141-154.

Suppressive effects of the chemokine (macrophage inflammatory protein) family of cytokines on proliferation of normal and leukemia myeloid cell proliferation

Hal E. Broxmeyer[1,2,3], Lisa Benninger[1,3], Nancy Hague[1,3], Paul Hendrie[2,3], Scott Cooper[1,3], Charlie Mantel[1,3], Kenneth Cornetta[1], Saroj Vadhan-Raj[4], Andreas Sarris[5] and Li Lu[1,3]

Departments of [1]Medicine (Hematology/Oncology), [2]Microbiology/Immunology, and [3]the Walther Oncology Center, Indiana University School of Medicine, 975 West Walnut Street, IB 501, Indianapolis, IN 46202-5121, and Departments of [4]Clinical Immunology and Biological Therapy, and [5]Hematology (Lymphoma Section), the University of Texas MD Anderson Cancer Center, 1515 Holcombe Boulevard, Houston, TX 77030, USA

ABSTRACT

A number of members of the chemokine family of molecules have suppressive activity on proliferation of stem cells and early subsets of multipotential, erythroid, and granulocyte-macrophage progenitor cells. This includes macrophage inflammatory protein (MIP)-1α, MIP-2α, platelet factor 4, interleukin-8, macrophage chemotactic and activating factor and interferon-inducible protein-10. This information is reviewed in the context of active vs. inactive molecules and in vitro and in vivo studies. Data is presented to show that patients with acute and chronic myeloid leukemia are heterogenous with regards to clonogenic progenitor cells responsive to inhibition by MIP-1α in vitro. Moreover, it is shown that the human factor-dependent cell line, MO7e, whose cells are responsive to enhanced stimulation by the combination of steel factor with either granulocyte-macrophage colony stimulating factor or interleukin-3, are responsive in a reversible fashion to suppression of cell-cycling by MIP-1α. These results may respectively be of use for future clinical trials with chemokines and for a cell line to evaluate possible intracellular mechanisms of action of chemokines.

KEY WORDS

Chemokines, Macrophage Inflammatory Proteins, Cytokines, Suppressor Molecules, Growth Factors, Stem Cells, Progenitor Cells, Leukemia Cells.

INTRODUCTION

Myeloid blood cell production is regulated by a complex interacting network of hematopoietic stem and progenitor cells, and the accessory-cell derived cytokines which can stimulate, enhance, and/or suppress stem and progenitor cells (Broxmeyer, 1992a, 1992b, 1993). These effects can be mediated directly on the stem and progenitor cells, or indirectly by an action on accessory cells. A number of molecules have been implicated in negative regulation of myelopoiesis (Table 1).

Table 1: Full Length Molecules with Known Suppressor Activities on Myeloid Progenitor Cells[a]

       Lactoferrin
       H-Ferritin
       Prostaglandins E_1 and E_2
       Tumor Necrosis Factors-alpha and beta (=lymphotoxin)
       Interferons-alpha, -beta, and -gamma
       Transforming Growth Factor-beta
       Inhibin
       Chemokines: Macrophage Inflammatory Protein-1 alpha
                               Macrophage Inflammatory Protein-2 alpha
                                 Interleukin-8
                                 Platelet Factor 4
                                 Macrophage Chemotactic and Activating Factor
                                 Interferon Inducible Protein-10

Activities Reviewed in: Broxmeyer 1992a, 1992b, 1993; Broxmeyer et al. 1993; Sarris et al. 1993.

Among these suppressor molecules are members of the chemokine family (Table 2) (Oppenheim et al., 1991; Sarris et al., 1993; Schall, 1991; Wolpe and Cerami, 1989) which include macrophage inflammatory protein (MIP)-1α (Bodine et al., 1991; Broxmeyer et al., 1990, 1991b, 1993; Clements et al., 1992; Cooper et al., 1993; Dunlop et al., 1992; Graham et al., 1990; Lord et al., 1992; Lu et al., 1993; Mantel et al., 1993; Maze et al. 1992; Queseniaux, 1993), MIP-2α, platelet factor 4 (PF4), interleukin-8 (IL-8), macrophage chemotactic and activating factor (MCAF) (Broxmeyer et al., 1993), and interferon inducible protein-10 (IP-10) (Sarris et al., 1993). The actions of natural and recombinant murine (rmu) MIP-1α, and recombinant human (rhu) MIP-1α, have been noted in vitro and in vivo. The first actions were detected on subsets of pluripotential stem cells, as assessed by the colony forming unit-spleen (CFU-S) assay, (Graham et al., 1990) and subsequently effects were also detected on early subsets of myeloid progenitor cells, including the multipotential (CFU-GEMM), erythroid (BFU-E), and granulocyte-macrophage (CFU-GM) progenitors. The myeloid progenitors that responded to the suppressive activity of MIP-1α were those that proliferated when stimulated by combinations of growth stimulating molecules, especially if steel factor (SLF, also termed c-kit ligand, stem cell factor, mast cell growth factor) was one of the molecules used for

Table 2: Members of the Human Chemokine Family

Beta Subfamily (MIP-1): coded for on human chromosome 17; contains position invariant cysteine-cysteine motif.
    Includes: MIP-1 alpha (= LD 78)
                    MIP-1 beta (= ACT-2)
                    RANTES
                    Macrophage Chemotactic and
                    Activating Factor (MCAF) = murine JE

Alpha Subfamily (MIP-2): coded for on human chromosome 4; contains position invariant cysteine - x - cysteine motif
    Includes: GRO-alpha = Melanoma Growth Stimulating Factor
                        (MGSA)=murine KC)
                    MIP-2 alpha = GRO-beta
                    MIP-2 beta = GRO-gamma
                    Platelet Factor 4 (PF4)
                    Interleukin-8 (IL-8) = Neutrophil Activating Peptide (NAP)-1
                    NAP-2
                    Interferon Inducible Protein-10 (IP-10)

Information adapted from: Oppenheim et al. 1991; Sarris et al. 1993; Schall 1991; Wolpe and Cerami 1989.

stimulation (Broxmeyer et al., 1990, 1991b, 1993). SLF is a potent co-stimulating cytokine that allows detection of early subsets of progenitors (Broxmeyer et al., 1991a). MIP-1α suppressed the SLF-enhanced colony formation of rhu granulocyte-macrophage colony-stimulating factor (GM-CSF)-dependent CFU-GM and the SLF-enhanced colony formation of rhu erythropoietin (Epo)-dependent BFU-E and CFU-GEMM (Broxmeyer et al., 1990, 1991b 1993). MIP-1α did not suppress proliferation of more mature CFU-GM stimulated with only GM-CSF or of more mature BFU-E stimulated with only Epo. It was subsequently demonstrated that other members of the chemokine family: MIP-2α, PF4, IL-8, MCAF (Broxmeyer et al., 1993), and IP-10 (Sarris et al.) had a suppressive action on early subjects of myeloid progenitors, similar to that of MIP1α. The suppressive effects of the chemokines in vitro were shown to be direct acting ones on early progenitors. This was demonstrated using purified populations of CD34^{+++} sorted normal human bone marrow cells (Broxmeyer et al., 1990, 1993; Sarris et al., 1993) and also at the level of a single isolated progenitor cell induced to proliferate in a single well in the presence of multiple growth factors (Lu et al., 1993).

    Cytokines are known to act synergistically in a growth-enhancing or -suppressing manner (Broxmeyer 1992a, 1992b, 1993). In this context, it was shown that very low and inactive concentrations of either MIP-1α, MIP-2α, PF4, IL-8, MCAF or IP-10, when used in combination with low concentrations of any other of these particular chemokines synergistically acted to suppress myeloid progenitor cell proliferation (Broxmeyer et al., 1993; Sarris et al., 1993). While MIP-1β, MIP-2β and GRO-α had no suppressive activity alone, and did not synergize to suppress progenitor cell proliferation, MIP-1β at a five-fold excess to MIP-1α blocked the suppressive effects of MIP-1α, and MIP-2β or GRO-α, at five-fold excess to either PF4 or IL-8, each blocked the suppressive effects of PF4 and IL-8. The mechanisms involved in suppression, synergistic suppression, or blocking of suppression are not yet known, but at least the initiation of these events must occur at the level of the cell surface receptors for the chemokines. A number of studies have demonstrated the isolation, purification and gene cloning of chemokine receptors, and some of the chemokines appear to compete for the same receptors (Cerretti et al., 1993; Gao et al., 1993; Graham et al., 1993; Grop et al., 1990; Holmes et al., 1991; Lee et al., 1992; Moser et al., 1991; Mukaida et al., 1991; Murphy et al., 1991; Neote et al., 1993; Oh et al., 1991; Schumacker et al., 1992; Van Ripper et al., 1993; Wang et al., 1993).

Preclinical in vivo suppressive and myeloprotective effects have been reported for MIP-1α in mice (Dunlop et al., 1992; Lord et al., 1992; Maze et al., 1992). This suggests the potential for possible clinical evaluation of MIP-1α. However, during biochemical isolation, polymerized forms of MIP-1α have been described (Davatelis et al., 1988; Sherry et al., 1988; Wolpe et al., 1988). While analyzing this phenomenon we found that polymerized rmuMIP-1α was inactive in vitro as a suppressor molecule of immature myeloid progenitors, but polymerized rmuMIP-1α didn't interfere with the action of monomeric rmuMIP-1α, the myelosuppressive effector molecule whose potency had been underestimated (Mantel et al., 1993). MIP-1α was in monomeric form when in an acetonitrile (ACN) solution (Mantel et al., 1993), a fact confirmed by others (Patel et al., 1993). It was found that polymerization was diluent- and concentration-dependent and monomeric MIP-1α had about 1000-fold higher specific activity in vitro than previously reported by us (Broxmeyer et al., 1990, 1991b) and others using preparations of MIP-1α that we believe to have been mainly in a polymerized non-active form. At least in our own studies we know this to be true. It was subsequently demonstrated that the in vivo effects of monomeric rmuMIP-1α were also about 1000-fold higher in specific activity (Cooper et al., 1992) than the MIP-1α we now know to be mainly polymerized (Maze et al., 1992). The in vivo effects of rmuMIP-1α were dose-dependent, time-related, and reversible with suppression of cycling rates of marrow and splenic CFU-GM, BFU-E and CFU-GEMM noted by 3 hrs. after injection of MIP-1α into C3H/HeJ mice, suppression of absolute numbers of marrow and splenic CFU-GM, BFU-E and CFU-GEMM noted by 24 hrs., and suppression of circulating levels of blood neutrophils decreased by 48 hrs. (Maze et al., 1992; Cooper et al., 1992). The enhanced specific activity of monomeric rmu and rhuMIP-1α (Broxmeyer, Mantel, Cooper, unpublished observations) is of relevance for potential clinical trials utilizing MIP-1α as a myelosuppressive/myeloprotective cytokine. It is not known how MIP-1α is exerting its effects in vivo. In vitro, the effects appear to be direct acting ones on the progenitor cells themselves (Broxmeyer et al., 1990, 1993; Lu et al., 1993). However, it is known that many cytokines have pleiotrophic effects (Broxmeyer 1992a, 1992b, 1993), and MIP-1α has effects in vitro on other cell types that are known to act as accessory cells for myeloid stem and progenitor cell proliferation and differentiation (Alam et al., 1992; Fahey et al., 1992; Oh et al., 1991; Schall et al., 1993; Zhou et al., 1993). Therefore, we cannot rule out the possibility, perhaps even more apparent in an in vivo situation, that MIP-1α is exerting some of its myelosuppressive effects by actions on accessory cells. We have evidence that MIP-2α, PF4, MCAF, IL-8 and IP-10 can also suppress myelopoiesis in mice in a manner analogous to that of MIP-1α (Broxmeyer, Cooper, Mantel, Sarris, unpublished observations) and these effects may also be mediated both directly and indirectly on the stem/progenitor cells. Actions of IP-10 on accessory cells have been noted by others (Taub et al., 1993).

There are no reports yet regarding the responsiveness in vitro of leukemic clonogenic cells to MIP-1α, and little or no information is available regarding how the effects of MIP-1α on progenitor cells are mediated at an intracellular level. To begin to address these issues we present information on responsiveness in vitro of CFU-GM from patients with leukemia to the effects of rhuMIP-1α, and responsiveness of the human factor-dependent cell line, MO7e (Hendrie et al.) to rmuMIP-1α. The latter study was done to identify a cell line responsive to inhibition by MIP-1α so that intracellular signalling mechanisms for the effects of MIP-1α could begin to be investigated.

MATERIALS AND METHODS

Cells. Samples from the blood and bone marrow of patients with acute myelogenous leukemia (AML) and chronic myelogenous leukemia (CML) were obtained from the Indiana University School of Medicine and the MD Anderson Cancer Center after the patients had given informal consent. Blood and bone marrow cells were subjected to a density cut on Ficoll-Hypaque to obtain the low density (<1.077 gm/cm^3) cells. MO7e cells were a gift from Genetics Institute (Boston, MA). MO7e cell cultures were maintained in rhuGM-CSF as described (Hendrie et al., 1991).
Colony Assays. Low density blood and bone marrow cells from patients with leukemia were plated at the concentrations shown in Table 3 in a 0.3% agar-culture medium in the presence of 100U/ml rhuGM-CSF (Immunex Corporation, Seattle, WA) with or without 50ng/ml rhuSLF (Immunex Corporation) as described for normal marrow cells (Broxmeyer et al., 1991b, 1993) and as briefly noted in the legend to Table 3. The tritiated thymidine kill technique was used to estimate the percentage of clonogenic cells in S-phase at the time of cell culture (Mantel et al., 1993). MO7e cells were evaluated after pulse exposure to control media, control diluent, or MIP-1α present in ACN and diluted in phosphate buffered saline (Mantel et al., 1993), washing, pulse treatment with control medium or high specific activity tritiated thymidine (20 Ci mmole, 50 uCi/ml, New England Nuclear, Boston, MA) (Mantel et al., 1993), washing and plating in 0.3% agar culture medium (Hendrie et al., 1991) in the presence of rhuGM-CSF or rhuIL-3 (Immunex Corporation) with or without rhuSLF as noted in the legend to Table 4. The tritiated thymidine kill technique allows an estimate of the percentage of MO7e cells in S-phase of the cell cycle immediately after pulse exposure to control medium, control diluent or MIP-1α. rmuMIP-1α and rhuMIP-1α were purchased from R and D Systems (Minneapolis, MN) in a 30% acetonitrile/0.1% trifluroacetic acid (ACN) solution.

Statistics. Results shown are the mean ± 1 S.E.M. for 3 plates per point for each experiment. Significant differences are based on student's t test (2-tailed).

RESULTS

Effects of MIP-1α on colony and cluster formation by CFU-GM from the blood and marrow of patients with leukemia. rhuMIP-1α in ACN was diluted into phosphate buffered saline (PBS) solution at a concentration of <20ng/ml MIP-1α in order to maintain the MIP-1α preparation in a monomeric and active form (Mantel et al., 1993). This preparation was used at 1ng/ml to assess effects on blood and marrow CFU-GM. It has previously been shown that the effects of rmuMIP-1α are initiated during S-phase of the cell cycle for myeloid progenitor cells (Mantel et al., 1993). Cells not in S-phase at the time MIP-1α is added are not suppressed by MIP-1α (Mantel et al., 1993). Therefore, it was important to know if the clonogenic cells from patients were in S-phase. The results of studies for five different patients are shown in Table 3. The results are grouped as those patients whose cells responded to the suppressive effects of MIP-1α (Table 3A: Responders) and those patients whose cells did not respond to the suppressive effects of MIP-1α (Table 3B: Non-Responders). The blood of a patient with AML and the marrow of a patient with CML had clonogenic cells in cycle which responded to the suppressive effects of MIP-1α. These clonogenic cells were stimulated in the presence of GM-CSF and SLF. Not shown is that the clonogenic cells from these patients stimulated by only GM-CSF had cells also in rapid cell cycle (50-60%) but these single factor stimulated cells were not responsive to the suppressive effects of MIP-1α, an effect consistent with the lack of effects of MIP-1α on colony/cluster formation of CFU-GM from normal donors stimulated by only one cytokine such as GM-CSF (Broxmeyer et al., 1990, 1991b, 1993). In contrast to these responders, the blood of two differentpatients with AML and the marrow of one patient with CML had clonogenic cells that were not suppressed by MIP-1α. For one of the patients with AML, the clonogenic cells responsive to stimulation by GM-CSF and SLF were not in S-phase, and therefore would not have been expected to be responsive to suppression. However, the other patient with AML and the patient with CML had 52-63% of their clonogenic cells in S-phase, yet they did not respond to inhibition with MIP-1α either. The GM-CSFstimulated clonogenic cells of these latter three patients were also not responsive to inhibition by MIP-1α.

Effects of MIP-1α on clonogenic MO7e cells. MO7e cells have been shown to be a good cell model for analysis of the synergistic stimulating effects of GM-CSF or IL-3 plus SLF (Hendrie et al., 1991). Initial attempts to demonstrate inhibition of MO7e colony formation by MIP-1α were not successful. After 6-14 days of incubation there were as many colonies and clusters in plates stimulated by GM-CSF plus SLF with or without 1-5ng monomeric rmuMIP-1α/10^3 cells/ml. With the possibility that the effects of MIP-1α on MO7e clonogenic cells might be rapidly reversible, we tested whether pulse-exposure of MO7e cells to MIP-1α would decreased the percentage of cells at that specific time that were in S-phase. This was done by first pulsing cells with control medium, control diluent or MIP-1α and then pulsing the cells with or without high specific activity tritiated thymidine. As shown in Table 4, MIP-1α significantly suppressed the percentage of MO7e responsive to stimulation by SLF plus either GM-CSF or IL-3 that were in S-phase, but not the percentage of MO7e cells responsive to stimulation by only GM-CSF or IL-3. Thus, the MO7e cells responsive to inhibition by MIP-1α are similar to normal bone marrow progenitors in that they proliferate in response to stimulation by more than one growth factor.

Table 3: Influence of Monomeric rhuMIP-1α In Vitro on Colony and Cluster Formation by CFU-GM from Patients with Leukemia Stimulated by rhuGM-CSF plus rhuSLF.[a]

Patient	Treatment Status	Cells	Control CFU-GM	% CFU-GM in S-phase	% Inhibition of CFU-GM by rhuMIP-1α
A) Responders					
AML	Untreated	Blood	1880 ± 95 / 5x10^4 cells	45[b]	50[b]
CML	Untreated	Marrow	15 ± 1 / 1x10^5 cells	57[b]	58[b]
B) Non-Responders					
AML	Untreated	Blood	1395 ± 53 / 5x10^4 cells	7	2
AML	Untreated	Blood	45 ± 6 / 5x10^5 cells	52[b]	11
CML	Treated w/ hydroxyurea	Marrow	75 ± 3 / 5x10^4 cells	63[b]	10

[a] *Cells were plated in the presence of 100 U/ml rhuGM-CSF plus 50 ng/ml rhuSLF, and in the presence of either control (McCoy's) Medium, control diluent (containing dilution of ACN present in MIP-1α preparation) or 1 ng/ml rhuMIP-1α. The control diluent had no significant effect compared to control (McCoy's) Medium; this is not shown. Cultures were scored after 14 days of incubation at 5% CO_2 and lowered (5%) O_2 tension.*

[b] *Significant % of CFU-GM in S-phase or inhibited by rhuMIP-1α, $p < 0.01$; other values are not significantly different from control, $p > 0.05$.*

Table 4: Influence of Monomeric rmuMIP-1α on Cycling Rates In Vitro of M07e Cells.[a]

	Percentage of M07e cells in S-phase After Pulse Exposure of Cells to rmuMIP-1α			
	Cells Stimulated With:			
Cells Pulsed with:	rhuGM-CSF	rhuIL-3	rhuGM-CSF + rhuSLF	rhuIL-3 + rhuSLF
Control (RPMI) Medium	67 ± 6	71 ± 6	60 ± 7	49 ± 6
Control Diluent	69 ± 6	74 ± 5	57 ± 8	49 ± 8
rmuMIP-1α	73 ± 4	76 ± 2	13 ± 7[b]	12 ± 12[b]

[a] M07e cells were pulse-exposed to control (RPMI) medium, control diluent (containing dilution of ACN present in MIP-1α preparation) or 5 ng monomeric rmuMIP-1α per 5×10^3 M07e cells at 37 °C for 1 hr. M07e cells were washed 2x and plated in 0.3% agar-culture medium (10^3 cells/ml/plate) in the presence of either 500 U/ml rhuGM-CSF or rhuIL-3, or 500 U/ml rhuGM-CSF or rhuIL-3 plus 50 ng/ml rhuSLF. Cultures were scored after 7 days of incubation at 5% CO_2 and lowered (5%) O_2 tension. Results shown are the mean ± 1 SEM of 3,2,5 and 3 experiments each respectively for cells stimulated with GM-CSF, IL-3, GM-CSF plus SLF, and IL-3 + SLF.

[b] Significant decrease compared to control (RPMI) medium-pulsed cells, $p < 0.001$; other values are not significantly different from control

DISCUSSION

MIP-1α and other chemokines such as MIP-2α, PF4, IL-8, MCAF and IP-10 suppress proliferation of progenitor cells in vitro that respond to stimulation by more than one growth factor (Broxmeyer et al. 1990, 1991b, 1993). Not all patients with leukemia have progenitor cells whose growth is enhanced by SLF (Maze et al. 1993). While the patients whose cells we evaluated had progenitor cells responsive to the enhancing effects of SLF (data not shown), there was heterogeneity amongst these patients as to the presence of clonogenic cells that could respond to MIP-1α. At this time it is not clear why clonogenic cells from some patients did not respond to inhibition by MIP-1α, but it did not relate to SLF-responsive cells, to the cycling status of these cells or apparently to the type of disease (acute vs. chronic leukemia) or to the treatment of patients (there were responsive and non-responsive cells from patients previously untreated). In vivo studies in mice (Dunlop et al. 1992, Lord et al. 1992, Maze et al. 1992) suggest that MIP-1α may eventually be tested in the clinic. Thus it is of potential importance to realize that not all patients may have cells responsive to inhibition by MIP-1α. Of interest in this context is that the patients who had clonogenic cells responding to inhibition by MIP-1α, also had clonogenic cells responding to inhibition by MIP-2α, PF4, IL-8, MCAF and IP-10, and the patients who had clonogenic cells not responding to inhibition by MIP-1α, also had clonogenic cells not responding to inhibition by MIP-2α, PF4, IL-8, MCAF and IP-10 (unpublished observations). It would seem appropriate that future patients being considered for clinical trials evaluating the efficacy of action in vivo of MIP-1α or other chemokines have their marrow and blood progenitors first assessed in vitro for responsiveness to inhibition by these chemokines. It can be envisioned that in a hyperproliferative condition in patients with chemokine-responsive clonogenic cells, the chemokines might be able to be used to dampen cell production. In a leukemic condition in which patients have leukemic clonogenic cells not responsive to chemokines, the chemokines may be able to be used to place the normal progenitor cells in those patients into a slow-proliferating and potentially more protected state so that chemotherapy can be used more efficaciously.

Unfortunately, because of the rareness of hematopoietic stem and progenitor cells and the difficulty of isolating enough of these cells at high yield and high purity, it is not possible to investigate most intracellular signalling mechanisms in these cells. These types of studies require very large numbers of cells. While cell lines established to grow in culture in the absence or presence of growth factors are not normal cells (normal cells do not appear to grow indefinitely even in the presence of growth factors), the presence of a human growth factor-dependent cell line such as M07e, which can respond to inhibition by MIP-1α, offers the possibility of beginning to understand some of the intracellular signalling mechanisms mediating the suppressive effects of MIP-1α on cell proliferation. The M07e cell line responds somewhat in a normal manner to stimulation by GM-CSF, IL-3, alone and in combination with SLF (proliferative effects are normal, but so far these cells don't show signs of much differentiation) (Hendrie et al. 1991). Receptor as well as intracellular tyrosine- and serine-protein phosphorylation events and immediate gene responses for c-myc, c-fos, junB, and c-egr have been evaluated in M07e cells responding to stimulation by SLF, GM-CSF and IL-3, alone and in combination (Miyazawa et al. 1991, 1992; Horie et al. 1993). Useful information may derive from similar studies evaluating responsiveness of M07e cells to MIP-1α and other chemokines.

ACKNOWLEDGEMENTS

These studies were supported by U.S. Public Health Service Grants R01 HL49202, R37 CA36464, and R01 HL46549 from the National Institutes of Health and the National Cancer Institute to H.E.B. We thank Rebecca Miller and Linda Cheung for typing this manuscript.

REFERENCES

Alam, R., Forsythe, P.A., Stafford, S., Lett-Brown, M.A., Grant, J.A. (1992): Macrophage inflammatory protein-1α activates basophils and mast cells. J. Exp. Med. 176: 781-786.

Bodine, D.M., Crosier, P.S., Clark, S.C. (1991): Effects of hematopoietic growth factors on the survival of primitive stem cells in liquid suspension culture. Blood 78:914-920.

Broxmeyer, H.E. (1992a): Suppressor cytokines and regulation of myelopoiesis: biology and possible clinical uses. Am. J. Ped. Hematol./Oncol. 14:22-30.

Broxmeyer, H.E. (1992b): Biomolecule-cell interactions and the regulation of myelopoiesis: an update. In: Concise Reviews in Clinical and Experimental Hematology. Murphy, Jr., M.J., ed. Alpha Med Press, Dayton, OH, pp. 119-149.

Broxmeyer, H.E. (1993): Role of cytokines in hematopoiesis. In: Clinical Applications of Cytokines: Role in Pathogenesis, Diagnosis and Therapy. Opphenheim, J.J., Rossio, J. Gearing, A., eds., Oxford University Press, New York, Chapter 27, in press.

Broxmeyer, H.E., Sherry, B., Lu, L., Cooper, S., Oh, K.O., Tekamp-Olson, P., Kwon, B.S., Cerami, A. (1990): Enhancing and suppressing effects of recombinant murine macrophage inflammatory proteins on colony formation in vitro by bone marrow myeloid progenitor cells. Blood 76:1110-1116.

Broxmeyer, H.E., Maze, R., Miyazawa, K., Carow, C., Hendrie, P.C., Cooper, S., Hangoc, G., Vadhan-Raj, S., Lu, L. (1991a): The Kit receptor and its ligand, steel factor, as regulators of hemopoiesis. Cancer Cells 3:480-487.

Broxmeyer, H.E., Sherry, B., Cooper, S., Ruscetti, F.W., Williams, D.E., Arosio, P., Kwon, B.S., Cerami, A. (1991b): Macrophage inflammatory protein (MIP)-1β abrogates the capacity of MIP-1α to suppress myeloid progenitor cell growth. J. Immunol. 147:2586-2594

Broxmeyer, H.E., Sherry, B., Cooper, S., Lu, L., Maze, R., Beckmann, M.P., Cerami, A., Ralph, P. (1993): Comparative analysis of the human macrophage inflammatory protein family of cytokines (chemokines) on proliferation of human myeloid progenitor cells. Interacting effects involving suppression, synergistic suppression, and blocking of suppression. J. Immunol. 150:3448-3458.

Cerretti, D.P., Kozlosky, C.J., Vonenbos, T., Nelson, N., Gearing, D.P., Beckmann, M.P. (1993): Molecular characterization of receptors for human interleukin-8, GRO/melanoma growth stimulatory activity and neutrophil activating peptide-2. Mol. Immunology 30:359-367.

Clements, J.M., Craig, A., Gearing, A.J.H., Hunter, M.G., Heyworth, C.M., Dexter, T.M., Lord, B.I. (1992) Biological and structural properties of MIP-1 alpha expressed in yeast. Cytokine 4:76-82.

Cooper, S., Mantel, C., Broxmeyer, H.E. (1992): Myelosuppressive effects in vivo with very low dosage of monomeric recombinant murine macrophage inflammatory protein-1α. Blood 80 (Suppl 1):346a (abstract). [Manuscript submitted for publication]

Davatelis, G., Tekamp-Olson, P., Wolpe, S.D., Hermsen, K., Luedke, C., Gallegos, C., Cort, D., Merryweather, J., Cerami, A. (1988): Cloning and characterization of a cDNA for murine macrophage inflammatory protein (MIP), a novel monokine with inflammatory and chemokinetic properties. J. Exp. Med. 167:1939-1944.

Dunlop, D.J., Wright, E.G., Lorimore, S., Graham, G.J., Holyoake, T., Kerr, D.J., Wolpe, S.D., Pragnell, I.B. (1992) Demonstration of stem cell inhibition and myeloprotective effects of SCI/rhMIP-1α in vivo. Blood 79:2221-2225.

Fahey, T.J., Tracey, K.J., Tekamp-Olson, P., Cousens, L.S., Jones, W.G., Shires, G.T., Cerami, A., Sherry, B. (1992): Macrophage inflammatory protein 1 modulates macrophage function. J. Immunol. 148:2764-2769.

Gao, J-L., Kuhns, G.B., Tiffany, H.L., McDermott, D., Li, X., Francke, U., Murphy, P.M. (1993): Structure and functional expression of the human macrophage inflammatory protein 1α/RANTES receptor. J. Exp. Med. 177:1421-1427.

Graham, G.J., Wright, E.G., Hewick, R., Wolpe, S.D., Wilkie, N.M., Donaldson, D., Lorimore, S., Pragnell, I.B. (1990): Identification and characterization of an inhibitor of haematopoietic stem cell proliferation. Nature 344:442-444.

Graham, G.J., Zhou, L., Weatherbee, J.A., Tsang, M.L-S., Napolitano, M., Leonard, W.J., Pragnell, I.B. (1993): Characterization of a receptor for macrophage inflammatory protein 1α and related proteins on human and murine cells. Cell Growth and Different. 4;137-146.

Grob, P.M., David, E., Warren, T.C., DeLeon, R.P., Farina, P.R., Homon, C.A. (1990): Characterization of a receptor for human monocyte-derived neutrophil chemotactic factor/interleukin-8. J. Biol. Chem. 265:8311-8316.

Hendrie, P.C., Miyazawa, K., Yang, Y-C., Langefeld, C.D., Broxmeyer, H.E. (1991): Mast cell growth factor (c-kit ligand) enhances cytokine stimulation of proliferation of human factor dependent cell line, M07e. Exp. Hematol. 19:1031-1037.

Holmes, W.E., Lee, J., Kuang, W.-J., Rice, G.C., Wood, W.I. (1991): Structure and functional expression of a human interleukin-8 receptor. Science 253:1278-1280.

Horie, M., Broxmeyer, H.E. (1993): Involvement of immediate-early gene expression in the synergistic effects of steel factor in combination with granulocyte-macrophage colony stimulating factor or interleukin-3 on proliferation of a human factor-dependent cell line. J. Biol. Chem. 268:968-973.

Lee, J., Horuk, R., Rice, G.C., Bennett, G.L., Camerato, T., Wood, W.I. (1992): Characterization of two high affinity human interleukin-8 receptors. J. Biol. Chem. 267:16283-16287.

Lord, B.I., Dexter, T.M., Clements, J.M., Hunter, M.A., Gearing, A.J.H. (1992): Macrophage-inflammatory protein protects multipotent hematopoietic cells from the cytotoxic effects of hydroxyurea in vivo. Blood 79:2605-2609.

Lu, L., Xiao, M., Grigsby, A., Wong, W.X., Wu, B., Shen, R.N., Broxmeyer H.E. (1993): Comparative effects of suppressive cytokines on isolated single CD34^{+++} stem/progenitor cells from human bone marrow and umbilical cord blood plated with and without serum. Exp. Hematol., in press.

Mantel, C., Kim, Y.J., Cooper, S., Kwon, S., Broxmeyer, H.E. (1993): Polymerization of murine macrophage inflammatory protein 1α inactivates its myelosuppressive effects in vitro: the active form is monomer. Proc. Natl. Acad. Sci. USA 90:2232-2236.

Maze, R., Sherry, B., Kwon, B.S., Cerami, A., Broxmeyer, H.E. (1992): Myelosuppressive effects in vivo of purified recombinant murine macrophage inflammatory protein-1α. J. Immunol. 149:1004-1009.

Maze, R., Horie, M., Hendrie, P., Vadhan-Raj, S., Tricot, G., Gordon, M., Nemunaitis, J., Ashman, L.K., Broxmeyer, H.E. (1993): Differential responses of myeloid progenitor cells from patients with myeloid leukemia and myelodysplasia to the costimulating effects of steel factor in vitro. Exp. Hematol. 21:545-551.

Miyazawa, K., Hendrie, P.C., Mantel, C., Wood, K., Ashman, L.K., and Broxmeyer, H.E. (1991): Comparative analysis of signaling pathways between mast cell growth factor (c-kit ligand) and granulocyte-macrophage colony stimulating factor in a human factor-dependent myeloid cell line involves phosphorylation of Raf-1, GTPase-activating protein and mitogen-activated protein kinase. Exp. Hematol. 19:1110-1123.

Miyazawa, K., Hendrie, P.C., Yang, Y-C., Kim, Y-J., Mantel, C., Broxmeyer, H.E. (1992): Recombinant human IL-9 induces protein tyrosine phosphorylation and synergizes with steel factor to stimulate proliferation of the human factor dependent cell line, M07e. Blood 80:1685-1692.

Moser, B., Schumacher, C., von Tscharner, V., Clark-Lewis, I., Baggiolini, M. (1991): Neutrophil-activating peptide 2 and gro/melanoma growth-stimulatory activity interact with neutrophil-activating peptide. J. Biol. Chem. 266:10666-10671.

Mukaida, N., Hishinuma, A., Zachariae, C.O.C., Oppenheim, J.J., Matsushima, K. (1991): Regulation of human interleukin-8 gene expression and binding of several other members of the intercrine family to receptors for interleukin-8. In: Chemotactic Cytokines, Westwich, J., Lindley, I.J.D., Kunkel, S.L. eds., Plenum Press, NY pp. 31-38.

Murphy, P.M., Tiffany, J.L. (1991): Cloning of complementary DNA encoding a functional human interleukin-8 receptor. Science 253:1280-1283.

Neote, K., DiGregorio, D., Mak, J.Y., Horuk, R., Schall, T.J. (1993): Molecular cloning, funational expression, and signalling characteristics of a C-C chemokine receptor. Cell

72:415-425.

Oh, K.O., Zhou, A., Kim, K.K., Samanta, H., Fraser, M., Kim, Y.J., Broxmeyer, H.E., Kwon, B.S. (1991): Identification of cell surface receptors for murine macrophage inflammatory protein-1α. J. Immunol. 147:2978-2983.

Oppenheim, J.J., Zachariae, C.O.C., Mukaida, N., Matsushima, K. (1991): Properties of the novel proinflammatory supergene "intercrine" cytokine family. Annu. Rev. Immunol. 9:617-648.

Patel, S.R., Evans, S., Dunne, K., Knight, G.C., Morgan, P.J., Varley, P.G., Craig, S. (1993): Characterization of the quaternary structure and conformational properties of the human stem cell inhibitor protein LD78 in solution. Biochem., in press.

Quesniaux, V.F.J., Graham, G.J., Pragnell, I., Donaldson, D., Wolpe, S.D., Iscove, N.N, Fagg, B. (1993): Use of 5-fluoruracil to analyze the effect of macrophage inflammatory protein-1α on long term reconstituting stem cells in vivo. Blood 81:1497-1504.

Sarris, A.H., Broxmeyer, H.E., Wirthmueller, U., Karasavvas, N., Cooper, S., Lu, L., Krueger, J., Ravetch, J.V. (1993): Human IP-10: expression and purification of recombinant protein demonstrates inhibition of early human hematopoietic progenitors. J. Exp. Med., in press.

Schall, T. (1991): Biology of the rantes/sis cytokine family. Cytokine 3:165-183.

Schall, T.J., Bacon, K., Camp, R.D.R., Kaspari, J.W., Goedell, D.V. (1993): Human macrophage inflammatory protein 1α (MIP-1α) and MIP-1β chemokines attract distinct populations of lymphocytes. J. Exp. Med. 177:1821-1825.

Schumacher, C., Clark-Lewis, I., Baggiolini, M., Moser, B. (1992): High- and low-affinity binding of GROα and neutrophil-activating peptide 2 to interleukin-8 receptors on human neutrophils. Proc. Natl. Acad. Sci. USA 89:10542-10546.

Sherry, B., Tekamp-Olson, P., Gallegos, C., Bauer, D., Davatelis, G., Wolpe, S.D., Masiarz, F., Cort, D., Cerami, A. (1988): Resolution of the two components of macrophage inflammatory protein 1, and cloning and characterization of one of those components, macrophage inflammatory protein 1β. J. Exp. Med. 168:2251-2259.

Taub, D.D., Lloyd, A.R., Conlon, K., Wang, J.M., Ortaldo, J.R., Harada, A., Matsushima, K., Kelvin, D.J., Oppenheim, J.J. (1993): Recombinant human interferon-inducible protein 10 is a chemoattractant for human monocytes and T-lymphocytes and promotes T cell adhesion to endothelial cells. J. Exp. Med. 177:1809-1814.

Van Ripper, G., Siciliano, S., Fischer, P.A., Meurer, R., Springer, M.S., Rosen, H. (1993): Characterization and species distribution of high affinity GTP-coupled receptors for human Rantes and monocyte chemoattractant protein 1. J. Exp. Med. 177:851-856.

Wang, J.M., McVicar, D.W., Oppenheim, J.J., Kelvin, D.J. (1993): Identification of RANTES receptors on human monocytic cells: competition for binding and desensitization by homologous chemotactic cytokine. J. Exp. Med. 177:699-705.

Wolpe, S D., Cerami, A. (1989): Macrophage inflammatory proteins 1 and 2: members of a novel superfamily of cytokines. FASEB J. 3:2565-2573.

Wolpe, S.D., Davatelis, G., Sherry, B., Beutler, B., Hesse, D.G., Nguyen, H.T., Moldawer, L.L, Nathan, C.F., Lowry, S.F., Cerami, A. (1988): Macrophages secrete a novel heparin-binding protein with inflammatory and neutrophil chemokinetic properties. J. Exp. Med. 167:570-581.

Zhou, Z., Kim, Y-J., Pollak, K., Lee, J-K., Broxmeyer, H.E., Kwon, B.S. (1993): Macrophage inflammatory protein-1α inhibits the anti-CD3 mab-mediated proliferation of T-lymphoyctes. J. Immunol., in press.

Résumé

Un certain nombre de molécules de la famille des chemokines a une activité suppressive sur la prolifération des cellules souches et des sous-populations précoces de progéniteurs multipotents, érythroides et granulo-macrophagiques. Parmi elles, la macrophage inflammatory protein MIP-1α, MIP-2α, la platelet facteur 4, l'interleukine-8, les facteurs chémotactique et activateur des macrophages et la protéine-10 inductible par l'interféron. Cette information est passée en revue dans le contexte de molécules actives et inactives dans des études in vitro et in vivo. Les données présentées indiquent que les patients atteints de leucémie aigue ou chronique sont hétérogènes dans la réponse de leur progéniteurs clonogéniques au MIP1α in vitro. De plus, il est montré que la lignée humaine facteur-dépendante MO7e, dont les cellules sont stimulées par la combinaison de steel factor avec le GM-CSF ou l'IL-3, répond de façon réversible à la suppression du cycle par le MIP1α. Ces résultats peuvent être utiles d'une part pour de futurs essais cliniques avec les chemokines et d'autre part pour disposer d'une lignée cellulaire permettant d'évaluer les mécanismes d'action intracellulaires des chemokines.

Macrophage inhibitory protein 1 (MIP-1α) and leukemia inhibitory factor (LIF) increase the number of colony-forming cells (CFC) in long-term cultures of human CD34+ cord blood cells

Brigitte Durand*, Babru Samal***, Anna Rita Migliaccio**, Giovanni Migliaccio** and John W. Adamson**

*Hôpital de la Croix-Rousse, Lyon, France; **New York Blood Center, New York, NY; ***Amgen, Thousand Oaks, CA, USA

INTRODUCTION

MIP-1α (Graham et al., 1990) and LIF (Gearing et al., 1987; Moreau et al., 1988) are two recently identified hematopoietic factors which share the following features; they are produced in the marrow microenvironment and they inhibit differentiation of stem cells. MIP-1α reversibly blocks the proliferation of day 12 CFU-S (Graham et al., 1990; Bodine et al., 1991) while LIF blocks the differentiation (but not the proliferation) of embryonic stem cells (Williams et al., 1988). LIF synchronizes the formation of blast cell colonies induced by IL-3 (Leary et al., 1990) and increases the efficiency of retroviral infection of stem cells in vitro (Fletcher et al., 1990). These properties make MIP-1α and LIF two potential regulators of stem cell biology.

The aim of this study was to identify the effect of MIP-1α and LIF on the generation of colony-forming cells (CFC) from pre-CFC, which may include stem cells, in liquid cultures of CD34+ cord blood cells.

MATERIALS AND METHODS

Separation of cord blood cells.

CD34+ cells were purified from human cord blood according to Migliaccio et al. (1992). Briefly, the light-density fraction was separated by Ficoll gradient centrifugation. Adherent cells and T-lymphocytes were depleted by soybean agglutination and the CD34+ cell fraction was then purified by panning on a flask coated with anti-CD34 antibody (Applied ImmuneSciences, Menlo Park, CA). The standard colony-forming efficiency of CD34+ cells enriched by this technique ranged from 0.5 to 2%.

Long-term cultures of CD34+ cord blood cells.

CD34+ (2.5×10^4/flask) cord blood cells were incubated in liquid culture for over 1 month under serum-deprived conditions as described (Migliaccio et al., 1992). The cultures were stimulated with IL-3 (10 U/ml) and SCF (1 μg/ml) and/or LIF (100 ng/ml) (all from Amgen, Thousand Oaks, CA) or MIP-1α (30 ng/ml) (Genetics Institute, Cambridge, MA).

Every 5 days, starting with day 0, the cultures were monitored for the presence of differentiated cells and CFC by taking an aliquot (20%) of replicate cultures, counting the Trypan blue negative cells and plating them in semisolid medium plus appropriate growth factors (Migliaccio et al., 1991) in the presence of fetal bovine serum (30% v/v).

RESULTS

We have previously shown that the generation of CFC from pre-CFC in liquid culture depends on the presence of SCF plus a second growth factor, such as IL-3, Epo or G-CSF (Migliaccio et al., 1992). SCF plus IL-3 increased the total number of cells for up to a month and generated large numbers of CFC including BFU-E, GM-CFC, and mixed-cell CFC.

We have used this liquid culture system of CD34+ cord blood cells to investigate the effects of MIP-1α and LIF on the recruitment of CFC from pre-CFC. MIP-1α and LIF had no effect either alone or in combination with SCF, IL-3, Epo or G-CSF (not shown). However, when LIF or MIP-1α was added in combination with SCF plus IL-3, while they had no effect on the total number of cells generated, they increased the total number of CFC generated in the culture (Table 1). MIP-1α increased CFC numbers at day 20, while LIF was already effective by day 5. All types of progenitor cell analyzed increased in number.

The modest effects of MIP-1α and LIF, and the great variability of the results (Table 1) led to the hypothesis that MIP-1α and LIF could be produced at high levels in our cultures.

To address this point, we studied by PCR-amplification the expression of LIF, of its receptor and of MIP-1α in CD34+ cord blood cells at day 0 and after 3 days of liquid culture stimulated with SCF and IL-3 (Fig. 1). Amplified bands of the expected sizes were seen for actin, LIF and for MIP-1α. LIF mRNA was present in the cultured cells at day 3 but not at day 0. MIP-1α, in contrast to LIF, is expressed at high levels at day 0 and its expression declines at day 3. So it is possible, that the CD34+ cells themselves produce MIP-1α. As early as day 3 of culture - as CD34+ cells proliferate - MIP-1α is switched off and LIF is switched on.

TABLE 1. TOTAL NUMBER OF PROGENITOR CELLS/CULTURE*

Factor Added	Day 0	Day 5	Day 10	Day 20
None	202±68	272±24	54±25	0
SCF + IL-3	202±68	1808±776	3642±1670	4677±2004
SCF + IL-3 + MIP-1α	202±68	1715±807	4600±2204	14756±7404
SCF + IL-3 + LIF	202±68	3077±677	5441±2268	12814±6432

* Total number of progenitor cells includes the number of BFU-E, GM-CFC and mixed-cell CFC. Mean (±SEM) of four experiments performed in duplicate.

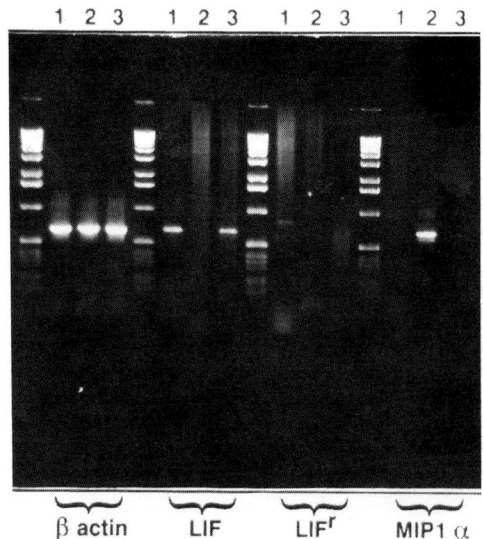

Figure 1. Analysis of ß-actin, LIF, LIF-receptor and MIP-1α in mRNA from CD34+ cord blood cells (10^4 cells/lane) reverse transcribed and amplified by PCR. Cells analyzed at onset or after 3 days of liquid culture in the presence of SCF and IL-3 are presented in the lanes labelled 2 and 3, respectively. RNA from 5637 cells (lane 1) was used as positive controls. Unlabelled lanes contain the markers for DNA length. The gel was stained by ethidium bromide.

CONCLUSIONS

In conclusion, both MIP-1α and LIF stimulated the recruitment of CFC from pre-CFC under defined culture conditions and in the presence of SCF plus IL-3. These results are compatible with the negative action on hematopoietic differentiation previously ascribed to these factors. We hypothesize that exposure of pre-CFC to MIP-1α or LIF results in one or two non-differentiative cell divisions which would be reflected in the accumulation of more CFC in the culture. If this is correct, then the definition of an "inhibitory action" on hematopoiesis would require redefinition. These proteins may be factors which keep the proliferative and differentiative potential of the target cells high.

Furthermore, based on the kinetics of the effects exerted by MIP-1α and LIF, on the differences of the amplification achieved, and on the expression of these factors, we propose that these two factors work hierarchically. The target cells for MIP-1α could be less differentiated than the target cells for LIF. This could be the counterpart in the "inhibitory" world of the hierarchy of stimulatory hematopoietic growth factors.

ACKNOWLEDGEMENTS

This study was supported by research grant DK-41937 from the National Institutes of Health, DHHS, institutional funds of the Lindsley F. Kimball Research Institute of the New York Blood Center, by Progetti Finalizzati CNR "Ingegneria Genetica" and "Applicazioni Cliniche della Ricerca sul Cancro".

REFERENCES

Bodine, D.M., Crosier, P.S., and Clark, S.C. (1991): Effects of hematopoietic growth factors on the survival of primitive stem cells in liquid suspension culture. Blood 78:914-920.

Fletcher, F.A., Williams, D.E., Maliszewski, C., Anderson, D., Rives, M., and Belmont, J.W. (1990): Murine leukemia inhibitory factor enhances retroviral-vector infection efficiency of hematopoietic progenitors. Blood 76:1098-1103.

Gearing, D.P., Gough, N.M., King, J.A., Hilton, D.J., Nicola, N.A., Simpson, R.J., Nice, E.C., Kelso, A., and Metcalf, D. (1987): Molecular cloning and expression of cDNA encoding a murine myeloid leukemia inhibitory factor (LIF). EMBO J. 6:3995-4002.

Graham, G.J., Wright, E.G., Hewick, R., Wolpe, S.D., Wilkie, N.M., Donaldson, D., Lorimore, S., and Pragnell, I.B. (1990): Identification and characterization of an inhibitor of haemopoietic stem cell proliferation. Nature 344:442-444.

Leary, A.G., Wong, G.G., Clark, S.C., Smith, A.G., and Ogawa, M. (1990): Leukemia inhibitory factor differentiation-inhibiting activity/human interleukin for DA cells augments proliferation of human hematopoietic stem cells. Blood 75:1960-1964.

Migliaccio, G., Migliaccio, A.R., Druzin, M.L., Giardina, P.-J.V., Zsebo, K.M., and Adamson, J.W. (1991): Effects of recombinant human stem cell factor (SCF) on the growth of human progenitor cells in vitro. J. Cell. Physiol. 148:503-509.

Migliaccio, G., Migliaccio, A.R., Druzin, M.L., Giardina, P.-J.V., Zsebo, K.M., and Adamson, J.W. (1992): Long-term generation of colony-forming cells in liquid culture of $CD34^+$ cord blood cells in the presence of recombinant human stem cell factor. Blood 79:2620-2627.

Moreau, J.-F., Donaldson, D.D., Bennett, F., Witek-Giannotti, J., Clark, S.C., and Wong, G.G. (1988): Leukaemia inhibitory factor is identical to the myeloid growth factor human interleukin for DA cells. Nature 336:690-692.

Williams, R.L., Hilton, D.J., Pease, S., Willson, T.A., Stewart, C.L., Gearing, D.P., Wagner, E.F, Metcalf, D., Nicola, N.A., and Gough, N.M. (1988): Myeloid leukaemia inhibitory factor maintains the developmental potential of embryonic stem cells. Nature (London) 336:684-687.

Contrasting effects of rh-MIP-1α and TGF-$β_1$ on chronic myeloid leukaemia progenitors *in vitro*

T.L. Holyoake*, M.G. Freshney*, W.P. Steward*, D.J. Dunlop*,
E. Fitzsimons** and I.B. Pragnell*

*CRC Beatson Laboratories, Glasgow, G61 1BD; **Monklands Hospital, Airdrie, Scotland, UK

In chronic myeloid leukaemia (CML) an abnormality at the stem cell level results in unregulated expansion of myeloid progenitors. The mechanism underlying this uncontrolled proliferation remains unclear. The murine CFU-A assay, which detects primitive progenitors with properties similar to CFU-S day 12, was used to purify the reversible inhibitor of stem cell proliferation, MIP-1α. Direct addition of MIP-1α or TGF-$β_1$ to murine CFU-A cultures is known to inhibit the formation of macroscopic CFU-A colonies in a reproducible fashion. Recently reported data (Eaves, 1992) offers preliminary evidence that in CML the primitive Philadelphia positive progenitors are inhibited by TGF-$β_1$ in a normal fashion, but appear insensitive to the ability of MIP-1α to block the entry into S-phase of their normal counterparts. We have recently described an in vitro clonogenic assay which detects the human counterpart of the murine CFU-A/CFU-S day 12. The aims of this study were firstly to demonstrate that primitive CML progenitors proliferate in the human CFU-A assay and that the *bcr/abl* status of individual colonies may be simply determined and secondly to document the effects of both MIP-1α and TGF-$β_1$ on the proliferation of CFU-A from normal control and CML chronic phase bone marrow, both by direct addition of the inhibitors to the CFU-A assay and by cytosine arabinoside suicide assays after 24 hour incubation with the inhibitors.

CML bone marrow samples were found to proliferate in the CFU-A assay, producing colonies morphologically indistinguishable from normal controls. The *bcr/abl* transcripts were sought in the RNA from individual colonies using the polymerase chain reaction (PCR). In order to ensure that sufficient RNA was isolated from each colony tested, the *abl* gene product was also sought by PCR as a positive control for each colony. For the CML samples tested to date, 100% of CFU-A colonies at diagnosis or in early chronic phase were found to be *bcr/abl* positive.

For normal controls both MIP-1α and TGF-$β_1$ inhibited the proliferation of CFU-A colonies when directly added to the assay (Table 1). Interestingly, CML progenitors responded normally to TGF-$β_1$, but showed no response to MIP-1α.

Table 1	Normal Controls		Chronic Myeloid Leukaemia	
	rh-MIP-1α	rh-TGF-$β_1$	rh-MIP-1α	rh-TGF-$β_1$
% inhibition of CFU-A colony number	52 ± 7.2 (n=7)	97 ± 3.0 (n=3)	0.7 ± 2.2 (n=15)	100 (n=4)
% reduction in number of cells per CFU-A colony	50 ± 5.5 (n=6)	54 ± 8.1 (n=3)	-5 ± 3.9 (n=7)	63 ± 13.9 (n=4)

In suicide assays, for five normal bone marrow samples, CFU-A progenitors induced into S-phase by incubation with IL-11 and Kit-ligand returned to a quiescent state after treatment with MIP-1α (% in S-phase: baseline 5.8 ± 3.8: after cytokine induction 39 ± 11.7: post MIP-1α 19.8 ± 12.8, n=5). In contrast, CML progenitors showed inherently high cycle status (% in S-phase: 55 ± 22.5, n=3) which failed to respond to the effects of MIP-1α during over night incubation at 37°C. TGF-β_1, however, resulted in quiescence of the CML progenitors after the same overnight incubation.

In conclusion, the primitive progenitors from CML samples were inhibited normally by TGF-β_1 but showed no response to MIP-1α. MIP-1α should therefore have clinical potential as a protective agent during chemotherapeutic purging of CML bone marrow grafts allowing selective killing of the malignant population.

References.
Eaves CJ (1992): Biology of Chronic Myeloid Leukaemia. *24th congress of the International Society of Haematology.* Abstract booklet, 19-20.

An assessment of the cell biological and biochemical effects of the stem cell inhibitor, MIP-1α, on the multipotent cell line, FDCP-Mix A4

Clare M. Heyworth*, Jane Owen-Lynch**, Mark A. Pearson*, Michael G. Hunter***, Stewart Craig***, Brian I. Lord*, Anthony D. Whetton** and T. Michael Dexter*

*CRC Department of Experimental Haematology, Paterson Institute for Cancer Research, Manchester M20 9BX; **LRF Group, UMIST, Manchester M60 1QD; *** British Biotechnology, Oxford OX4 5LY, UK

Macrophage Inflammatory Protein-1α (MIP-1α) is a potent inhibitor of pluripotent haemopoietic cell proliferation. However, the attempts to characterize the biochemical basis of the action of this cytokine by the lack of an available in vitro system to study the inhibition of primitive haemopoietic cell proliferation. We have shown that the proliferation of the multipotent, IL-3 dependent cell line FDCP-Mix A4 (A4) can be inhibited in a dose dependent fashion by MIP-1α, this therefore provides a system with which the biochemical effects of this cytokine can be elucidated.

(^{125}I)-MIP-1α was employed to assess the number of binding sites present on A4 cells. A single class of binding sites numbering 30,000 in total were present, with a K_d of 4nM. The MIP-1α receptor has recently been shown to be coupled to a G protein, suggesting that G protein linked second messenger pathways such as inositol lipid hydrolysis or cyclic AMP production may be important in MIP-1 signal transduction. There was, however, no effect of MIP-1α on A4 cell cyclic AMP levels, although prostaglandin E_2 stimulated a marked increase in cyclic AMP and inhibited A4 cell proliferation. MIP-1α did, however, stimulate a rapid (peaking after 5-10 seconds) increase in inositol 1,4,5 triphosphate levels in A4 cells. However, this effect was transient ($InsP_3$ levels were increased for <1 minute) and occurred at higher concentrations than that at which MIP-1α-mediated inhibition of A4 cell proliferation was observed. This suggests that inositol lipid hydrolysis may not be associated with growth inhibition and that there may be alternative signalling events coupling a discrete pool of as yet uncharacterized MIP-1α receptors to the molecular mechanisms associated with decreased proliferation in multipotent A4 cells.

TNFα enhances the CSF dependent growth of human hematopoietic progenitors and induces their differentiation into dendritic cells

Christophe Caux and Jacques Banchereau

Schering-Plough, Laboratory for Immunological Research, 27, chemin des Peupliers, 69571 Dardilly, France

Summary

TNFα enhances the early IL-3 and GM-CSF dependent proliferation of human CD34+ hematopoietic progenitor cells. Concomitantly, it enhances their differentiation into dendritic cells while inhibiting granulopoiesis. The generated dendritic cells display the phenotype of skin Langerhans cells and act as powerful antigen presenting cells.

I - Introduction

Dendritic cells comprise a system of highly efficient antigen presenting cells which initiate immune responses such as sensitization of MHC-restricted T cells, rejection of organ transplants and formation of T cell dependent antibodies. Found in many non lymphoid tissues, such as skin (Langerhans cells) and mucosa, dendritic cells, following antigen capture, migrate via the afferent lymph or the blood stream to lymphoid organs where they efficiently present antigen to T cells (Steinman, 1991). Isolation of dendritic cells from skin, blood or lymphoid organs is tedious and yields only low numbers of viable cells. Although they are known to originate from bone marrow, their site of generation and the conditions that direct their growth and differentiation are still poorly characterized. We thus sought for culture conditions allowing the differentiation of human CD34+ hematopoietic progenitors into dendritic cells. Combinations of GM-CSF (or IL-3) and TNFα were found to result in the generation of dendritic cells with potent antigen presenting capacity as shown by the induction of a strong proliferation of allogeneic T lymphocytes.

II - TNFα enhances GM-CSF and IL-3 induced-growth of CD34 cells.

CD34+ cells, isolated by panning from cord blood or bone marrow mononuclear cells, were grown in the presence of various cytokine combinations. Tumor necrosis factors (α and ß) strongly potentiated the proliferative effects of IL-3 and GM-CSF on CD34+ HPC (Caux, 1990). Within 8 days, the number of cells generated in TNFα-supplemented liquid cultures of CD34+ HPC increased three times when compared to cultures carried out with IL-3 alone. This reflected a true potentiating effect as TNFα per se does not act as a growth factor for CD34+ HPC. As shown by secondary liquid cultures, the observed potentiation was not the consequence of an IL-3-dependent generation of cells proliferating in response to TNFα alone. Rather, colony assays and limiting dilution studies indicated that the effect of TNFα resulted from both an increased frequency and an increased clone size of IL-3 responding cells. Furthermore, limiting dilution analysis demonstrated a direct effect of TNFα on its target cells. Three observations indicated that the enhancing effect of TNFα on early myelopoiesis was due to the recruitment of primitive hematopoietic progenitor cells hyporesponsive to IL-3. First, by separating IL-3-sensitive and IL-3-hyposensitive CD34+ HPC according to transferrin receptor (TfR) expression following a 48 h pulse of CD34+ cells in IL-3, TNFα was found to allow specifically the recruitment of TfR- cells, which proliferated weakly in response to IL-3 alone. Second, preincubation experiments have indicated that TNFα was required only during the initiation of the proliferative process, thus demonstrating that TNFα prepared cells to respond to IL-3. Third, morphologic and phenotypic characterization of clones generated in the presence of noninhibitory concentrations of TNFα revealed that high proliferative potential cells selectively recruited by TNFα were in fact granulocyte/monocyte bipotent progenitors (Caux, 1993).

Finally, culturing CD34+ HPC in the presence of TNFα resulted in an upregulation of the expression of the ß chain common to the IL-3/IL-5/GM-CSF receptor complexes thus providing a simple explanation for the induction by TNFα of an enhanced proliferative response of early HPC to IL-3 and GM-CSF (Sato, 1993).

III - TNFα alters the differentiation of CD34+ cells.

In presence of TNFα, no further CSF dependent growth of CD34+ cells could be observed after 12 days and the number of cells eventually declined after 18 to 20 days. This result contrasted with cultures performed with IL-3 alone, in which cell growth persisted for at least 30 days. The inhibitory effects of TNFα on IL-3-dependent development of CD34+ HPC could also be observed in semisolid cultures, following an early potentiation of proliferation of clonogenic cells. TNFα strongly blocked the generation of day 14 IL-3-dependent BFU-e, G, and GM colonies. In fact, TNFα was found to inhibit granulocytic development via two mechanisms: 1) by inhibition of differentiation of blast cells towards granulocytes; 2) by inhibition of proliferation of granulocyte-committed cells (Caux, 1991).

TNFα did not decrease the number of recruited CFU-M and even potentiated IL-3-dependent development of the monocytic/macrophage lineage in liquid cultures of CD34+ HPC. However, the monocytic cells that acquired CD14 became slowly proliferating cells. Therefore the TNFα -dependent early potentiation of monocyte development subsequently contributes to the arrest of growth progression observed in late cultures. CD34+ HPC cultured for 12 days with either IL-3 or GM-CSF and TNFα, generated both adherent and non-adherent cells with a dendritic morphology and delicate membrane projections. Freshly isolated HPC do not bear CD1a (a molecule specificily expressed on Langerhans cells, and thymocytes), and culturing them for 12 days with GM-CSF resulted in the generation of 5-15% CD1a cells. Adding TNFα to GM-CSF cultures increased the total cell number by 5 to 10 fold and the proportion of CD1a+ cells up to 70% immunoprecipitation demonstrated a typical 49 kDa CD1a molecule. Examined by electron microscopy, HPC derived CD1a+

cells exhibited a lobulated nucleus and a villous surface with dendritic projections. Furthermore, 1 out of 5 CD1a+ cells showed Birbeck granules, i.e. organelles with double membrane joinings, which are characteristic of Langerhans cells. CD1a+ cells expressed antigens (such as CD4, CD40, B7/BB1 and HLA class II) previously identified on dendritic cells isolated from either tissues or blood (Caux, 1992).

IV - Functions of dendritic cells .

One way to study the antigen presenting capacity of a given cell type is to measure its ability to induce the proliferation of allogeneic CD4+ T lymphocytes. CD34+ cells cultured with GM-CSF alone or with GM-CSF + TNFα induced a strong proliferation of allogeneic CD4+ T cells. 6500 and 60 cells, from respectively GM-CSF and GM-CSF + TNFα cultures, increased by 50 fold the tritiated thymidine incorporation of 2.5×10^4 CD4+ T cells. Allogeneic CD4+ T cells derived from either cord or adult blood were equally stimulated while syngeneic CD4+ T cells were only weakly, although significantly stimulated. As few as 10 CD1a+ cells induced a strong proliferation of allogeneic CD4+ T cells, whereas CD1a- cells were poor antigen presenting cells. Allogeneic CD4+ T cells formed rosettes with GM-CSF + TNFα cells within 24 h and took on a blast morphology after 96 h coculture. In fact, allogeneic CD8+ T lymphocytes were also able to proliferate when cultured in the presence of dendritic cells generated *in vitro*. Thus, cells cultured with GM-CSF + TNFα showed high antigen presenting capacity to T cells and allospecific T cell lines and T cell clones could be generated by repeated culturing on dendritic cells.

V - Conclusion.

The complex stimulatory and inhibitory effects of TNFα on hematopoiesis, which were revealed in the past few years, can now be put together in the enclosed figure and into a general *in vivo* context which, we believe, pertains to the generation of an efficient response to pathogen invasion. At the site of injury, TNFα is produced in conjunction with CSFs by non-antigen specific cells, such as monocytes, granulocytes or keratinocytes. The released cytokines are transported to the bone marrow where TNFα stimulates early central hematopoiesis and CSFs stimulate the generation and the migration to the site of injury, of granulocytes and monocytes involved in non-antigen specific defense mechanisms. The strong inhibition of erythropoiesis by TNFα temporarily permits to devote the anabolic resources to myelopoiesis and possibly lymphopoiesis. Concomitantly, CSFs trigger the recirculation of pluripotent hematopoietic progenitors which are chemoattracted at site of injury. The localized production of TNFα and GM-CSF then induces an extramedullary myelopoiesis, yielding dendritic/Langerhans cells. Once loaded with invading antigen, the dendritic/Langerhans cells migrate to the secondary lymphoid organs and initiate antigen specific immune responses. Thus, the local production of TNFα, at the site of tissue injury may represent a bridge between non-antigen specific defences and antigen specific adaptive immune responses. In keeping with this later notion, *in vivo* administration of neutralizing anti-TNFα antibodies has been shown to inhibit the development of secondary lymphoid organs (de Kossodo, 1992).

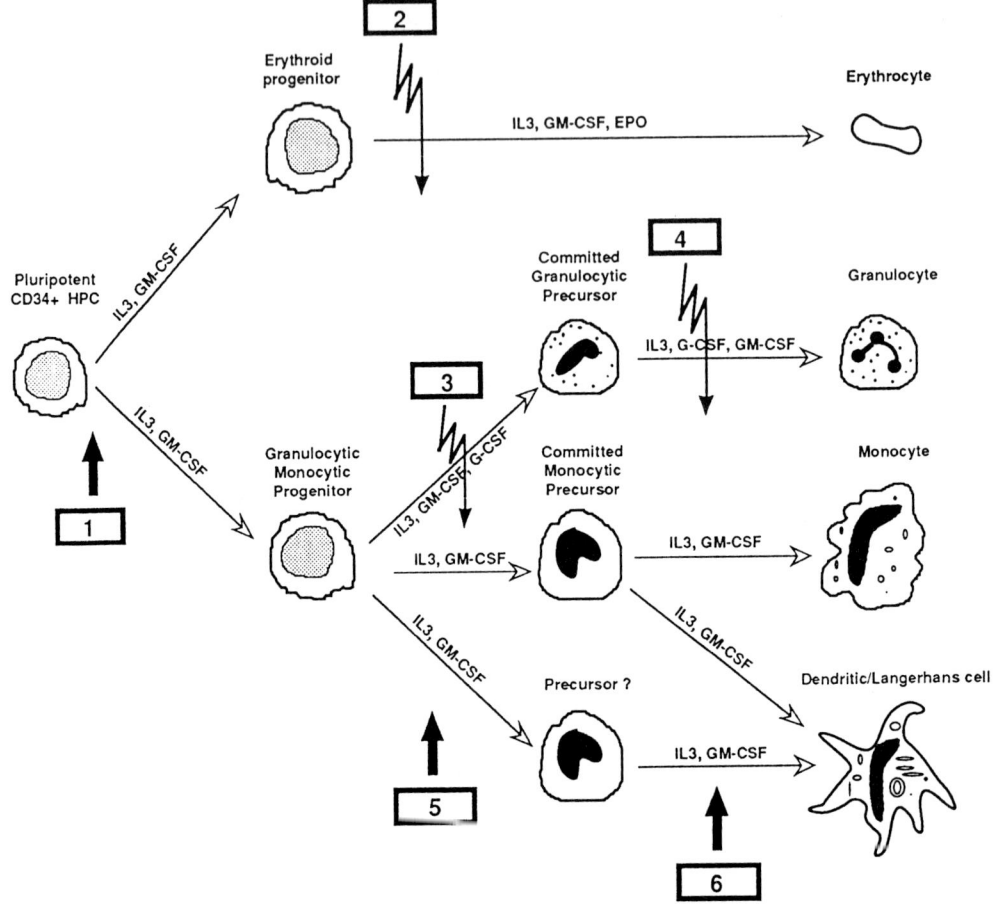

Figure Legend

Régulation of myelopoiesis by TNFα.

Positive effects are indicated by bold arrows, negative effects are indicated by angular arrows.

TNFα directly acts on primitive pluripotent progenitors by upregulation of the ß chain of the IL-3/GM-CSF receptor complex, thus inducing CSFs responsiveness (phase 1). Then, TNFα blocks erythropoiesis (phase 2) and affects granulopoiesis by inhibition of granulocytic diferentiation of uncommitted progenitors (phase 3) and inhibition of proliferation of immature granulocytic cells (phase 4). Concomitantly, TNFα specifically induces the generation of functionally mature dendritic/Langerhans cells (phase 5 and 6).

References.

Caux, C., Dezutter-Dambuyant, C. et al. (1992) : GM-CSF and TNF-α cooperate in the generation of dendritic Langerhans cells. Nature. 360:258-261.

Caux, C., Durand, I. et al. (1993) : TNFα cooperates with IL-3 in the recruitment of a primitive subset of human CD34$^+$ progenitors. J. Exp. Med. (in press).

Caux, C., Favre, C. et al. (1991) : Potentiation of early hematopoiesis by Tumor Necrosis Factor-α is followed by inhibition of granulopoietic differentiation and proliferation. Blood. 78:635-644.

Caux, C., Saeland, S. et al. (1990) : Tumor necrosis factor-alpha strongly potentiates interleukin-3 and granulocyte-macrophage colony-stimulating factor-induced proliferation of human CD34$^+$ hematopoietic progenitor cells. Blood. 75:2292-2298.

de Kossodo, S., Grau, G. E. et al. (1992) : Tumor necrosis factor α is involved in mouse growth and lymphoid tissue development. J. Exp. Med. 176:1259-1564.

Sato, N., Caux, C. et al. (1993) : Expression and factor dependent modulation of the IL-3 receptor subunits on human hematopoietic cells. Blood. (in press).

Steinman, R. M. (1991) : The dendritic cell system and its role in immunogenicity. Annu. Rev. Immunol. 9:271-296.

Résumé

Le TNFα stimule la prolifération des progéniteurs hématopoiétiques humains CD34$^+$, induite par l'IL-3 et le GM-CSF. Parallèlement, le TNFα favorise la différenciation des cellules CD34$^+$ en cellules dendritiques, alors qu'il inhibe la granulopoïèse. Les cellules dendritiques ainsi produites ont le phénotype des cellules de Langerhans de la peau et agissent comme de puissantes cellules présentatrices d'antigènes.

Effects of the tetrapeptide AcSDKP (Seraspenide) on normal and malignant hematopoietic cells

D. Bonnet, F.M. Lemoine, L. Liu, S. Pontvert-Delucq, C. Baillou, A. Najman and M. Guigon

Laboratory of Hematology, CHU Saint-Antoine, Paris, France

SUMMARY

The tetrapeptide AcSDKP (Seraspenide), an inhibitor of murine pluripotent stem cells (CFU-S), reduces the number and the percentage in DNA synthesis of normal progenitors from human mononuclear cells. In order to precise the target cells of the peptide, a study on purified CD34+ cells was performed. This study demonstrates that AcSDKP inhibits both the clonogenicity and the proliferation of the CD34+ cells and that it acts directly in a dose-dependent and reversible way. This effect was also obtained on more primitive cells (HPP-CFC) which are present in the CD34++HLA-DR^{low} subpopulation. On the contrary, AcSDKP has no effect on the studied leukemic cells. It did not reduce the clonogenicity and the proliferation of both leukemic cell lines and progenitors from CML patients. It has also no effect on the proliferation of blast cells from AML or ALL patients. This differential effect and the fact that in long term cultures (LTCs), AcSDKP did not impair the formation of the stromal layer are of major importance for the potential clinical application of the peptide.

INTRODUCTION

The tetrapeptide Acetyl-N-Ser-Asp-Lys-Pro (AcSDKP), isolated from fetal calf marrow (Frindel et al., 1977, Lenfant et al., 1989 a) is now chemically synthetized (Lenfant et al., 1989 b). The entire sequence of the peptide is included in two other known molecules, thymosin β4 and TNF α :

AcSDKPMAEIEKFDKSKLKKTETQEKNPLPSKETIQETIEQ... Thymosin β4
VRSSSRTPSDKPVAHVVANPQGELQWLDRRANALLANG.... TNFα

AcSDKP inhibits the entry into DNA synthesis of murine pluripotent stem cells (spleen-colony-forming-units, CFU-S) after cytosine arabinoside treatment (Frindel et al., 1977). Moreover, its administration increases the survival of mice to high doses of chemotherapy treatment (Guigon et al., 1982, 1987, Bogden et al., 1991). The fact that this molecule, originally isolated from bovine marrow, is also active on murine cells, suggests that it has no species - specificity.
Because these properties seem to be important in view of a clinical potential application of this molecule as a marrow protector, we were interested to study its action on human hematopoietic progenitors.

EFFECT OF AcSDKP ON HUMAN NORMAL PROGENITORS

Using "in vitro" clonogenic assays, we have previously shown that AcSDKP inhibited the growth of human granulo-macrophagic (CFU-GM) and erythroid (BFU-E) progenitors from normal bone marrow and decreased their percentage in DNA synthesis (Guigon et al., 1990). These first experiments were performed using human placenta conditioned medium (HPCM) as a source of stimulating factor in our clonogenic assay. In order to improve the reproducibility of the culture conditions, we used different combinations of recombinant growth factors. It should be pointed out that whatever the combinations used, the same range of inhibition of the progenitors was obtained (Table 1), confirming our first results. The inhibition was obtained after a 20 hour-preincubation of mononuclear cells from normal bone marrow as well as from cord blood, with nanomolar concentration of the peptide (Guigon et al., 1990, Bonnet et al., 1992c). The inhibition was not higher than 50% and concerned cycling progenitors. This latter point has been confirmed by the fact that the inhibition was no longer obtained when the cells were pretreated with tritiated thymidine at high specific activity, which kill cells in S-phase, and then incubated with the peptide (fig.1).

Table 1 : Percentage of inhibition of the number of progenitors

Growth factors added	BFU-E	CFU-GM	n
HPCM + Epo	31 ± 6 %	28 ± 6 %	8
IL-3 + GM-CSF + Epo	39 ± 11 %	37 ± 11 %	7
IL-3 + GM-CSF + Epo + SCF	30 ± 7 %	21 ± 3 %	6

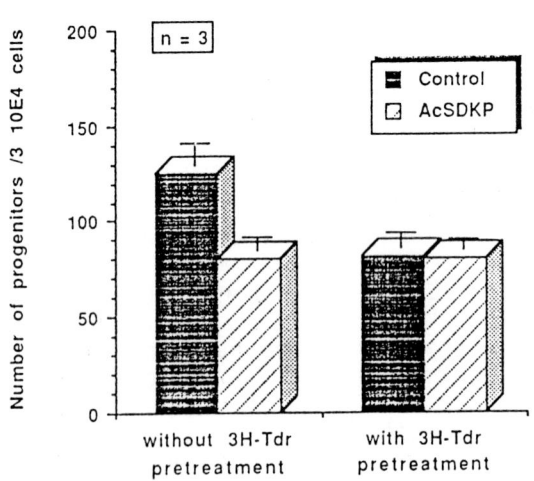

Fig 1 : Effect of AcSDKP on the number of progenitors after pretreatment with tritiated thymidine

EFFECT OF AcSDKP ON PURIFIED CD34+ PROGENITORS

In order to precise the target cells of the peptide, experiments were performed on purified human progenitors. To this end, we highly purified by flow cytometer CD34+ HLA-DR^{high} and CD34++ HLA-DR^{low} cell fractions. Both CD34+ subpopulations were stimulated by combinations of growth factors (GFs), with or without AcSDKP and tested both in clonogenic and proliferative assays. The results showed that AcSDKP inhibited the response of the two CD34+ cell subsets to GFs by reducing both the clonogenicity and the cell proliferation (Bonnet et al., submitted). The inhibitory effect of AcSDKP observed at low concentrations (10^{-10} to 10^{-9}M), disappeared at higher concentrations. Such a bell-shaped curve, which has been reported on unfractionated bone marrow, may indicate that the peptide has an indirect effect through accessory cells, as suggested by Cashman et al. (1992) and that high concentrations of AcSDKP induce accessory cells to release factor(s) able to counterbalance its suppressive effect.

EFFECT OF AcSDKP AT A SINGLE CELL LEVEL

To determine whether or not AcSDKP act directly, we have performed studies at a single cell level using limiting dilution assays. We have shown that AcSDKP acts directly on CD34+ cells at nanomolar concentration and that its inhibitory effect is reversible (Bonnet et al., submitted). The fact that even at a single cell level, we still obtained a bell-shaped curve indicates that accessory cells are not responsible for this phenomenon. Although the mechanism underlying this biphasic effect is not known, it should be pointed out that similar dose-response curves has been reported for other small peptides (Cannon et al.,1986). A possible explanation would involve a down-regulation of receptor at high concentration as reported for TNFα (Ishikura et al., 1989) but up to now, the receptor of the peptide has not been yet identified.

EFFECT OF AcSDKP ON PRIMITIVE PROGENITORS

High proliferative potential colony forming cells (HPP-CFC) and long term initiating cells (LTC-IC) (McNiece et al., 1989, Sutherland et al., 1989) represent the most primitive hematopoietic progenitor cells that can be presently assayed in vitro in the human hematopoietic system. HPP-CFC and LTC-IC can be further enriched by sorting CD34++HLA-DR^{low} cells. Thus, in view of demonstrating whether or not AcSDKP may act on very primitive stem cells, we studied its effects on purified CD34++HLA-DR^{low} cells using both HPP-CFC and LTC-IC assays. Treatment of CD34++HLA-DR^{low} cells with nanomolar concentration of the peptide reduced significantly the number of HPP-CFC. The effect of AcSDKP on LTC-IC which may represent cells capable of reconstituting hematopoiesis in vivo (Berenson et al., 1988) is now under study (Bonnet et al., in press).

EFFECT OF AcSDKP IN LONG TERM BONE MARROW CULTURE

Long-term bone marrow cultures (LTBMCs) composed of both an adherent stromal layer (AL) containing primitive progenitors and a non-adherent layer (NAL) represent the most physiological in vitro system and thus provide a good strategy to further investigate the role of AcSDKP. Therefore, LTBMCs were established from normal marrows and AcSDKP was added at the onset of the culture and then added either weekly or daily for 4 and 6 weeks. We have shown (Bonnet et al., 1992a) that the weekly addition of the peptide did not induce any significant change either on the cellularity or on the number of progenitors in the adherent and non-adherent compartments whatever the doses tested. This phenomenon can be explained by the very short lifespan of AcSDKP in LTBMC medium as shown by immunoenzymatic and biochemical studies (Pradelles et al., 1990, Grillon et al., 1993). Further experiments performed by adding AcSDKP at 10^{-10}M daily allowed us to demonstrate that AcSDKP; 1) decreased the number of progenitors present in the NAL; this inhibition was reversible, since after stopping the addition of the peptide the number of progenitors returned to control level within a week confirming our data on CD34+ purified cells; 2) conversely, AcSDKP did not show any effect on both the cellularity and the number of progenitors present in the AL. The difference of sensitivity of these two populations of progenitors to AcSDKP could be explained by the difference in their cell cycle status. Indeed, contrary to nonadherent progenitors, the very primitive hematopoietic cells which are present in the adherent layer are quiescent and are induced to proliferate for 24-48 hr after medium change, then returning to quiescence until the next feeding (Cashman et al., 1985). This hypothesis was confirmed by the same group (Cashman et al., 1992) who demonstrated that AcSDKP could reversibly block the entry of adherent progenitors into S-phase induced by medium change; 3) AcSDKP did not modify the formation of the stromal layer nor did it induce the secretion of different cytokines (TNFα, IL-6, GM-CSF) detected by immunoassays. This latter point indicates that the effects of AcSDKP on long term hematopoiesis differed from that of TNFα which contains the sequence of the peptide (Khoury et al., 1992). Taken together, our data indicate the absence of long-term toxicity of AcSDKP on human hematopoiesis at least at nanomolar concentration, which is an important point in view of its clinical application.

EFFECT OF AcSDKP ON LEUKEMIC CELLS

It has already been reported that the peptide had no effect on the proliferation of different leukemic cell lines (Lauret et al., 1989, Guigon et al., 1991, Bonnet et al., 1992 b). We have studied its effect on fresh malignant cells from patients with chronic myeloid leukemia (CML). This clonal disease is characterized by the presence of a chromosomal translocation bcr/abl which can be detected by PCR analysis and by the fact that leukemic progenitors can give rise to colonies as normal progenitors. We have demonstrated that, whatever the concentration used, AcSDKP did not inhibit the cell cycle of these leukemic progenitors nor their clonogenicity (Bonnet et al., 1992b, 1992c). These data has been confirmed recently by Cashman et al. (1992). Using the MTT assay, we have also shown that AcSDKP has no effect on the proliferation of blasts from 7 acute myeloid leukemia (AML) patients. These results were confirmed on an extending study including 9 AML and 4 acute lymphoid leukemia (ALL) cases. As shown in fig. 2, we have stimulated the proliferation of leukemic cells by different combinations of growth factors during 7 days and we have studied whether or not a daily addition of AcSDKP could decrease this proliferative response. Table 2 shows that the index of stimulation of leukemic cells varied from 1 to 63, but despite of this variability, the tetrapeptide did not modify the proliferative response whatever the doses (10^{-6} to 10^{-12} M) tested. Thus, contrary to normal CD34+ cells, leukemic cells expressing the CD34+ antigen seem to be resistant to the action of the peptide confirming our previous results on the differential effect of AcSDKP on normal and leukemic cells (Guigon et al., 1990, Bonnet et al., 1992 b, 1992 c).

Fig2 : proliferative assay

Blast cells from AML and ALL patients
↓
+ recombinant growth factors
AML 7GFs : IL1β + IL3 + IL6 + Epo + GM-CSF + G-CSF + SCF
ALL 4GFs : IL3 + IL7 + LIF + SCF
↓
± daily addition of AcSDKP
↓ 7 days
3H-Tdr incorporation assay

Table 2 : Effect of AcSDKP on the proliferative response of AML and ALL blast cells induced by growth factors (GFs)

FAB classification	Expression of CD34 antigen	Proliferative response to GFs / control	Response to AcSDKP (10^{-6} to 10^{-12} M)
AMLs			
AML 2	+	x 1.1 (a) x 1.2 (b)	no no
AML 2	+	x 6.6 (b)	no
AML 4	+	x 11.5 (a)	no
AML 4	+	x 63 (a)	no
AML 4	+	x 4.0 (a)	no
AML 4	+	x 2.2 (a) x 2.1 (b)	no no
AML 4	+	x 2.0 (a)	no
AML 5	+	x 1.5 (a)	no
AML 6	+	x 2.0 (a)	no
ALLs			
ALL-preB	+	x 4.0 (a) x 3.0 (b)	no no
ALL - pre B	+	x 45 (a)	no
ALL- pre B	-	x 10 (b)	no
ALL-T	-	x 3.0 (b)	no

legend : (a) bone marrow ; (b) peripheral blood

CONCLUSION

Because AcSDKP appears to be a physiological negative regulator in mice (Frindel et al., 1989, Lauret et al., 1989, Wdzieczack-Bakala et al., 1990) and probably also in human beings (Pradelles et al., 1990, Lopez et al., 1991), its role in the leukemic process can be questioned. Preliminary data showing an increase of AcSDKP level in the serum of some CML patients (Liozon et al., 1990) could also suggest that this molecule may play a role in the proliferative advantage of the leukemic clone. Studies on the molecular mechanism of action of the tetrapeptide, which are now in progress, will certainly be useful to explain the differential effect of AcSDKP on normal and leukemic cells. Besides, the differential effect of AcSDKP on normal and leukemic cells allows its clinical use as a marrow protector. Indeed, experiments reported in this book indicate an in vitro protective effect during marrow purging (P.K. Wierenga et al; C.Coutton et al., M. Guigon et al.). Furthermore, preliminary results from a clinical trial, indicate that it reduces the myelotoxicity of systemic chemotherapy (Carde et al., this book).

ACKNOWLEDGMENTS

D.Bonnet is a recipient from CIFRE award. We are grateful to Dr E.Deschamps de Paillette and Dr F.Thomas (Ipsen-Biotech, France) for supporting this work and providing AcSDKP.
We thank ARC, Beecham Institute and the Groupe des Assurances Nationales (GAN Company, France) for their financial contribution to the cell sorter EPICS Elite (Coultronics, France).
We are grateful to Dr K.Zsebo (Amgen,USA), Dr S.Clarke (Genetics Institute, USA), Dr L.Arden (Amsterdam, The Netherlands) for providing growth factors and Dr P.Lansdorp (Terry Fox Laboratory, Vancouver, Canada) for 8G12 FITC monoclonal antibody.

REFERENCES

Berenson RJ, Andrews RG, Bensinger WI, Kalamasz D, Knitter G, Buckner D, Bernstein ID. (1988) CD34+ marrow cells engraft lethally irradiated baboons. J. Clin. Invest 01 : 951
Bogden AE, Carde P, Deschamps de Paillette E, Moreau JP, Tubiana M, Frindel E.(1991) Amelioration of chemotherapy induced toxicity by cotreatment with a hematopoiesis-inhibiting tetrapeptide AcSDKP. Ann NY Acad Sci, 628 : 126
Bonnet D, Lemoine F.M, Khoury E, Pradelles P, Najman A, Guigon M. (1992a) Reversible inhibitory effects and absence of toxicity of the tetrapeptide Acetyl-N-Ser-Asp-Lys-Pro (AcSDKP) in human long-term bone marrow culture. Exp. Hematol, 20 : 1165
Bonnet D, Césaire R, Lemoine F.M, Aoudjhane M, Najman A, Guigon M.(1992b) The tetrapeptide AcSDKP, an inhibitor of the cell cycle status for normal human bone marrow progenitors, has no effect on leukemic cells. Exp Hematol 20 : 251
Bonnet D, Fabrega S, Lemoine F.M, Aoudjhane M, Najman A, Guigon M.(1992c) Differential effect of the tetrapeptide AcSDKP on human normal and leukemic blood progenitors. Int J Cell Cloning 10 : Supp 1, 185
Bonnet D, Lemoine F.M, Pontvert-Delucq S, Baillou C, Najman A, Guigon M. (1993) Direct and reversible inhibitory effect of the tetrapeptide AcSDKP (Seraspenide) on the growth of human CD34+ subpopulations in response to growth factors. Blood : in press.
Cannon J.G, Tatro J.B, Reichlin S, Dinarello C.A.(1986) α melanocyte stimulating hormone inhibits immunostimulatory and inflammatory actions of interleukin 1. J. Immunol. 137 : 2232
Cashman J, Eaves AC, Eaves CJ. (1985) Regulated proliferation of primitive hematopoietic progenitor cells in long-term human marrow cultures. Blood 66, 1002
Cashman JD, Eaves AC, Eaves CJ. (1992) Evidence for an indirect mechanism mediating the inhibitory effect of the tetrapeptide AcSDKP on primitive human hematopoietic cell proliferation. J.Cell.Biochem, Suppl 16C, 95 (abstr)

Frindel E, Guigon M. (1977) Inhibition of CFU entry into cycle by a bone marrow extract. Exp Hematol 5 : 74

Frindel E, Montpezat E. (1989) The physiological role of the endogenous colony-forming-units spleen (CFU-S) inhibitor Acetyl-N-Ser-Asp-Lys-Pro (AcSDKP)Leukemia 3 : 753

Grillon C, Bakala J, Lenfant M. (1993) Optimization of cell culture conditions for the evaluation of biological activities of AcSDKP, a natural inhibitory molecules. Growth factors : in press.

Guigon M, Mary JY, Enouf J, Frindel E. (1982) Protection of mice against lethal doses of 1-β-arabinofuranosyl-cytosine by pluripotent stem cell inhibitors. Cancer Res 42 : 638

Guigon M.(1987) Biological properties of low molecular weight pluripotent stem cell (CFU-S) inhibitors. In : Najman A, Guigon M, Gorin NC, Mary JY. (eds) : The inhibitors of Hematopoiesis, vol.162. John Libbey Eurotext, 241

Guigon M, Bonnet D, Lemoine F.M, Kobari L, Parmentier C, Mary JY, Najman A.(1990) Inhibition of human bone marrow progenitors by the synthetic tetrapeptide AcSDKP. Exp Hematol 18 : 1112

Guigon M, Bonnet D, Césaire R, Lemoine F.M, Najman A.(1991) Effects of small peptidic inhibitors of murine stem cells on human normal and malignant cells. Ann NY Acad Sci, 628 : 105

Ishikura H, Hori K, Bloch A.(1989) Differential biologic effects resulting from bimodal binding of recombinant human tumor necrosis factor to myeloid leukemia cells. Blood 73 : 419

Khoury E, Lemoine F.M, Baillou C, Kobari L, Deloux J, Guigon M and Najman A. (1992) Tumor Necrosis Factor alpha in human long term bone marrow cultures : Distinct effects on non-adherent and adherent progenitors. Exp. Hematol 20 : 991

Lauret E, Dumenil D, Miyanomae T, Sainteny F. (1989a) Further studies on the biological activities of the CFU-S inhibitory tetrapeptide AcSDKP. II. Unresponsiveness of isolated adult rat hepatocytes, 3T3, FDC-P2 and K562 cell lines to AcSDKP : possible involment of intermediary cell(s) in the mechanism of AcSDKP action. Exp Hematol 17 : 1081

Lauret E, Miyanomae T, Frindel E, Troalen F, Sotty D. (1989b) Abrogation of the biological activity of the inhibitor AcSDKP by a polyclonal antiserum. Leukemia 3 : 315

Lenfant M, Wdzieczak-Bakala J, Guittet E, Promé JC, Sotty D, Frindel E. (1989a) Sequence determination of an inhibitor of hemopoietic pluripotent stem cell proliferation. Proc Natl Acad Sci USA 86 : 779

Lenfant M, Itoh K, Sakoda M, Sotty D, Sasaki NA, Wdzieczak-Bakala J, Mori KJ.(1989b) Enhancement of the adherence of hematopoietic stem cells to mouse bone-marrow derived stromal cell line MS-1-T by the tetrapeptide acetyl-N-Ser-Asp-Lys-Pro. Exp Hematol 17 : 898

Lopez M, Wdzieczak-Bakala J, Pradelles P, Frindel E. (1991) Human placenta low molecular weight factors inhibit the entry into cell cycle of murine pluripotent stem cells. Leukemia 5 :101

McNiece I.K, Stewart F.M, Deacon D.M, Temeles D.S, Zsebo K.M, Clark S.C, Quesenberry P.J, Bernstein I.D. (1989) Detection of human CFC with a high proliferative potential. Blood 74 : 609

Pradelles P, Frobert Y, Créminon C, Liozon E, Massé A, Frindel E. (1990) Negative regulator of pluripotent hematopoietic stem cell proliferation in human white blood cells and plasma as analysed by enzyme immunoassay. Biochem Biophys Res Communic 170 : 986

Sutherland HJ, Eaves CJ, Eaves AC, Dragowska W, Lansdorp PM. (1989) Characterization and partial purification of human marrow cells capable of initiating long-term hematopoiesis in vitro. Blood 74 : 1563

Wdzieczak-Bakala J, Fache MP, Lenfant M, Frindel E, Sainteny F. (1990) AcSDKP, an inhibitor of CFU-S proliferation, is synthesized in mice under steady state conditions and secreted by bone marrow in long term culture. Leukemia 4 : 235

Résumé

Le tétrapeptide AcSDKP (Seraspenide), un inhibiteur des cellules souches pluripotentes de la souris (CFU-S), réduit le nombre et le pourcentage en synthèse d'ADN de progéniteurs provenant des cellules mononucléées de la moelle humaine normale. Dans le but de préciser les cellules cibles du peptide, nous avons étudié son action sur les progéniteurs purifiés CD34+ de la moelle osseuse normale. Nous avons montré qu'AcSDKP agit à la fois sur la clonogénicité et la prolifération des progéniteurs CD34+ et ceci de façon directe, réversible et dose-dépendante. Cet effet inhibiteur a également été observé sur les "high proliferative potential cells" (HPP-CFCs) population plus immature présente dans la fraction CD34++ HLA-DR^{low}. Par contre, AcSDKP n'a pas d'effet sur les cellules leucémiques. En effet, il ne réduit ni la clonogénicité, ni la prolifération des cellules de lignées leucémiques ou des progéniteurs de leucémie myéloïde chronique. Il n'a pas d'effet non plus sur la réponse proliférative des blastes leucémiques de patients atteint de leucémie aigüe myéloïde ou de leucémie aigüe lymphoïde. L'effet différentiel du tétrapeptide sur les cellules normales ouvre des perspectives d'utilisation clinique comme chimioprotecteur en hémato-cancérologie d'autant plus qu'AcSDKP ne semble pas modifier la formation du microenvironnement médullaire comme le montrent nos expériences de culture à long terme.

Comparison of the inhibitory effect of Seraspenide (AcSDKP), Tumor Necrosis Factor α (TNFα), Thymosin β4, Transforming Growth Factor β1 (TGFβ1) and Macrophage Inflammatory Molecule 1α (MIP1α) on CD34+ cell growth

D. Bonnet, F.M. Lemoine, A. Najman and M. Guigon

Department of Hematology, CHU Saint-Antoine, Paris, France

We have previously shown that the tetrapeptide Acetyl-N-Ser-Asp-Lys-Pro, an inhibitor of murine CFU-S, inhibited the response of purified hematopoietic stem cells (HSC) to a combination of growth factors (Bonnet et al., 1993). Here, we compared the effects of AcSDKP to both TNFα and thymosin β4 which contain the sequence of the peptide, and to TGFβ1 and MIP1α known to act at a very primitive level of the hematopoietic system.

For this purpose, HSC from normal bone marrow were sorted by flow cytometer on the basis of their light scatter properties and fluorescence intensities for CD34 and HLA-DR labeled monoclonal antibodies.

Sorted CD34+ HLA-DR^{high} and CD34++ HLA-DR^{low} cell fractions were seeded in liquid culture containing 5% fetal calf serum and a combination of growth factors : IL-3, IL-Iβ, IL-6, G-CSF, GM-CSF, Epo and SCF used at optimal concentration. AcSDKP, TNFα, Thymosin β4, TGFβ1 and MIP1α were added either at day 0 or daily during 6 days. Clonogenic assays were performed either after 20hrs or 6 days of liquid culture and cell proliferation was tested at day 6 by ^{3}H-Thymidine incorporation.

Results show that 1) AcSDKP, Thymosin β4, TNFα, TGFβ1 and MIP1α used at optimal concentrations reduced the number of colonies from both CD34+ cell fractions by about 30% and 65 % after 20hrs and 6 days incubation respectively. 2) Whereas a single addition at day 0 of TNFα and TGFβ1 was sufficient to reduce the proliferation of both CD34+ cell subsets, a daily addition of AcSDKP, thymosin β4 and MIP1α was absolutely required to decrease cell proliferation. 3) Besides, with TNFα and TGFβ1, we obtained a dose-dependent inhibition whereas with AcSDKP and thymosin β4 a bell-shaped curve was observed.

Our data indicate that all of these molecules reduced the response of both CD34+ HLA-DR^{high} and CD34++ HLA-DR^{low} cell subsets to growth factors and suggest some differences in their mechanism of action.

Reference :

Bonnet D, Lemoine F.M, Pontvert-Delucq S, Baillou C, Najman A, Guigon M. Direct and reversible inhibitory effect of the tetrapeptide AcSDKP (Seraspenide) on the growth of human CD34+ subpopulations in response to growth factors. Blood in press.

The specificity of action of the tetrapeptide Acetyl-N-Ser-Asp-Lys-Pro (AcSDKP) in the control of haematopoietic stem cell proliferation

Simon Robinson*, Maryse Lenfant**, Joanna Wdzieczak-Bakala** and Andrew Riches*

*School of Biological and Medical Sciences, University of St. Andrews, Scotland, UK; **Institut de Chimie des Substances Naturelles, Centre National de la Recherche Scientifique, Gif-sur-Yvette, France

A tetrapeptide of amino acid sequence Acetyl-N-Ser-Asp-Lys-Pro (AcSDKP, M_r = 487amu) has been isolated in dialysates of foetal bovine liver and bone marrow extracts (Lenfant et al., 1989). AcSDKP acts to prevent the G_0-G_1 recruitment of haematopoietic stem cells into S-phase and its administration to mice following the administration of an S-phase-specific, cytotoxic drug, cytosine arabinoside, significantly reduces the degree of haematopoietic toxicity observed (Lenfant et al., 1989). Haematopoietic stem cells in late G_1, or S-phase, are insensitive to the action of AcSDKP (Robinson et al., 1992), which suggests that the molecule is not a direct-acting, inhibitor of haematopoietic stem cell proliferation.

The proportion of the haematopoietic stem cell population in S-phase is modified by the presence of direct-acting, haematopoietic stem cell-specific, proliferation regulators. An inhibitor of haematopoietic stem cell proliferation, present in extracts of normal haematopoietic tissues and subsequently characterized as MIP1α, acts directly to reduce the proportion of the haematopoietic stem cell population in S-phase; while a stimulator of haematopoietic stem cell proliferation, present in extracts of "regenerating", or foetal haematopoietic tissues, acts directly to increase the proportion of the haematopoietic stem cell population in S-phase. In the presence of AcSDKP, the activity of the stimulator of haematopoietic stem cell proliferation was 'blocked' (Robinson et al., 1992). It is thus proposed that AcSDKP may act to 'fine tune' the proportion of the haematopoietic stem cell population in S-phase by modulating the activity of the direct-acting, stimulator of haematopoietic stem cell proliferation although the precise mechanism of action of the tetrapeptide is presently unclear.

The molecular specificity of the tetrapeptide was investigated using a number of analogues of AcSDKP and an in vitro assay of a primitive murine haematopoietic precursor characterised by a high proliferative potential and behavioural and regulatory properties similar to those reported for the in vivo spleen colony-forming unit (CFU-S) and for the in vitro "CFU-A" (Lorimore et al., 1990). The high proliferative potential colony-forming cell (HPP-CFC) population is considered to be a component of the haematopoietic stem cell compartment (Robinson & Riches, 1991). While AcSDKP acted to 'block' the action of the stimulator of haematopoietic stem cell proliferation, two structurally-distinct analogues of the molecule (AcS$_D$DKP and AcSD$_β$KP) and the tripeptide Ala-Asp-Lys (ADK) did not. However, the tripeptide Ser-Asp-Lys (SDK) did show an AcSDKP-like 'block' of stimulator activity. It would appear that the haematoregulatory role of AcSDKP is specific and that the tripeptide sequence 'SDK' may be a significant component of the molecule.

Supported by the [1]Leukaemia Research Fund, [2]INSERM and [1]IPSEN International Ltd.

References

Lenfant, M., Wdzieczak-Bakala, J., Guittet, E. et al. (1989): Inhibitor of hematopoietic pluripotent stem cell proliferation: Purification and determination of its structure. Proc. Natl. Acad. Sci. USA 86, 779-782.

Lorimore, S.,Pragnell, I., Eckmann, L. et al. (1990): Synergistic interactions allow colony formation in vitro by murine haemopoietic stem cells. Leuk. Res. 14, 481-489.

Robinson, S., Lenfant, M., Wdzieczak-Bakala, J. et al. (1992): The mechanism of action of the tetrapeptide Acetyl-N-Ser-Asp-Lys-Pro (AcSDKP) in the control of haematopoietic stem cell proliferation. Cell Prolif. 25, 623-632.

Robinson, S. & Riches, A. (1991): Haematopoietic stem cell proliferation regulators investigated using an in vitro assay. J. Anat. 174, 153-162.

Involvement of human plasma angiotensin converting enzyme in the degradation of the tetrapeptide N-Ac-Ser-Asp-Lys-Pro (AcSDKP), an inhibitor of hematopoietic stem cell (CFU-S) proliferation

Klaus-Jörg Rieger*,**, Nathalie Saez-Servent*, Joanna Wdzieczak-Bakala*, Anne Rousseau*, Wolfgang Voelter** and Maryse Lenfant*

*Institut de Chimie des Substances Naturelles, CNRS, 91198 Gif-sur-Yvette, France; **Physiologisch-chemisches Institut der Universität Tübingen, 7400 Tübingen, Germany

The tetrapeptide NAc-Ser-Asp-Lys-Pro (AcSDKP) isolated from fetal calf bone marrow has been shown to prevent in vivo the entry of hematopoietic pluripotent stem cell (CFU-S) into S-phase (Lenfant et al., 1989). AcSDKP administration increases the survival of mice treated with lethal doses of cytotoxic drugs, indicating a potential therapeutic application for this molecule (Bogden et al., 1991). Clinical trials already in progress demonstrate that AcSDKP is devoid of toxicity and exhibits significative protection against chemotherapy induced myelotoxicity (Carde et al., 1992). Clinical use of the myeloprotective tetrapeptide led us to investigate its sensitivity to proteolytic enzymes present in biological fluids.

The degradation of AcSDKP by enzymes present in human plasma has been investigated using the tetrapeptide specifically radiolabelled in the lysyl residue. 1 mM [^{3}H]-AcSDKP was completely metabolized in human plasma with a half-life of 80 min, leading exclusively to the formation of radiolabelled lysine. The cleavage of AcSDKP was completely blocked by specific inhibitors of angiotensin converting enzyme (ACE; kininase II; peptidyldipeptide hydrolase, EC 3.4.15.1), showing that the first step of the hydrolysis was indeed due to ACE. Hydrolysis of AcSDKP by rabbit lung and pure human kidney ACE generated the COOH-terminal dipeptide Lys-Pro. Thus, ACE cleaves AcSDKP by a dipeptidyl carboxypeptidase activity. In fact the formation of Lys-Pro was observed when AcSDKP was incubated in human plasma in the presence of $HgCl_2$, a cysteyl protease inhibitor.

These results suggest that ACE is involved in the first limiting step of AcSDKP degradation in human plasma. The second step seems to be under the control of a cathepsin H - like, cysteyl protease. This might suggest a possible new function of human ACE as an enzyme implicated in the degradation of this regulator of hematopoiesis.

REFERENCES

Lenfant, M., Wdzieczak-Bakala, J., Guittet, E., Prome, J.C., Sotty, D., and Frindel, E. (1989): Inhibitor of hematopoietic pluripotent stem cell proliferation: purification and determination of its structure. *Proc. Natl. Acad. Sci. USA* 86: 779-782.
Bogden, A., E, Carde P., Dechamps de Paillette, E., Moreau, J.P., Tubiana, M., and Frindel, E. (1991): Amelioration of chemotheraphy induced toxicity by co-treatment with AcSDKP a tetrapeptide inhibitor of hematopoietic stem cell proliferation. *Ann NY Ac Sci* 628: 126-139.
Carde, P., Chastang, C., Goncalves, E., Mathieu-Tubiana, N., and Vuillemin, E. (1992): Séraspenide (acetylSDKP): étude en phase I-II d'un inhibiteur de l'hématopoïèse la protégant de la toxicité de monochimiothérapies aracytine et ifosfamide. *C.R.Acad.Sci.Paris* t.315, Série III: 545-550.

Degradation rate of the tetrapeptide N-Ac-Ser-Asp-Lys-Pro (AcSDKP), an inhibitor of hematopoietic stem cell (CFU-S) proliferation, in tissues from normal and leukemic mice

Joanna Wdzieczak-Bakala*, Catherine Grillon*, Simon Robinson**, Andrew Riches**, Patrice Carde*** and Maryse Lenfant*

*Institut de Chimie des Substances Naturelles, CNRS, 91198 Gif-sur-Yvette, France; **School of Biological and Medical Sciences, University of St. Andrews, Fife, KY16 9TS, UK; ***Institut Gustave-Roussy, 94800 Villejuif, France

Elevated levels of proteolytic activity associated with malignant proliferation has been recognized for over 70 years. The possibility that the catabolism of the hemoregulatory peptide N-Ac-Ser-Asp-Lys-Pro (AcSDKP), an inhibitor of hematopoietic pluripotent stem cell proliferation (Lenfant et al., 1989) could be modified in the case of hematopoietic disorders, was worthwhile evaluating. In fact a rise in various proteolytic activities has been previously reported in leukemia (Foon & Todd, 1986; Nakamura et al., 1992).

The comparative degradation of AcSDKP was studied in vitro following incubation with plasma, bone marrow and spleen cells from normal mice and mice bearing a transplantable SA8 myeloid leukemia. Using the tetrapeptide specifically radiolabelled in the lysyl residue, degradation of [^{3}H]AcSDKP was followed by measurement of [^{3}H]Lys formation resulting from its catabolism. It was shown that already after 1 hour the degradation of AcSDKP in plasma from leukemic mice was higher compared to that following incubation in plasma from normal mice, whereas incubation with bone marrow cells exhibit a small difference only after 4 hours incubation. However no increase of AcSDKP catabolic activity was observed following incubation with spleen cells from leukemic animals when compared with incubation with normal spleen cells. Preliminary studies indicated that similar results are obtained using human plasma and bone marrow from leukemic patients.

These results suggested that malignant cells which are present both in spleen and bone marrow are not directly implicated in the increased catabolism of AcSDKP observed in the leukemic plasma. However, the higher activity of proteolytic enzymes involved in the degradation of AcSDKP in the plasma of SA8 leukemic mice, should lead to a decreased concentration of the circulating AcSDKP and in consequence may contribute to the disturbance of the regulation of hematopoietic cell proliferation which is observed in leukemia.

REFERENCES

Lenfant, M., Wdzieczak-Bakala, J., Guittet, E., Prome, J.C., Sotty, D., and Frindel, E. (1989): Inhibitor of hematopoietic pluripotent stem cell proliferation: purification and determination of its structure. Proc. Natl. Acad. Sci. USA 86: 779-782.
Foon, K.A., Todd, R.F. (1986): Immunologic clasification of leukemia and lymphoma. Blood 68: 1-31.
Nakamura, N., Tsuru, A., Hirayoshi, K., Nagata, K. (1992): Purification and characterisation of a vimentin-specific protease in mouse myeloid leukemia cells. Regulation during differentiation and identity with cathepsin G. Eur J Biochem 205: 947-954.

The tetrapeptide AcSDKP-preventing stimulation of stem cells into cycle

J. Godden*, A. Riches, G. Graham and I. Pragnell

*Beatson Institute for Cancer Research, Bearsden, Glasgow, UK; *Department of Biology and Preclinical Medicine, University of St. Andrews, UK*

The tetrapeptide Acetyl-N-Ser-Asp-Lys-Pro (AcSDKP), first identified as a dialysable extract of foetal calf bone marrow by Frindel and colleagues (1), has been shown to inhibit spleen colony forming unit (CFU-S) entry into cell cycle in cytosine arabinose treated mice (2). The protective mechanism of the tetrapeptide has been postulated to be due to the inhibition of recruitment of haematopoietic stem cells into the cell cycle.

The lack of *in vitro* data concerning the tetrapeptide has left our understanding of its mechanisms very restricted. Using the CFU-A assay, which is an *in vitro* clonogenic assay system able to detect a multipotent haematopoietic stem cell similar to the CFU-S at day 12 (3), we were able to detect *in vitro* tetrapeptide activity.

Normal bone marrow (NBM) from B6D2F1 mice was incubated in serum free Fischer's medium for 2Hrs in the presence of a stimulator which was KLS cell conditioned medium containing high levels of *kit* ligand or the tetrapeptide inhibitor. Using an ARA-C suicide technique the proliferative status of the CFU-A cells was assessed following growth of CFU-A colonies. After incubation with stimulator (KLS) the cycling status was increased from 15% (Control) to 35% CFU-A cells in S phase (shown in Figure 1). When cells were incubated with tetrapeptide alone (AcS) at 2×10^{-9}M AcSDKP there was no detectable change in the cycling status. Incubating NBM in the presence of AcSDKP (2×10^{-9}M) and stimulator (KLS/AcS) meant the cycling status remained low at 17%. This represented a significant reduction in the cycling status from treatment with stimulator alone (P<0.05) using a paired t-test.

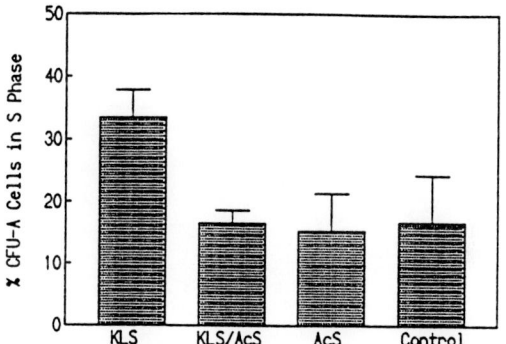

Figure 1. The proportion of CFU-A cells in S phase in normal bone marrow incubated for 3Hrs in serum free Fischer's medium (Control). Bone marrow treated for 2Hrs with 10% KLS CM (KLS), 10% KLS CM + 2×10^{-9}M AcSDKP (KLS/AcS) and 2×10^{-9}M AcSDKP (AcS). The percentage of S phase cells in the presence of KLS CM was significantly reduced with the addition of AcSDKP (P<0.05). Mean +/- SEM of four experiments. Each sample comprising of 10 replicate plates.

These results suggest the tetrapeptide AcSDKP has no direct effect on either the number or proportion of CFU-A cells in S phase. The tetrapeptide appears to function by inhibiting the effect of stimulators of stem cell proliferation, which is in accord with the proposal that tetrapeptide acts by preventing recruitment of stem cells into cycle.

References
1. Frindel,E. & Guigon,M.,(1977),Exp.Hematol.,5,74.
2. Guigon,M.*et al*,(1982), Cancer Research,42,638.
3. Pragnell,I.*et al*,(1988), Blood,72,196.

(These studies are supported by Ipsen International Limited)

Effects of interleukin 4 (IL-4) and the tetrapeptide Acetyl Ser-Asp-Lys-Pro (AcSDKP) on the proliferation of multipotent FDCP-Mix cell line

Roland Bourette, Martine Guigon*, Jean-Paul Blanchet and Guy Mouchiroud

*Centre de Génétique Moléculaire et Cellulaire, UMR CNRS 106, Université Claude-Bernard Lyon I, 69622 Villeurbanne Cedex, France; *Department of Haematology, CHU St-Antoine, 75012 Paris Cedex, France*

IL-4 inhibits the interleukin 3 (IL-3)-dependent colony formation from human and murine multipotent progenitor cells whereas AcSDKP prevents the entry into S phase of murine pluripotent stem cells and inhibits growth of human CD34+ progenitors. Here, we have looked at the effects of IL-4 and AcSDKP on the growth of the murine IL-3-dependent multipotential cell line FDCP-Mix.

Using liquid cultures and clonal assays, we have shown that IL-4 inhibits the IL-3-driven proliferation of FDCP-Mix cells in a dose-dependent manner. This effect was neutralized by the monoclonal anti-IL-4 antibody 11B11 (Fig. A). In contrast, AcSDKP did not inhibit FDCP-Mix cell proliferation under conditions in which IL-4 was effective. Because AcSDKP acts specifically on resting stem cells *in vivo*, we have induced the transition to G0/G1 of FDCP-Mix cells by cultivating them for 12hours in the absence of IL-3. Under these conditions, IL-3 and Stem Cell Factor (SCF) induced 3H-thymidine uptake by FDCP-Mix cells. In the presence of 10^{-9} M AcSDKP, the positive effect of SCF was abolished, whereas IL-3-induced DNA synthesis was not altered (Fig. B). Thus, inhibitory effects of AcSDKP seem to be related to the cell cycle status of FDCP-Mix cells, and these results are consistent with the known effects of AcSDKP on stem cells *in vivo*.

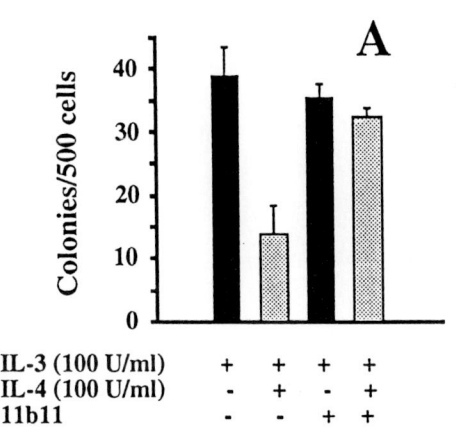
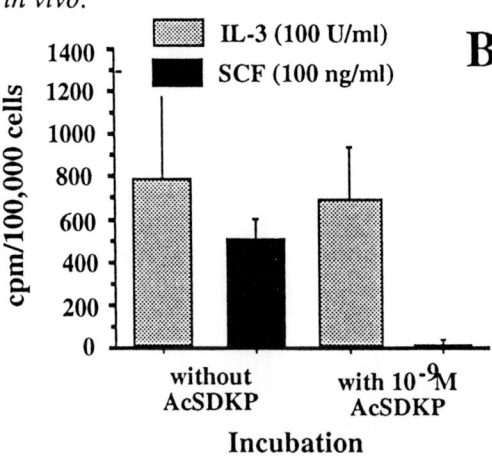

In vivo effects of the tetrapeptide N-Acetyl-Ser-Asp-Lys-Pro (Seraspenide) on hematopoiesis of normal primates and mice

F. Hérodin, N. Grenier, J.C. Mestries, E. Deschamps de Paillette* and F. Thomas*

*Department of Radiobiology, Centre de Recherches du Service de Santé des Armées, 38702 La Tronche-Grenoble, France; *Groupe Beaufour/Ipsen-Biotech, Paris, France*

Seraspenide (AcSDKP) is a synthetic tetrapeptide known to be effective in preventing murine CFU-S entry into the cell cycle and protecting mice against the toxicity of high doses of chemotherapy. Seraspenide has also been shown to inhibit the in vitro growth of human hematopoietic progenitors and to decrease the proportion of cycling cells.

Here, we report the in vivo effects of AcSDKP on hematopoiesis of normal primates and mice. We have treated four adult male monkeys (Macaca fascicularis) with two daily intramuscular injections of 10 and 20 µg Seraspenide/kg/injection for 2.5 days and 8.5 days respectively. The bone marrow (BM) progenitors were estimated 1 hour, 2 days and 7 days after the last injection (performed at 9 a.m.). Plasma AcSDKP levels were assayed using a specific enzyme immuno-assay. Twenty minutes after a 10 µg/kg injection, the mean plasma AcSDKP level was 22.8 ± 6.8 nM (base line value: 3.2 ± 0.3 nM). After 2.5 days of treatment, a moderate and reversible decrease in the number (≤ 45 per cent) and the proliferation of hematopoietic progenitors was observed. Seraspenide administered for 8.5 days exerted a strong and durable inhibition of both the number (up to 80 per cent) and the percentage in S phase (up to 90 per cent) of GEMM-CFU, BFUe and GM-CFU, noticeable as soon as 1 hour after the last injection.

We also report the effects of Seraspenide in mice treated with a single intraperitoneal (at 9 a.m.) administration. Eighty adult female CBA mice were used. Three single doses of Seraspenide were tested (5, 12.5 and 62.5 µg/kg). The control group mice received the vehicle of the injection. The hematopoietic parameters were estimated 3 hours, 1, 2 and 7 days after the injection. Twenty minutes after the 12.5 µg/kg injection, the mean plasma AcSDKP level was 27.8 nM (base line value: 8.1 nM). The doses of 5 and 12.5 µg/kg induced a significant decrease in peripheral WBC count ($p<.01$ on day 2 and $p<.02$ on day 7 [12.5µg], $p<.001$ on day 7 [5µg]), consisting in a decrease in neutrophils as well as in lymphocyte count. A significant decrease in the absolute number of spleen GEMM-CFU was observed on day 1 at 12.5 and 62.5 µg ($p<.01$).

Further experiments are now in progress to study the effects of a continuous subcutaneous perfusion of Seraspenide in mice, in order to assess the reversibility of its suppressive effects on normal cycling progenitor cells. Taken together, our results suggest that the tetrapeptide AcSDKP has myelosuppressive activity in vivo. It could be a useful hematoprotective adjunct to treatments involving cytotoxic drugs.

Seraspenide: a peptide with hematopoietic regulatory activities. Pharmacokinetics in animals and man

J.M. Grognet[1], P. Carde[2], F. Isnard[3], X. Morge[1], F. Herodin[4], E. Ezan[1], P. Pradelles[1], F. Thomas[5] and E. Deschamps de Paillette[5]

[1]*Service de Pharmacologie et d'Immunologie, CEA/Saclay, 91191 Gif-sur-Yvette Cedex;* [2]*Institut Gustave-Roussy, 94805 Villejuif Cedex;* [3]*Department of Hematology, CHU St-Antoine, 75012 Paris;* [4]*CRSSA, 38702 La Tronche Cedex;* [5]*Laboratoires Beaufour and Ipsen-Biotech, 75015 Paris, France*

NacSer-Asp-Lys-Pro (acSDKP or seraspenide) is a tetrapeptide first isolated from fetal calf bone marrow and subsequently purified and identified (Frindel and Guigon, 1977; Lenfant et al., 1989). In mammals, acSDKP is naturally present in bone-marrow, lymphoids organs and T-lymphocytes (Pradelles et al., 1990). It inhibits normal hematopoiesis progenitors. Preclinical data support that this peptide may be used in human to protect *in vivo* hematopoiesis from anticancer agents. Preliminary evaluation of pharmacokinetic parameters was investigated in order to conduct phase I-II studies in man.

Seraspenide was administered to 4 beagles dogs (100 and 200 µg/kg) using intravenous route (IV) and to 4 cynomolgus monkeys (9.3 µg/kg) using IV, intramuscular (IM) and subcutaneous (SC) routes. Seraspenide was also given by continuous infusion during 48 hours (25, 50, 125 and 250 µg/kg) in 6 men under clinical evaluation. Seraspenide plasma levels have been assayed using an enzyme immunoassay (3).

Main plasma pharmacokinetic parameters have been calculated. $T_{1/2\beta}$ values (0.19 h in dog, 0.36 h in monkeys and 0.55 h in man) show the rapid elimination of the peptide from the central compartment. Clearance values (1.57 $l.h^{-1}.kg^{-1}$ in dog, 0.70 $l.h^{-1}.kg^{-1}$ in monkeys and 0.56 $l.h^{-1}.kg^{-1}$ in man) suggest an important metabolisation of the peptide. Seraspenide is rapidely resorbed after IM or SC administration as demonstrated by Tmax values (SC : 0.40 h and IM : 0.17 h).

In conclusion, a multi-center clinical phase I-II study of seraspenide is underway in France. Moreover, evaluation of pharmacokinetic data in human volunteers is under investigation after intravenous, intramuscular and subcutaneous administration.

Frindel E. and Guigon M. (1977) Inhibition of CFU-S entry into cycle by a bone marrow extract. Exp. Hematol., **5**, 74-76,
Lenfant M., Wdzieczak-Bakala J., Guittet E., Prome J.C., Sotty D. and Frindel E.(1989) Sequence determination of an inhibitor of hematopoietic pluripotent stem cell proliferation Proc. Natl. Acad. Sci. USA, **86**, 770-782
Pradelles Ph., Frobert Y., Creminon C., Liozon E., Masse A. and Frindel E. (1990) Negative regulator of pluripotent hematopoietic stem cell proliferation in human white blood cells and plasma as analyzed by enzyme immunoassay. Biochem. Biopphys. Res. Comm. **170**, 986-993

Novel hematoregulatory peptides: monomeric and dimeric forms determine opposite biological activity

Louis M. Pelus, Peter DeMarsh, Andrew King, Carrie Frey and Pradip Bhatnagar

Departments of Anti-Infectives and Peptidomimetic Research, SmithKline Beecham Pharmaceuticals, 709 Swedeland Road, King of Prussia, PA 19406, USA

Almost twenty five years ago, a crude granulocyte extracts were shown to have myelopoietic inhibitory activity [Rytomma & Kiviniemi, 1968]. In 1982 Paukovits and Laerum identified a synthetic pentapeptide (pGlu-Glu-Asp-Cys-Lys) with the same hematopoietic properties and named it hematoregulatory peptide (HP5) [Paukovits, 1982; Laerum & Paukovits 1984; Laerum & Paukovits 1985]. Confirmation of the structure and biological activity of this peptide has come from several laboratories [reviewed in Boll *et al.* 1979; Foa *et al.*, 1982; Foa *et al.*, 1987; Lu *et al.*, 1989].

During the purification and synthesis of HP-5 it was determined that monomeric peptide was easily oxidized to a dimeric compound which stimulated rather than inhibited myelopoiesis in vitro [Laerum *et al.*, 1988]. Stimulation of hematopoiesis required suboptimal concentration of colony stimulating activity (CSA) suggesting an effect on accessory cells. Recent studies have confirmed the indirect effect of HP5 dimer and indicate a mechanism of action associated with the upregulation of hematopoietic cytokine production by stromal cells, particularly fibroblasts [Langen *et al.*, 1992; Veiby *et al.*, 1992; King *et al.*, 1992].

The ability of the HP5 monomer to dimerize to a compound with opposite biological activity, and the possibility that the HP5 dimer can be reduced to the monomeric peptide particularly in vivo is a significant drawback to the analysis of the therapeutic potential of these compounds. SK&F107647 is a new synthetic peptide which contains a dimethylene carbon bridge replacement of the disulfide bridge in HP5 dimer. In addition, SK&F108636 is a novel synthetic monomeric pentapeptide incapable of dimerization. These compounds and have undergone extensive analysis for potential therapeutic application.

Upregulation of hematopoietic growth factor production

SK&F 107647 stimulates marrow fibroblasts to produce CSA. SK&F 107647 (EC=5 pg/ml) is more potent than HP-5 dimer (EC 100 pg/ml). SK&F 108636 is an antagonist of SK&F 107647. The effects of SK&F 107647 are specific; both the SK&F 108636 and HP-5 monomers exhibits dose dependent antagonistic activity towards the CSA stimulatory activity of SK&F 107647 but not Interleukin-1 (IL1) or lipopolysaccharide (LPS) (Figure 1). An IC_{50} for inhibitory activity of the monomer was estimated at 0.1 ng/ml. HP-5 monomer at 100 ng/ml did not reduce CSA stimulated by increasing concentrations of IL1 (FIGURE 1 Center) or LPS (FIGURE 1 Right), and was without effect on constitutive CSA production by C6 cells.

The CSA stimulated by SK&F 107647 was completely neutralized by anti-M-CSF. Treatment of C6 stromal cells with SK&F 107647 results in a rapid, short-lived induction of M-CSF mRNA as determined by Northern blot analysis. The rapidity of the induction (5-30 min) as compared to the half-life of the mRNA (30 min) suggest that transcriptional activation is probably responsible for the observed increase.

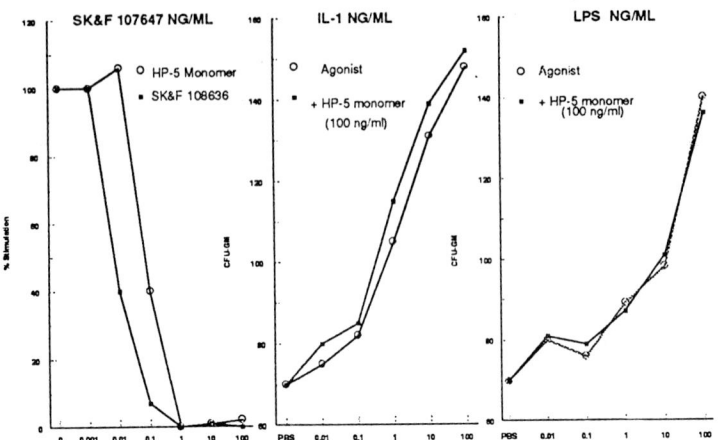

FIGURE 1: HP-5 MONOMER ANTAGONIZES THE EFFECTS OF SK&F 107647 BUT NOT IL1 OR LPS. C6 fibroblasts were incubated with agonist (SK&F 107647, Interleukin-1 or LPS) and HP-5 simultaneously for 1 hour, washed and supernates analyzed for CSA on mouse marrow CFU-GM.

Stimulation of hematopoietic precursor cell proliferation

SK&F 107647 is a potent stimulator of hematopoietic progenitor (precursor) cell proliferation. Administration of SK&F 107647 QDx4, i.p. to mice results in an increase in the proportion of CFU-GM, erythroid progenitors (BFU-E) and multilineage progenitors (CFU-GEMM) in S-phase of the cell cycle (Figure 2). Similar results were observed following administration of peptide QDx2-QDx4 to mice by other parenteral routes (sc, iv) or orally (po). The stimulatory activity of SK&F 107647 in this assay spans at least 5-7 log doses depending upon route of administration. Administration of SK&F 108636 monomer resulted in inhibition of cell cycle rate for CFU-GM, BFU-E and CFU-GEMM.

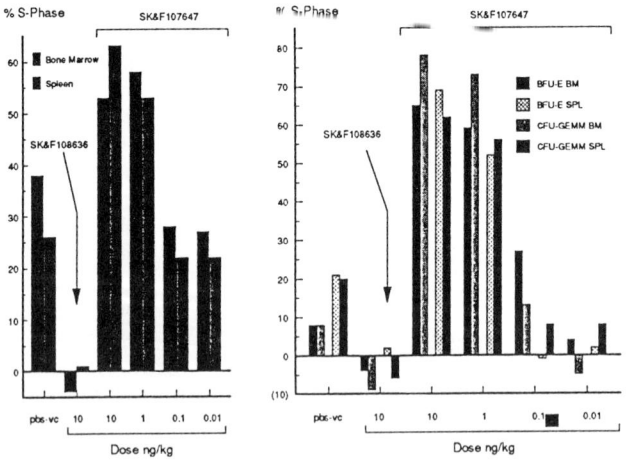

Figure 2: Effects of SK&F 107647 and SK&F 108636 administration QDx4, I.P on cell cycle rates of marrow and spleen CFU-GM, BFU-E and CFU-GEMM.

We have also determined that SK&F 107647 administration to mice elevates CFU-S, an early multilineage myeloid stem cell able to form colonies in the spleens of lethally irradiated mice (Data not shown). The CFU-S assayed in vivo is closely related to the CFU-GEMM precursor assayed in vitro or ex vivo.

These data indicate that SK&F 107647 has broad activity to stimulate proliferation of hematopoietic precursor cells. This pattern of activity is broader than has been demonstrated by single natural hematopoietic cytokines to date, and encompasses activities ascribed to several cytokines presently in clinical development.

In order to investigate the effects of SK&F 107647 on human marrow precursor cell populations, normal untreated bone marrow cells were pulsed with compounds for 2 hours followed by washing and determination of the number of S-phase progenitor cells. Pulse exposure of low density (<1.077 g/ml) human marrow cells to SK&F107647 stimulated CFU-GM, BFU-E and CFU-GEMM to enter S-Phase. Stromal cell depletion by adherence to plastic (NALD) resulted in abolition of the effect of SK&F 107647 These results indicate that the effect of SK&F 107647 on precursor cell proliferation was mediated by the stromal cell population, rather than directly on the progenitor cell population confirming a multi-step mechanism of action of this.

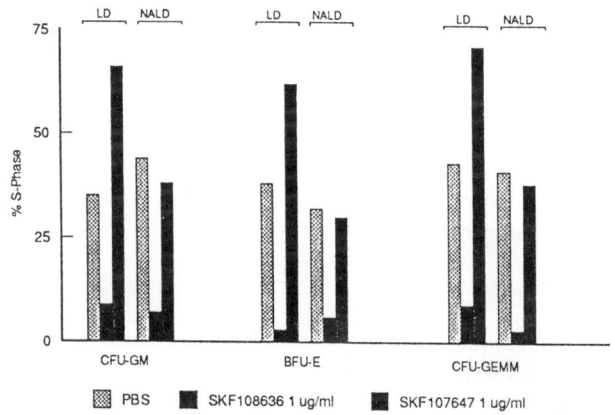

FIGURE 3: EFFECTS OF SK&F107647 ON S-PHASE OF CFU-GM, BFU-E AND CFU-GEMM FROM LOW DENSITY AND NON-ADHERENT LOW DENSITY HUMAN BONE MARROW.

Peripheral Blood Studies

SK&F 107647 increases peripheral blood counts from γ irradiated immunosuppressed mice when dosed either i.p. or by oral gavage. Elevation of PB counts were moderate but statistically significant (p<0.05 T-test), consisting of increases in neutrophils (105%), monocytes (86%), RBC (16%) and platelets (15%) by day 21 after irradiation. Due to the limitations of repeated blood sampling from mice SK&F 107647 was also evaluated in femorally cannulated normal rats. Fourteen day Alzet pump were implanted s.c. infusing either dilution buffer or 10 ng/kg/day of SK&F 107647. Repetitive peripheral blood samples were taken from the femoral cannula and counted on a Technicon hematology analyzer. Overall, total WBC was increased 30-40%. Neutrophil counts were significantly increased (103%) by day 9 with the counts returning to normal on day 14, while monocyte counts increased (50%) between days 9-16 and returned to normal within 8 days of presumed cessation of compound delivery.

Candida albicans infected mice

SK&F 107647 has been extensively evaluated in immunosuppressed and normal Balb/c mice lethally challenged with Candida albicans. In all studies, mice received 7 days prophylactic treatment with SK&F 107647, cytokines or control vehicle prior to infection and then dosed daily throughout the period of study. In radiation immunosuppression studies, treatments began 2 hrs after irradiation. Seven days post irradiation the mice were infected IV with 4.0×10^4 C. albicans. Normal mice were treated on the same schedule but infected with 1.2×10^5 yeasts.

Survival of immunosuppressed and immune intact mice

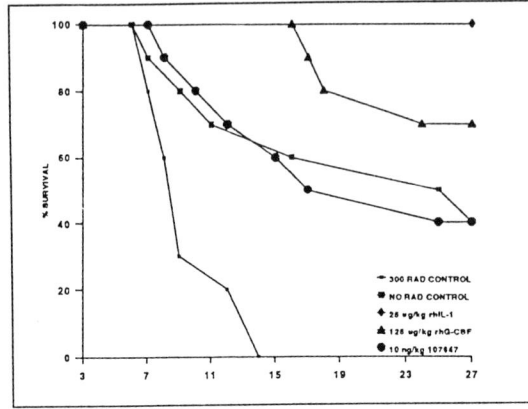

FIGURE 4: EFFICACY OF SK&F 107647 IN IRRADIATED IMMUNOSUPPRESSED AND NORMAL MICE INFECTED WITH C. ALBICANS. SK&F 107647 demonstrated protection in Candida infected mice at 10 ng/kg (p<0.05) over the controls. G-CSF also demonstrated significant protection (p<0.001).

Irradiated mice receiving no treatment and infected with C. albicans die within 7 days of infection (Figure 4). Normal immune intact mice infected with three times the number of yeast are more resistant but approximately 50% will succumb to Candida infection within 4 weeks. Treatment of irradiated mice with 10 ng/kg SK&F 107647 provides significant protection from lethal Candida infection. These mice appear to now be as resistant as immune intact mice. Treatment with G-CSF or Il1 are also protective in this model.

Combination therapy with amphotericin B

SK&F 107647 has also been evaluated as adjunct therapy in combination with amphotericin B (Figure 5). In order to ensure death in the amphotericin B treated animals, the dose of Candida was escalated even further. SK&F107647 alone significantly protected the mice from Candida lethality, as did Amphotericin B treatment at 0.3 mg/kg (the maximally tolerated dose in mice). When amphotericin was given therapeutically to mice that were prophylactically treated with 10 ng/kg SK&F107647 the combined therapy was significantly superior to either agent used alone. Similar results were observed using Fluconazole instead of amphotericin B (not shown).

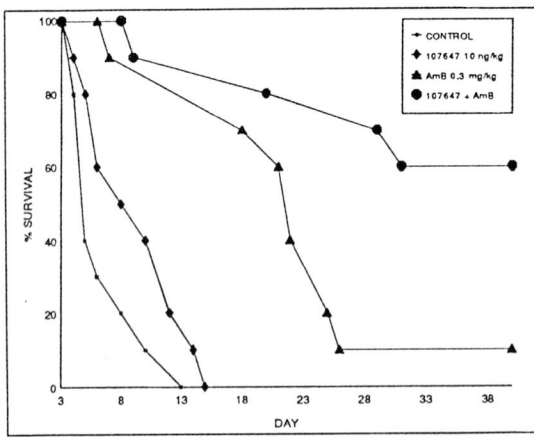

FIGURE 5: EFFICACY OF SK&F 107647 ON NORMAL MICE INFECTED WITH C. ALBICANS TREATED WITH AMPHOTERICIN. Control mice received daily IP injections of dilution buffer starting on day -7, control mice were also dosed IV on day +2. Experimental mice were dosed daily with SK&F 107647 with or without Amphotericin dosed at 0.3 mg/kg IV on day +2.

The protection seen with SK&F107647 in normal, immunosuppressed and the combination therapy with anti-fungals indicates that SK&F107647 augments non-specific host defenses against C. albicans. The activity of SK&F107647 in mouse models of candidiasis would indicate that the hematoregulatory activities of SK&F107647 could lead to improved survival in patients with systemic candidiasis.

Survival of Herpes type II infected mice

SK&F 107647 was evaluated in normal non-immunosuppressed mice challenged with Herpes type II virus in an encephalitis model. The mice were prophylactically treated for seven days then infected in the rear foot pads with Herpes type II virus strain MS. The SK&F 107647 treatments continued for the duration of the study. In this model neither IL-1 nor G-CSF demonstrated any activity. SK&F 107647 gave significant activity at 10 ng/kg (P<0.05) (Figure 6).

Combination therapy with acyclovir

SK&F 1076476 usage in the clinic for fungal or Herpes infections would almost certainly be in conjunction with conventional therapy. In the case of Herpes the conventional therapy would be acyclovir. To determine the effect of the combination of prophylactic SK&F 107647 with therapeutic acyclovir we pretreated mice for 7 days then infected with HSV II. On days +2, 3, 4, 5 and 6 after infection the mice were treated SC with acyclovir at 30 mg/kg (maximally tolerated dose). Acyclovir dosed s.c. at 30 mg/kg failed to give significant protection. However, when SK&F 107647 and acyclovir were given in combination, there was a significant increase in protection (p<0.01) over the acyclovir or SK&F 107647 alone treated animals.

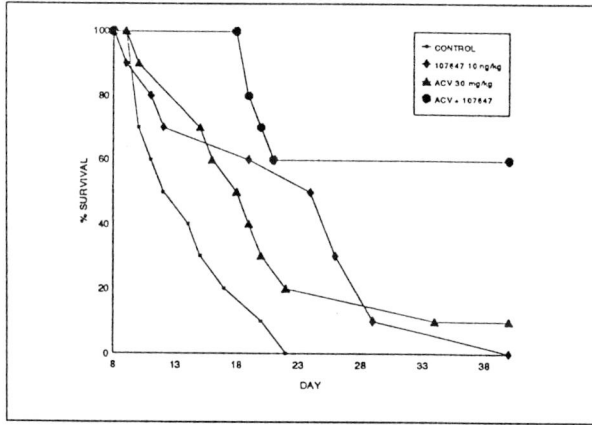

FIGURE 6: THE EFFECT OF SK&F 107647 ON MICE INFECTED WITH HERPES TREATED WITH ACYCLOVIR. Control mice received daily IP injections of dilution buffer and SC injections of dilution buffer on days +2, 3, 4, 5, and 6. Acyclovir was dosed SC at 30 mg/kg on days +2, 3, 4, 5, and 6 after infection. SK&F 107647 was dosed IP daily starting day -7 to infection at 10 ng/kg. SK&F 107647 at 10 ng/kg IP significantly protected the mice (p<0.05). Acyclovir dosed SC at 30 mg/kg failed to give significant protection. SK&F 107647 in combination with acyclovir significantly protected over either the acyclovir alone treated mice (p<0.01) or the SK&F 107647 alone treated mice (p<0.001).

The mechanism by which SK&F 107647 protects mice challenged with Herpes virus is unclear. A recent study has demonstrated that M-CSF had efficacy in a guinea pig model of vaginal Herpes [Ho *et al.*, 1991]. Furthermore, activated macrophages have been shown to provide significant protection against CMV in a murine marrow transplant model [Agah *et al.*, 1991]. Given the ability of SK&F 107647 to elevate M-CSF levels and enhance macrophage superoxide production the activity of this compound in our Herpes model is not without rationale.

Herpes is a major problem in bone marrow transplant patients, fungal infections are serious but somewhat less frequent. Efficacy in our Herpes model together with the candida mouse data and the bone marrow transplant data would indicate that SK&F107647 could have a major impact on the successful engraftment and anti-fungal/anti-viral usage in bone marrow transplant patients.

Effects of SK&F 107647 on host defense effector cells.

Evaluation of hematopoietic effector cell function was initiated to determine possible mechanism(s) of action responsible for the observed anti-infective therapeutic efficacy of SK&F 107647. Macrophage and polymorphonuclear neutrophil (PMN) effector cell functions were studied under conditions where efficacy in infectious disease models could be demonstrated (Figure 7).

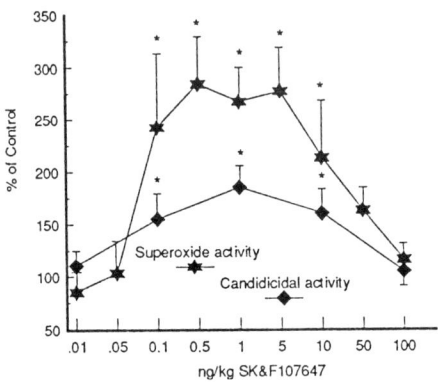

FIGURE 7. THE EFFECT OF SK&F 107647 TREATMENT ON THE EX VIVO SUPEROXIDE AND CANDIDACIDAL ACTIVITY OF MURINE PEC/PEM. Mice were administered SK&F 107647 IP daily for 8 days. Resident PEC from treated mice displayed elevated ex vivo respiratory burst and candidacidal responses. * $p<0.05$.

In neutropenic conditions, monocytes and macrophages become an important part of non-specific host defense against invading pathogens [Fromtling & Shadomy, 1986]. SK&F 107647 enhances the candidacidal potential of resident PEM by 50-60% over the responses observed with PEM from control animals. The increase in the measured ex vivo phagocytic response observed in SK&F 107647 treated animals is not sufficient in itself to account for the increase in candidacidal activity observed following SK&F 107647 treatment. Additional experimentation has indicated a possible role for increased respiratory burst (determrined by measurement of superoxide production) activity in the enhanced cidal efficiency of PEM from treated mice. Superoxide is the major initial product of oxygen reduction in the respiratory burst of activated effector cells and the various species of oxygen radicals generated during this process are cytotoxic to invading pathogens [Babior et al., 1970]. PEC/PEM from SK&F 107647 treated mice generated a superoxide response that was 1.5 - 3 times greater than that observed with cells from control mice (depending upon in vitro stimulus). This increase in superoxide as well as candidacidal activity is observed over a 3 log dose range (0.1 - 10 ng/kg) of SK&F 107647. To date we have not observed any direct in vitro effect of SK&F 107647 on monocyte or macrophage candidacidal, phagocytic or superoxide activity.

SK&F 107647 treatment significantly prolongs the survival of C. albicans-infected mice and since PMNs are the first-line defense against these organisms [Crislip & Edwards, 1989] the assessment of ex vivo candidacidal activity of PMNs from treated animals is important in determining the mode of action of SK&F 107647 in this animal model. Studies utilizing PMNs isolated from peripheral blood of treated mice indicated no differences in the Candidacidal capacity of cells from control and SK&F 107647-treated mice. However, ex vivo assays did indicate a significantly elevated respiratory burst response to the soluble stimulus PMA. Additional studies utilizing PMNs from an alternative treated animal species are required to determine the effect of SK&F 107647 therapy on the function of this effector cell population.

Summary

In Vitro	Ex Vivo	In Vitro
SK&F 107647:		
• Increases CSA production from mouse, rat, rabbit and human marrow stromal cells.	• Increases number of hematopoietic progenitor cells.	• Increases PB WBC and RBC in mice and rats.
• Upregulates M-CSF specific mRNA.	• Increases cell cycle rates of CFU GM, BFU-E and CFU-GEMM.	• Increases survival in murine models of lethal fungal and viral infection.
	• Increases serum CSA.	
	• Increases efficiency of successful BMT.	• Synergizes with empirical anti-fungal and anti-viral drugs.
	• Increases superoxide and Candidacidal activity of effector cells.	
SK&F 108636:		
• Antagonizes CSA induced by SK&F 107647 and HP5 dimer, but not LPS or IL1.	• Decreases cell cycle rates of CFU-GM, BFU-E and CFU-GEMM.	

References:

Agah, R., Charak, B.S., Chen, V., Mazumder, A. (1991): Adoptive transfer of anti-cytomegalovirus effect of interleukin-2-activated bone marrow: Potential application in transplantation. *Blood* 78: 720-727.

Babior, B.M., Kipnes, R.S. and Curnutte, J.J. (1970): Biological defense mechanisms. The production by leukocytes of superoxide, a potential bactericidal agent. *J. Clin. Invest.* 52:791-798.

Boll, I.T.M., Sterry, K., Maurer, H.R. (1979): Evidence for a rat granulocyte chalone effect on the proliferation of normal human bone marrow and of myeloid leukemias. *Acta Haemat.* 61:130-137.

Crislip, M.A. and Edwards, J.E., Jr. (1989): Candidiasis. *Infect. Dis. Clin. North Am.* 3:103-133.

Foa, P., Maiolo, T., Lombardi, L., Rytomaa ,T., Polli, E.E. (1982): Effect of granulocytic chalone on the growth rate of continuous cell lines propagated in vitro. *Scand. J.. Haematol.* 29:257-264.

Foa, P., Chillemi, F., Lombardi, L., Lonati, S., Maiolo, A.T., Polli, E.E. (1987): Inhibitory activity of a synthetic pentapeptide on leukemic myelopoiesis both in vitro and in vivo in rats. *Eur. J. Haematol.* 39:399-403.

Fromtling, R.A. and Shadomym H.J. (1986): An overview of macrophage fungal interactions. *Mycopathologia* 93:77-93.

Ho, R.J.Y., Chong, K.T., Merigan, T.C. (1991): Anti viral activity and dose optimum of recombinant macrophage colony-stimulating factor on Herpes Simplex genitalis in guinea pigs. *J. Immunol.* 146: 3578-3582.

King, A.G., Talmadge, J.E., Badger, A.M. and Pelus, L.M. (1992): Regulation of colony stimulating activity production from bone marrow stromal cells by the hemoregulatory peptide HP-5. *Exp. Hematol.* 20:223-228.

Laerum, O.D. & Paukovits, W.R. (1984): Modulation of murine hemopoiesis in vivo by a synthetic hemoregulatory pentapeptide (HP-5b). *Differentiation* 27:106-112.

Laerum, O.D. & Paukovits, W.R. (1985): Peripheral blood leukocyte alterations in mice induced by a hemoregulatory pentapeptide (HP-5b). *Leukemia Res.* 9:1075-1084.

Laerum, O.D., Sletvold, O., Bjerkness, R., Eriksen, J.A., Johansen, J.H., Schanche, J-S., Tveteras, T., Paukovits, W.R. (1988): The dimer of hemoregulatory pentapeptide (HP-5b) stimulates mouse and human myelopoiesis in vitro. *Exp. Hematol.* 16:274-280.

Langen, P., Schunck, H., Hunger, B., Schutt, M. and Laerum, O.D. (1992): Adherent-cell-dependent stimulation of CFU-GM by nucleobases, nucleosides, their analogues and the hemoregulatory peptide dimer. *Exp. Hematol.* 20:196-200.

Lu, L., Foa, P., Chillemi, F., Shen, R-N., Lin, Z.H., Carow., C., Broxmeyer, H.E. (1989): Suppressive biological activity of a synthetic pentapeptide on highly enriched human and murine marrow hematopoietic progenitors: Synergism with recombinant human tumor necrosis factor-alpha and interferon-gamma. *Exp. Hematol.* 17:935-941.

Paukovits, W.R. (1982): Isolation and synthesis of a hemoregulatory peptide. *Z. Naturforsch* 37C:1297- 1300.

Rytomma, T. & Kiviniemi, K. (1968): Control of granulocyte production. I. Chalone and anti-chalone, two specific humoral regulators. *CellTissue Kinet.* 1:329-334.

Veiby, O.P., Lovhaug, D., Fjerdingstad, H. and Engelsen, S.J. (1992): Indirect stimulation of hemopoiesis by hemoregulatory peptide (HP-5b) dimer in murine long term bone marrow cultures. *Exp. Hematol.* 20:192-195.

Résumé

	in vitro	ex vivo	in vivo
SK&F 107647 :	- augmente la production de CSA à partir de cellules stromales de souris, rat, lapin et homme - augmente l'ARNm spécifique du M-CSF	- augmente le nombre de progéniteurs hématopoïétiques - augmente le cycle des CFU-GM, BFU-E et CFU-GEMM - augmente le CSA dans le sérum - augmente l'efficacité de transplantation de moelle - augmente l'activité superoxyde et Candidacide des cellules effectrices	- augmente le nombre de globules blancs et de globules rouges dans le sang de souris et rats - augmente la survie dans des modèles murins d'infection fungique et virale - agit en synergie avec des agents antifungiques et antiviraux
SK&F 108636 :	- antagoniste de la production de CSA induite par SK&F107647 et le dimère HP5, mais pas par LPS ou IL-1	- diminue le cycle des CFU-GM, BFU-E et CFU-GEMM	

The negative regulation of hematopoiesis. Ed. M. Guigon et al. Colloque INSERM/John Libbey Eurotext Ldt.
© 1993, Vol. 229, pp. 201-211.

Physical interaction of a negative regulator of erythropoiesis with the membrane lipid bilayer

Nicholas Dainiak, Amala Guha and R. Preston Mason

University of Connecticut School of Medicine, Departments of Medicine, Laboratory Medicine and Radiology, Biomolecular Structure Analysis Center, Connecticut Cancer Institute, Farmington, CT, USA

ABSTRACT

Plasma membranes are organized into a two-dimensional mosaic of phospholipids and proteins. Interactions between membrane proteins and lipids may alter the function of either membrane component. Recently, plasma membrane-bound growth factors for hematopoietic cells have been described. To determine the influence of membrane lipid composition on the expression of the erythroid-directed growth factor, membrane-bound erythroid burst-promoting activity (mBPA), we incubated normal human splenic B-cells with an emulsion of lipids that included Liposin II. Rather than stimulating human marrow erythroid burst formation, plasma membranes isolated from the "fluidized" (i.e., treated) cells suppressed erythroid burst formation in a concentration-dependent fashion. The inhibitory activity was immunoprecipitated by prior incubation with a monoclonal antibody (D3-E4) that was previously shown to specifically recognize mBPA. To assess whether these changes in mBPA expression could be explained on the basis of alteration in membrane structure, small-angle x-ray diffraction was performed on intact plasma membranes and extra cellular vesicles shed from normal and "fluidized" B-cells. This approach determines the electron density of matter across the width of the membrane. We observed that plasma membranes and shed vesicles have distinct x-ray diffraction patterns, consistent with distinct lipid/protein compositions. In addition to demonstrating dilution of cholesterol and diminished electron density in the area of terminal methyl groups at the center of bilayer, electron density profiles were consistent with outward displacement of the phosphate head groups. The latter displacement resulted in an overall increase in plasma membrane width of approximately 4 Angstroms. We hypothesize that alteration in plasma membrane width may affect the function of mBPA. This hypothesis is consistent with the notion that the width of plasma membranes is a variable that may play a role in the expression of membrane-associated growth factors and/or their receptors.

KEY WORDS
Membrane-bound growth factors, x-ray diffraction, erythropoiesis.

INTRODUCTION

The basic structural unit of biological membranes is the phospholipid bilayer. Phospholipids are amphipathic, thereby permitting hydrophobic interactions that result in the formation of two layers of phospholipid molecules whose polar head groups are oriented outward (i.e., toward surrounding water molecules) (Singer and Nicolson, 1972). Membrane proteins are permitted to interact with the lipid bilayer in several ways. Integral membrane proteins contain amino acid residues having hydrophobic side chains that interact with fatty acyl groups of the membrane phospholipids. These proteins can be removed from the membrane with detergents (Dainiak, 1990). In contrast, peripheral membrane proteins do not enter the hydrophobic core of the phospholipid bilayer. Rather, they are bound to the membrane indirectly via interactions with integral membrane protiens (Darnell, Lodish and Baltimore, 1990). Peripheral proteins may be removed from the membrane by incubation with high ionic strength solutions.

Membrane proteins and lipids are laterally mobile on the cell surface (Gupte et al, 1984). While interactions between membrane molecules occur on nearly all scales of space and time (Edidin, 1990), little information is available regarding the functional consequences of such interactions. Insertion of proteins into the lipid bilayer may result in disordering of the structure of lipids located adjacent to the protein (Marsh, 1985). Alternatively, insertion of lipids into the bilayer may alter the function of membrane proteins (Smith and Stubbs, 1987).

We have recently observed that plasma membranes and extracellular vesicles shed from the surface of B-cells that are lipid loaded by incubation with Liposin II (containing 95% WT/WT free fatty acids) suppress rather than stimulate the proliferation of human erythroid burst-forming units (BFU-Es) in serum-depleted marrow culture (Dainiak et al, 1991a). Here, we report additional findings suggesting that the inhibitory factor shares antigenic determinants with the growth promoting factor. Although we have previously demonstrated that incubation of intact human peripheral blood mononuclear cells with Liposin II results in "fluidization" of the plasma membrane (accompanied by decreased fluorescence polarization and increased anisotropy; Armstrong, et al, 1988), virtually no information is available regarding structural changes induced by this treatment.

To characterize potential molecular interactions of the components of Liposin II with intact plasma membranes, small angle x-ray diffraction was utilized. This procedure assesses electron density of partially dehydrated multi-bilayers that are prepared from test and control plasma membranes and shed extracellular vesicles. Our results indicate that the x-ray diffraction patterns of extracellular vesicles and intact plasma membranes are distinct, consistent with our previous observation that extracellular vesiculation (i.e., shedding or exfoliation) is a directed process wherein specific regions of the membrane are released from the surface. Furthermore, we report that profound changes in electron density profile and overall membrane width occur following treatment with Liposin II. We hypothesize that alterations in membrane width may be responsible for the profound effect of fluidization on mBPA function.

METHODS AND RESULTS

Immunologic Characterization of the Negative Regulator.

To probe the relationship of the membrane-associated "negative" regulator of BFU-E proliferation to mBPA, a monoclonal antibody (D3-E4) that specifically recognizes mBPA (Dainiak et al, 1988) was employed. D3-E4 or control mouse anti-human IgG (0-200 ug) was incubated, 16 hr, 4°C, with 10 ug beta-D-octylglucoside (OG) extracts of intact plasma membranes or shed extracellular vesicles. An excess amount of 10% staphylococcal protein A solution was added, and antigen-antibody complexes were pelleted, as described previously (Dainiak et al, 1985a). As shown in Figure 1, inhibitory activity expressed by plasma membranes and shed vesicles was removed by immunoadsorption with D3-E4. These results suggest that the inhibitor and the growth factor mBPA share antigenic determinants and raise the possibility that they are identical molecules.

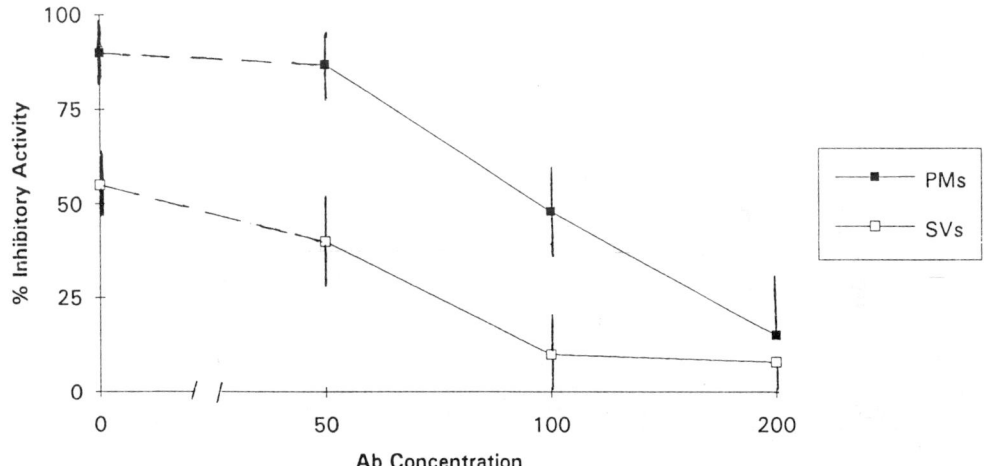

Figure 1: Immunoadsorption of "negative" regulator with D3-E4 from lipid treated B-cell OG extracts. Antibody was incubated at the indicated concentrations with OG-extracted plasma membranes (PMs) or shed vesicles (SVs). Dose-dependent removal of inhibitory activity is evident in the immunoadsorbed extracts. Cultures were established under serum-free conditions with recombinant human erythropoietin. Baseline colony formation (0% inhibition) was 80 ± 7 erythroid bursts/5×10^4 bone marrow mononuclear cells.

Effects of Lipid Treatment on Membrane Structure.

Plasma membrane order is determined by the free cholesterol:phospholipid ratio, relative saturation of phospholipid acyl chains and various experimental factors that include temperature, hydration (i.e., bulkiness) of phospholipid head groups, surface pressure and solute partitioning into the plasma membrane. Taking these factors into mind, we varied the free cholesterol:phospholipid ratio by incubating B-lymphocytes with known concentrations of free fatty acid, using carefully defined experimental conditions. Briefly, test and control plasma membranes and shed vesicles were centrifuged onto an aluminum foil substrate. Samples were partially dehydrated over a saturated salt solution to define relative humidity, and mounted onto a curved support. The multi-bilayer sample was then placed into a temperature-controlled water jacket of an Elliot GX-18 Rotating Anode X-ray Generator (ENRAF Nonius, NY). As shown in Figure 2, the x-ray beam was directed at discreet Bragg's angles so that angles of incidence defined by Bragg's law would occur simultaneously. The incident unreflected beam was adsorbed by the metal beam stop, while the Bragg reflections were collected on a one-dimensional, position-sensitive electronic detector or on x-ray film. Areas under the x-ray diffraction peaks were then calculated and a Fourrier analysis of the data was performed. Figure 2 depicts the conditions for these small angle x-ray diffraction experiments.

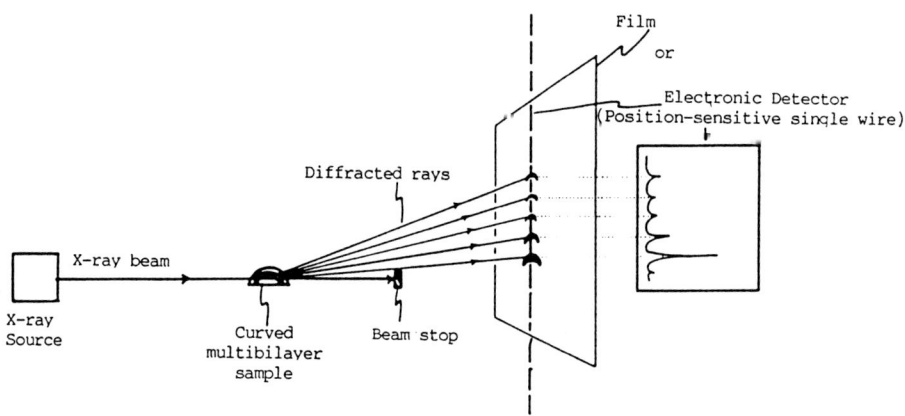

Figure 2: Small angle x-ray diffraction. Adapted from Mason, et al, 1990.

As shown in Figure 3, distinct diffraction patterns were obtained for plasma membranes and extracellular vesicles shed from normal human B-lymphocytes. Whereas a series of x-ray peaks was readily evident in native plasma membranes, a single peak in x-ray counts was observed for extracellular vesicles. These results indicate the presence of a population of shed vesicles that is highly enriched in protein (or relatively depleted in lipid). Biochemical analysis of the shed vesicles is underway. Overall, the finding of differences in lipid and protein composition is consistent with our previous hypothesis that exfoliation from the cell surface is an energy-dependent process wherein specific regions of the membrane are selectively released (Armstrong et al 1988; Dainiak and Sorba, 1991b).

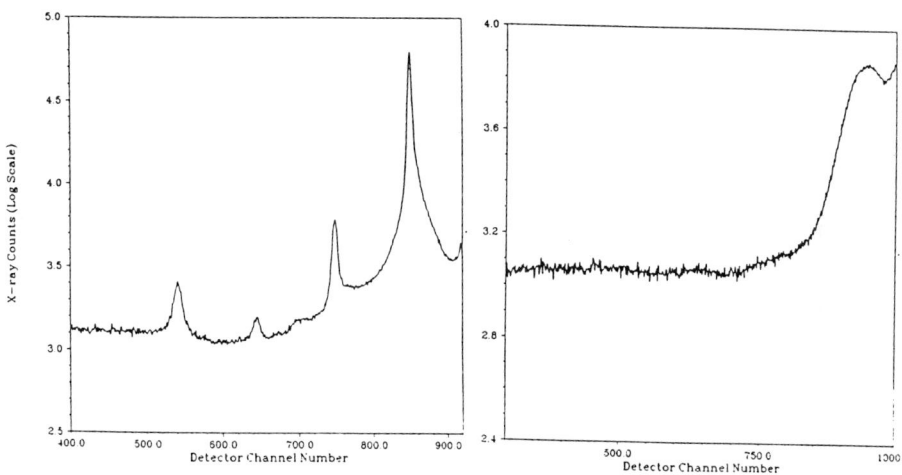

Figure 3: Diffraction patterns of intact plasma membranes **(left)** and extracellular vesicles **(right)** released from the surface of normal human B-cells.

In order to assess whether the plasma membrane structure of lipid-treated cells was different from that of untreated cells, electron density profiles of test and control cells were compared. Figure 4 shows that marked differences in the electron density profiles are readily evident, with widening of the peaks and troughs (consistent with a more disordered membrane), a decrease in the cholesterol "hump", and diminished electron density of terminal methyl groups at the center of the bilayer. In addition, an unexpected finding was outward displacement by approximately 4 Angstroms in the position of the phosphate head groups. The latter finding is consistent with the interaction of glycerol (representing approximately 3% WT/WT of Liposin II) with the phosphate head groups.

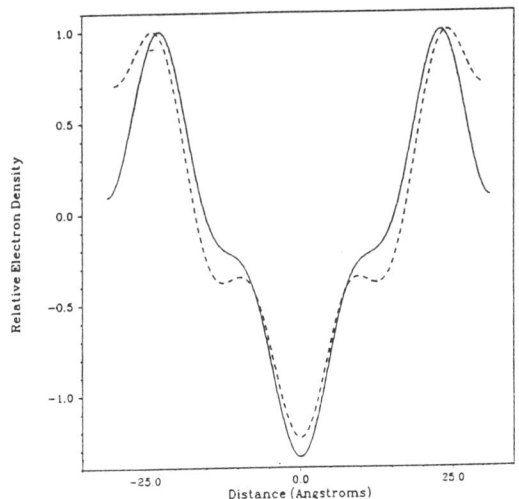

Figure 4: Electron density profiles of plasma membranes from normal and fluidized B-cells. Plasma membranes prepared from Liposin II-treated cells (dashed line) and control cells (solid line) were analyzed by small angle x-ray diffraction. Shown above are the electron density profiles resulting from Fourrier transformation and "phasing" of the data. Note that the overall width of the membrane (from peak to peak) is increased for test (relative to control) cells. In addition, the cholesterol "hump" is decreased on either side of the center of the bilayer, while the electron density in the center of the bilayer is distinctly abnormal.

DISCUSSION

The role of membrane-associated growth factors in hematopoietic cell differentiation and growth has recently been reviewed (Dainiak, 1991c). Stem cell factor (growth factor kit-ligand or mast cell growth factor), colony-stimulating factor-1, epidermal growth factor, tumor necrosis factor and CD23 are some examples of integral membrane proteins that have biological activity in vitro. mBPA is a heat stable ($56°C$, 10 min), 28,000 dalton glycoprotein that has been purified from plasma membranes and extracellular vesicles from human B-cells (Dainiak, et al, 1987, Feldman, et al, 1987). This factor partitions into the detergent (i.e., amphiphilic) phase after extraction of plasma membranes and shed vesicles with Triton X-114, followed by temperature elevation from $0°C$ to $37°C$ (Dainiak, 1990). Accordingly, mBPA is deeply embedded in the bilayer, raising the potential for extensive interactions of the protein with adjacent phospholipids.

Shedding of plasma membrane components from the cell surface is a directed process that is undergone by virtually all mammalian cell types (Beaudoin and Grondin, 1991; Dainiak, 1991b). Since mBPA is expressed at a concentration of approximately two-fold greater on shed vesicle vs. intact plasma membranes, we explored the potential of fluidization to accelerate exfoliation. However, while exfoliation was indeed augmented by lipid treatment, fluidized plasma membranes and shed extracellular vesicles suppressed rather than stimulated human erythroid burst formation (Dainiak et al, 1991a). Here, we demonstrate that the inhibitory activity shares antigenic determinants with mBPA. In addition, we explore the influence of fluidization on plasma membrane structure.

Our results indicate that the concentration of lipids in the ambient environment of mononuclear cells directly affects plasma membrane composition and structure. Profound alterations in x-ray diffraction patterns and electron density profiles are observed following fluidization of B-cells with free fatty acids. In addition, significant alterations in membrane width are also evident. Our results are consistent with the hypothesis that alterations in plasma membrane width may in turn, result in altered function of integral membrane proteins. This hypothesis is diagrammed in Figure 5.

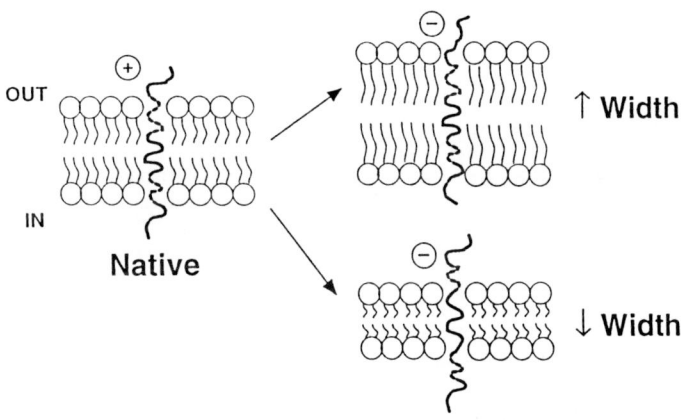

Figure 5: Protein-membrane interactions. Shown above is a schematic that illustrates how alteration in membrane width may lead to differential exposure of transmembrane proteins at either the external or internal surface of the plasma membrane. The schematic shows the conversion of a "positive" regulator of hematopoietic cell growth to a "negative" regulator by either increasing or decreasing plasma membrane width. This may result in alterations in the quartinary structure of the putative regulator, and lead to abnormal protein function.

Since transmembrane proteins traverse the plasma membrane with a specific number of alpha-helical turns (for example, 7 turns for rhodopsin, one of the most intensively studied membrane proteins; Dratz and Hargrave, 1983), each requiring 3.5 amino acids per turn with an average distance of 1.54 Angstroms between amino acids, and average total rise along the helical axis of 5.4 Angstroms (Branden and Tooze, 1991); the distance within the membrane through which a transmembrane protein must traverse may become critical with respect to the portion and amount of the molecule exposed at either the extracellular or cytoplasmic surface of the membrane. Interestingly, of all regions of integral membrane proteins, the transmembrane domain appears to be the most highly conserved throughout evolution. Accordingly, an increase (or decrease) in overall membrane width of 4 Angstroms is highly significant (i.e., nearly a full turn of an alpha-helix). Such a change may lead to alteration in the quartinary structure of the protein, the net result of which is altered protein function. Examples of altered function may include covering (or burying) of receptor proteins for growth factors as well as abrogation of the function of membrane bound growth factors.

In summary, our results raise potentially important questions with regard to the ambient environment in which mammalian cells are cultured. Since lipoproteins are transfered passively between the plasma membrane and ambient plasma lipid pools as a direct consequence of first-order kinetics (Phillips, et al, 1987; Johnson et al, 1988), varying the lipid composition of additives to the tissue culture mixture in which hematopoietic cells are maintained and/or are induced to differentiate, may have profound effects on the expression of membrane bound growth factors and/or their receptors. This variable may be particularly important when carefully defined culture conditions are employed wherein the influence of paracrine signalling among stromal and hematopoietic cells may be greatest (Dainiak et al, 1985; Peleraux and Eliason, 1989; Otsuka et al, 1991). Our results demonstrate that treatment of B-cells with phospholipids abrogates mBPA expression and results in elaboration of a "negative" regulator of erythropoiesis at the cell surface. The hypothesis that abnormal protein function may be the result of altered surface membrane structure is presented. Additional studies are required to confirm/deny this hypothesis.

REFERENCES

Armstrong MJ, Storch J and Dainiak N (1988): Structurally distinct plasma membrane regions give rise to extracellular membrane vesicles in normal and transformed lymphocytes. Biochim Biophys Acta 946:106-112.

Beaudoin AR and Grondin G (1991): Shedding of vesicular material from the cell surface of eukaryotic cells: different cellular phenomena. Biochim Biophys Acta 203-219.

Branden C, Tooze J (1991): Introduction to Protein Structure, Garland Publishing, Inc., p. 12.

Edidin M (1990): Molecular associations and membrane domains. In: Claudio T (ed.) Protein-Membrane Interactions, Vol 36, Current Topics in Memebranes and Transport, Academic Press, 81-96.

Dainiak N, Feldman L, and Cohen CM (1985a): Neutralization of erythroid burst-promoting activity in vitro with antimembrane antibodies. Blood 65:877-885.

Dainiak N, Kreczko S, Cohen A, Pannell R and Lawler J (1985b) Primary human marrow cultures for erythroid bursts in a serum-substituted system. Exp Hematol 13:1073-1079.

Dainiak N, Warren G, Sutter D, Kreczko S and Howard D (1988): A monoclonal antibody to exfoliated surface vesicles that recognizes a membrane-associated erythroid burst-promoting activity. Blood 72:989-994.

Dainiak N, Najman A, Kreczko S, Baillou C, Mier J, Feldman L, Gorin NC, and Duhamel G (1987): B-lymphocytes as a source of cell surface growth-promoting factors for hemopoietic progenitors. Exp Hematol 15:1086-1096.

Dainiak N (1990): Cell membrane family of growth regulatory factors. In: Dainiak N, Cronkite EP, McCaffrey R, Shadduck R (eds.) The Biology of Hematopoiesis, Prog Clin Biol Res 352, Wiley-Liss, Inc., p. 49-61.

Dainiak N, Guha A, Silva M, Sorba and Armstrong MJ (1991a): Expression of a negative regulator of human erythropoiesis by fluidized lymphocyte plasma membranes. Ann NY Acad Sci 628:212-221.

Dainiak N, and Sorba S (1991b): Intracellular regulation of the production and release of human erythroid-directed lymphokines. J Clin Invest 87:213-220.

Dainiak N (1991c): Surface membrane-associated regulation of cell assembly, differentiation and growth. Blood 78:264-276.

Darnell J, Lodish H, Baltimore D (1990): Molecular Cell Biology, 2nd Ed, Scientific American Books, Inc., p. 500.

Dratz EA, Hargrave PA (1993): The structure of rhodopsin and the rod outer segment disk membrane. Trends Biochem Sci 8:128-131.

Feldman L, Cohen CM, Riordan MA and Dainiak N (1987): Purification of a membrane-derived human erythroid growth factor. Proc Natl Acad Sci USA 84:6775-6779.

Gupte S, Wu E-S, Hoechli L, Hoechli M, Jacobson K, Sowers AE and Hackenbrock C (1984): Relationship between lateral diffusion, collision frequency, and electron transfer of mitochondrial inner membrane oxidation-reduction components. Proc Natl Acad Sci USA 81:2606-2610.

Johnson WJ, Mahlberg FH, Chacko GK, Phillips MC and Rothblat GH (1988): The influence of cellular and lipoprotein cholesterol contents on the flux of cholesterol between fibroblasts and high density lipoprotein. J Biol Chem 263:14099.

Mason PR, Moring J, Herbette LG (1990): A molecular model involving the membrane bilayer in the binding of lipid soluble drugs to their receptors in heart and brain. Nucl Med Biol 17:13-33.

Marsh D, (1985): In: Watts A, de Pont JJHHM (eds). Progress in Protein-Lipid Interactions, Vol 1, Elsevier, 143-172.

Otsuka T, Thacker D, Eaves CJ and Hogge DE (1991): Differential effects of microenvironmentally presented interleukin-3 versus soluble growth factor on primitive human hematopoietic cells. J Clin Invest 88:417-422.

Phillips MC, Johnson WJ, Rothblat GH (1987): Mechanisms and consequence of cellular cholesterol exchange and transfer. Biochim Biophys Acta 906:223.

Peleraux A, and Eliason JF (1980): Proliferation of single hemopoietic progenitor cells in the absence of colony-stimulating factors and serum. Exp Hematol 17:1032-1037.

Singer SJ, Nicolson GL (1972): The fluid mosaic model of the structure of cell membranes. Science 175:720-731.

Smith AD, Stubbs CD (1987): Modulation of membrane protein by bilayer lipids. Basic Res Cardiol 82:93-97.

Résumé

La membrane cytoplasmique est organisée en une mosaïque bidimensionnelle de phospholipides et de protéines. Des interactions entre les protéines et les lipides membranaires peuvent altérer les fonctions de chaque composant de la membrane. On a décrit récemment des facteurs de croissance hématopoiétiques liés à la membrane. Pour déterminer l'influence de la composition lipidique de la membrane sur l'expression du facteur de croissance érythroide, "erythroid-burst-promoting activité liée à la membrane (mBPA)", nous avons incubé des cellules spléniques humaines normales B avec une émulsion de lipides incluant de la Liposine II. Plutôt que de stimuler la formation de bursts érythroides de moelle humaine, les membranes plasmatiques isolées à partir des cellules "fluidifiées" (c.a.d traitées), entrainaient une suppression de la formation des bursts de façon dose-dépendante. L'activité inhibitrice a été immunoprécipitée par préincubation avec un anticorps monocloncal (D3-E4) qui reconnait de façon spécifique le mBPA. Pour déterminer si ces modifications dans l'expression du mBPA pouvait être expliquées sur la base d'une altération de la structure de la membrane, une étude par diffraction aux rayons X avec petit angle a été effectuée sur des membranes plasmatiques intactes et sur des vésicules extra-cellulaires produites à partir de cellules B normales ou "fluidifiées". Cette approche permet de déterminer la densité électronique de la matière dans la largeur de la membrane. Nous avons observé que les membranes plasmatiques et les vésicules produites avaient des images de diffraction aux rayons X différentes, corrélées avec des compositions en lipides/protéines différentes. Outre qu'ils démontrent une dilution du cholestérol et une diminution de la densité électronique dans la zone des groupements méthyles terminaux du centre de la double couche, les profils de densité électronique sont en faveur d'un déplacement vers l'extérieur des groupes phosphates. Ce dernier déplacement entraine une augmentation globale de l'épaisseur de la membrane plasmatique d'environ 4 Angstroms. Notre hypothèse est que l'altération de la membrane cytoplasmique pourrait affecter la fonction du mBPA. Cette hypothèse est en accord avec la notion que l'épaisseur de la membrane est une variable qui pourrait jouer un rôle dans l'expression des facteurs de croissance associés à la membrane et /ou de leurs récepteurs.

Apoptosis as a regulatory mechanism in the late stages of hematopoiesis

Mark J. Koury, Linda L. Kelley and Maurice C. Bondurant

Vanderbilt University and Veterans Administration, Medical Centers, Nashville, TN 37232, USA

CHARACTERISTICS OF APOPTOSIS

One physiological mechanism which plays a role in the regulation of hematopoiesis is apoptosis. Apoptosis is the name given to a process of cellular death which can be distinguished from necrosis, the other well recognized form of cell death. Acute severe injury such as that accompanying sudden anoxia, thermal injury, or chemical toxicity often causes necrosis. The cells within necrotic tissues are swollen with disrupted plasma membranes and indistinct cytological appearances. Apoptosis, on the other hand, is associated with specific morphological and biochemical events. For more detailed information about apoptosis and the biological events described in this section, the reader is referred to recent reviews (Lockshin and Zakeri, 1990; Arends and Wyllie, 1991; Kerr and Harmon, 1991). Cells undergoing apoptosis are smaller than their viable counterparts due to the combined effects of decreased cell water and loss of membrane-bound cytoplasmic blebs. The nuclei of apoptotic cells are homogeneously condensed and often fragmented. The plasma membrane remains intact, at least temporarily, and energy stores are not rapidly depleted. Thus, when examined with trypan blue dye the apoptotic cell appears to be viable since it excludes the dye. *In vitro*, the integrity of the plasma membrane is eventually lost and the cells then stain with trypan blue. *In vivo*, apoptotic cells are rapidly recognized by macrophages which ingest them. The recognition appears to be mediated through the surface display of specific molecules. The major biochemical event in apoptosis is double-stranded cleavage of nuclear DNA at internucleosomal sites. This cleavage results in a characteristic "ladder" pattern of DNA fragments, which are multiples of approximately 180 base pairs, following electrophoresis in neutral agarose gels.

Apoptosis is widespread and plays a major role in many developmental and physiological processes. Apoptosis has been long recognized in embryonic development. Morphogenesis by the apoptosis of specific cells or groups of cells has been identified in the formation of the nervous system, the reproductive system, the heart, and the limbs. In the development of the immune system in post-natal life, apoptosis provides the means for selection and differentiation of both T lymphocytes and B lymphocytes which do not have autoimmune activity. Some hormones and cytokines can directly induce apoptosis in their target cells. An extensively studied example is the glucocorticoid-mediated death of thymocytes. Another is the apoptosis induced in a variety of cell types by tumor necrosis factor-α. Other hormones and growth factors inhibit apoptosis and thereby function as trophic factors for their target tissues. Thus, apoptosis plays a role in the endometrial involution in normal menstruation, the reduction in prostate size following castration, and the adrenal atrophy indirectly mediated through reduced ACTH production in prolonged corticosteroid therapy.

APOPTOSIS IN HEMATOPOIESIS

Although the blood cells are turning over constantly, the production of new cells to maintain the normal blood cell count must change rapidly in response to specific stimuli. These changes occur in specific types of cells and when the stimulus to produce more cells is removed, the production rate must return to normal in a prompt but well-controlled manner. In bacterial infections granulocyte numbers in the blood increase while the erythrocyte and platelet numbers remain stable. Conversely, when erythrocytes are being repleted after blood loss their numbers increase while granulocyte and platelet counts remain in the normal range. Since all of the cell types in the blood are derived from pluripotent stem cells, any hematopoietic growth factor that regulates the numbers of one specific type of blood cell must act on lineage-committed progenitor cells. Among the lineage-specific hematopoietic growth factors many have been shown to maintain the viability of their target cells. For several of these factors, the mode of action has been shown to be the prevention of apoptosis. Specific examples include the action of GM-CSF on granulocytic-macrophage progenitors (Williams et al., 1990) interleukin-5 on eosinophils (Yamaguchi et al., 1991), and erythropoietin (EPO) on erythroid progenitor cells (Koury and Bondurant, 1990). Of these hematopoietic factors, EPO has several advantages for understanding how prevention of apoptosis regulates blood cell production. EPO's sites of production are confined to specific types of cells in the kidneys (Koury et al., 1988; Lacombe et al., 1988) and liver (Koury et al., 1991) while its target cells appear to be restricted to erythroid progenitor cells. The physiological stimulus that induces the production of EPO is tissue hypoxia which, in the large majority of cases, is a function of the number of circulating erythrocytes (Erslev, 1991). Furthermore, as a hormone, concentrations of EPO in the serum can be easily determined. These factors have permitted studies in mice that led to the development of a model of erythrocyte production based upon apoptosis of erythroid progenitor cells and its prevention by EPO (Koury and Bondurant, 1990, 1992).

ERYTHROPOIETIN AND APOPTOSIS

When mice are made anemic by bleeding, the EPO-producing cells in the kidneys sense the resultant hypoxia, and they are induced to make and secrete EPO. These cells produce EPO in an all-or-none fashion with more cells recruited to produce EPO when the anemia is more severe (Koury et al., 1989). The number of renal cells producing EPO and the resultant serum concentration of EPO are inversely and exponentially related to the hematocrit. Thus, in the range of significantly low hematocrits a small decrease in hematocrit results in a large increase in serum EPO concentration. When the hematocrits of mice or humans are reduced to less than 20 percent, the serum EPO concentration increases more than 100 times the normal level. This wide range of EPO concentrations between normal and severely anemic states allows the fine control of erythrocyte production that results in prompt reticulocytosis and erythrocytic repletion without overshoot polycythemia.

The control of erythrocyte production by EPO is mediated through the ability of EPO to prevent apoptosis in erythroid progenitor cells. Our studies using proerythroblasts isolated from mice infected with the "anemia-inducing" strain of Friend leukemia virus (FVA cells) (Koury and Bondurant, 1990; Kelley et al., 1992, 1993) and other studies using either purified, regenerating CFU-E (Boyer et al., 1992) or fetal liver erythroid progenitors (Yu et al., 1993) have demonstrated that EPO-deprivation *in vitro* leads to apoptosis of these erythroid cells. This dependence upon EPO to prevent apoptosis begins at least by the CFU-E stage of erythroid differentiation (Landschulz et al., 1992) and lasts until just before the erythroblast begins hemoglobin synthesis (Kelley et al., 1993). In this EPO-dependent period, an individual erythroid progenitor cell dies when the EPO concentration is insufficient to prevent apoptosis. If, however, the EPO concentration is sufficient to prevent apoptosis, then that progenitor cell lives and continues to differentiate. Although terminal erythroid differentiation appears morphologically similar to apoptosis at the light microscopic level, these two processes are distinctly different (Kelley et al., 1993). The loss in cell size in EPO-deprivation induced apoptosis occurs rapidly and before the more gradual decrease in size that characterizes terminal erythroid differentiation. The nucleus in apoptotic erythroid progenitor cells is uniformly condensed and has no envelope while the nuclear condensation in terminal differentiation preserves areas of euchromatin and heterochromatin as well as the nuclear envelope.

HETEROGENEITY OF PROGENITOR CELLS

The proposed model of erythropoiesis based on modulation of apoptosis by EPO (Koury and Bondurant, 1990, 1992) requires that populations of erythroid progenitors are heterogeneous in EPO requirement. Although apoptosis is an all-or-none event for each individual erythroid progenitor cell, the finely controlled graded responses that occur in repletion of lost erythrocytes indicate that the erythroid progenitor cell population does not live or die as a group. Rather, a wide range of responses must exist within the population. Indeed, a continuum of EPO responsiveness must extend from erythroid progenitor cells which only require the normal serum EPO concentration or less for survival through those progenitors which survive only when EPO concentration reaches the high levels found in severe anemia. Support for this idea of heterogeneity of individual cells in EPO requirement existed in earlier studies although the evidence was indirect. Early studies of CFU-E growth *in vitro* demonstrated that a dose-response relationship between EPO and colony development encompassed the wide range of EPO concentrations found in normal and anemic individuals (Eaves et al., 1979). More recently using recombinant, purified human EPO and FVA cells, an even wider range of dose-responsivity has been found for the DNA cleavage associated with apoptosis (Kelley et al., 1993). This result and other new direct evidence presented below constitute a strong argument that, by some unknown cellular mechanism, the amount of EPO required to prevent apoptosis can vary more than 100-fold in a population of erythroid progenitor cells that are otherwise similar.

FVA cells can be used to isolate subpopulations of progenitors with differing EPO requirements (Kelley et al., 1993). Such subpopulations should be useful in elucidating the biochemical basis for the differing requirement. The reason that such separations are possible is that there is a correlation between the length of time that individual FVA cells can survive in culture without added EPO and the EPO dose requirement of those cells. That is, those cells with the greatest intrinsic sensitivity to EPO appear to live longest in cultures without added EPO. Kelley et al. (1993) isolated a non-apoptotic, viable subpopulation of FVA cells by density gradient centrifugation of cells previously cultured for 20 h without EPO. In experiments in which they were recultured with or without EPO, this EPO-deprived subpopulation was compared to an isodense subpopulation isolated from freshly procured, previously uncultured FVA cells. The EPO-deprived subpopulation survived and completed differentiation with one-tenth of the EPO required by the previously uncultured subpopulation (Kelley et al., 1993). However, when the EPO-deprived subpopulation was recultured without EPO all of the cells underwent apoptosis within 24 hours. Thus, the EPO-deprived subpopulation had greatly increased EPO sensitivity, but it was not totally independent of EPO. The EPO-deprived subpopulation reached its maximal response within 24 hours of reculture with EPO. This time was significantly shorter than for freshly isolated FVA cells. Also, this subpopulation only doubled its cell number with reculture, but it still achieved the same differentiation (as measured by ^{59}Fe incorporation into heme per cell) as the previously uncultured cells (Kelley et al., 1993). Thus, the EPO-deprived subpopulation had continued their differentiation during the 20 hours of EPO deprivation, indicating that they had the same differentiation time course and were at the same differentiation stage as the general population of FVA cells from which they were originally selected.

In retrospect, several previous results had indicated that heterogeneity existed in FVA cells such that individual cells have a wide range of time over which they can remain cultured without EPO and not undergo apoptosis. The longer the delay in the addition of EPO to cultures of FVA cells (Koury and Bondurant, 1988) the more decreased the cellular proliferation and differentiation are. The same phenomenon was observed for regenerating CFU-E (Landschulz et al., 1989). With FVA cells, a minor percentage was known to survive 24 h in EPO-deprived cultures and still respond by proliferating and differentiating into reticulocytes in the same time that they would have had they received EPO at the initiation of culture (Koury and Bondurant, 1988). Finally, when EPO-deprived FVA cells are examined for apoptosis, the percentage of apoptotic cells accumulates with time in culture (Kelley et al., 1993). It was only later appreciated that the cells which survived for extended periods in culture without EPO had much higher than average EPO sensitivity.

BASIS OF PROGENITOR HETEROGENEITY

How does such a great variability in sensitivity to a growth factor develop in a target cell population that appears to be otherwise similar? In the case of FVA cells and EPO, and presumably other types of committed hematopoietic progenitors and their lineage-specific growth factors, several possible sources of this variability exist. The affinities or numbers of growth factor receptors may be different for one cell than for another. In this respect, the EPO-deprived FVA cell subpopulation and the previously uncultured FVA cells have similar numbers of EPO receptors with similar affinities (Kelley et al., 1993). Similarly, other components of the receptor complex which do not directly bind the growth factor but which may mediate signal transduction may vary from one cell to another. Many of the receptors for the hematopoietic growth factors have such secondary components (Taga and Kishimoto, 1992) but none has yet been identified in the erythropoietin receptor. A possible secondary component in the special case of FVA cells is the viral gp55 envelope protein which has been shown to interact with the EPO receptor (Li et al., 1990). However, when gp55 mRNA and protein levels were examined in the EPO-deprived subpopulation described above and the previously uncultured FVA cells, they were the same (Kelley et al., 1993). A third possibility is that the signal transduction mechanism that leads to the prevention of apoptosis may have varying input requirements before it is triggered.

HEMATOPOIETIC REGULATION BY APOPTOSIS

A model of regulation of erythrocyte production based upon apoptosis and its prevention by EPO in late stage erythroid progenitor cells has been previously proposed (Koury and Bondurant, 1990, 1992). The extension of this model to other hematopoietic lineages and their respective lineage-specific growth factors has also been proposed (Koury, 1992). Although apoptosis in this model is due to insufficient concentrations of a "positive" hematopoietic factor, increased amounts of a lineage-specific "negative" factor could have the same effect. In the model based upon EPO, the normal individual produces low levels of EPO which maintain the slight daily turnover of erythrocytes. In the normal state, large numbers erythroid progenitor cells are lost to apoptosis. However, when the individual becomes anemic, the physiological increase in EPO concentrations prevent the apoptosis of many of these erythroid progenitor cells that would have been eliminated by apoptosis in normal concentrations of EPO. As the blood erythrocytes are repleted, the hematocrit and the attendant tissue oxygen increase, EPO levels decline and the incidence of apoptosis increases toward the normal level. With further understanding about the physiological control of other hematopoietic growth factors, the extension of this model beyond erythropoiesis to other areas of hematopoiesis may be made.

ACKNOWLEDGEMENTS

The authors are supported by grants from the National Institutes of Health, the Veteran's Administration and the National Blood Foundation.

REFERENCES

Arends, M.J., and Wyllie, A.H. (1991) Apoptosis: Mechanisms and roles in pathology. Int. Rev. Exp. Pathol., *32*:223-255.

Boyer, S.H., Bishop, T.R., Rogers, O., Noyes, A.N., Frelin, L.P., and Hobbs, S. (1992) Roles of erythropoietin, insulin-like growth factor 1, and unidentified serum factors in promoting maturation of purified murine erythroid colony-forming units. Blood, *80*:2503-2512.

Eaves, C.J., Humphries, R.K., and Eaves, A.C. (1979) In vitro characterization of erythroid precursor cells and the erythropoietic differentiation process. In Cellular and Molecular Regulation of Hemoglobin Switching. G. Stamatoyannopoulos, and A.W. Nienhuis, eds. Grune and Stratton, New York, pp. 251-273.

Erslev, A.J. (1991) Erythropoietin. N. Engl. J. Med., *324*:1339-1344.

Kelley, L.L., Koury, M.J., and Bondurant, M.C. (1992) Regulation of programmed death in erythroid progenitor cells by erythropoietin: Effects of calcium and of protein and RNA syntheses. J. Cell. Physiol., *151*:487-496.

Kelley, L.L., Koury, M.J., Bondurant, M.C., Koury, S.T., Sawyer, S.T., and Wickrema, A. (1993) Survival or death of individual proerythroblasts results from differing erythropoietin sensitivities: A mechanism for controlled rates of erythrocyte production. Blood, in press.

Kerr, J.F.R., and Harmon, B.V. (1991) Definition and incidence of apoptosis: An historical perspective. In: Apoptosis: The Molecular Basis of Cell Death. L.O. Tomei, and F.O. Cope, eds. Cold Spring Harbor Laboratory Press, New York, pp. 5-29.

Koury, M.J., and Bondurant, M.C. (1988) The maintenance by erythropoietin of viability, proliferation and maturation of murine erythroid precursor cells. J. Cell. Physiol., *137*:65-74.

Koury, M.J., and Bondurant, M.C. (1990) Erythropoietin retards DNA breakdown and prevents programmed death in erythroid progenitor cells. Science, *248*:378-381.

Koury, M.J., and Bondurant, M.C. (1992) The molecular mechanism of erythropoietin action. Eur. J. Biochem., *210*:649-663.

Koury, S.T., Bondurant, M.C., and Koury, M.J. (1988) Localization of erythropoietin synthesizing cells in murine kidneys by *in situ* hybridization. Blood *71*:524-527.

Koury, S.T., Koury, M.J., Bondurant, M.C., Caro, J., and Graber, S.E. (1989) Quantitation of erythropoietin producing cells in kidneys of mice by *in situ* hybridization: Correlation with hematocrit, renal erythropoietin mRNA and serum erythropoietin concentration. Blood, *74*:645-651.

Koury, S.T., Bondurant, M.C., Koury, M.J., and Semenza, G.L. (1991) Localization of cells producing erythropoietin in murine liver by *in situ* hybridization. Blood, *77*:2497-2503.

Lacombe, C., DaSilva, J.L., Bruneval, P., Fournier, J.G., Wendling, F., Casadevall, N., Camilleri, J.P., Bariety, J., Varet, B., and Tambourin, P. (1988) Peritubular cells are the site of erythropoietin synthesis in the murine hypoxic kidney. J. Clin. Invest., *81*:620-623.

Landschulz, K.T., Noyes, A.N., Rogers, O., and Boyer, S.H. (1989) Erythropoietin receptors on murine erythroid colony-forming units: Natural history. Blood, *73*:1476-1486.

Landschulz, K.T., Boyer, S.H., Noyes, A.N., Rogers, O.C., and Frelin, L.P. (1992) Onset of erythropoietin response in murine erythroid colony-forming units: Assignment to early S-phase in a specific cell generation. Blood, *79*:2749-2758.

Li, J.-P., D'Andrea, A., Lodish, H., and Baltimore, D. (1990) Activation of cell growth by binding of Friend spleen focus-forming virus gp55 glycoprotein to the erythropoietin receptor. Nature, *343*:762-674.

Lockshin, R.A., and Zakeri, Z.F. (1990) Programmed cell death: New thoughts and relevance to aging. J. Gerontol., *45*:B135-140.

Taga, T., and Kishimoto, T. (1992) Cytokine receptors and signal transduction. FASEB J., *6*:3387-3396.

Williams, G.T., Smith, C.A., Spooncer, E., Dexter, T.M., and Taylor, D.R. (1990) Haemopoietic colony stimulating factors promote cell survival by suppressing apoptosis. Nature, *343*:76-79.

Yamaguchi, Y., Suda, T., Ohta, S., Tominaga, K., Miura, Y., and Kasahara, T. (1991) Analysis of the survival of mature human eosinophils: Interleukin-5 prevents apoptosis in mature human eosinophils. Blood, *78*:2542-2547.

Yu, H., Bauer, G., Lipke, G.K., Phillips, R.L., and VanZant, G. (1993) Apoptosis and hematopoiesis in murine fetal liver. Blood, *81*:373-384.

Résumé

L'apoptose, une forme de mort cellulaire programmée, joue un rôle important dans de nombreux processus physiologiques et pathologiques, incluant l'hématopoièse. Des expériences réalisées sur les progéniteurs érythroides et l'érythropoiétine ont fourni la base d'un modèle dans lequel la production des cellules sanguines est controlée par l'apoptose des progéniteurs aux stades tardifs de leur différenciation. Les contraintes de ce modèle sont 1) une période pendant la différenciation de progéniteurs hématopoiétiques spécifiques d'une lignée au cours de laquelle ils deviennent dépendants d'un facteur de croissance hématopoiétique pour empêcher leur apoptose; 2) une hétérogénéité significative dans la sensibilité individuelle des progéniteurs au facteur de croissance. Plusieurs facteurs de croissance spécifiques d'une lignée empêchent l'apoptose des cellules cibles concernées. Nous avons montré l'existence d'une hétérogénéité de la sensibilité à l'érythropoiétine des progéniteurs érythroides individuels au sein d'une population érythropoiétine-dépendante. Des résultats préliminaires montrent que ces différences de sensibilité à l'érythropoiétine sont liées plutôt aux mécanismes de signaux de transduction intracellulaires qu'aux récepteurs à l'érythropoiétine.

Summary

Apoptosis, a form of programmed cell death, plays an important role in many physiological and pathological processes, including hematopoiesis. Data from experiments with erythroid progenitor cells and the hormone erythropoietin provide the basis for a model in which blood cell production is controlled by apoptosis of hematopoietic progenitor cells in the late stages of their differentiation. Two requirements for this model are: 1) a period during the differentiation of lineage-specific hematopoietic progenitor cells in which they become dependent upon a hematopoietic growth factor to prevent their apoptosis; 2) a significant heterogeneity in the sensitivity of individual progenitor cells to the hematopoietic growth factor. Several lineage-specific hematopoietic growth factors prevent apoptosis in their respective hematopoietic target cells. Evidence is presented for heterogeneity in the erythropoietin sensitivity of individual erythroid progenitor cells within a erythropoietin-dependent population. Preliminary results indicate that these differences in erythropoietin sensitivity reside in the intracellular signal transduction mechanism rather than in the receptors for erythropoietin.

Structural characterization of recombinant LD78 the human Stem Cell Inhibitor protein

S. Craig, S. Patel, D. Brotherton, S. Evans, L. Czaplewski, N. Woods, L. Howard, R. Gilbert, J. Fisher* and P. Morgan**

*British Bio-Technology, Cowley, Oxford, OX4 5LY; *School of Chemistry, University of Leeds, LS2 9JT; **National Centre for Macromolecular Hydrodynamics, Nottingham University, LE12 5RD, UK*

Recombinant human Stem Cell Inhibitor (SCI) protein LD78 has been purified from a synthetic gene expressed in Saccharomyces cerevisiae. At physiological ionic strength, this 7.8 kDa protein exists in solution as large, soluble, heterogeneous multimers with molecular weight ranging from 8 - 1000 kDa when analyzed by size exclusion chromatography and analytical ultracentrifugation. In conditions of high ionic strength or low pH, the protein is observed to exist only as a tetrameric complex of mass 32 kDa. Ionic interactions between charged side-chains, therefore, stabilize the high molecular weight complexes. In the acidic, non-polar solvent 30% acetonitrile, 0.1% TFA, LD78 exists as a homogeneous monomer. N.m.r. and circular dichroism (c.d.) studies demonstrate that the monomer is folded with defined tertiary structure under these solution conditions. Hydrophobic interactions, therefore, provide a major stabilizing force for the formation of LD78 tetramers. Spectroscopic analysis of wild type and single site variants of LD78 demonstrate that the interaction of Tyr-61 with another aromatic residue (probably Trp-57) provides a major contribution to the near u.v. c.d. spectrum. Fluorescence emission and c.d. spectroscopic analyses also reveal that the local environment of the single tryptophan residue (-57) is highly sensitive to the multimeric state of the molecule (Patel et al, 1993). The nature of the observed spectroscopic changes are entirely consistent with the formation of ionic interactions proximal to the tryptophan during the multimerisation process. The specific environment of aromatic residues (in particular) Trp-57, therefore, provide a sensitive probe for the quaternary structure of the human SCI in solution.

REFERENCES

Patel, S.R., Evans, S., Dunne, K., Knight, G.,Morgan, P.J., Varley, P.G. and Craig, S. (1993): Characterization of the quaternary structure and conformational properties of the human Stem Cell Inhibitor protein LD78 in solution. *Biochemistry*. In press.

Characterisation of granulopoietic inhibitory activity

E.C. Stevenson, A.E. Irvine, T.C.M. Morris and G.J. Graham*

*Department of Haematology, The Queen's University of Belfast, BT12 6BA, N. Ireland; *The Beatson Institute for Cancer Research, Glasgow, UK*

We have previously shown that normal unstimulated lymphocytes co-cultured *in vitro* with normal human bone marrow will inhibit the formation of 7d CFU-C (Irvine et al., 1991). Furthermore we have shown that there is a molecular basis for this inhibition and that the granulopoietic inhibitory activity (GIA) is a glycoprotein with a molecular weight greater than 100 000 daltons. Neutralising antibodies to α and γ interferon, TNF and acidic and basic isoferritins do not diminish its action.

In our original reports we prepared GIA using RPMI-10% FCS. GIA produced in the following serum-depleted media consistently showed inhibition in the myeloid colony assay (Pike and Robinson, 1970), although at a reduced level:- HL-1, Optimem, and Biorich (Table 1). We established the conditions to produce GIA in serum-free media which allowed us to further characterise the activity.

GIA was previously demonstrated to inhibit human bone marrow 7d CFU-C formation. We have now shown that it also acts on an earlier progenitor cell using the CFU-A assay (Pragnell et al., 1988). Bone marrow from phenylhydrazine treated mice was cultured with L-929 and AF1-19T cell conditioned media. GIA produced a dose-related inhibition of colony formation at 11d (Fig. 1). Further ELISA characterisation showed that this activity was not due to LD78, the human homologue of MIP 1α.

GIA ASSAY OF SERUM-FREE CULTURES

MEDIUM	SERUM CONTENT	NUMBER OF GIA BATCHES TESTED	MEAN % INHIBITION + S.D.
RPMI	+10% FCS	9	63 + 12
OPTIMEM	0	7	52 + 10
HL - 1	0	7	47 + 11
BIORICH	0	6	43 + 20

TABLE 1

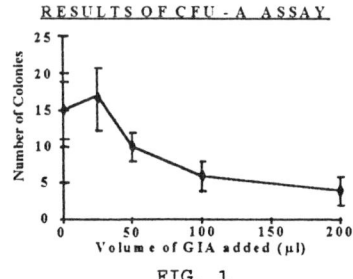

RESULTS OF CFU-A ASSAY

FIG. 1

References

Irvine, A.E., French, M.A., Bridges, J.M., Crockard, A.D., Desai, Z.R. and Morris, T.C.M. (1991), Exp. Hematol., 19, 106-109.

Pike, B.L. and Robinson, W.A. (1970), J. Cell Physiol., 76, 77-84.

Pragnell, I.B., Wright, E.G., Lorimore, S.A., Adam, J., Rosendaal, M., DeLarter, J.F., Freshney,M., Eckmann,L., Sproul,A. and Wilkie,N. (1988), Blood, 72, 196-201.

Negative regulation of megakaryocytopoiesis

Zhong Chao Han, Isabelle Lebeurier and Jacques Philippe Caen

Institut des Vaisseaux et du Sang, Hôpital Lariboisière, 8, rue Guy-Pantin, 75010 Paris, France

Abstract   Megakaryocytopoiesis is negatively regulated mainly by megakaryocyte-platelet derived inhibitors including TGFß, platelet factor 4 and its related molecules. Other factors such as interferon-α and -γ, thrombin and prothrombin also inhibit megakaryocytopoiesis. The activity of most inhibitors is negatively modified by heparin and other glycosaminoglycans. A model of negative regulation of megakaryocytopoiesis is proposed on the basis of these observations

Megakaryocytopoiesis is a complex cellular and biological process which can broadly be divided into several stages by means of various relevant assays (Table 1). Normal megakaryocytopoiesis, like granulopoiesis and erythropoiesis, is controlled by both positive and negative regulators. The positive regulators stimulate the growth of megakaryocyte precursors and ultimately result in cell death through apoptosis process. The negative regulators maintain megakaryocyte-platelet lineage at constant size by inhibiting cell growth and prevent both the loss of megakaryocyte precursors through apoptosis and the disorders caused by the increase of number of platelets. It has been shown that almost no inhibitor of hematopoiesis is restricted to act only on one hematopoietic lineage and most described inhibitors of hematopoiesis also play a role in megakaryocytopoiesis. However, the negative regulation of megakaryocytopoiesis, unlike that of granulopoiesis or erythropoiesis, is carried out mainly by the inhibitors produced by megakaryocyte-platelets themselves, a regulation fashion called negative feedback. The inhibitors produced by megakaryocyte-platelets act predominantly on megakaryocytic lineage but also on the growth of multipotent (mCFU-MK) and committed granulocytic (CFU-GM) and erythroid (BFU-E) progenitors (Hoffman, 1989; Han et al, 1991b). These observations are of particular interest since the inhibitors, in addition to their general significance in the understanding of the regulation of hematopoiesis, may also have profound clinical implications.

Megakaryocyte derived inhibitors of megakaryocytopoiesis

Megakaryocytes produce several factors capable of inhibiting hematopoiesis, including transforming growth factor ß (TGFß), platelet factor 4 (PF4), ß-thromboglobulin (ßTG) and its precursor, the connective tissue-activating peptide III (CTAP-III). The TGF-ß is produced not only by megakaryocytes (Fava et al, 1990) but also by various other cell types (Moses & Yang, 1990), whereas PF4, ßTG and CTAP-III are specific for megakaryocytic lineage (Holt et al, 1986).

TGFß

TGFßs are multifunctional cytokines and have five isoforms designated TGFß1 to TGFß5. Much more information is available for TGFß1 because it was the first isoform isolated and studied and is present in large amounts in platelets. TGFß1 is known as a growth-stimulating factor for mesenchymal cells and matrix formation and a growth inhibitor for normal hematopoietic progenitor cells and leukemic cells (Moses & Yang, 1990).

TGF-ß1 is, to our knowledge, the most potent inhibitor of megakaryocytopoiesis. It acts on normal megakaryocytopoiesis, probably erythropoiesis, at picomol concentrations but on granulopoiesis at nanomol levels. It inhibits the proliferation of megakaryocyte progenitors at different stages and also the endomitosis of megakaryocytes (Ishibashi 1987; Kuter et al, 1992; Han et al 1992a). Interestingly, TGFß1 inhibits mainly the colony formation induced by interleukin-3 (IL3) (Keller et al, 1989; 1991; Han et al, 1992) and by aplastic anemia serum. The effect of other factors such as GM-CSF and IL6 on megakaryocyte growth is hardly modified by TGFß1 (Fig. 1). These observations are paralleled with several recent studies that TGF-ß1 enhanced granulocyte and macrophage proliferation in the presence GM-CSF and M-CSF (Keller et al, 1991; Fan et al, 1992).

Table 1. Developmental process of megakaryocytic cells and relevant assays

Cell compartment	Main relevant assay
Stem cells - HPP-mCFU-MK	Bone marrow culture
Multipotent progenitor - mCFU-MK or CFU-GEMM	Bone marrow & blood culture
Committed progenitors - BFU-MK - CFU-MK	Bone marrow & blood culture Bone marrow & blood culture
Megakaryoblasts & megakaryocytes	Morphology Cell diameter & ploidy detection Purification by various methods Glycoprotein analysis Specific enzyme histochemistry Proplatelet formation Incorporation of thymidine & amino acids
Platelets	Number & volume Function Ultrastructure Specific protein analysis

The in vivo studies of the effect of TGF-ß1 on haematopoiesis in mice have further suggested that TGFß1 acts on hematopoiesis either as an inhibitor or promotor. TGF-ß1, when injected locoregionally to the bone marrow, inhibited the proliferation of early haematopoietic progenitors (Goey et al, 1989). Subcutaneous daily injection of TGF-ß1 for 14 days resulted in a decrease in the count of platelets and red blood cells, and an increase in the number of white blood cells and granulopoieisis in the spleen and the bone marrow (Carlino et al, 1990). Furthermore, Bursuker et al (1992) showed that injection of either natural or recombinant TGF-ß1 into mice caused an increase in the number of progenitors in the bone marrow that gave rise to granulocytes and macrophages in response to M-CSF and GM-CSF. These contradictory results indicate the complexity of the in vivo activity of TGFß1. As TGFß has multiactivities, what is its physiological function in bone marrow and hematopoiesis? As TGFß is produced in latent forms, what is the in vivo mechanism of its activation? Will all isoforms of TGFß be inhibitor of megakaryocytopoiesis in vitro and in vivo? These questions remain to be answered in the future.

PF4 and related peptides PF4 and related peptides (ßTG and CTAP-III) are synthesized in the golgisomes of megakaryocytes, transfered and stored in alpha granules and released after cell activation. PF4 was identified and purified on the basis of its role in the related processes of coagulation (Hermoson et al, 1977). PF4 is now recognized as having a variety of biological activities including heparin-binding, immunoregulatory, chemotactic, anticollagenase and anti-angiogenic effects (Zucker & Katz, 1991).

PF4, ßTG and CTAP-III inhibit in vitro human megakaryocytopoiesis (Gewirtz et al, 1989; Han et al, 1990a,1990b,1990c) and might be useful agents for suppression of the growth of megakaryocyte progenitor cells (CFU-MK) of essential thrombocythemia (Han et al, 1990b). In addition, PF4 and ßTG have been shown to be capable of inhibiting the growth of a megakaryocytic cell line, the human erythroleukemia cells (Han et al, 1992b).

Based on the ability of PF4 to inhibit megakaryocyte colony formation in vitro, it was anticipated that PF4 would be a negative regulator of megakaryocyte and platelet production in vivo. The in vivo effect of human PF4 on murine megakaryocytopoiesis and thrombopoiesis was studied subsequently by our laboratory. Administration twice daily for 4 days of PF4 (0.1-1 µg/mouse) into 8 week-old Balb/c mice induced a dose-dependent decrease in the numbers of megakaryocytes and their progenitors (CFU-MK), continuing for 1 week after injection. However, the size of megakaryocytes and their colonies was not changed following PF4 injection. Platelet counts were significantly reduced at day 3-4. The number of CFU-GM was decreased at day 1-2. White blood cells and hemoglobin were unaffected by PF4. These results indicate that PF4 inhibits megakaryocyte and platelet production in vivo by acting on the early stage of megakaryocyte development (Han ZC, 1991a). Similar results were also obtained when ßTG was tested in vitro and in vivo in mice (Data not shown).

Because PF4, ßTG and CTAP-III share amino acid sequence homology, we hypothesized that their homologous region might be involved in their inhibitory activity. Several synthetic peptides corresponding to the homologous region of PF4 and related peptides were therefore studied for their effects on megakaryocytopoiesis in vitro and in vivo. A dodecapeptide, -Asn-Gly-Arg-Lys-Ile-Cys-Leu-Asp-Leu-Ala-Pro-, was found to be a potent inhibitor of megakaryocytopoiesis in vitro and in vivo in mice. This dodecapeptide also is able to inhibit the growth of human megakaryocytic-like cells (HEL, Dami and Meg-01). When this peptide was prepared in monomeric form or its cystein residue was replaced with arginine, it was no longer active on megakaryocytopoiesis (Results are shown in the separate Abstract presented by our laboratory).

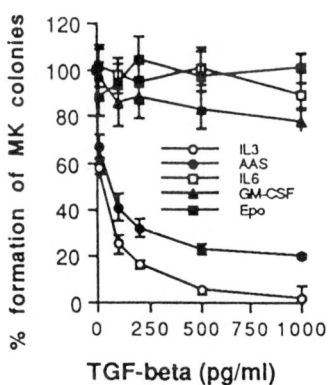

Fig.1. Effect of TGFß1 on the formation of murine megakaryocyte (MK) colonies stimulated by pig aplastic anemia serum (AAS; 10%, v/v)) or various growth factors including interleukin (IL)-3 (100 u/ml), IL-6 (20 ng/ml), erythropoietin (Epo, 1 u/ml)) and GM-CSF (5 ng/ml). Data represent the mean ± SEM of quadruplicate determinations from three separate experiments. The 100% (control) corresponds to 132 ± 8, 78 ± 7, 26 ± 3, 28 ± 2 and 11 ± 2 MK colonies/ 2 X 10^5 cells for cultures stimulated by AAS, IL3, IL6, GM-CSF and Epo, respectively.

Other inhibitors of megakaryocytopoiesis

Megakaryocytopoiesis is negatively regulated not only by megakaryocytes/platelets themselves, but also by the accessary cell population and some serum factors.

Interferon   The experimental studies have shown that alpha and gamma interferons are able to inhibit the formation of megakaryocyte colonies (Ganser et al, 1987; Han et al, 1987). Clinical studies in essential thrombocythemia have further confirmed the role of alpha interferon in suppressing megakaryocytopoiesis in vivo (Wadenvik et al, 1991).

Thrombin & prothrombin

Platelet formation from megakaryocytes is a unique process through which anucleate platelets are formed by fragmentation of megakaryocyte cytoplasma. This process is still incompletely understood, due in part to the difficulty to study it in vivo. The interaction of megakaryocytes and extracellular matrix (ECM), especially megakaryocyte vitronectin receptor and its action with vitronectin in ECM, seems to promote the proplatelet formation. The tetrapeptide Arg-Gly-Asp-Ser (RGDS) and a monoclonal antibody against vitronectin receptor blocked proplatelet formation (Leven & Tablin, 1992). The megakaryocyte proplatelet formation process in vitro is inhibited by serum prothrombin. Thrombin directly prevents the formation of new processes and also induces retraction of existing processes. (Hunt et al, 1993).

It has also been shown that thrombin at a concentration of 7.5 u/ml can cause a reduction in the proliferation of Meg-01 cells, without affecting their differentiation stage as determined by the expression of platelet glycoproteins GPIIb/IIIa and GPIb, FVIII-related-antigen and cell size measurement (Vittet et al, 1992).

Other factors

Sonda et al. (1993) have recently reported that interleukin-4 (IL4) strongly inhibited pure and mixed megakaryocyte colony formation in a dose-dependent manner. This inhibitory action is selective for megakaryocytic as well as monocytic lineage. In their experiments, megakaryocyte colonies were identified in situ on an inverted microscope by analyzing the morphology of megakaryocytes. Previously, IL4 has been shown to have no effect (Bruno et al, 1989) or to stimulate megakaryocyte colony formation alone or in combination with other cytokines in murine cuture system (Peschel et al, 1987). Whether IL4 plays an important role in regulating megakaryocytopoiesis remains to be further determined.

It is not yet known whether other previously described hematopoietic inhibitors such as the Tetrapeptide AcSDKP and macrophage inflammatory protein (MIP) play a role in megakaryocytopoiesis. The tetrapeptide AcSDKP is a inhibitor of murine pluripotent stem cells (Frindel & Guigon, 1977; Lenfant et al, 1989), and subsequently found to inhibit the growth of human CFU-GM and BFU-E, and to decrease their percentage in DNA synthesis, but not to inhibit the growth of leukemia cells (Bonnet et al, 1992). MIP-1 is known to be a potent inhibitor of the proliferation of stem cells (Broxmeyer et al, 1990; Dunlop et al, 1992).

Megakaryocytopoietic inhibitors and glycosaminoglycans (GAGs)

One important point in the negative regulation of megakaryocytopoiesis is that most inhibitors described above are heparin-binding proteins and their inhibitory action can be modified by heparin and other GAGs. The inhibitory effect of PF4 on HEL cell growth can be blocked by heparin and heparan sulfate (Han et al, 1992). The prothrombin induced inhibition of proplatelet process formation can also be blocked by matrix-bound GAGs (Hunt et al, 1993). Figure 2 illustrates the effect of heparin on the action of TGFß1, PF4 and related dodecapeptide in vitro in mice, and indicates that heparin neutralizes completely the inhibitory effect of PF4 and related dodecapeptide and partially that of TGFß1 on megakaryocytopoiesis. Similar results were also obtained in our laboratory when heparin and PF4 or dodecapeptide were simultaneously injected into mice (data not shown).

Mechanism of action of inhibitors

The mechanism of action of the inhibitors in regulating megakaryocytopoiesis is still incompletely understood. There may be several action pathways for the inhibitors to execute their activities. One is the inhibitor-receptor mediated mechanism. The binding of an inhibitor to its specific receptor initiates a signal transduction pathway which crosses the cell membrane, passes through the cytoplasma into nucleus, resulting in altered transcription of certain genes and arrest of cell division. Another pathway is that the inhibitors may interfere with growth factors responsible for megakaryocyte development. In addition, the inhibitors may affect megakaryocytopoiesis on the basis of their other activities. These different mechanisms may coexist.

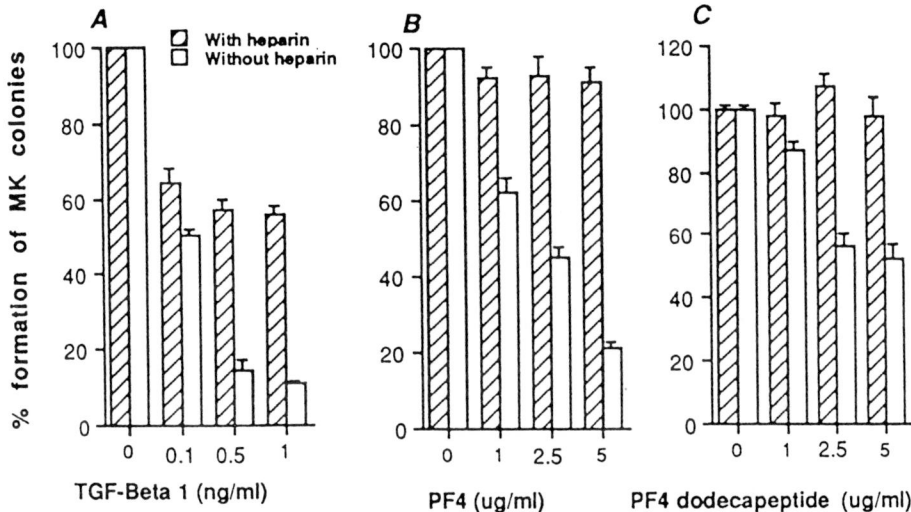

Fig. 2. Effect of heparin (5 u/ml) on the action of TGFß1 (A), PF4 (B) and dodecapeptide of PF4 (C) in murine megakaryocyte colony formation in vitro. Bone marrow cells from Balb/c mice were cultured at concentration of 2×10^5 nucleated cells per dish in plasma clot system in the presence of 10% aplastic anemia serum of irradiated pig (Han et al, 1991a). Data represent the mean (± SEM) of triplicate determinations from three separate experiments. The 100% (control) corresponds to 132 ± 5 (A), 124 ± 4 (B) or 136 ± 7 (C) megakaryocyte (MK) colonies / 2×10^5 nucleated cells.

TGFß receptors (of which there are three types) are present in almost every cell type (Massague et al, 1990). A single cell type can express multiple TGFß receptors which are distinguishable by their molecular weights, their affinities for TGFß and their polysaccharide components. TGFß inhibition of c-myc may be a central event in TGFß growth arrest. Arrest of c-myc expression would prevent the cell from proceeding past the G1/S boundary. A nuclear protein, pRB, probably is the cellular factor mediating TGFß effects on c-myc transcription. Addition of TGFß in mid/late G1 interrupts phosphorylation of pRB (Laiho et al, 1990). Factor-selective action of TGFß may explain why it acts on megakaryocytopoiesis at picomol levels. TGF-ß1 inhibits IL3 or AAS-induced colony formation but not colony formation due to GM-CSF, G-CSF, IL6 or Epo (Keller et al, 1988, 1989; Han et al, 1992a; Fig. 1). Among all the well characterized hematopoietic growth factors, IL3 is the most potent stimulator of megakaryocyte colony formation. AAS has long been thought to contain a unique factor, termed megakaryocyte colony-stimulating factor (MK-CSF), which is different from IL-3, IL6, Epo and GM-CSF (Hoffman, 1989; Han et al, 1991b).

The mechanism of action of PF4 has recently been proposed to be associated with its binding to the heparin-like molecules of target cells (Han et al, 1992). This hypothesis was based on the following observations: a) The binding of PF4 to cells could be almost completely blocked by heparin and heparan sulfate, b) The inhibitory activity of PF4 and related peptides on megakaryocytopoiesis in vitro and in vivo could be completely neutralized by heparin and heparan sulfate (Han et al, 1992). Heparin-like molecules on cell surface are known to be involved in the binding to cells and the growth-stimulating activity of some growth factors such as basic fibroblast growth factor (bFGF) (Yayon et al, 1991). These heparin-like molecules may be the binding sites of PF4 and related peptides. The binding of PF4 (a high concentration is required for saturating binding sites) to these molecules may induce cell growth inhibition by blocking the binding of megakaryocyte growth-stimulating factor(s) to cells, as suggested by recent observation that PF4 blocks the binding of bFGF to the receptor (Sato et al, 1991).

Thrombin-induced inhibition of Meg-01 proliferation has been suggested to be due to increased cAMP formation but not to thrombin effect on calcium mobilization. However, whether such a mechanism occurs at normal bone marrow megakaryocyte level remains to be confirmed.

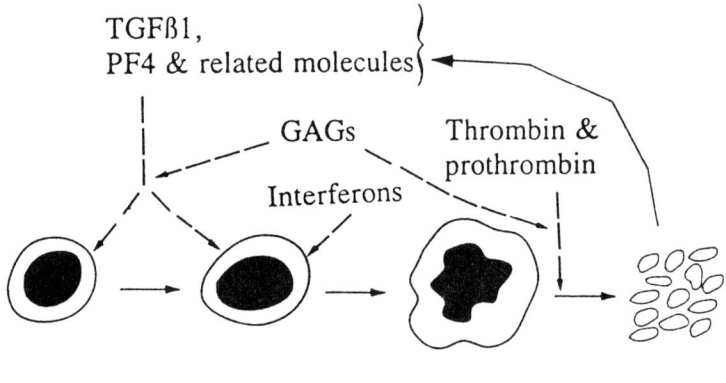

Fig. 2. Schematic presentation of megakaryocytopoiesis and its negative regulation. Interferons include alpha and gamma interferon. PF4 related molecules include ß-thromboglobulin and its precursor, connective tissue activating peptide III, and their fragments of C-terminal domain. MK, megakaryocyte; GAGs, glycosaminoglycans.

Concluding remarks

Taken together the data collected here may be summarized in Fig.3. Megakaryocytopoiesis is multi-stage cellular process. Negative regulation of megakaryocytopoiesis is controlled essentially by megakaryocyte-platelet products but also controlled to some degree by molecules such as interferons and thrombin and its precursor. Most inhibitors are heparin-binding factors and their inhibitory action can negatively be modified by heparin and other GAGs.

Whether such a model meets physiologic requirements remains to be further confirmed. However, some possible clinical implications may be drawn up on the basis of this model. For example, the observations that PF4 and related peptides as well as TGFß are capable of inhibiting normal hematopoietic progenitor and leukemic cell proliferation will probably lead to an application of these inhibitors in the therapeutic management of myeloproliferative disorders. In addition, because the inhibitory effect of megakaryocyte-platelet derived inhibitors can be abrogated by heparin and other GAGs, an application of heparin or other GAGs may be therefore envisaged in the treatment or at least improvement of thrombocytopenia resulting from platelet destruction.

Acknowledgement   The authors thank HY Wan and AM Maurer for experimental assistance, and Dr. J Amiral for providing PF4 and synthetic peptides.

References

Bonnet, D., Césaire, R., Lemoine, F. et al (1992): The tetrapeptide AcSDKP, an inhibitor of the cell cycle status for normal human bone marrow progenitors, has no effect on leukemic cells. Exp Hematol. 20:251-255

Broxmeyer, H.E., Sherry, B., Cooper, S. et al (1991): Macrophage inflammatory protein (MIP)-1ß abrogates the capacity of MIP-1 to suppress myeloid progenitor cell growth. J Immunol 147:2586-2594

Bruno, E., Briddle, R., Hoffman, R. (1988): Effect of recombinant and purified hematopoietic growth factors on human megakaryocyte colony formation. Exp Hematol   16:371-377

Bursuker, I., Neddermann, K.M., Petty, B.A. et al (1992): In vivo regulation of hemopoiesis by transforming growth factor beta 1: stimulation of GM-CSF- and M-CSF-dependent murine bone marrow precursors. Exp Hematol 20:431-435

Carlino, J.A., Higley, H.R., Avis, P.D. & Ellingsworth, L.E. (1990): Hematologic and hematopoietic changes induced by systemic administration of TGF-ß1. Annals New York Acad Sci   593:326-333

Dunlop, D.J., Wright, E.G., Lorimore, S. et al (1992): Demonstration of stem cell inhibition and myeloprotective effects of SCI/rhMIP 1 in vivo. Blood 79:2221-2225

Fan, K., Ruan, Q., Sensenbrenner, L. & Chen, B. (1992): Transforming growth factor-ß1 bifunctionally regulates murine macrophage proliferation. Blood 79:1679-1685

Frindel, E & Guigon, M. (1977): Inhibition of CFU entry into cycle by a bone marrow extract.   Exp Hematol 5:74-79

Ganser, A., Carlo-Stella, C., Greher, J. et al (1987): Effect of recombinant interferon alpha and gamma on human bone marrow-derived megakaryocytic progenitor cells. Blood 70:1173-9

Gewirtz, A.M., Calabretta, B., Niewiarowski, S., Xu, W.Y. (1989): Inhibition of human megakaryocytopoiesis in vitro by platelet factor 4 (PF4) and a synthetic COOH-terminal peptide. J. Clin. Invest. 83:1477-1486.

Goey, H., Keller, J.R., Back, T. et al. (1989): Inhibition of early murine hemopoetic progenitor cell proliferation after in vivo locoregional administration of tranforming growth factor ß1. J Immunology 143:877-880

Han, Z.C., Briere, J., Abgrall, J.F. et al (1987): Effects of recombinant human interferon alpha on human megakaryocyte and fibroblast colony formation.   J Bio Regul Homeo Agents. 1:195-200

Han ZC, Sensebe L, Abgrall JF & Brière J (1990a) Platelet factor 4 inhibits human megakaryocytopoiesis in vitro. Blood. 75:1234-1239

Han, Z.C., Bellucci, S., Walz, A., Baggiolini, M., and Caen, J.P. (1990b): Negative regulation of human megakaryocytopoiesis by human platelet factor 4 (PF4) and connective tissue-activating peptide III (CTAP-III). Int. J. Cell. Cloning. 8:253-259.

Han, Z.C., Bellucci, S., Tenza, D., and Caen, J.P. (1990c): Negative regulation of human megakaryocytopoiesis by human platelet factor 4 and beta-thromboglubulin: comparative analysis in bone marrow cultures from normal individuals and patients with essential thrombocythaemia and immune thrombocytopenia. Br. J Haematol. 74:395-401.

Han, Z.C., Bellucci, S., Bodevin, E et al (1991a): In vivo inhibition of megakaryocyte and platelet production by Platelet factor 4 in mice. C R Acad Sci Paris 313:553-558

Han ZC, Bellucci S, Caen JP (1991b) Megakaryocytopoiesis: characterization and regulation in normal and pathologic states. Int J Hematol 54:3-14.

Han, Z.C., Bellucci, S., Wan, H.Y. & Caen, J.P. (1992a): New insights into the regulation of megakaryocytopoiesis by haematopoietic and fibroblast growth factors and transforming growth factor ß1. Br J Haematol 81:1-5

Han, Z.C., Maurer, A.M., Bellucci, S et al (1992b): Inhibitory effect of platelet factor 4 (PF4) on the growth of human erythroleukemia cells: proposed mechanism of action of PF4. J Lab Clin Med   120:645-660.

Hermoson, M., Schmer, G., & Kurachi, K.(1977). Isolation, characterization and primary amino acid sequence of human platelet factor 4. J. Biol. Chem. 252:6276-6279.

Hoffman, R. (1989): Regulation of megakaryocytopoiesis. Blood 74:1196-1212

Holt, J.C., Harris, M.E., Holt, A.M., Lange, E. Henschen, A. & Niewiarowski, S (1986): Characterization of human platelet basic protein, a precursor form of low-affinity platelet factor 4 and ß-thromboglobunin. Biochemistry 25:1988-1996

Hunt, P., Hokom, M.M., Wiemann, B., Leven, R.M. & Arakawa, T. (1993): Megakaryocyte proplatelet-like process formation in vitro is inhibited by serum prothrombin, a process which is blocked by matrix-bound glycosaminoglycans. Exp Hematol 21:372-381

Ishibashi, T., Miller, S. & Burstein, S.A. (1987): Type B transforming growth factor is a potent inhibitor of murine megakaryocytopoiesis in vitro. Blood 69:1737-1741

Keller, J.R., Mantel, C., Sing, G.K. et al (1988): Transforming growth factor ß1 selectively regulates early murine hematopoietic progenitors and inhibits the growth of IL3-dependent myeloid leukemia cell lines. J Exp Med 168:737-750

Keller, J.R., Sing, G.K., Ellingsworth, L.R. & Ruscetti, F.W. (1989): Transforming growth factor ß: possible roles in the regulation of normal and leukemic hematopoietic cell growth. J Cell Biochem 39:175-180

Keller, J.R., Jacobsen, S.E.W., Sill, K.T. et al (1991): Stimulation of granulopoiesis by transforming growth factor ß: Synergy with granulocyte/macrophage colony-stimulating factor. Proc Natl Acad Sci USA 88:7190-7194

Kuter, D.J., Gminski, D.M. & Rosenberg, R.D. (1992): Transforming growth factor ß inhibits megakaryocyte growth and endomitosis. Blood 79:619-626

Laiho, M., DeCaprio, J.A., Ludlow, J.W., Livingston, D.M., Massagué, J. (1990): Growth inhibition by TGFß linked to suppression of retinoblastoma protein phosphorylation. Cell 62:175-179

Lenfant, M., Wdzieczak-Bakala, J., Guittet, E. et al (1989): Inhibitor of hematopoietic pluripotent stem cell proliferation: purification and determination of its structure. Proc Natl Aca Sci USA. 6:779-782

Massagué, J. (1990): The transforming growth factor family. Annu Rev Cell Biol 6:597-602

Maze, R., Sherry, B., Know, B.S. et al (1992): Myelosuppressive effects in vivo of purified recombinant murine macrophage inflammatory protein-1. J Immunology 149:1004-1009

Moore, M.A.S. (1991) Clinical implications of positive and negative hematopoietic stem cell regulators. Blood 78:1-19

Moses, H.L. & Yang, E.Y. (1990): TGF-ß stimulation and inhibition of cell proliferation: new mechanistic insights. Cell 63:245-247, 1990

Peschel, C., Paul, W.E., Ohara, J., Green, I. (1987): Effect of B cell stimulatory factor-1/interleukin 4 on hematopoietic progenitor cells. Blood 70:254-259

Ruoslahti E & Yamaguchi Y (1991) Proteoglycans as modulators of growth factor activities. Cell 64:867-869

Sato, Y., Abe, M., Takaki, R. (1991): Platelet factor 4 blocks the binding of basic fibroblast growth factor to the receptor and inhibits the spontaneous migration of vascular endothelial cells. Biochem Biophys Res commun 172:595-600

Sonada, Y., Kuzuyama, Y., Tanaka, S., Yokota, S., Maekawa, T., Clark, S.C. & Abe, T. (1993): Human interleukin-4 inhibits proliferation of megakaryocyte progenitor cells in culture. Blood 81:624-630

Vittet, D., Mathieu, M.N., Launay, J.M. & Chevillard, C. (1992): Thrombin inhibits proliferation of the human megakaryoblastic MEG-01 cell line: a possible involvement of a cyclic-AMP dependent mechanism. J Cell Physiol 150:65-75

Yayon, A., Klagsbrun, M., Esko J.D., Leder, P., Ornitz, D.M. (1991): Cell surface, the heparin-like molecules are required for binding of basic fibroblast growth factor to its high affinity receptor. Cell 64:841-848

Wadenvik, H., Kutti, J., Ridell, B et al (1991): The effect of -interferon on bone marrow megakaryocytes and platelet production rate in essential thrombocythemia Blood 77:2103-2108

Zucker, M.B. & Katz, I.R. (1991): Platelet factor 4: production, structure, and physiologic and immunologic action. Proc Soci Exp Med. 693-702

Résumé

La mégacaryocytopoïèse est régulée négativement principalement par des inhibiteurs dérivés des mégacaryocytes et plaquettes incluant le TGFb, le platelet factor 4 et les molécules apparentées. D'autres facteurs tels que les interférons a et g, la thrombine et la prothrombine inhibent aussi la mégacaryocytopoïèse. L'activité de la plupart de ces inhibiteurs est supprimée par l'héparine et d'autres glycosaminoglycans. Nous proposons un modèle de régulation négative de la mégacaryocytopoïèse basé sur ces observations.

Inhibition of megakaryocytopoiesis by molecules related to the platelet factor 4 (PF4)

Isabelle Lebeurier, Jean Amiral*, Jacques P. Caen and Zhong C. Han

*Institut des Vaisseaux et du Sang, Hôpital Lariboisière, Paris, *Serbio, Gennevilliers, France*

The platelet factor 4 (PF4), a protein specific for megakaryocyte-platelet lineage, is an inhibitor of the human and murine megakaryocytopoiesis. PF4 and its related molecules: the beta-thromboglobulin (β-TG) and the interleukin 8 (IL 8) are members of a multigene family: the small inducible gene (SIG). In this study, we studied the effect of these molecules on the growth of three human megakaryocytic cell lines: HEL, Dami and MEG-01 and one granulocytic cell line, the HL60. The results showed that all these molecules had, although at different degree, an inhibitory activity on the proliferation of HEL, Dami and MEG-01 but not HL60 cell line, suggesting a predominant effect of these molecules on megakaryocytic lineage. Several synthetic peptides corresponding to the homologous region of these factors were therefore tested in order to know if the C-terminal sequence homology existing among the different molecules was implicated in the inhibitory activity. The synthetic peptide C1-13 of PF4 inhibited the in vitro HEL and MEG-01 cell growth but this peptide was less potent than the peptide C13-24 of PF4 in dimeric form. The C13-24 monomeric type and modified type (cystein residue was replaced with arginin) had no inhibitory activity. In vivo, peptide C13-24 of PF4 in dimeric form also inhibited the murine megakaryocytopoiesis. This activity was predominant on the megakaryocyte progenitors (CFU-MK). The megakaryocyte and the platelets were slightly affected. The leucocytes, white blood cells and hemoglobin were not affected by this synthetic peptide. Interestingly, the inhibitory effect of PF4, β-TG and C13-24 on megakaryocytopoiesis in vitro and in vivo all could be abrogated by heparin. These results suggest that the homologous region of C-terminal of PF4 and related proteins may be responsible for their inhibitory activity on megakaryocytopoiesis and that a specific dimeric structure seems to be necessary for peptide C13-24 to execute its inhibitory activity. The peptide C13-24 of PF4, like PF4, could be a potential component for the treatment of diseases characterized by an exessive proliferation of megacaryocytic lineage.

Cyclic AMP inhibits cell growth and glycoprotein Ib expression on a human megakaryocytic cell line

Daniel Vittet, Christophe Duperray* and Claude Chevillard

*INSERM U 300, Faculté de Pharmacie, 34060 Montpellier, France; *INSERM U 291, Service commun de cytométrie en flux, 99, rue Puech Villa, 34090 Montpellier, France*

Megakaryocytopoiesis is a process characterized by proliferation of megakaryocytic precursors and differentiation of their progeny. The regulation of megakaryocyte development, in mammals, has been shown to be influenced by a complex network of interacting cytokines (for review, see Hoffman, 1989). However, intracellular processes activated during megakaryocytopoiesis remain unclear. Cyclic AMP has been implicated in the regulation of growth and differentiation in a variety of normal and malignant cells including human leukemia cell lines (Dumont et al., 1989 ; Cho-Chung, 1990). A few studies have suggested the possible involvement of the cyclic AMP pathway in intracellular mechanisms whereby megakaryocytic growth factors regulate cellular development (Long, 1989). To further explore cyclic AMP action on megakaryoblastic cells we have investigated the role of cyclic AMP analogs and of the water soluble derivative of forskolin, L858051 (7ß-desacetyl-7ß-[γ-(N-methyl piperazino)-butyryl] forskolin) on the regulation of growth and differentiation of the human megakaryocytic-like DAMI cell line.

Our results show that exposure of DAMI cells to cyclic AMP analogs or to L858051 yielded dose-dependent inhibition of cell growth. In order to examine a possible effect of the cyclic AMP pathway on DAMI cell differentiation, we analyzed membrane expression of two platelet glycoproteins (GPIb and GPIIb/IIIa) whose expression are enhanced during megakaryocytic differentiation. Cyclic AMP analogs and L858051 were shown to selectively inhibit GPIb expression without affecting GPIIb/IIIa expression as determined by immunofluorescence and flow cytometry analysis. All these effects could be attributed to the cyclic AMP entity, since the stable Sp-phosphorothioate isomer of cyclic AMP, which exhibits high resistance toward phosphodiesterases and no metabolic side effects, exerts similar effect on both cell proliferation and GPIb expression.

Our results are then consistent with a potential negative regulatory role of the cyclic AMP pathway on the proliferation and the differentiation of megakaryocytic cells. However, analysis of other megakaryocyte differentiation markers, such as DNA ploidy, and experiments using normal megakaryocytic cells remain to be performed to clearly establish this role.

REFERENCES

Cho-Chung, Y.S. (1990) : Role of cyclic AMP receptor proteins in growth, differentiation, and suppression of malignancy : new approaches to therapy. *Cancer Res.* 50, 7093-7100.
Dumont, J.E., Jauniaux, J.C., and Roger P.P. (1989) : The cyclic AMP-mediated stimulation of cell proliferation. *Trends Biochem. Sci.* 14, 67-71.
Hoffman, R. (1989) : Regulation of megakaryocytopoiesis. *Blood* 74, 1196-1212.
Long, M.W. (1989) : Signal transduction events in in vitro megakaryocytopoiesis. *Blood cells* 15, 205-229.

Inhibition of actin polymerization induces endomitosis of human megakaryocyte cell lines

Christian Chatelain, Sarah Baatout, Philippe Staquet, Bernard Chatelain and Christine B. Chatelain

Laboratory of Experimental Hematology and Oncology, Oncology Unit Catholic University of Louvain, UCL Mont-Godinne, Av. G. Therasse 1, B-5530 Yvoir, Belgium

INTRODUCTION. Previous experiments have suggested that mouse megakaryocyte polyploidization is due to inhibition of cytoplasmic separation or cytodieresis [1]. In order to detect a cytoskeleton protein involved in cytodieresis, cytochalasin B (an inhibitor of actin polymerization) was added to cultures of Dami, K-562, MEG-01 and HEL leukemic cell lines, known to differentiate into the megakaryocyte pathway in the presence of phorbol myristate acetate.

RESULTS. Cytochalasin B alone (2μg/ml) augmented the diameter of Dami, K-562, MEG-01 and HEL cells obtained after 4 days of culture and measured on spun smears. Cells cultured in the presence of Cytochalasin B looked like megakaryocytes on a morphological standpoint. The DNA content of these cells was evaluated by fluorescence photometry on spun smears stained with chromomycin A3 [2]. The proportion of polyploid cells was increased after addition of cytochalasin B. Furthermore, cell size (Forward Scatter, FSC) and ploidy (propidium iodide fluorescence) were measured by flow cytometry. In response to cytochalasin B, Dami, K-562 and HEL cells, normally diploid, became substantially polyploid with a mode of 4N and appearance of higher ploidy classes. Concomitantly, the cell size distribution measured by FSC was significantly shifted to higher values.

CONCLUSION. The data suggest that inhibition of actin polymerization induces and enhances endomitosis of megakaryocyte leukemic cells, probably by preventing the cytoskeleton-related annular constriction of the cytoplasm at the diploid level.

REFERENCES.
1. Chatelain C, De Bast M, Symann M: Enhancement of megakaryocyte polyploidization by actin inhibition. Blood 80:497a, 1992 (abstr.)

2. Chatelain C, Burstein SA: Fluorescence cytophotometric analysis of megakaryocytic ploidy in culture: studies of normal and thrombocytopenic mice. Blood 64:1193, 1984

Cellular architecture of the foetal liver with respect to the production of haemopoietic growth factors and inhibitors
Expression of TGFβ, MIP1α, SCF and IL-3 detected by *in situ* hybridization

Theonne de Kretser

Peter MacCallum Cancer Institute, 481 Little Lonsdale Street, Melbourne, Victoria 3000, Australia

Regulation of haemopoietic stem cell proliferation and differentiation is essential in maintaining the ability of the haemopoietic system to respond to physiological insult throughout life. Candidate factors for the regulation of haemopoietic stem cells are gradually being discovered. The overall aim of our laboratory is to study the molecular mechanisms by which two of these, transforming growth factor β1 (TGFβ1) and macrophage inflammatory protein 1α (MIP1α), act in regulating haemopoietic stem cells. In order to establish that, *in vivo*, these two factors are potential candidates to influence haemopoietic stem cells, and whether their pattern of elaboration within haemopoietic organs held any clues as to the manner by which they might interact with haemopoietic stem cells, we have studied the expression of these factors in murine foetal liver.

In the mouse embryo, the liver is the principal haemopoietic organ from day 10 until birth. We have examined the expression of TGFβ1, MIP1α, interleukin-3 (IL-3) and stem cell factor (SCF) mRNAs in 5μm sections of foetal livers (day 11, day 13 and day 15 gestation) using dioxygenin-labelled, polymerase chain reaction-generated cDNA as probes, detected using a modification of the peroxidase-antiperoxidase immunohistochemical reaction.

In day 11 foetal liver, individual cells expressing readily detectable levels of TGFβ1 were observed scattered throughout the central areas of the liver lobes. A smaller number of MIP1α-expressing cells were seen but these were in most instances located in the periphery of the lobes. IL-3- and SCF- expressing cells were present, widely scattered throughout the liver in similar numbers to TGFβ1-expressing cells.

At day 13, the number of TGFβ1-producing cells peaked. At this point, small clusters of TGFβ1-expressing cells (2-3 cells) were found, although single cells were still present. The TGFβ1-positive cells stained more strongly than at day 11 and were widely scattered through the liver. MIP1α-expressing cells were also present in increased numbers and with more intense staining, but still predominantly in the outer regions of the lobes. The number of IL-3- and SCF-producing cells and their distribution within the liver had not significantly altered from that observed at day 11. However, unlike day 11, at day 13 the number of TGFβ1 and MIP1α-producing cells was greater than IL-3- and SCF- producing cells.

By day 15, the relative amounts of inhibitor to stimulator produced in the liver had altered. TGFβ1 expression was down, with TGFβ1-positive cells more frequently found in the outer regions of the liver lobes. MIP1α expression was not detected. Conversely, both IL-3 and SCF- producing cells had increased both in number and intensity of staining, with both cell types still distributed throughout the liver.

While essentially qualitative, our results show that both TGFβ1 and MIP1α are produced in the same cellular compartment in which generation and maintenance, and thus, regulation of haemopoietic stem cells occur. The different patterns of distribution of TGFβ1- and MIP1α- producing cells suggest that these inhibitors cannot not play identical or redundant roles in stem cell regulation. The patterns observed for cells expressing the four factors we have studied (individual cells or small clusters spread throughout the liver) raises intriguing possibilities as to how regulatory niches in foetal liver are organized. It was noted that, particularly at day 13, TGFβ1-expressing cells were found in close contact with small cells with dense, compacted nuclei. In order to further study the intra-organ spatial relationship between cells expressing these factors, as well as the characteristics of those non-factor-expressing cells in close contact with them, we are extending this work to combine *in situ* hybridization with standard antibody-based immunohistochemistry. Preliminary experiments have shown that many of the antigens currently used to define haemopoietic cell populations are not expressed at sufficiently high density in foetal liver to allow immunohistochemical detection. Thus, we are presently evaluating novel reagents to the appropriate receptors, in order to simultaneously localize cells capable of responding to, and those producing, the factors above.

The restrictive role of stromal factors in hemopoiesis

Yael Gothelf, Amnon Peled, Naama Brosh, Judy Honigwachs-Shaanani, Byeong-Chel Lee, Dalia Sternberg, Pnina Carmi, Allan Levy, Jeki Toledo and Dov Zipori

Department of Cell Biology, The Weizmann Institute of Science, Rehovot 76100, Israel

The structure and function of the hematopoietic system is discussed in view of the proposed "theory of restrictins". It is argued that amongst the regulators of hemopoiesis, antagonists of differentiation are renewal factors that differ functionally from well known hemopoietic cytokines. It is further suggested that a set of lineage specific killer molecules are involved in hemopoietic tissue organization.

The constant supply of mature blood cells to the periphery, depends on the regulation of the processes by which the stem cells differentiate into committed progenitors which in turn differentiate into mature progeny. The major steps in the hemopoietic process are: renewal, differentiation, and spatial organization. In the current models of hemopoiesis, differentiation inducing cytokines are suggested to be involved in all the above aspects of hemopoiesis. These models are based on data derived mostly from the use of colony formation assays for the study of cytokines. Those molecules that cause a reduction either in size or in the number of colonies formed, were subsequently defined as inhibitors. This definition introduces some difficulties since similar effects may be caused by totally unrelated mechanisms. For example, both cell lysis and accelerated differentiation, may lead to the lack of colony formation. An additional problem is that some cytokines can be inhibitory to one cell type and at the same time have a stimulatory effect on another. Thus, an inhibitor can be defined only in a certain context and never in general terms. We have recently reviewed this issue in detail (Zipori & Honigwachs-Shaanani, 1992).

The difficulties we pointed out in the current models of hemopoiesis may be overcome if these models are modified to include information derived from the study of stroma dependent hemopoiesis. We found that primary stromal cells from the bone marrow, that comprise a mixture of various cell types, support hemopoiesis by restraining the differentiation flow. Primary stromal cells antagonized the effects of various differentiation inducing cytokines. We demonstrated that in co-cultures of stromal cells and hemopoietic cells, an inverse relationship existed between the production of mature cells and the maintenance of progenitors. High concentrations of stromal cells markedly reduced differentiation into mature progeny but promoted the production of progenitors (Zipori, 1981, Zipori et al., 1981, Zipori & Sasson, 1980).

Several stromal cell lines have been developed and characterized in our laboratory (Zipori et al. 1984, Zipori et al. 1985). These were classified as belonging to several subtypes (Zipori, 1989). 14F1.1, an endothelial-adypocyte cell line, was found to support stem cell renewal and long term hemopoiesis. When 14F1.1 cells were grown to confluence and bone marrow or enriched stem cell populations were seeded onto the stromal cell layer, long term hemopoiesis developed accompanied by the formation of cobblestone-like areas. Several weeks later stem cells, that were able to reconstitute lethally irradiated mice, could be recovered from these cultures (Zipori, 1988). Other functions of the 14F1.1 stromal cell line include: induction of myelopoiesis, long term pre-B cell growth (Zipori & Lee, 1988), long term growth of T lymphocyte precursor cells (Tamir et al., 1990) and stimulation of granulocyte, macrophage and megakaryocyte colony formation in short term cultures (Kalai & Zipori, 1989). The 14F1.1 line has been shown to express low levels of CSF-1 (Zipori et al., 1984), c-kit ligand, and IL-6 (Peled et al. unpublished results) and no detectable levels of GM-CSF, IL-3, G-CSF or IL-4 (Zipori & Lee, 1988). Other stromal cell lines that also express the above cytokine genes do not however support stem cell renewal. It was therefore concluded that as yet unidentified stromal factors regulate stem cell renewal.

The MBA-2.1 endothelial-like cell line is one of those stromal cell lines which are devoid of any ability to support hemopoiesis. This cell line produces a protein which has a unique ability to kill plasmacytoma cells. This protein (designated restrictin-P) is highly specific to plasmacytoma cells and does not affect the growth of any of the cell types studied thus far. We predict that this lineage and differentiation stage specific factor is one of a family of inhibitors that are target cell specific and may account for tissue organization.

The available data about hemopoeitic cell growth and differentiation, together with our current knowledge about stroma dependent hemopoiesis, indicate necessary changes in models of hemopoiesis. Indeed, we suggest a different classification for hemopoietic cytokines: all cytokines "stimulators" and "inhibitors" alike are designated "restrictins". The term restrictin describes the ability of the cytokine to limit the cell in its choice of a differentiation pathway, thus restricting the options a given cell can choose (Zipori, 1992). Furthermore, restrictins are classified according to their ability to affect differentiation: they can be either inducers or antagonists of differentiation.

Inducers of differentiation include classical CSFs, that stimulate proliferation associated with differentiation. This category also includes molecules like IFNγ and TNF which are well known for their inhibitory effects on hemopoietic cells (Craig & Buchan, 1989). These molecules often induce differentiation which is not coupled with proliferation, thus contributing to the reduction of the pool of mature cells. This latter function may also be achieved via induction of programmed cell death (apoptosis). Some of these molecules are lineage and stage specific. The other category, antagonists of differentiation, includes maintenance and renewal factors. These are not necessarily inhibitory molecules, but rather cytokines that block the differentiation flow thus leaving the cell with the option for self renewal only.

We hereby describe the structure of the hemopoietic system based on the "theory of restrictins". The obligatory step in this process is stem cell renewal, which does not necessarily involve rapid proliferation, and is mediated by yet undefined factors that antagonize differentiation processes. Renewal may also occur at later stages of hematopoietic differentiation. Stem cell renewal is dependent upon intimate interactions between stromal cells and hemopoietic cells. Such interactions can be seen in the bone marrow microenvironment. The quiescence of the stem cell pool is probably part of the safeguard mechanism that forms a barrier protecting the stem cell niches from events occurring in the blood stream and peripheral organs. The first step towards differentiation must involve therefore an escape of the stem cell from G_0 of the cell cycle.

Some cytokines act directly on stem cells in the G_0 phase "awakening" them. These cytokines probably shorten the G_0 phase. Mushasi et al. (1991), for example, studied the development of blast cell colonies under the influence of IL-3: colony formation from individual blast cells in the presence of IL-3, requires an average of 9 days. The addition of either IL-11, IL-6, G-CSF or c-kit ligand leads to colony formation within 5 days only. Thus the early stem cell may form colonies in vitro, but this process occurs earlier when the stem cells are presented with certain cytokine mixtures.

The next step in the differentiation cascade requires growth and differentiation factors. These are classified as those that (a) increase the number of mature cells and induce proliferation coupled with differentiation like CSFs or (b) induce terminal differentiation without growth, potentiating the function of mature differentiated cells, like IFNγ (Perussia et al., 1987) and (c) factors that cause cytolysis like TNF (Craig & Buchan 1989). Some differentiation factors contribute to the reduction of the pool of mature cells. The process of differentiation is self perpetuating since the mature progeny, like macrophages and lymphocytes elaborate ample amounts of differentiation inducing cytokines. This might be dangerous as the differentiation flow could eventually create a depletion of the stem cell pool. To prevent such an outcome, a potent feedback mechanism must function to dampen the differentiation signal. Feedback regulators of differentiation are in fact produced by mature hemopoietic cells and inhibit cell proliferation and cycling of stem cells by reducing the number of cells in S phase: these include cytokines like TGFβ and MIP1α (SCI) and regulatory peptides like pEEDCK and AcSDKP (reviewed in Zipori & Honigwachs-Shaanani, 1992).

The hemopoietic microenvironment within the bone marrow involves some very characteristic tissue organization. No satisfactory molecular explanation is available to date to explain this level of organization within the hemopoietic tissue, which has been suggested to take an active part in the regulation of hemopoieis. The defined microorganization of cells could be due to factors locally expressed by the stroma. Cytokines bound to extracellular matrix components or expressed on the cell surface could also be involved in the formation of such microenvironments. However, the spatial organization of hematopoietic tissues cannot be explained by inducer cytokines only, as these can have a variety of targets. Such cytokines are therefore insufficient by themselves to account for the formation of domains within the tissue in which specific cell types reside. The "theory of restrictins" predicts the existence of lineage specific inhibitors characterized by a limited target cell range, similar to restrictin-P mentioned above. We propose that each hemopoietic cell type, belonging to a specific lineage or stage of differentiation, bears cell surface molecules that have corresponding receptors on stromal cells. The stromal cell receptor (restrictin) is a molecule capable of signalling cell death. Thus the result of an encounter between a specific hemopoietic cell type and a stromal cell bearing its corresponding restrictin is the death of the hemopoietic cell. Some stromal cells may bear one type of restrictin that will limit the accumulation of a single cell type while other stromal cells may bear a whole range of restrictive molecules that will not allow any hemopoiesis to occur despite the presence of differentiation inducing cytokines. This last situation may exist in non-hemopoietic organs such as the uterus or the lung which in fact harbor high titers of hemopoietic cytokines. However, some restrictins may not be lineage specific and could be more abundant in non hemopoietic organs.

Restrictin-P has a unique capability to inhibit the proliferation of plasmacytoma cells while having no inhibitory effect on any other type of hemopoietic or non hemopoietic cell tested thus far. Addition of restrictin-P to the plasmacytoma MPC-11 causes an inhibition in cell proliferation within 10-15 hours. This correlates with an early decrease in cell size, decrease in DNA synthesis and a shift in the G_0-G_1 distribution of the cell cycle. The viability of MPC-11 plasmacytoma cells treated with restrictin-P decreases significantly within 24 hours. We found that restrictin-P induces apoptosis in its target cells. In scanning electron microscopy we observed a loss of microvilli, detectable as early as 6 hours following addition of restrictin-P. This was then followed by characteristic membrane changes. Furthermore restrictin-P caused DNA fragmentation within 5-10 hours of its addition. All these processes were independent of *de novo* protein synthesis. Restrictin-P may be one representative of an entire class of yet unknown lineage specific inhibitors of stromal origin that kill in a cell lineage and differentiation specific manner.

Figure 1 outlines the major claims of the "theory of restrictins". Panel A shows that stem cells under in vitro conditions respond to the effect of differentiation factors by utilizing their spectrum of differentiation

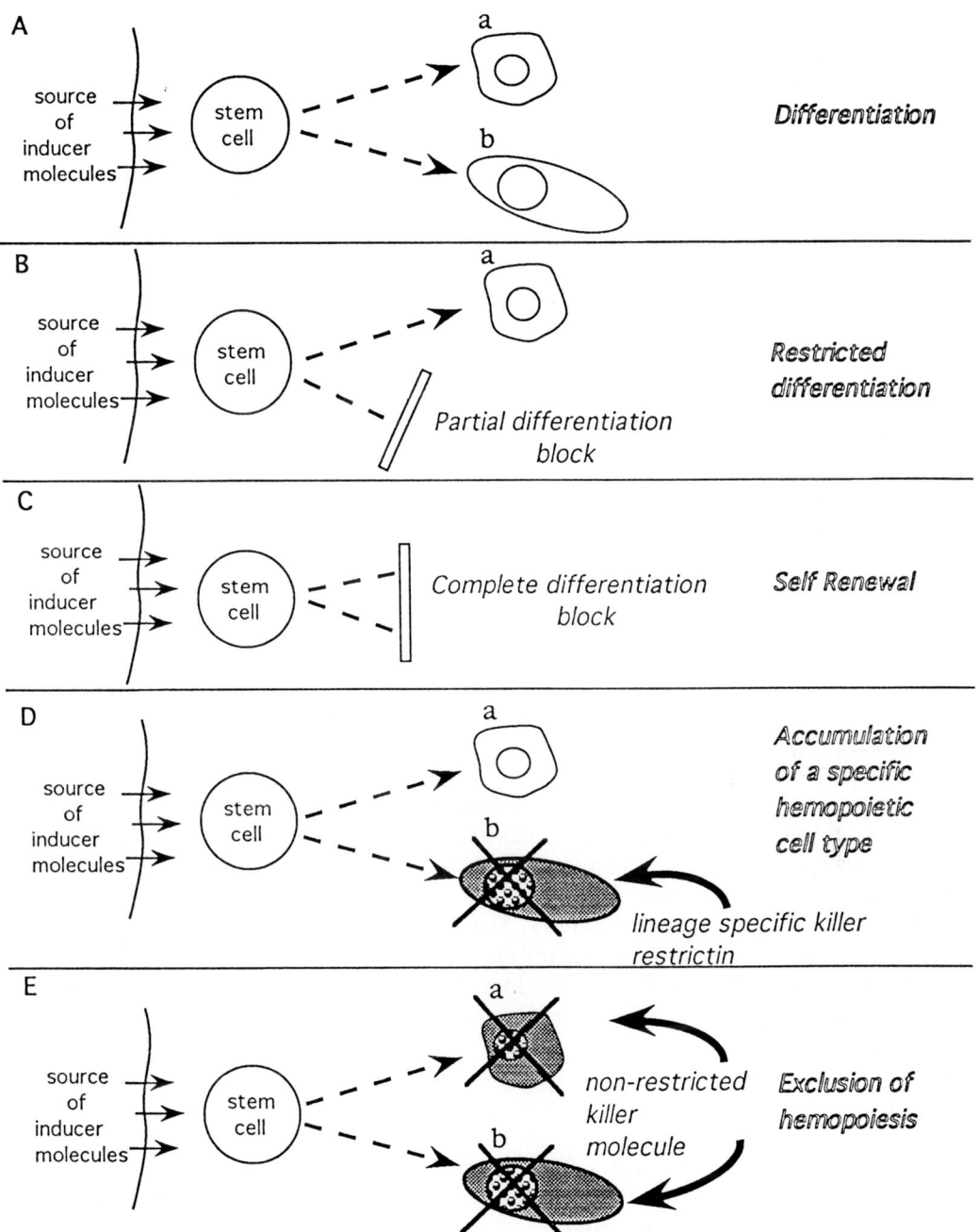

Fig. 1: Schematic view of some of the major steps in the regulation of hemopoiesis. The hemopoietic stem cell is shown to be capable of differentiating along various differentiation pathways leading to mature cells of different lineages as exemplified by cells **a** and **b**. Small arrows indicate activity of differentiation inducing cytokines. Dashed arrows or lines indicate the direction of differentiation. Empty bars represent a block in differentiation. Thick arrows indicate activity of killer molecules. Cell death is demonstrated by dark crossed cells. Further explanations are detailed in the text.

capabilities which enable them to differentiate for example along differentiation pathways a and b. This unrestricted differentiation is clearly not what happens under most physiological conditions. Indeed, panel B indicates that stem cells are usually restricted. This restriction is performed by a molecule that blocks the flow of differentiation in direction b and leaves the stem cell with the option to differentiate in the direction of cell a only. A complete differentiation block shown in panel C results in the lack of production of mature progeny. The stem cell is thus left with the alternative of self renewal only. It is the restraint of differentiation that leads to self renewal rather than a specific induction of renewal that accompanies differentiation processes. This point is crucial and was deduced from experiments in which it was shown that there is an inverse relationship between differentiation and stem cell renewal in short term bone marrow cultures (Zipori, 1981, Zipori et al., 1981, Zipori & Sasson, 1980). Figure 1B demonstrates how a single cell type can accumulate at a specific site due to restriction in differentiation imposed on the stem cell. The same result can be achieved by a completely different mechanism, as shown in Fig.1D. Here, a killer restrictin, specific to cell b eliminates this cell type, thus allowing the accumulation of cell type a only. If the restriction is imposed on all types of mature progeny, then hemopoietic cell accumulation is completely blocked, as happens in non-hemopoietic organs. This process, in contrast to the one shown in Fig.1C, is not accompanied by the renewal of the stem cells which are lost during the differentiation process.

The various aspects of hemopoiesis can be summarized as follows: the earliest stem cell is endowed with high proliferative and repopulating potential. Its quiescent state and slow renewal is maintained by stromal factors of yet undefined nature. Some stem cells can be "awakened" by specific cytokines. After this step stem cells become cycling and are committed to differentiation. The lineage choice and rate of differentiation will depend on the quantity and quality of stimulators encountered. Strong differentiation and proliferation signals are countered by the secretion of feedback regulators from mature cells. The overall organization and localization of the cells are directed by lineage specific restrictins that create specific microenvironments.

References

Craig, R.W., and Buchan, H.L. (1989): Differentiation-inducing and cytotoxic effects of tumor necrosis factor and interferon-gamma in myeloblastic ML-1 cells. *J.Cell. Physiol.* 141, 46-52.

Graham, G.J., Wright,E.G., Hewick,W.R., Wolpe,S.D., and Wilkie, N..M., Donaldson, D., Lorimore,S., & Pragnell,I.B. (1990): Identification and characterization of an inhibitor of hematopoietic stem cell proliferation. *Nature* 344, 442-444.

Kalai, M., & Zipori, D. (1989): A quantitative assay for stromal dependant hemopoiesis In *Experimental Hematology today,1988*, ed. Baum, S.J., Dicke,K.A. & Pluznik D.H. New York Springer-Verlag 25-30.

Musashi, M., Yang,Y.-C., Paul,S.R., Clark,S.C., Sudo,T., and Ogawa,M. (1991): Direct and synergistic effects of interleukin 11 on murine hemopoiesis in culture. *Proc.Nat.Acad.Sci. USA* 88,765-769.

Perussia, B., Kobayashi, M., Rossi, M.E., et al. (1987): Immune interferon enhances functional properties of human granulocytes: the role of Fc receptors and effect of lymphotoxin, tumor necrosis factor and granulocyte-macrophage colony-stimulating factor. *J.Immunology* 38, 765-774.

Tamir, M., Eren.,R., Globerson,A., Kedar,E., Epstein,E., Trainin,N., & Zipori,D. (1990): Selective accumulation of lymphocyte precursosr cells mediated by stromal cells of hemopoietic origin. *Exp. Hematol.* 18,332-340.

Zipori, D. (1981): Cell interactions in the bone marrow microenvironment: Role of endogenous colony-stimulating activity. *J. Supramolecular structure and cellular biochemistry* 17,347-357.

Zipori, D. (1988): Modulation of hemopoiesis by novel stromal cell factors. *Leukemia* 2,9S-15S.

Zipori, D. (1989): Cultured stromal cell lines from hemopoietic tissues. In *Handbook of the hemopoietic microenvironment*, ed. M. Tavassoli, pp. 287-333. Clifton NJ: The Humana Press, Inc.

Zipori, D. (1992): The renewal and differentiation of hemopoietic stem cells. *FASEB J.* 6,2691-2697.

Zipori, D., Duksin,D., Tamir,M., Argaman,A., Toledo,J. & Malik,Z. (1985): Cultured mouse marrow stromal cell lines. II. Distinct subtypes differing in morphology, collagen types, myelopoietic factors and leukemic cell growth modulating activities. *J.Cell.Physiol.* 122,81-90.

Zipori, D., Friedman,A., Tamir,M., Silverberd,D., & Malik,Z. (1984): Cultured mouse marrow cell lines: interactions between fibroblastoid cells and monocytes. *J.Cell. Physiol.* 118,143-152.

Zipori, D. & Honigwachs-Shaanani, J. (1992): In *Growth restrictions in the regulation of hemopoiesis.* Lord,B.I. & Dexter T.E.M. Bailliere's Clinical Hematology 5, 741-752.

Zipori, D. & Sasson, T. (1980): Adherent cells from mouse bone marrow inhibit the formation of colony stimulating factor (CSF) induced myeloid colonies. *Exp. Hematol.* 8, 816-817.

Zipori, D., Sasson, T. & Frenkel, A. (1981): Myelopoiesis in the presence of stromal cells from mouse bone marrow: I. Monosaccharides regulate colony formation. *Exp. Hematol.* 9, 656-663.

Zipori, D. & Lee, F. (1988): Introduction of interleukin-3 gene into stromal cells from the bone marrow alters hemopoietic differentiation but does not modify stem cell renewal. *Blood* 71, 586-596.

Résumé

La structure et la fonction du système hématopoiètique sont présentées en vue de proposer la théorie des 'restrictines". Il est avancé que parmi les régulateurs de l'hématopoièse, il existe des antagonistes de la différenciation agissant comme facteurs d'autorenouvellement et fonctionnellement différents des cytokines bien connues de l'hématopoïèse. Il est alors suggéré que tout un ensemble d'inhibiteurs spécifiques de lignée est impliqué dans l'organisation du tissu hématopoïètique.

Restrictive microenvironments: the mechanism controlling stromal dependent myelopoiesis

Amnon Peled and Dov Zipori

Department of Cell Biology, The Weizmann Institute of Science, Rehovot, Israel

In-vivo and in vitro studies indicate that early events in the hemopoietic process are stroma-dependent. In order to study the mechanism by which stroma-dependent myelopoiesis is controlled, a series of stromal cell lines were intercompared for their ability to secrete factors that stimulate/inhibit hemopoiesis. Endothelial-adipocytes (14F1.1) were previously shown to support long-term hemopoiesis in culture, whereas osteoblasts (MBA-15), fibroblasts (MBA-1.1.1) and endothelial-like cells (MBA-2.1) exhibit various functions, but were unable to support long-term hemopoiesis. In correlation with this observation, conditioned medium (CM) from 14F1.1 cells was specifically capable of supporting long-term maintenance of myeloid progenitors, as well as stem cells from 5FU-treated bone marrow (BM). CM from all the other cell lines induced massive macrophage differentiation. When colony stimulating activity (CSA) in these conditioned media was measured, using a standard bone marrow assay, we observed an inverse realationship between the ability of the stromal cells to support hemopoiesis and the titer of the stimulating factors produced. This finding was further demonstrated by use of CSF-1 dependent cell line 14M1.4. In this assay 14F1.1 cells were the weakest producers of CSF-1. MBA-2.1 and MBA-1.1.1 stromal cells that produced high titers of CSA and CSF-1 elaborated high titers of inhibitors that yielded, in a standard bone marrow assay, an inhibitory dose response curve. The same CM also inhibited the growth of M1 myeloid cells, ABLS-8 pre-B cells, IL-3 dependent cell line MC-9 and MPC-11 plasmacytoma cells, whereas CM from 14F1.1 and MBA-15 cells were not inhibitory to both hemopoietic cell lines and BM cells. Neutralizing antibody to TGF-β abolished the inhibitory effect of MBA-2.1 CM, both on 5FU treated BM cells and M1 cells. Although TGF-β was also found in concentrated conditioned medium from 14F1.1, its concentration seemed to be very low. All BM stromal cell lines produced colony stimulating activities, but only those that had low inhibitory activities and poor capacity to induce differentiation could maintain long-term hemopoiesis. It is possible that an additional differentiation restraining activity also contributes to the phenomena observed. We suggest that stem cell renewal and expansion of progenitors occurs in microenvironments which are unrestrictive to cell growth and in which the stimuli driving cells to differentiate are poorly expressed and/or restrained by factor(s) released from the stroma.

A stroma-derived protein induces apoptotic cell death of plasmacytoma cell lines

N. Brosh, J. Honigwachs-Sha'anani, D. Sternberg, A. Levi and D. Zipori

Department of Cell Biology, The Weizmann Institute of Science, Rehovot, Israel

The natural process of hemopoiesis is carried out in close contact with cellular elements collectively called stroma. These cells stimulate the complicated events resulting in the proliferation and differentiation of blood and immune cells. Evidence has accumulated suggesting that stroma controls hemopoiesis by producing signals which impose growth restrictions on hemopoietic cells. We suggest that stromal cells produce inhibitory molecules which specifically inhibit the growth of certain cell lineages in a given stage of their differentiation. These inhibitory signals are called restrictins. One such stroma-derived molecule was shown to be a protein which exclusively restricts the growth of B cell tumors at a late stage of their differentiation, plasmacytomas and hybridomas. The protein, which we named restrictin-P does not affect B cell tumors representing an earlier maturation stages, nor any other tested lymphoid or myeloid tumor cell lines. Restrictin-P was purified by conventional biochemical methods including anion exchange chromatography, gel filtration and reversed phase HPLC. The mechanism by which restrictin-P arrests the growth of its target cells was elucidated. The viability of the cells was reduced within 16 hours of the exposure of the cells to the protein. Membrane integrity was shown to be altered after 8-17 hours. DNA synthesis was decreased within 12-16 hours. Analysis of this damage by gel electrophoresis revealed the typical ladder of DNA fragments. These events were shown to be independent of *de novo* protein synthesis. It was also shown that the protein induced plasmacytoma cells to accumulate at the G_0/G_1 stage of the cell cycle. Electron microscopy analysis of the exposed cells validated the data suggesting that restrictin-P killed its target cells in an apoptotic manner. The above properties of restrictin-P are compatible with the proposed role ascribed to restrictins as regulators of tissue organization.

Effects of interferons on IL-6 receptor expression in multiple myeloma cells

Ahmed Lasfar, Jeanne Wietzerbin and Christian Billard

INSERM U 365, Institut Curie, Section de Biologie, Pavillon Pasteur, 26, rue d'Ulm, 75231 Paris Cedex 05, France

Interferon-alpha (IFN-α) has shown efficacy in the therapy of several B cell malignancies, such as hairy cell leukemia and multiple myeloma (MM) and previous studies suggested that IFN-α may act by interacting with other cytokines. Interleukin-6 (IL-6) is the major tumor growth factor in MM. Interleukin-6 mediates pleiotropic functions through specific receptors (IL-6R) composed of an 80 kDa binding protein, associated with a non-ligand binding protein (gp130) which transduces the signal. Little is known on the regulations of IL-6R in B cells. Therefore, we investigated the regulation of IL-6R in two human MM cell lines.

Binding experiments with [^{125}I]-IL-6 showed that IL-6R were expressed at a high density on RPMI-8226 cells (15,000 receptors/cell), but no specific binding was detected on XG-1 cells, whose growth depends on the presence of exogenous IL-6. However, when IL-6 was removed from the culture medium, high affinity IL-6R appeared on the surface of XG-1 cells (5,300 sites/cell). Treatment of RPMI-8226 cells with IL-6 reduced the number of IL-6R without changing their affinity. This reduction was dose-dependent and was not affected by acid treatment which dissociates ligand-receptor complexes. Cross-linking experiments showed that the formation of one IL-6/receptor complex of 160 kDa markedly decreased upon IL-6 treatment, while an other complex of 190 kDa became undetectable. These data provide evidence for ligand-induced down-regulation of membrane IL-6R expression in myeloma cells. Treatment of RPMI-8226 cells with IFN-α, which inhibits the growth of these cells, stimulated IL-6R expression and increased the formation of the 160 kDa IL-6/receptor complex. This stimulation was specific for IFN-α, since IFN-γ reduced the number of IL-6R.

These data indicate that, in myeloma cells, IL-6R are differentially regulated by IL-6 and IFN-α. They also suggest that interferons may interact with IL-6 in malignant B cells via modulation of IL-6 receptors.

A recombinant soluble receptor for IgG inhibits G_1 progression of B cells stimulated by LPS or anti IgM

Caroline Bouchard, Daniel Choquet*, Wolf Herman Fridman and Catherine Sautès

*INSERM U 255, Institut Curie, 26, rue d'Ulm, 75231 Paris Cedex 05; *INSERM U 261, Institut Pasteur, 28, rue du Docteur-Roux, 75724 Paris Cedex 15, France*

Soluble forms of membrane receptors of hematopoietic cells play important roles in immunoregulation. Both in man and in mouse, soluble forms of low affinity receptors for the Fc portion of IgG (sFcγR) have been described in supernatants of cells of the immune system (T cells, B cells, macrophages and Langherans cells) and in biological fluids such as serum and saliva (Fridman et al., 1992). The so-called "sFcγRII" are produced by proteolysis of the extracellular portion of the corresponding membrane receptor.
We have obtained a murine recombinant sFcγRII containing the two extracellular domains of FcγRII, by transfection of eukaryotic cells with a cDNA mutated by insertion of a stop codon (Varin et al., 1989). After purification to homogeneity, we showed that sFcγRII inhibited in a dose-dependent way cell proliferation -as measured by [^{3}H] Thymidine incorporation- of resting B lymphocytes stimulated by LPS or by Fab'2 fragments of anti-IgM antibodies. IgM and IgG$_3$ secretion by LPS stimulated B cells was also decreased in presence of sFcγRII. Kinetics experiments showed that sFcγRII could be added till 24 hours after initiation of the culture to exert its inhibitory effect. The effect(s) of sFcγRII on the early activation events induced by LPS or Fab'2 fragments of anti-IgM antibodies were studied. sFcγRII had no significant effect on the phosphorylation of proteins on tyrosine residues and the transient rise of intracellular calcium - measured at a single cell level- which occurs after stimulation by anti-IgM. The entry of B cells in cell cycle was analyzed by FACS by measuring cell size and expression of cell surface markers of the G_0 and G_1 phases of the cell cycle on LPS or anti-IgM stimulated B lymphocytes (Amigorena et al. 1989). sFcγRII did not affect the increase of expression of class II antigens which occurs before the activated B cells enter G_1. However it inhibited the cell enlargement and the induction of Transferrin receptor, a marker of the late G_1 phase. Thus, sFcγR inhibits B cell proliferation by blocking early events of the G_1 phase of the cell cycle.

REFERENCES

Amigorena S., Bonnerot C., Choquet D., Fridman W.H. and Teillaud J.L. (1989) : FcγRII expression in resting and activated B lymphocytes. Eur. J. Immunol. 19 : 1379-1385.
Fridman W.H., Bonnerot C., Daëron M., Amigorena S., Teillaud J.L. and Sautès C. (1992) : Structural bases of Fcγ receptor functions. Immunological Reviews. 125 : 49-76.
Varin N., Sautès C., Galinha A., Even J., Hogarth P.M. and Fridman W.H. (1989) : Recombinant soluble receptors for the Fcγ portion inhibit antibody production *in vitro*. Eur. J. Immunol. 19 : 2263-2268.

Regulation of myeloblastin expression by retinoic acid in acute promyelocytic leukemia

Paola Ballerini, Catherine Labbaye and Yvon E. Cayre*

*Departments of Physiology and Medicine, Columbia University, New York, NY 10032, USA; *Département d'Hématologie, Hôpital Saint-Antoine, Paris, France*

Myeloblastin (mbn) is a serine protease involved in the control of growth and differentiation of human leukemic cells (Bories et al., 1989). In the promyelocyte-like human leukemia cell line HL-60, this protease is inhibited during retinoic acid (RA) - induced differentiation. RA has been shown to play a major role in myeloid differentiation and achieve temporary remissions in patients with by acute promyelocytic leukemia (APL) (Huang et al., 1988; Castaigne et al., 1990). These remissions are probably due to the differentiation of the leukemic cells in vivo. One of the genes that mediates the RA effects in vivo is the RA-Receptor alpha (RAR α) which is located on chromosome 17q21. The t(15;17) translocation specific of the APL involves the RARα gene (Borrow et al., 1990) and creates a PML-RARα fusion gene between the C terminus of the Retinoic Receptor alpha (RAR α) and the amino terminal region of a new locus PML (Kakizuka et al., 1991; de Thé et al., 1991). It has been suggested that the PML-RARα fusion proteins might be incapable of responding to normal levels of RA for regulating genes that are critical to differentiation. The hybrid gene transcripts are highly heterogeneous. Two major isoforms, type A and type B, have been sofar cloned and used for in vitro assays of the transactivating capability on known response elements (Kastner et al., 1992). A few observations indicate that the transactivating properties of the chimeric proteins may differ from the wild type RAR α product. A certain variability of effects was observed depending on the cells and the promoter elements tested. Until now we lack information about the effects of the PML-RAR translocation on endogenous genes that are involved

in the control of cell growth and differentiation of myeloid leukemia cells. Expression of mbn is inhibited by RA treatment in different human leukemia cell lines. The mechanism(s) of such down-regulation have been investigated in two RA-sensitive cell lines, HL-60 and NB4; the latter unlike HL-60 cells carries the t(15;17). We observed that mbn expression was differently regulated in these two cell lines. In HL-60 cells mbn was transcriptionally down-regulated within 2 hours of treatment by RA. In NB4 cells, though mbn expression was down-regulated after 6 days, this regulation was not transcriptional (Labbaye et al., 1993). In both HL-60 and NB4 cells, mbn was down-regulated with the same efficiency and within approximately the same time of treatment by other inducers of differentiation such as DMSO, vit.D3 and PMA. To verify whether the absence of the transcriptional regulation of mbn by RA in NB4 cells was due to an altered transregulating function of the translocated receptor PML-RAR α, we cloned a genomic sequence of 800 nucleotides upstream of the mbn initiation codon containing a putative promoter activity. This sequence revealed the presence of two potential CAAT boxes (position -28 et -56), a TATA box (position -24), a well conserved SP1 binding site and a few palindromic sequences scattered from position -150 to -600. The putative mbn promoter was subcloned in a CAT expression vector (pMBN-CAT) and the functional activity was subsequently tested in transfection experiments using different cell types with very low endogenous levels of RARα (COS, CV1 or Pyt21). These cells were also co-transfected with wild type RARα, PML-RARα receptors type A and B or PML expression vectors. A weak CAT activity was detected in the cells transfected with the pMBN-CAT alone, which was not significantly modulated by treatment with 10^{-7}M or 10^{-6}M RA concentrations. The cotransfection of pMBN-CAT and RARα resulted in a weak and inconstant up-regulation of the CAT activity weakly inhibited by the addiction of RA. On the contrary, a strong stimulation of CAT activity (up to 20 or 40 folds, depending of the cell type) was observed when cells were co-transfected by type A PML-RARα and treated with various concentrations of RA. The co-transfection of pMBN-CAT and PML alone did not affect CAT activity. Taken together these data favor the hypothesis that the differences in the regulation of mbn observed in HL-60 and NB4 cells could be due to the existence of the PML-RAR in the latter cells. To give further support to this hypothesis we tried to identify potential RA response elements (RARE) which are specific binding sites for the RARα on the genomic DNA. A few palindromic sequences were selected according to homologies with other response elements and tested in mobility gel shift assays with in vitro-transcribed and translated receptors, RAR α, RXRα, PML-RARα/A, PML-RARα /B et PML. One of them, RE-2, exhibiting a palindromic motif, was able to fix specifically RARα/RXRα heterodimers or RXR homodimers and to compete with a known RARE (DR5G) for the binding to the same receptors. Conversely, the PML-RAR type A and type B , alone or mixed with equimolar amount of RAR α or RXR receptors, were not able to bind (type A) or bind weakly (type B) to the RE-2 motif, while they retained the ability to bind efficiently to the DR5G. Moreover, the introduction of point mutations in the half of the palindromic sequence RE-2 resulted in the loss of the binding activity of RARα/RXRα. Further investigations are being developed to verify the function of this putative mbn-RARE in vivo and to confirm whether the absence of transcriptional control of mbn in NB4 cells rely on the abnormal binding of PML-RAR fusion product to RE-2.

AKNOWLEDGEMENTS: CAT and gel-shift assays will not have been possible without the generous help, lab space and material provided by Prof. P. Chambon and Dr. M.P. Gaub.

REFERENCES

Bories D., Raynal M.C. et al. (1989): Down regulation of a serine protease, myeloblastin, causes growth arrest and differentiation of promyelocytic leukemia cells. *Cell*, 59, 959-968.

Borrow J., Goddard A.D. et al. (1990): Molecular analysis of acute promyelocytic leukemia breakpoint cluster regions on chromosome 17. *Science*, 249, 1577-1580;

Castaigne S., Chomienne C. et al. (1990): All-trans retinoic acid as a differentiation theraphy for acute promyelocytic leukemia. I. Clinical results. *Blood*, 76, 1704-1709.

de Thé H., Laveau C. et al. (1991): The PML-RARα fusion mRNA generated by the t(15;17) translocation in acute promyelocytic leukemia encodes a functionally altered RAR. *Cell*, 66, 675-684.

Huang M.E., Ye Y.C. et al. (1988): Use of all-trans retinoic acid in the treatment of acute promyelocytic leukemia. *Blood*, 72, 567-572.

Labbaye C., Zhang J. et al. (1993): Regulation of myeloblastin messanger RNA expression in myeloid leukemia cells treated with all-trans retinoic acid. *Blood*, 81, 475-481.

Kakizuka A., Miller W.H. et al. (1991) Chromosomal translocation t(15;17) in human promyelocytic leukemia fuses RARa with a novel putative transcription factor. Cell, 66, 663-674.

Kastner P., Perez A. et al. (1992). Structure, localization and transcriptional properties of two classes of retinoic acid receptor a fusion proteins in acute promyelocytic leukemia (APL): structural similarities with a new family of oncoproteins. Embo J., 11, 629-642.

Résumé
La Myéloblastine (mbn) est une sérine protéase impliquée dans le contrôle de la croissance et de la différenciation de cellules leucémiques humaines. L'expression de cette protéase est inhibibée transcriptionnellement au cours de l'induction de la différenciation par l'acide rétinoïque. Dans le cellules NB4 qui possèdent la t(15;17) caractéristiques des leucémie aiguës promyélocytaires, la régulation transcriptionnelle de la mbn n'est pas observée. Afin de vérifier si l'absence de régulation transcriptionnelle dans les cellules NB4 est due à l'expression du gène de fusion PML-RARa qui résulte de la t(15;17) nous avons étudié l'activité fonctionnelle et la présence éventuelle de séquence de liaison du RARα sur le promoteur de la mbn.

Divergent regulation of cell surface proteases on human myeloblastic (HL-60) cells differentiated into macrophages or neutrophils. Comparison to normal myeloid cells

Amale Laouar and Brigitte Bauvois

INSERM U 365 «Interférons et Cytokines», Institut Curie, Section de Biologie, Pavillon Pasteur, 26, rue d'Ulm, 75231 Paris Cedex 05, France

Because there is recent evidence for a multifunctional role of cell surface proteases in cell differentiation and invasion (Bauvois, 1992), the regulation of such enzymes seems to be an important issue. The results of our investigation provide new information about the regulation of cell surface protease expression on myeloblastic precursors and their blood mature counterparts. We used the HL-60 cell line as a model of immature myeloid precursors, rHuGM-CSF treated HL-60 cells as a model of differentiated macrophages and retinoic acid treated HL-60 cells as a model of differentiated granulocytes. Using a spectrophotometric assay, we demonstrated this assay to be a sensitive and specific indicator of cell surface N-aminopeptidases and serine dipeptidyl peptidase IV (DPP IV) expression in HL-60 cell line as well as on normal myeloid cells. Both classes of protease activities significantly increased following granulocyte-macrophage colony stimulating factor (GM-CSF)-induced maturation of the HL-60 cell line and normal monocytes toward the macrophage pathway. This up-regulation was mainly due to the enhancement in cell surface protease number. In contrast, N-aminopeptidase expression was mainly down-regulated on HL-60 cells differentiated into granulocytes with retinoic acid and low affinity was paralleled with that expressed by normal blood granulocytes. Retinoic acid treatment also reduced soluble protease activity *in vitro* indicating that down-regulation of membrane aminopeptidases was not due to their proteolytic clip. HL-60 maturation into the granulocyte lineage however did not cause any modulation in DPP IV expression. Our data indicate that myeloid cell surface protease up and down cell surface expression correlates with the acquisition of a differentiated phenotype. Dysregulation in blood cell associated protease levels may be important in malignant pathology.

REFERENCE

Bauvois,B. (1992). Ectopeptidases, enzymes plurifonctionnelles. In Med. Sci. 8: 441-447.

In vitro, sensitivity of hematopoietic progenitors to NKH-1 positive cells

Dina Fardoun, Elisabeth Sumereau-Dassin, Gorette Lebrun-Texeira, Bernard Lenormand and Jean-Pierre Vannier

Department of Research on Hematopoietic Stem Cells and Extracelllar Matrix Molecules, School of Medecine, BP 97, 76803 St-Etienne-du-Rouvray Cedex, France

Summary - Peripheral blood mononuclear cells (PBMNCs) were depleted of NK cells by fluorescence activated cell sorter (FACS). Progenitor assay of NK-depleted cells revealed a significant decrease of CFU-GM ($p<0.01\%$). Addition of NK cells in the mixed culture suppressed the growth of CFU-GM, BFU-E and CFU-Mix in a dose response fashion. The role of NK cells in hematopoiesis seems to be more than accessory.

Natural Killer (NK) cells have been defined as large granular lymphocytes that do not express on their surface the CD3 antigen or any of the known T cell receptor chains (Lanier and al,1992) but do express CD16 and NKH-1 cell surface markers (Ellis and al, 1989). Upon activation, NK cells produce GM-CSF, IL-3 (Limb and al, 1989) and in certain conditions TNF α or INF δ (Kasahara and al, 1983; Robertson and al, 1992). The relationship between NK cells and hematopoietic precursors still remain unclear. Data are somewhat confusing, and both stimulation (Niemeyer and al, 1989) and inhibition (Miller and al, 1991) of various colony forming unit (CFU) have been described.To gain additional informations on the role played by NK cells in regulation of hematopoiesis, nonadherent (NA) PBMNCs were depleted of NK cells by FACS. For five experiments, FACS analysis indicated that NA PBMNCs were 16% ± 6% NKH-1+ (mean ± SD). Sorted NKH-1+ were 93% ± 5% on reanalysis by FACS. Both NKH-1+ and NK-depleted cells were collected and plated in a progenitor assay. Depletion of NK cells suppressed CFU-GM colony growth up to 67%. No significant effect on either BFU-E or CFU-Mix was observed. To test the possibility of a NK stimulation effect on CFU-GM, NK cells were added to NK-depleted cells in a dose response fashion. A progressive increase of CFU-GM numbers occured upon addition of NK cells. The maximal effect of NK cells on CFU-GM colony growth was obtained at NK/NK-depleted cells ratio of 0.15/1 (17±5 versus 4±5/10^5 cells; $p<0.01$). Decrease of CFU-GM (9±2; $p<0.05$), BFU-E (48±9 versus 88±43/10^5 cells) and CFU-Mix (1±1 versus 6±5/10^5 cells) was observed at ratio corresponding to 0.5/1. Our results suggest that stimulation or inhibition of colony growth depend on the number of NK cells present in the mixed culture. These are of importance considering the fact that after a bone marrow transplantation the number of NK cells is often increased (Yabe and al, 1990).

REFERENCES

Ellis TM, Fisher IM. (1989): Functional heterogeneity of Leu 19$^{bright+}$ and Leu 19^{dim+} lymphokine-activated killer cells. J.Immunol. 142, 2949-2954.

Kasahara T, Djeu JY and al. (1983): Capacity of lymphokine-activated killer to produce multiple lymphokines: IL-2, INFδ and colony stimulating factor (CSF). J.Immunol. 131, 2379-2385.

Lanier LL, Spits H and al. (1992): The developmental relationship between NK cells and T cells. Immunol.Today. 13 (10), 392-395.

Limb GA, Meager A and al. (1989): Release of cytokines during generation of lymphokine-activated killer cells by IL-2. Immunol. 68, 514-519.

Miller SJ, Verfaillie C and al. (1991): Adherent lumphokine-activated killer cells suppress autologous human normal bone marrow progenitors. Blood. 77, 2389-2395.

Niemeyer MC, Sieff AC and al. (1989): Hematopoiesis in vitro coexists with NK cells. Blood. 74, 2376-2382.

Robertson MJ, Soiffer JR and al. (1992): Response of NK cell to NK cell Stimulatory Factor (NKSF) : cytolytic activity and proliferation of NK cells are differentialy regulated by NKSF. J.Exp.Med. 75, 779-788.

Yabe H, Yabe M and al. (1990): Increased numbers of CD8+CD11+, CD8+CD11- and CD8+Leu7+ cells in patients with chronic graft-versus-horst desease after allogenic bone marrow transplantation. Bone Marrow Transplant. 5 (5), 295-300.

This work is supported by ARC and CDLCC (Seine-Maritime et Eure).

The role of Natural Killer (NK) Cells on human umbilical cord blood erythropoiesis

M. Hamood, F. Corazza, W. Bujan, E. Sariban and P. Fondu

Laboratory of Experimental Hematology, Brugmann University Hospital and Hematology/Oncology Unit, Pediatric University Hospital, Free University of Brussels, Brussels, Belgium

The Natural Killer cells are defined phenotypically as lymphocytes expressing the antigen CD56 and lacking CD3. From previous studies they are assumed to play a role on bone marrow erythropoiesis (Mangan et al, 1984). Although lymphocytes in human umbilical cord (HUC) blood have been shown immunologically naive, little is known about the role of NK cells on HUC erythropoiesis.

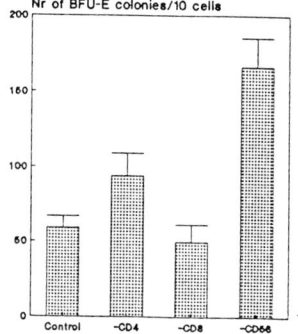

Fig 1: BFU-E growth before and after cell depletion (mean +/- SEM).

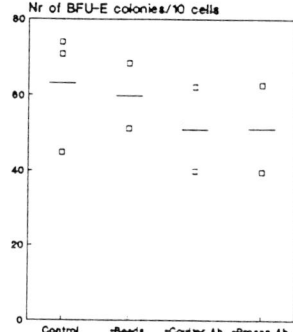

Fig 2: No significant effect of beads or antibodies on BFU-E growth.

In this study, we investigated the role of NK cells on HUC blood early erythropoietic progenitors (BFU-E). Twenty-five normal HUC blood samples were depleted of CD56+ cells by immunomagnetic beads coated with CD56 monoclonal antibodies (MoAb). The CD56+ cell depleted preparations demonstrated about 2 fold increase in BFU-E over non-depleted preparations when stimulated by erythropoietin to form colonies in plasma-clot medium (fig1). This stimulatory effect is apparently not due to artifact stimulation of other accessory cells by the MoAb or by the beads used in depletion procedure since the addition of these materials separately to culture did not exert any stimulatory effect on BFU-E growth (fig2). Whether the inhibitory effect of CD56+ cells on BFU-E was mediated by lymphokines or by cell to cell interaction is a matter of question. Further investigations are undergoing in our laboratory in this respect.

References: Mangan KF, Hartnett ME, Matis SA, Winkelstein A, Abo T. 1984: Natural Killer cells suppress human erythroid stem cell proliferation in vitro. Blood, 63: 260-269.

This work was supported by a grant of the Fonds National de la Recherche Scientifique (Télévie).

Effects of cultured medium derived from purified lymphokine-activated killer cells(pLAK-CM) on survival and growth of normal blood cells and leukemia cells

Ning Mao, Cheng-Yong Jiang, Min-Wei Zhang, Xiu-Shen Li, Fei-Zi Jiang, De-Lin Du and Pei-Hsien Tang

Department of Experimental Hematology-Oncology, Institute of Basic Medical Sciences, Beijing 100850, PO Box 130, PR China

The mechanisms of negative effects of lymphokine-activated killer(LAK) cells on leukemia cells were not known well so far. Whether LAK cells could secrete some factors/moleculars and what were its effects on normal blood cells and leukemia cells? To anwser these questions, human LAK cells were purified by a semisolid-liquid cloning-culture procedure with IL-2 stimulation and pLAK-CM was obtained by centrifugation from supernatant of single clone at 3-day during 4-6 weeks of long-term culture. The biological activity of pLAK-CM against target cells was studied with MTT assay, ^{3}H-TdR up-take assay, clonogenic assay and nitroblue tetrazolium(NBT) test. After 3-day culture with $2-4 \times 10^4$/well in 96-well flat-bottomed plates, pLAK-CM showed different degrees of cytotoxicity to a variety of leukemia cell lines including NK-sensitive K562 and NK-resistant HL-60, KG-1, U937, Raji, JM-1, HPB-ALL, Molt-4 and CCRF-CEM cells, but not normal human bone marrow mononuclear cells, peripheral blood mononuclear cells, and autologous/allogenic LAK cells in MTT assay. The HL-60 cells were more sensitive to cytotoxicity of pLAK-CM and the cytotoxic level was increased as the increasing of both the final concentrations of pLAK-CM from 2.5% to 50% and the numbers of HL-60 cells plated between $1-10 \times 10^4$/wells. The cytotoxicity was up to 80%($p<0.01$) with addition of 50% pLAK-CM. Moreover, a dose-dependent inhibition on colony-formation of HL-60 cells in agar semisolid cultures was observed when pLAK-CM incubated with HL-60 cells for 6-day at the concentrations of 5-40%. In contrast, the growth of CFU-GM from normal bone marrow was increased by 35% and inhibited by 50% when pLAK-CM was added at final concentrations of 5-15% and 40, respectively. The results of ^{3}H-TdR up-take also showed that pLAK-CM inhibited the DNA synthesis of HL-60 cells. The IFN-γ activity(100-300 U/ml), TNF activity(<10pg/ml) and GM-CSF activity(10-40 U/ml) were found in pLAK-CM. Besides, pLAK-CM could increase the number of NBT-positive cells of HL-60 by 50-80%,which reflected the tendency of cell-differentiation to maturation. Our results suggested that pLAK-CM had the potential of selective cyotoxicity to leukemia cells and their progenitors, and inducing them differenciation to maturation, which mediated by one or more known and unknown different soluble factors and which may be an important part of mechanisms of LAK cells killing/inhibiting leukemia cells. The secretion of soluble factors by pLAK cells did not depend on the interaction between LAK cells and target cells.

In vitro effect of non-steroidal anti-estrogenic drugs on B lymphocytes from chronic lymphocytic leukemic patients

M. Gorodinsky, A. Dvilansky and I. Nathan

Faculty of Health Sciences, Clinical Biochemistry Unit and Department of Hematology, Ben-Gurion University, Beer-Sheva, Israel

Chronic lymphocytic leukemia (CLL) is the most common disease of the elderly with an incidence of 3 per 100,000 persons in western countries. In over 95 percent of the cases, the clonal expansion is of B-cell lineage. Triphenylethylene non-steroidal antiestrogens are known inhibitors of mammary cancer cell proliferation. Recent evidence indicates that the antiestrogen receptors, as distinct from the estrogen receptors, are present in normal and neoplastic tissues. Moreover, these drugs may affect cell proliferation through an estrogen-independent mechanism. The molecular basis of this anti-tumor activity is not as yet clear. The present study was undertaken in order to evaluate the sensitivity of CLL cells to the various non-steroidal antiestrogen derivatives.

Cells were isolated from patients, and were incubated with and without drugs for 4 days. Thereafter, cell viability was evaluated using the 3-[4,5-dimethylthiazol-2-yl]-2,5-diphenyltetrasolium bromide (MTT) assay in vitro.

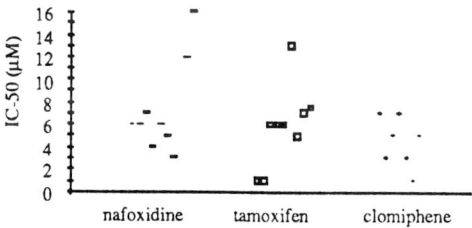
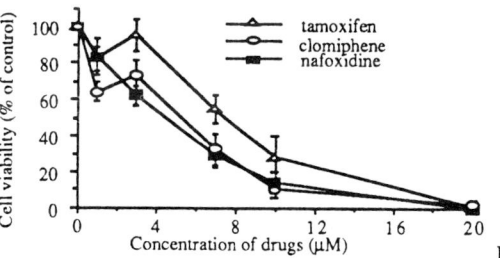

Fig 1. IC_{50} values of B CLL cells.　　　Fig 2. Effect of antiestrogens on B CLL cells.

Figure 1 depicts the IC_{50} values of the various drugs of B-CLL cells. Cells derived from the various patients revealed different sensitivity to each derivative. For most of the patients, IC_{50} values were in the range of 3-7.5 μM. Three patients had IC_{50} values in the 1μM range while three cases were relatively resistant (IC_{50} above 10μM.). Nevertheless, most of the patients were more sensitive to nafoxidine and clomiphene than to tamoxifen at the range up to 20μM (Figure 2). The effect of the drugs on cell viability in the presence and absence of estradiol at 10^{-8} M concentration was not significantly changed. This suggests that the cytotoxic effect was mediated mainly through the non-estrogenic receptor. 12-O-tetradecanoylphorbol 13-acetate (TPA), a well known activator of protein kinase C (PKC), at a concentration of 10^{-9} M antagonized the cytotoxic effect of nafoxidine and clomiphene. Sensitivity to the T lymphocytic human line (MOLT-3) to the drugs was also examined. The results showed that cells were less sensitive to tamoxifen than to nafoxidine and clomiphene, similar to the B CLL studies already reported. In the present study, we report that tamoxifen, a non-steroidal antiestrogen, revealed a cytotoxic effect on CLL cells through a non-estrogen receptor mediated mechanism. This is supported by two facts: 1) estradiol did not abrogate the effect and 2) two other triphenylethylene antiestrogens, nafoxidine and chlomiphene, are known to be less effective in their antiestrogenic activity. However, they were more effective in their cytotoxic activity on CLL cells, again supporting the notion that this effect was not mediated through the estrogen receptor. Our results, showing that TPA antagonizes the effect of the drugs, may suggest that PKC can down regulate the antiestrogen receptor or signal transduction mechanism, as shown for other growth factors. The effective concentrations against CLL cells are at the same therapeutic levels, thus implying therapeutic potential for their use.

IV. Molecular approaches for the manipulation of hematopoietic progenitors

IV. Approches moléculaires pour la manipulation des progéniteurs hématopoïétiques

The c-kit protooncogene in normal and malignant human hematopoiesis

Alan M. Gewirtz

Hematology/Oncology and Molecular Diagnosis Sections, Departments of Pathology and Internal Medicine, University of Pennsylvania School of Medicine, Philadelphia, Pennsylvania 19104, USA

INTRODUCTION

The c-kit protooncogene is the normal cellular homologue of v-kit, the Harvey-Zuckerman 4 feline sarcoma virus oncogene (Besmer et al, 1986). c-kit encodes a dimeric transmembrane tyrosine kinase receptor. The receptor belongs to the immunoglobulin supergene family and is homologous to the colony stimulating factor (CSF)-1 and platelet derived growth factor (PDGF) receptors (Yarden et al, 1987). The mouse c-kit gene maps to chromosome 5 and is allelic to the white spotting locus (W) (Geissler et al, 1988). W mice are sterile, have abnormal coat coloration, and manifest abnormal hematopoiesis characterized by a profound macrocytic anemia (Russel, 1979). The marrows of these animals have greatly decreased numbers of erythroid burst forming units (BFU-E). These, and other observations established the importance of the c-kit receptor in murine hematopoietic stem cell development, and predicted its importance in human hematopoiesis as well (see Ratajczak et al, 1992a for recent review). Nevertheless, to gain a more complete understanding of the physiologic function of the KIT receptor in human hematopoiesis, we disrupted the gene's function in hematopoietic cells with antisense oligodeoxynucleotides (Ratajczak et al, 1992c). The results of our studies, which are summarized below, suggested that c-kit plays a predominant role in normal erythropoiesis, and that it might be of importance in regulating some types of malignant hematopoietic cell growth.

MATERIALS AND METHODS

Cells: Light density marrow mononuclear cells (MNC) were obtained from normal consenting donors, or patients with acute myelogenous (AML) or lymphatic (ALL) leukemia, Philadelphia chromosome-positive chronic myelogenous leukemia (CML), and polycythemia vera (PV). They were partially enriched for progenitor cells as previously described (Gewirtz and Calabretta, 1988 9; Ratajczak et al, 1992b and c).

Oligodeoxynucleotides: Unmodified, 18-nucleotide (nt) oligodeoxynucleotides (oligomers) were synthesized, purified and utilized as previously reported (Gewirtz and Calabretta, 1988; Ratajczak et al, 1992 b and c).

Cell Culture: Assays for hematopoietic progenitor cells of varying lineages were carried out as previously reported ((Gewirtz and Calabretta, 1988; Ratajczak et al, 1992b and c).

Reverse Transcription-Polymerase Chain Reaction (RT-PCR): Total RNA was extracted from cells, reverse transcribed, and detected by Southern Blotting as previously reported (Ratajczak et al, 1992b and c).

Statistical Analysis: Statistical significance of differences in numbers of colonies arising in the duplicate control, sense, or antisense plates was determined using the Student t test for unpaired samples. p values less than 0.05 were judged to be of statistical significance.

RESULTS

Expression of c-kit mRNA in normal MNC

We initially determined if c-kit expression was detectable in normal MNC, and potential relationships between c-kit gene expression and cellular metabolic activity. To carry out these studies A-T- MNC were made quiescent by culturing cells for 24 hours at 4°C in medium containing 2% normal human AB serum. The cells were then rapidly warmed to 37°C, and placed in serum containing medium (5 % v/v) supplemented with recombinant human interleukin-3 (20 U/ml) and erythropoietin (5 U/ml). Immediately (Time 0), and at 2, 8, 12, 24, 36 and 48 hour intervals thereafter, RNA was extracted from cells to detect c-kit expression using RT-PCR.

At 4°C, c-kit expression was not detectable using a single step PCR reactions suggesting that expression was minimal or non-existant. However, by two hours after warming in growth factor and serum replete medium, c-kit expression was unequivocally detectable. Further, expression appeared to increase over the ensuing 24 hours, at which time expression appeared to peak, and persisted through 48 hours. The results of these experiments suggested the possibility that c-kit expression might also be linked to cell cycle activity.

Effect of c-kit oligomers on Normal Human Hematopoietic Progenitor Cell Growth

We then determined if c-kit expression in MNC could be abrogated in a sequence specific manner by exposure to c-kit antisense oligodeoxynucleotides (Fig. 1). A-T- MNC (2 x10^6) were again incubated at 4°C for 24 hours in a low serum (2% v/v) containing medium. Cells were re-warmed to 37°C and exposed to growth factors in the presence of c-kit oligomers. After 36 hours, total RNA was extracted for RT-PCR detection of c-kit message. c-kit mRNA was expressed at very low level after 24 hours at 4°C. Thirty six hours after warming, c-kit mRNA was detected in control cells and in cells exposed to sense and scrambled sequence oligomers. In contrast, antisense treated cells had no detectable c-kit mRNA. Accordingly, inhibition of c-kit expression was highly efficient and sequence specific.

Figure 1: Expression of c-kit mRNA in normal MNC exposed to c-kit oligomers. A-T- MNC were cultured as described in the text. Oligomers were added at time zero (100 µg/ml), and 50% of the initial dose was added again 18 hours later. At Time 0 and thirty-six hours after the first addition of oligomers, total RNA was extracted for RT-PCR analysis of c-kit expression. Lane 1-Time 0 control. Lane 2-Thirty-six hour control. Lane 3- Cells exposed to c-kit sense oligomers. Lane 4- Cells exposed to c-kit antisense oligomers. Lane 5- Cells exposed to scramble sequence c-kit oligomers (Adapted from Ratajczak et al, 1992c).

The effect of c-kit antisense, and control sequence oligomers on normal hematopoietic progenitor cell cloning efficiency and development was then investigated by assessing the effect of oligomer exposure on CFU-E, BFU-E, CFU-GM, and CFU-MEG derived colony growth. As shown in Table 1, c-kit antisense oligomers inhibited erythroid colony formation in a dose dependent fashion. Inhibition was sequence specific as neither sense, nor scrambled sequence oligomers significantly affected colony growth in comparison to untreated controls. At the highest doses employed, inhibition of CFU-E and BFU-E, was ~75% and ~71% respectively. Residual colonies were also much smaller than those of the untreated controls. Nevertheless, maturation of the cellular constituents of the colonies appeared normal as assessed by the cells' ability to synthesize hemoglobins F and A. In contrast to the erythroid colony results, CFU-GM and CFU-MEG derived colony formation was unaffected by exposure to any of the oligomers, at any of the doses employed (Table 1). Similar results were obtained after exposure of CD 34+ cells to c-kit antisense oligomers.

Table 1: Effect of c-kit oligomers on A$^-$T$^-$ colony formation. A$^-$T$^-$MNC were cultured and exposed to the oligomers as described in text. Colony counts shown are from two or three individual studies at the highest concentration tested (Adapted from Ratajczak et al, 1992c).

Progenitor Cell Type	STUDY	Control	Sense	Scrambled Sequence	Antisense
CFU-E	1	182, 209	153, 142	119, 128	33, 59
	2	1943, 543	1635, 1135	627, 649	243, 213
	3	148, 110	129, 176	149, 206	97, 107
		Mean (±SE)= 522±291	Mean (±SE)= 562±268	Mean (±SE)= 313±103	Mean (±SE)= 125±34
BFU-E	1	133, 152	117, 106	94, 64	60, 149
	2	534, 392	601, 249	273, 246	126, 113
	3	206, 172	215, 258	162, 246	59, 51
		Mean (±SE)= 265±66	Mean (±SE)= 258±74	Mean (±SE)= 181±36	Mean (±SE)= 76±14
CFU-GM	1	212, 189	231, 179	282, 193	195, 220
	2	412, 408	395, 421	457, 384	407, 471
	3	217, 241	230, 237	201, 199	293, 187
		209, 246	Mean (±SE)= 282±41	Mean (±SE)= 286±46	Mean (±SE)= 296±49
		Mean (±SE)= 280±42			
CFU-MEG	1	114, 107	133, 117	154, 113	127, 112
	2	93, 100	58, 52	53, 40	47, 54
		Mean (±SE)= 104±5	Mean (±SE)= 90±20	Mean (±SE)= 90±27	Mean (±SE)= 85±20

To provide additional proof that c-kit antisense mediated inhibition of erythropoiesis was due to the absence of c-kit receptor function, we demonstrated that BFU-E responsiveness to kit ligand (KL) could be abolished in a sequence specific manner after exposure to c-kit oligomers. Accordingly CD34+ MNC (2×10^4) were cloned in the presence of erythropoietin (5U/ml) and KL (100ng/ml) alone, or with sense, antisense, or scrambled sequence c-kit oligomers (final concentration=150 µg/ml=~26µM). In four experiments, 191 ± 19 BFU-E (mean ± SD) were grown in the presence of the growth factors alone. These numbers were not statistically different from those cloned with sense (183 ± 29; p=.654) or scrambled sequence oligomers (180 ± 20; p=.758). In the presence of the c-kit antisense oligomers BFU-E derived colony formation was completely abolished (0.4 ± 0.7; p<.0001) demonstrating that kit receptor was no longer present to interact with its ligand.

Role of c-kit receptor in malignant hematopoietic cell growth

The data generated in the above described experiments suggested that the c-kit encoded receptor played an important role in normal hematopoiesis, especially for erythroid cell development. c-kit mRNA and protein has been detected in some acute leukemias well as suggesting a role for the kit encoded protein in malignant hematopoietic cell growth as well. Nevertheless, its importance in the pathogenesis of these diseases remains unclear since expression in primary leukemias is quite variable (Wang et al, 1989; Lerner et al, 1991) and, to our knowledge had not been examined in other malignant hematopoietic diseases such as the myeloproliferative disorders. To address this issue, A⁻T⁻ MNC were obtained from patients with a variety of hematologic malignancies, and exposed to the c-kit oligomers. The effect of oligomer exposure on the ability of malignant CFU-GM to form colonies in semi-solid media was then assessed (Table 2). A total of 20 patients were studied; three with acute lymphocytic leukemia, four with acute myelogenous leukemia, ten with chronic myelogenous leukemia, and four with polycythemia vera. CFU-GM from 10 of these patients were inhibited by exposure to the c-kit antisense oligomers. Of these, 8 were derived from patients with myeloproliferative disorders (all of whom had PCR documented expression of c-kit mRNA) suggesting that progenitor cells isolated from patients with these disorders were more dependent on c-kit function than those with acute leukemia. Five responding patients had chronic myelogenous leukemia (CML). Interestingly, when residual colonies were probed for bcr-abl expressing cells by RT-PCR, bcr-abl expression was substantially decreased or undetectable in cases where colony inhibition was observed. In patients with PV, BFU-E (92%±3) and, unexpectedly, CFU-GM derived colony formation were significantly inhibited by the c-kit antisense oligomers.

Table 2: Effect of c-kit oligomers on malignant hematopoietic cell colony growth (CFU-GM). Cells were exposed to oligomers and cultured as described in the text.

Disease Type	No. Pts Studied	No. Pts With Decrease in Colonies	% Decrease Colony Number
Acute lymphocytic leukemia	3	1	68%
Acute myelogenous leukemia	4	1	63%
Chronic myelogenous leukemia	10	5	65±26 (Mean±SD)
Polycythemia Vera (PV)	4	4	74%±24 (Mean±SD)
TOTAL	21	11	Av=68%

DISCUSSION

The importance of the c-kit receptor in murine hematopoiesis is undisputed. Inferential data that the receptor played a significant role in human hematopoiesis is also now substantial. The c-kit ligand appears to have little colony stimulating activity on its own, but instead acts synergistically with other growth factors, in particular with interleukin-3, to enhance colony formation (Broxmeyer et al, 1991). Nevertheless, the importance of the c-kit ligand's receptor in regulating normal human hematopoietic cell development was not directly demonstrated until these knockout studies were performed. The biological function (s) governed by the c-kit receptor are not known in great detail. However, given its similarity to other growth factor receptors, it is likely a component of a growth factor signal transduction apparatus. Other functions, such as serving as a cell surface adhesion molecule receptor, are also possible (Flanagan et al, 1990).

Though the exact function of the receptor remains undefined, our data provided some insight into factors which appear to regulate c-kit expression in normal bone marrow mononuclear cells. In this regard, we found that c-kit expression was virtually undetectable in cells that had been rendered quiescent by exposure to cold and a low serum environment, and that expression increased after cells were re-warmed and then stimulated with IL-3 and GM-CSF (Figure 1). Whether the augmented c-kit expression we found was due to increased metabolic activity of the cell, or was related to initiation of cell cycle activity is uncertain. Nevertheless, since c-kit expression appeared relatively early after stimulation (~2 hours) initial expression in G_0 cells would not appear dependent on activation of "late" cell cycle genes such as c-myb. Nevertheless, since perturbation of c-myb function led to downregulation of c-kit expression at thirty six hours, without affecting β-actin mRNA expression, c-kit may be regulated by different mechanisms in cells that are cycling, and in a manner independent of cellular metabolic activity. These regulatory

In addition to elucidating biological factors which regulate c-kit expression, the experimental approach employed has also generated some unexpected findings about the importance of c-kit expression in human hematopoietic cell development. First, the c-kit receptor does not appear to play a critical role in either myeloid or megakaryocyte development, at least at the early progenitor cell level. Whether development from even earlier progenitor cells than those assayed, or from cells with more "stem" like properties would also be unaffected is unclear. In marked contrast to these results however, erythroid colony development at both the BFU-E and CFU-E level appeared to be markedly dependent on c-kit function. This dependence appeared to be most critical during the early events of colony formation since maturation in residual colonies was otherwise normal. Second, the results also strongly imply the existence of two distinct subsets of erythroid progenitors with different dependence on the c-kit ligand, and therefore the c-kit receptor. These subsets may differ in their maturational state, and other as yet unidentified properties. Finally, the c-kit receptor may be of importance in regulating growth of some malignant hematopoietic cell types as well. In this regard, it was found that growth of granulocyte colony forming units from most cases of acute leukemia, either myeloid or lymphoid, were not inhibited by exposure to the c-kit antisense oligomers. In contrast, growth of granulocyte colony forming units derived from patients with chronic myelogenous leukemia and polycythemia vera were much more sensitive to the inhibitory effects of the c-kit antisense oligomers. The reason for this discrepancy is unclear. One possible explanation may be that CFU from patients with acute leukemia are blocked at a maturation level where c-kit function is relatively unimportant for cell growth. This situation would contrast with CFU from patients with myeloproliferative disorders who are likely to have progenitor cells at varied levels of maturation. Alternatively, acute leukemia CFU may have generated alternate ways of carrying out c-kit related functions so that receptor deprivation does not adversely effect cell growth. Accordingly, c-kit may subserve different functions in normal as opposed to malignant hematopoietic cells since CFU-GM derived granulocyte colony formation was unaffected by c-kit disruption in the former but led to inhibition of colony formation in the latter cell types. The possiblity that these differences can be exploited for therapeutic purposes is currently under investigation in our laboratory.

ACKNOWLEDGEMENTS: This work was supported in part by U.S. Public Health Service grants CA36896, CA01324, and CA 54384. Dr. Gewirtz is the recipient of a Research Career Development Award from the National Cancer Institute. The editorial assistance of Ms. Elizabeth R. Bien is gratefully acknowledged.

REFERENCES

Besmer, P., Murphy, J.E., George, P.C., Qiu, F., Bergold, P.J., Ledermann, L., Snyder, H.W., Brodeur, D., Zuckerman, E.E. & Hardy, W.D. (1986): A new acute transforming feline retrovirus and relationship of its oncogene v-kit with the protein kinase gene family. Nature 320, 415-421.

Broxmeyer, H.E., Cooper, S., Lu, L., Hagve, G., Anderson, D., Cosman, D., Lyman, S.D. & Williams, D.E. (1991): Effect of murine mast cell growth factor (c-kit proto-oncogene ligand) on colony formation by human marrow hematopoietic progenitor cells. Blood 77, 2142-2149.

Geissler, E.N., Ryan, M.A. & Housman, D.E. (1988): The dominant white spotting (W) locus of the mouse encodes the c-kit proto-oncogene. Cell 55, 185 - 192.

Gewirtz, A.M. & Calabretta, B.(1988): An anti-sense oligodeoxynucleotide to proto-oncogene c-myb inhibits normal human hematopoiesis in vitro. Science 242, 1303-1306.

Ratajczak M. & Gewirtz A. (1992a): Role of the c-kit protooncogene in normal and malignant human hematopoiesis. Int J Cell Cloning 10:205-215.

Ratajczak M.Z., Hijiya N., Catani L., DeRiel K., Luger S.L., McGlave P., & Gewirtz, A. (1992b): Acute and Chronic Phase Chronic Myelogenous Leukemia Colony Forming Units Are Highly Sensitive to the Growth Inhibitory Effects of C-myb Antisense Oligodeoxynucleotides. Blood 79, 1956-1961.

Ratajczak M.Z., Luger S.M., De Riel K., Abraham J, Calabretta B., Gewirtz A.M. (1992c): Role of the KIT protooncogene in normal and malignant human hematopoiesis. Proc Natl Acad Sci USA 89 : 1710.

Russel, E.S. (1979): Hereditary anemias of the mouse: A review for geneticists. Adv.Genet. 20, 357-459.

Yarden, Y., Kuang, W.J., Yang-Feng, T., Coussens, L., Munemitsu, S., Dull, T.J., Chen, E., Schlessinger, J., Francke, U. and Ullrich, A. (1987): Human proto-oncogene c-kit: a new cell surface receptor tyrosine kinase for an unidentified ligand. EMBO. J. 6, 3341-3851.

Wang, C., Curtis, J.E., Geissler, E.N., McCulloch, E.A. & Minden, M.D. (1989): The expression of the proto-oncogene c-kit in the blast cells of acute myeloblastic leukemia. Leukemia 3, 699-702.

Lerner, N.B., Nocka, K.H., Cole, S.R., Qiu, F., Strife, A., Ashman, L.K., and Besmer, P. (1991): Monoclonal antibody YB5:B8 identifies the human c=kit protein product. Blood 77, 1876-1883.

Flanagan, J.G., and Leder, P. (1990): The kit ligand: a cell surface molecule altered in steel mutant fibroblasts. Cell 63, 185-194.

Résumé

Le proto-oncogène c-kit est l'homologue cellulaire normal de v-kit, l'oncogène du virus du sarcome félin Harvey Zuckerman 4. c-kit code pour un récepteur tyrosine kinase transmembranaire dimérique qui fait partie de la superfamille des gènes des immunoglobulines. De nombreuses observations ont établi l'importance du récepteur de c-kit dans le développement des cellules souches hématopoiétiques murines et laissent présager de son importance dans l'hématopoïèse humaine. C'est pourquoi, afin de mieux comprendre la fonction physiologique du récepteur de c-kit dans l'hématopoïèse, nous avons bloqué la fonction du gène dans les cellules hématopoiétiques en utilisant des oligodéoxynucléotides antisenses. Les résultats de ces études suggèrent que c-kit joue un rôle prédominant dans l'érythropoïèse normale et qu'il pourrait être impliqué de façon importante dans la régulation de la croissance de certains types de cellules malignes.

Antisense oligonucleotides for inhibitors and tumor suppressor genes reveal the hematopoietic potential of quiescent progenitors

Jacques Hatzfeld, Ma-Lin Li, Angelo Cardoso, Pascal Batard, Jean-Pierre Levesque, Eugene Brown*, Hemchand Sookdeo*, Steven C. Clark* and Antoinette Hatzfeld

*CNRS Laboratory of Cell and Molecular Biology of Cytokines, GBGM, UPR 272, IRSC, 7, rue Guy-Môquet, BP 8, 94801 Villejuif Cedex, France; *Genetics Institute, Cambridge Park Drive, Cambridge, MA 02140, USA*

ABSTRACT AND INTRODUCTION

The hematopoietic stem cell compartment is mostly quiescent. To better understand its proliferative potential, we need to release its cells from a Go phase or some other less well characterized resting stage. Clonal or non-clonal long term culture (Dexter et al., 1977; Leary and Ogawa, 1987; Sutherland et al., 1990) have been developed to release early progenitors from quiescence, but in these assays, progenitor activation is generally unpredictable. A first attempt to better control progenitor cell activation from this compartment has been to characterize and clone the relevant cytokines (Yang et al., 1986; Hirano et al., 1986; Zsebo et al., 1990). Until now, it would appear that these cytokines have not been able to activate quiescent stem cell by a direct effect. We first considered the possibility that an excess of cytokines could selectively down-regulate receptors on early progenitors (Walker et al., 1985; Zhou et al., 1988). Indeed, this was the case (Panterne et al., 1993). Even after an extensive attempt to define the optimal combination of cytokines which would not down-regulate receptors on early progenitors, it was not possible to directly activate the stem cell compartment. Our laboratory has therefore examined the possibility that extracellular inhibitors and intracellular negative regulators could control Go phase by a mechanism that does not utilize known cytokines. To test this hypothesis, we are developing various antisense oligonucleotides (ODNs) to block the expression of inhibitors or tumor suppressor genes. From the work of numerous laboratories, it is reasonably clear that antisense ODNs prevent the expression of specific genes by blocking the translation of the corresponding mRNAs. We will discuss the advantages and disadvantages of this strategy and will use it to demonstrate that early progenitors produce TGF-ß1 in an autocrine fashion to control their Go phase. We will also show that blocking TGF-ß from umbilical cord blood CD34+ CD38- cells reveal their hematopoietic potential which demonstrates that regular umbilical cord blood has enough stem cells to engraft not only children but adults as well. We will give an example of complementation study which suggests that TGF-ß1 and Rb genes may belong to a common pathway that controls early progenitor quiescence. Antisense ODNs should, in the future, permit the study of any gene function in the same way that geneticists have used conditional mutants of bacteria, yeast or drosophila to study DNA replication, cell cycle or development.

DIFFICULTIES OF THE STRATEGY OF ANTISENSE ODN.

Antisense ODNs must have four important properties: 1) they must be stable to enzymatic degradation, 2) they must be relatively non toxic, 3) they must be able to enter cells and 4) they must be specific for their target sequence. Stability can be obtained by protecting the ODN phosphodiester bonds from nuclease digestion by replacing them by phosphorothioate linkages. However, a disadvantage of phosphorothioate ODNs is their apparent toxicity. Using 21 mers targeted to TGF-ß1 or the retinoblastoma susceptibility (Rb) genes (Hatzfeld et al., 1991), we observed a toxic effect at concentrations over 8µM. At this concentration, the full antisense potential has not been reached. A solution to this problem may be to utilize normal phosphodiester ODNs with modified internucleotides linkages at the ends that are resistant to 3' and 5' exonucleases. Indeed, it has been demonstrated by Dagle et al. (1990) that several phosphoramidate bonds at the 5' and 3' ends significantly increased the stability of ODNs when injected into embryos and oocytes. Various modifications of the 5' and 3' ends of antisense ODNs are under study to improve their stability and decrease their toxicity. The advantage of our cell assay for testing the effectiveness of a new type of ODNs is that it allows one to readily distinguish between the ODNs toxic and specific effects since the latter activates the cell by blocking a negative control. On the other hand, when the antisense ODN targets an oncogene, for example, cell growth function is inactivated which is difficult to distinghish from a non-specific toxic effect.
As non-toxic and stable antisense ODNs are developed, additional successes will be achieved by increasing the penetration of ODNs into cells. The use of polycationic ligands (Lipofectin) as described by Bennett et al.(1992) is under study.
It is anticipated that the increased stability, cellular penetration, and decreased toxicity will have two important consequences. First, we should be able to decrease by a factor of 10 to 100 the concentration of antisense ODN needed to block a specific mRNA. This will decrease the cost of their synthesis in a considerable way. Second, it should permit the simultaneous addition of several ODNs in order to determine if their corresponding genes function by a common pathway. We will provide in this chapter an example of this type of complementation study. The sequence of the antisense ODN must be chosen so that it will

ideally hybridize only to the desired target. For example, the sequence of the TGF-ß1 antisense ODN 21-mer (Hatzfeld et al., 1991) chosen after computer searches, did not reveal any significant sequence similarity between this oligonucleotide and sequences in GenBank, including TGF-ß2 and TGF-ß3.
We have shown (Cardoso, Li et al.,1993) that this antisense ODN is active mainly on immature progenitors (CFU-GEMM, early BFU-E and CFU-GM) such as CD34+ CD38- cells which represent 1/10,000 mononucleated cells in bone marrow. It is much less effective on CD34+ CD38+ cells which are more mature progenitors (CFU-M, CFU-G, CFU-E). Because the population of these immature progenitors is so low in the pool of mononucleated cells, it would be very difficult to purify enough of these cells to do a Northern Blot and follow the fate of the mRNA after treatment of the cell with the antisense ODN. Accordingly, we have employed reverse PCR to demonstrate that indeed the TGF-ß1 mRNA disappear when the antisense ODN is added to early progenitors. Additionally, the fact that the TGF-ß1 antisense ODN is active mainly on early progenitors and leads to an increase in colony number, further suggests that its effect is specific since the opposite effect was observed when TGF-ß protein was added to hematopoietic cultures (Ohta et al., 1987; Keller et al., 1989; Broxmeyer et al.,1988; Kishi et al., 1989; Ruscetti 1991). Thus, TGF-ß inhibits early progenitors and TGF-ß1 antisense ODN activates these same cells.

AUTOCRINE TGF-ß1 CONTROLS QUIESCENCE OF EARLY HEMATOPOIETIC PROGENITORS.

Our results with the TGF-ß1 ODN have allowed us to make two important observations about the controls of the quiescent state of early hematopoietic progenitors. First, we have been able to demonstrate that the early progenitors produce in an autocrine fashion TGF-ß1, which is sufficient to maintain them in a quiescent state in single cell culture. This implies that early progenitors did not necessarily require accessory cells to remain in this resting state. It further suggests that the hematopoietic stem cell could have its own machinery to stay quiescent. Second, the new mixed colonies and erythroid bursts obtained with TGF-ß1 antisense ODN are much larger than those obtained in the absence of the antisense compound. This suggests that early progenitors are being activated and they have a higher proliferative potential than those we can observe in culture conditions with cytokines alone. It may be that these early progenitors are similar to the Long Term Culture-Initiating Cells (LTC-IC) described by Sutherland et al.(1991). The advantage of our culture system is that it permits the exit from quiescence precisely at the begining of the culture, the time at which the TGF-ß1 antisense ODN is added. We know that the effect of the TGF-ß1 antisense compound occurs within the first day of the culture. It then takes 2 to 4 weeks for the colonies to develop into macroscopic mixed colonies which contain up to $1.5.10^5$ myeloid lineage cells which are readily recognized by eye.

HEMATOPOIETIC POTENTIAL OF UMBILICAL CORD BLOOD TO ENGRAFT AN ADULT

We decided to take advantage of this culture assay to compare the proliferative potential of CD34+ CD38- cells from human bone marrow and umbilical cord blood. It has been shown by Terstappen et al.(1991) that CD34+ CD38- cells represent about 1% of CD34+ cells and are the most immature progenitors, those which contain the stem cell compartment. In our study (Cardoso et al.,1993), we put 1000 CD34+ CD38- in liquid culture with IL3, IL6, GM-CSF, SLF together with a TGF-ß blocking antibody kindly provided by Anita Roberts and Michael Sporn (Danielpour et al., 1989). As TGF-ß is highly conserved in mammalian species, this blocking antibody was raised in the turkey. Every week for 7 weeks, the medium was partly replaced with fresh medium, cells were numerated and assayed in a semi-solid methyl cellulose medium for production of large erythroid bursts, large GM and mixed colonies. We observed that, when TGF-ß is neutralized, the cumulative production of CFU-GM and BFU-E by CD34+ CD38-cells present in a regular umbilical cord blood is respectively 3 and 2 times higher than in a standard bone marrow sample for adult transplantation. Production of CFU-mix was not significantly different. Therefore umbilical cord blood contain enough early progenitors to promote long term engraftment in adult.

TUMOR SUPPRESSEUR GENE AS NEGATIVE REGULATORS OF HEMATOPOIETIC STEM CELL.

TGF-ß1 is not the only negative control of hematopoietic progenitor proliferation. Other inhibitors are well detailed in this book. We are using the strategy of gene inhibition with antisense ODNs to identify tumor suppressor genes that could also play a role in this type of regulation. We have shown (Hatzfeld et al., 1991) that the retinoblastoma susceptibility gene (Rb) controls quiescence of early progenitors in a way similar to TGF-ß. The fact that anti-TGF-ß blocking antibodies, antisense TGF-ß1 or Rb ODNs all had similar effects and that no additive effects were observed when these reagents were combined suggested a common pathway of these two genes. For example, using colony formation by early CFU-GEMM as an indicator of release from quiescence, addition of anti-TGF-ß antibody to CD34+ enriched human bone marrow cells yielded the same enhancement of colony formation as was achieved with addition of either antisense TGF-ß1 or Rb ODNs. Likewise, the combination of the anti TGF-ß antibody and the Rb antisense ODN did not result in additional enhancement of colony formation. It must be noted that in these experiments, we were forced to employ a blocking antibody for TGF-ß and the Rb antisense ODN because the simultaneous addition of the two ODNs gave an overall oligonucleotide concentration at which nonspecific inhibition became an important factor in the culture assay. Thus, the goal of our current work is to develop new types of antisense oligonucleotides that are stable, less toxic and enter the cell more readily. It is anticipated that these developments will allow for the use of less concentrated solutions of antisense oligodeoxyribonucleotides and make the identification of the genes that maintain the hematopoietic stem cell in its quiescent state a more straight forward process.

Informations on stem cell culture media: tel (33 1) 47 26 60 99.

ACKNOWLEDGMENTS: this work was supported by Association de la Recherche sur le Cancer and by Direction des Recherches Etudes et Techniques (DRET). M.-L. Li was on leave from Kumning Medical College, Kumning, People's Republic of China and obtained a fellowship from the French government. A. Cardoso is a fellow from the Junta Nacional de Investigação Cientifica e Technológica (Portugal). P. Batard is a fellow of DRET.

REFERENCES

Dexter, T.M., Allen, T.D., and Lajtha, L.G. (1977): Conditions controlling the proliferation of haematopoietic stem cells in vitro. *J. Cell Physiol.* 91: 335-342.

Leary, A.G., and Ogawa, M. (1987): Blast cell colony assay for umbilical cord blood and adult bone marrow progenitors. *Blood.* 69: 953-956.

Sutherland, H.J., Lansdorp, P.M., Henkelman, D.H., Eaves, A.C., and Eaves, C.J. (1990): Functional characterization of individual human hematopoietic stem cells cultured at limiting dilution on supportive marrow stromal layers. *Proc. Natl. Acad. Sci. USA.* 87: 3584-3588.

Yang, Y.-C., Ciarletta, A.B., Temple, P.A., Chung, M.P., Kovacic, S.,Witek-Giannotti,J.S, Leary, A.C, Kriz, R., Donahue, R.E., Wong, G.G., and Clark, S.C. (1986): Human IL3 (Multi-CSF): Identification by expression cloning of a novel hematopoietic growth factor related to murine IL-3. *Cell.* 47: 3-10.

Hirano, T., Yasukawa, K., Harada, H., Taga, T., Watanabe, Y., Matsuda, T., Kashiwamura, S., Nakajima, K., Koyama, K., Iwamatu, A., Tsunasawa, S., Sakiyama, F., Matsui, H., Takahara, Y., Taniguchi, T., and Kishimoto, T. (1986): Complementary DNA for a novel human interleukin (BSF-2) that induces B lymphocytes to produce immunoglobulin. *Nature.* 324: 73-76.

Zsebo, K.M., Wypych, J., McNiece, I.K., Lu, H.S., Smith, K.A., Karkare, S.B., Sachdev, R.K., Yuschenkoff, V.N., Birkett, N.C., Williams, L.R., Satyagal, V.N., Tung, W.F., Bosselman, R.A., Mendiaz, E.A., and Langley, K.E. (1990): Identification, purification and biological characterization of hematopoietic stem cell factor from buffalo rat liver conditioned medium. *Cell,* 63: 195-200.

Walker, F., Nicola, N.A., Metcalf, D., and Burgess, A.W. (1988): Hierarchical down-modulation of hemopoietic growth factor receptors. *Cell.* 43: 269-276.

Zhou, Y.Q., Stanley, E.R., Clark, S.C., Hatzfeld, J.A., Levesque, J.-P., Federici, C., Watt, S.M., and Hatzfeld, A. (1988): Interleukin-3 and interleukin-1α allow earlier bone marrow progenitors to respond to human colony-stimulating factor 1. *Blood.* 72: 1870-1874.

Panterne, B., Zhou, Y.Q., Hatzfeld, J., Li, M.L., Levesque, J.-P., Clark, S.C. and Hatzfeld, A. (1993): CSF-1 control of c-fms expression in normal human bone marrow progenitors. *J. Cell. Physiol.* In press.

Hatzfeld, J., Li, M.L., Brown, E.L., Sookdeo, H., Levesque J.-P., O'Toole, T., Gurney, C., Clark S.C. and Hatzfeld, A. (1991): Release of early human hematopoietic progenitors from quiescence by antisense transforming growth factor-ß1 or rb-oligonucleotides. *J. Exp. Med.* 174: 925-929.

Dagle, J.M., Walder, J.A., and Weeks, D.L. (1990): Targeted degradation of mRNA in Xenopus oocyte and embryos directed by modified oligonucleotides: studies of An2 and cyclin in embryogenesis. *Nucleic Acids Res.* 18: 4751-4757.

Bennett, C.F., Chiang, M.-Y., Chan, H., Ellen, J., Shoemaker, J.E.E., and Mirabelli, C.K.. (1992): Cationic lipids enhance cellular uptake and activity of phodsphorothioate antisense oligonucleotides. *Mol. Pharmacol*, 41: 1023-1033.

Cardoso, A.A., Li, M.L., Hatzfeld, A., Brown, E.L., Levesque, J.-P., Sookdeo, H., Batard, P., Clark, S.C., and Hatzfeld, J. (1993): Release from quiescence of CD34+ CD38- human umbilical cord blood reveals their potentiality to engraft adults. *Proc. Natl. Acad. Sci. USA.* In press.

Ohta, M., Greenberger, J.S., Anklesaria, P., Bassols A., and Massague, J. (1987): Two forms of transforming growth factor-ß distinguished by multipotential haematopoietic progenitor cells. *Nature.* 329: 539-541.

Keller, J.R., Mantel, C., Sing, G.K., Ellingsworth, L.R., Ruscetti, S.K., and Ruscetti, F.W. (1988): Transforming growth factor ß1 selectively regulates early murine hematopoietic progenitors and inhibits the growth of IL-3-dependent myeloid leukemia cell lines. *J. Exp. Med.* 168: 737-750.

Broxmeyer, H.E., Lu, L., Cooper, S., Schwall, R.H., Mason, A.J., and Nikolics, K. (1988): Selective and indirect modulation of human multipotential and erythroid hematopoietic progenitor cell proliferation by recombinant human activin and inhibin. *Proc. Natl. Acad. Sci. USA,* 85: 9052- 9056.

Kishi, K., Ellingsworth, L.R., and Ogawa, M. (1989): The suppressive effectscof type ß transforming growth factor (TGFß) on primitive murine hemopoietic progenitors are abrogated by interleukin -6 and granulocyte colony-stimulating factor. *Leukemia.* 3: 687-691.

Ruscetti, F.W.. (1991): In vivo and in vitro effects of TGF-ß1 on normal and neoplastic haemopoiesis. In *Clinical applications of TGF-ß.* Ciba Found. Symp. 157: 212-222.

Sutherland, H.J., Eaves, C.J., Lansdorp, P.M., Thacker, J.D., and Hogge, D.E. (1991): Differential regulation of primitive human hematopoietic cells in long-term cultures maintained on genetically engineered murine stromal cells. *Blood.* 78: 666-672.

Terstappen, L.W.M.M., Huang, S., Safford, M., Lansdorp, P.M. and Loken, M.R. (1991): Sequential generations of hematopoietic colonies derived from single nonlineage-committed CD34+ CD38- progenitor cells. *Blood.* 77: 1218-1227.

Danielpour, D., Dart, L.L., Flanders, K.C., Roberts, A.B., and Sporn M.B. (1989): Immunodetection and quantitation of the two forms of transforming growth factor-ß (TGF-ß1 and TGF-ß2) secreted by cells in culture. *J. Cell. Physiol.* 138: 79-86.

Résumé

La cellule souche se trouve la plupart du temps à l'état quiescent. Aucune des cytokines actuellement clonées n'est capable de l'activer pour la sortir de la phase Go. Notre laboratoire développe des oligonucléotides antisens pour des gènes suppresseurs de tumeur et des inhibiteurs susceptibles de contrôler la phase Go de la cellule souche. Le but est d'activer la cellule souche en bloquant l'expression de ces gènes. Les oligonucléotides antisens doivent avoir 4 propriétés importantes : ils doivent 1) être stables ; 2) être non toxiques ; 3) diffuser facilement dans la cellule ; 4) être spcifiques pour la séquence de l'ARNm qu'ils doivent neutraliser.

Nous montrons l'intérêt d'un système dans lequel nous activons la cellule en bloquant des contrôles négatifs pour développer différents types d'oligonucléotides antisens plus efficaces. Des oligonucléotides antisens pour le gène du TGF-ß1 nous ont permis de débloquer des progéniteurs proches de la cellule souche, que nous ne pouvons pas mettre en évidence en culture à court terme avec des cytokines. Grâce à des cultures de cellules isolées, nous montrons que les progéniteurs les plus précoces produisent de façon autocrine du TGF-ß1 pour contrôler leur phase Go. En combinant plusieurs oligonucléotides antisens, nous réalisons des expériences de complémentation pour analyser les gènes qui contrôlent un même mécanisme de maintien en phase Go. Par exemple, nous montrons que le gène du TGF-ß1 et le gène suppresseur de tumeur Rb contrôlent une même voie de régulation. Nous avons utilisé cette stratégie pour évaluer le potentiel hématopoïétique d'un sang de cordon en sortant de la phase Go les cellules CD34+ CD38-(*). En comparant un sang de cordon ombilical avec un prélèvement de moelle osseuse permettant la transplantation d'un adulte de taille moyenne, nous constatons.qu'un sang de cordon correctement prélevé a un potentiel hématopoïétique supérieur à celui d'un prélèvement de moelle osseuse pour greffer un adulte.

Antisense approach to examining the role of RAF-1 in erythropoietin action

W. Stratford May and Michael P. Carroll

The Johns Hopkins Oncology Center, 424 N. Bond Street, Baltimore, MD, USA

Hematopoietic growth factors (HGF) such as Erythropoietin (EPO) and Interleukin-3 (IL-3) stimulate net cell growth by binding to specific receptors displayed on the surface of the responding hematopoietic cells. It is now well recognized that once a receptor is engaged there is an immediate triggering of a series of intracellular biochemical and molecular events, known collectively as transmembrane signaling, that is required to initiate and/or sustain cell growth and/or differentiation. The mechanism by which a signal is generated at the plasma membrane and how it is transmitted to the cell nucleus is not clear but appears to involve regulation via rapid and reversible phosphorylation of crucial intermediate proteins. For example, recent studies demonstrate that EPO and IL-3 rapidly stimulate the transient phosphorylation of membrane, cytosolic and nuclear proteins at both tyrosine and serine sites (Isfort et al., 1988; Fields et al., 1989; Quelle, 1991), supporting a regulatory role for protein phosphorylation. Much effort has been put forth recently to identify the specific molecular components responsible for signaling. One technique for accessing the potential role of an intracellular molecule as a signaling candidate is by eliminating and/or decreasing the expression of the specific protein candidate component. We have identified one such crucial cytosolic component, the 74KDa RAF-1 protein (Carroll et al., 1990,1991) encoded by the c-raf protooncogene (Rapp, 1988). RAF-1 is an ubiquitously expressed serine/threonine protein kinase. We and others have found that RAF-1 plays a central regulatory role in the transmission of a variety of hematopoietic mitogenic signals including IL-3, GM-CSF, IL-3, PDGF, CSF-1, EGF and insulin (Carroll, 1990, 1991; Turner, 1991; Morrison, 1988, Baccarin, et al., 1990; Kovacina et al., 1990; Blackshear et al., 1990; App et al., 1991). Antisense technology has proved useful in our studies. By synthetically producing antisense deoxyribonucleotides designed to target specific intracellular mRNA (i.e. sense strands to be translated into protein product) messages, the antisense oligo's can be added directly to cell culture (Gewirtz & Calabretta, 1988). These deoxyribonucleotides are short enough (ie usually 18-21 nucleotides) to be taken up by the cell and reach the intracellular milieu. Once having penetrated the cell, the antisense strands can specifically target and perturb the expression of the mRNA either prior to transcription or by interfering with processing at the nuclear level, or at the extranuclear level (i.e. cytoplasmic) by affecting the mRNA to be translated (Akhtar & Juliano, 1992). For example, as a result of this specific interaction a heteroduplex (RNA sense and DNA antisense) is formed which is recognized by the cell machinery as abnormal and is thus either degraded and/or not further processed or translated by the cell. In the absence of functional mRNA that can be translated into protein, and according to the half-life of the specific protein product, partial or complete depletion of the specific intracellular protein can be achieved. Analysis of the effect of a growth factor such as EPO or IL-3 on the process of cell growth can then be performed and the results should reflect the role of that specific protein component that has been depleted.

Using such an antisense (AS) approach, we have determined that RAF-1 expression is required for EPO or IL-3 mediated proliferative growth by IL-3 or EPO responsive murine FDC-P1/ER cells (Carroll, 1991). RAF-1 is an intracellular ser/thr protein kinase whose activity is rapidly activated by either EPO or IL-3 in association with ser phosphorylation at several protein kinase C (PKC) consensus sites on RAF-1 (Carroll et al., 1990 & 1991; Carroll et al., 1992). These data suggested that RAF-1 may be activated directly by PKC, another intracellular ser/thr protein kinase whose activity can be activated by EPO and IL-3 (Spivak et al., 1992; Fields et al., 1989). Recently several studies have found that both the IL-3 and GM-CSF ligands, members of HGF cytokine receptor superfamily (Bazan, 1990), can activate intracellular PKC (Durunio et al., 1989; Nishimura et al., 1992). The mechanism of HGF activation of PKC is thought to result from DAG generation, perhaps from phosphatidylcholine.

Recently we have discovered that a potent marine antineoplastic natural product Bryostatin (Pettit et al, 1982), can activate PKC (Kraft et al., 1989) and can mimic, in part, IL-3's mitogenic effect on murine FDC-P1 cells (Fields et al., 1989). Interestingly Bryostatin can also support growth of normal bone marrow progenitor cells (May et al., 1989) which normally require multipotential HGFs.

Since IL-3 and EPO or IL-2 can mediate the rapid ser and tyr phosphorylation of RAF-1 (Carroll, 1991; Turner et al., 1991), while tyrosine kinase containing receptors can be stimulated by their specific ligands (eg. EGF and insulin) to also mediate ser phosphorylation of RAF-1 (Blackshear et al., 1990; Morrison et al., 1988), we reasoned that either RAF-1 had a functional autokinase activity or a separate intracellular ser/thr protein kinase could act as an upstream activator of RAF-1. Since we have found PKC to be rapidly and functionally activated by IL-3, and others have shown EPO may activate PKC (Spivak et al, 1992), we evaluated a role for PKC in RAF activation. First, we used antisense RAF-1 as above (Carroll et al., 1992) and, again results indicated that at 10µM antisense Bryostatin proliferation is inhibited 60-70% (Fig. 1).

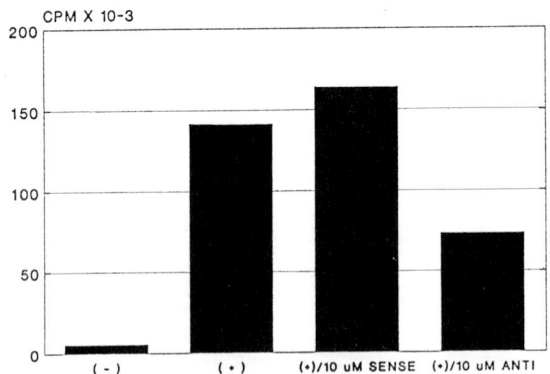

Figure 1: ANTISENSE RAF C-RAF INHIBITS BRYOSTATIN MEDIATED PROLIFERATION. FDC-P1 cells were incubated with 10µM antisense (made against the first six codovis in the C1 region of PKCα) or sense oligodeoxyribonucleotides for 16 hrs prior to addition the of 5nM Bryostatin. Proliferation was measured after an additional 24 hr using ^{3}H-thymidine incorporation as described (Carroll et al., 1991). (+) indicates addition of 5nM Bryostatin. (-) indicates the control without Bryostatin.

These data indicate that RAF-1 is necessary for Bryostatin mediated proliferation. Next, cells were equilibrated with ^{32}P-orthophosphoric acid and Bryostatin was added at various concentrations. Cells were lysed and RAF-1 was isolated by immunoprecipitation and the phosphorylation status examined by SDS-PAGE. Results reveal that RAF-1 is hyperphosphorylated in a dose-dependent manner (Fig. 2A).

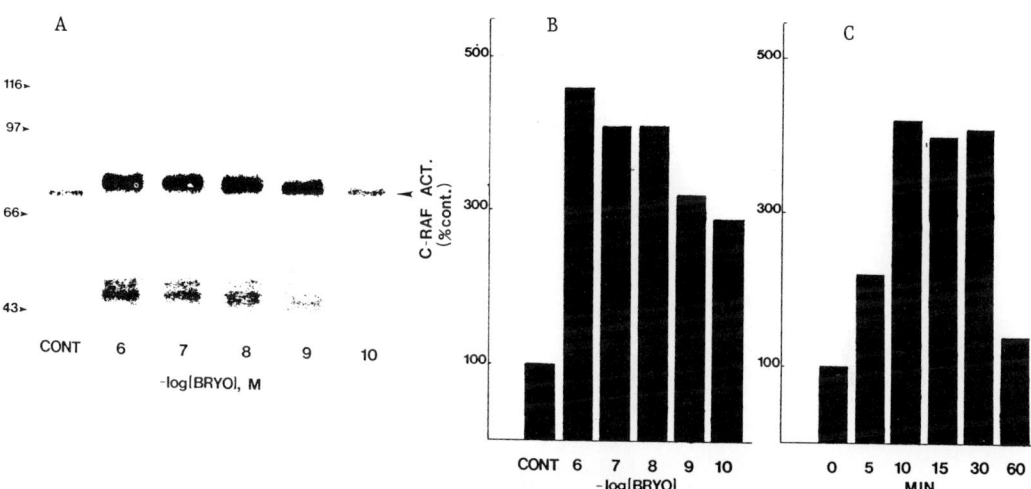

Figure 2: EFFECT OF BRYO ON RAF-1 PHOSPHORYLATION AND ACTIVATION IN FDC-P1 CELLS. (A) Bryostatin was added to cells at various concentrations and for various times and the effect on RAF-1 phosphorylation determined as described in the text. Arrow indicates RAF-1. Numbers at left indicate M_r in thousands. (B,C) Dose and time dependent activation of RAF-1 kinase activity in vitro was performed as previously described (Carroll, 1990).

Phosphorylation occurs maximally between 1 and 10 nM Bryostatin, concentrations that support maximal growth of FCD-P1 cells (Fields et al., 1989). Furthermore, phosphorylation of RAF-1 is stoichiomatic, indicated by the virtual complete shift from baseline phosphorylated RAF (Fig. 2), suggesting a functional consequence. Next, an in vitro kinase reaction was performed with anti-RAF immunopreciptates isolated from cells treated with Bryostatin. Results indicate that both a time and dose dependent enzymatic activation of RAF-1 was observed (Fig. 2C&D) indicating phosphorylation may activate RAF-1.

Finally, phosphoamino acid analysis indicated that ser was the only site(s) phosphorylated on RAF-1 following addition of Bryostatin indicating that enzymatic activation of RAF-1 can be achieved in association with ser phosphorylation.

Exploring further a role for PKC in RAF-1 activation, FDC-P1/ER cells were screened for PKC isoforms by Western blotting with isoform specific antibodies. We found that PKCα is quantitatively the major isoform expressed (data not shown). Therefore, we again turned to the antisense approach to assess whether PKC may be involved in EPO proliferative signaling. We found that both antisense PKCα as well as antisense pan PKC (i.e. antisense to the classic PKC isoforms α, βI,II,γ; kind gift of Craig Hooper, Centers for Disease Control, Atlanta, GA.) can inhibit EPO-mediated proliferation in a dose dependent manner by 60-70% as compared with sense or missense oligo controls (Figure 3).

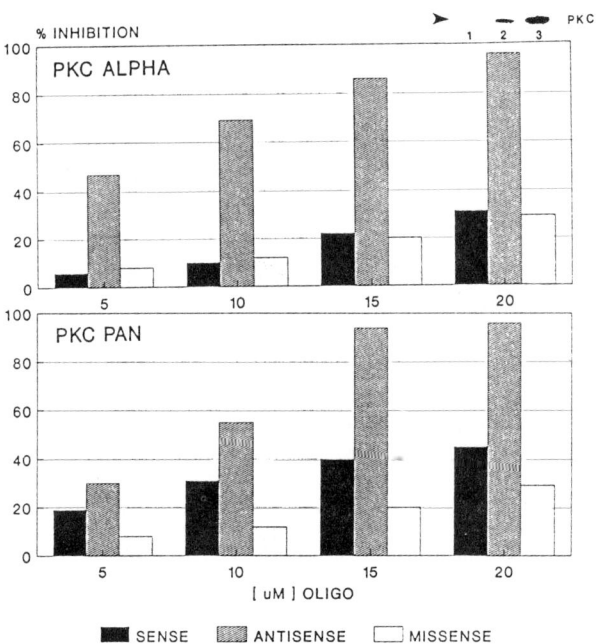

Figure 3: EFFECTS OF PKCα and PAN ANTISENSE OLIGOS ON GROWTH OF FDC-P1ER CELLS. Antisense, sense or missense oligos were added twice as described in the text to growth factor deprived FDC-P1ER cells to achieve the final concentration indicated. Synthetic murine IL-3 (gift of Dr. Ian Clark-Lewis) was then added and growth determined by ^{3}H-thymidine incorporation all as described (Carroll, 1991). Western blotting (insert) using an anti-PKCα antibody indicates that neither sense (lane 1) or missense (lane 2) affects PKCα, but antisense (lane 3) inhibits PKCα protein levels.

Figure 4: **BRYO ACTIVATED PURIFIED PKC DIRECTLY PHOSPHORYLATES BACULOVIRUS-INSECT PRODUCED, PURIFIED RAF-1**.

(A) A baculovirus containing a human c-raf cDNA insert (gift of Dr. Ulf Rapp) was used to transfect SF9 insect cells. After 48 hr, cells were harvested and RAF-1 purified by immunoprecipitation using a specific anti-RAF-1 Ab that we made against a c-terminal RAF-1 peptide (sp-63). Immunoprecipitates were mixed with purified rat brain PKC, Mg^2 and $^{32}P_\gamma$-ATP added and the reaction carried out for 30 min at RT. ^{32}P-labeled protein products were separated by SDS-PAGE on a 10% gel. PKC ~80kDa, RAF-1 ~72kDa. RAF-1 was isolated from untransfected (wt) cells or from RAF-1 transfected cells (RAF) and an immunoblot is shown in the left two lanes $\underline{in\ vitro}$ kinase reaction were performed. Lane 1 RAF-1 immunoprecipitate alone; lane 2 RAF-1 immunoprecipitate plus bryostatin, Ca^{++}, P/S and EGTA plus PKC, Lane 3 RAF-1 plus bryostatin, Ca^{2+}, P/S and PKC; lane 4 RAF-1 plus bryostatin, Ca^{2+}, P/S but without PKC; Lane 5 preimmune serum immunoprecipitate plus bryostatin, Ca^{2+}, P/S and PKC; Lane 6 autokine of purified rat brain PKC; Lane 7 as in lane 6 except with 2mM EGTA to demonstrate Ca^{++} dependence of PKC. Double arrows at left indicates RAF-1 and single arrow at right PKC migration.

(B) $\underline{In\ vitro}$ kinase reaction using PKC "primed" RAF-1. Immunoprecipitates of RAF-1 from SF-9 infected cells were incubated with (+) or without (-) PKC plus cofactors and activator for 10 or 30 min in the presence of unlabeled ATP (i.e. primed RAF). The immunoprecipitates were washed free of residual PKC and used to assess $\underline{in\ vitro}$ kinase activity of the primed RAF-1 using $^{32}\gamma$/ATP, Mg^2 and histone III (30μg/ml) as substrate (N = 3).

It is clear that only when the PKC cofactors plus an activator is added that RAF-1 becomes phosphorylated (Fig. 4A lane 3). RAF-1 incubated without PKC shows little auto phosphorylation, indicating that recombinant RAF-1 purified from insect cells is a poor autokinase $\underline{in\ vitro}$, at least without prephosphorylation and activation (lane 5). Furthermore, results with extensively washed RAF precipitates that had been incubated under the identical conditions except that unlabeled ATP was used instead of ^{32}P-labeled ATP, indicate that this "primed" RAF-1 is enzymatically active. Thus, RAF-1 kinase activity $\underline{in\ vitro}$ is activated in a time dependent manner following PKC "priming" (Fig. 4B). Collectively, these findings suggest that PKC, and specifically PKCα, may be an upstream activator of RAF-1 in murine hematopoietic FDC-P1 cells and that the HGFs, represented by EPO and IL-3, may activate RAF-1 through PKC serving as a RAF-1 kinase.kinase. Finally, in recent studies (Carroll & May, 1992), we find that ser phosphorylation of RAF by PKC occurs at residues ser 43, ser 497 and ser 619. Currently it is not clear whether one or all of these phosphorylation sites are required to be phosphorylated to achieve full enzymatic activation of RAF-1. Interestingly, ser 43 lies in the CR1 region of the amino terminal portion of RAF and, analogous to PKC regulation, may represent a pseudosubstrate site. If this were the case, phosphorylation of this site could be envisioned to relieve any inhibitory effect on the catalytic activity perhaps through altering inhibitory conformational constraints present when the sites are not phosphorylated. Whatever the mechanism of activation of RAF-1, ser phosphorylation appears to be sufficient.

Also, in support of a role for PKC as a RAF-1 activating kinase, a recent publication using the recombinant baculovirus system for PKC isoform production indicates that the conventional or classic PKC isoforms (α,β,γ) can phosphorylate RAF-1 intracellularly when recombinant baculovirus expressing both PKC and RAF-1 are used to coinfect SF-9 insect cells (Sazeri et al., 1992). Furthermore, recent findings indicate that ras may operate upstream of RAF-1 in development (Dickson et al., 1992) as well as IL-3 signaling (Boswell et al., 1991). This could suggest a post-receptor activation step for PKC.

Furthermore, when the PKCα content following oligo treatment is estimated by Western blotting it can be seen that this isoform is virtually depleted following AS treatment of cells but not by treatment using sense or missense oligos (Fig. 4, Insert lane 3).

Since PKC appears to be necessary for hematopoietic growth factor activation of RAF-1, we next determined whether PKC could directly phosphorylate RAF-1. We used highly purified rat brain PKC isolated as described (May & Tyler, 1987) to incubate with immunopurified RAF-1 isolated from SF-9 insect cells infected with a Baculovirus vector containing a human RAF-1 cDNA construct (kindly provided by Ulf Rapp). Experimentally, PKC and RAF-1 are mixed and the PKC cofactors Ca^{++} and phosphatidylserine plus Bryostatin as activator were added along with $\gamma^{32}P$-ATP and Mg^{++} for 10 or 30 min. Following incubation, the RAF immunoprecipitate was washed extensively to remove residual PKC and the phosphorylation status or enzymatic activation of purified RAF-1 was evaluated by SDS-PAGE (Fig. 4A).

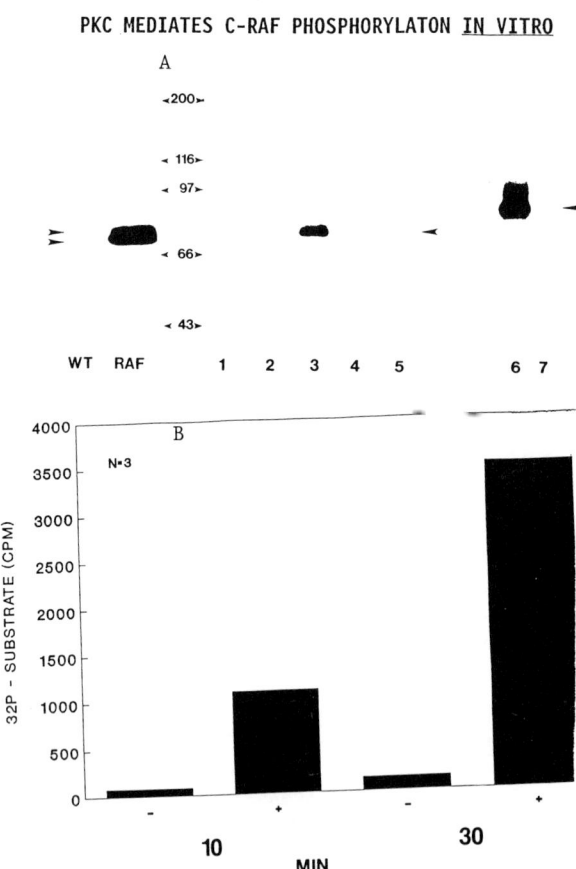

Interestingly, evidence has appeared that the PKC inhibitors H-7 and staurosporine can reverse the suppressive effect on apoptosis that HGF's, including IL-3 or GM-CSF have in human MO7-E factor dependent cells (Rajotte et al., 1992). We have recently found that Bryostatin is also a potent inhibitor of apoptosis in murine FDC-P1 cells whose growth is stimulated by this agent (unpublished data). Together these data also support a potential linker role for PKC in the regulation of both the apoptosis as well as the RAF-1 proliferation-associated pathways in hematopoietic cells (Fig. 5).

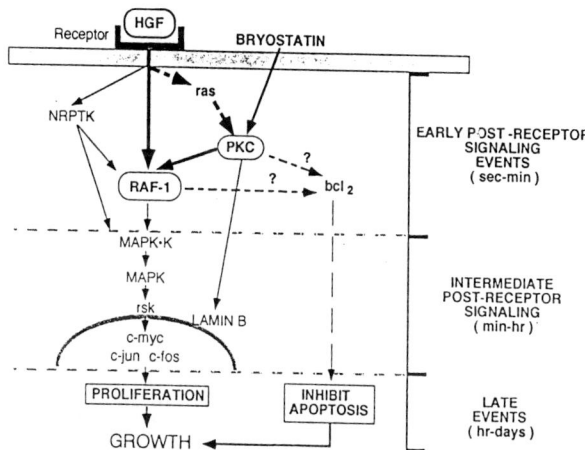

Figure 5: TEMPORAL EVENTS IN HEMATOPOIETIC CELL GROWTH. Hematopoietic growth factors (HGF), represented by EPO and IL-3, bind to their surface receptor and trigger the rapid and sequential activation of crucial intracellular enzymatic components of the transmembrane signaling pathways. Net growth is envisioned to occur as a collective result of stimulating a proliferation associated pathway involving RAF-1 and its distal components which include; MAPK.K (Mitogen Activated Protein Kinase.kinase), MAPK, rsk (ribosomal S-6 kinase), etc. PKC may be activated through either a ras or another pathway which is envisioned to be capable of directly activating RAF-1 in the proliferation pathway and also an anti-apoptosis pathway which likely involves the bcl2 protein (i.e. B-cell leukemia/lymphoma protooncogene product) which is known to be required for survival of hematopoietic cells.

Collectively our results indicate that studies with antisense deoxyribonucleotides have helped to demonstrate that RAF-1 and PKC are required for EPO and IL-3 mediated net cell growth in hematopoietic cells and that PKCα may be a direct upstream activator of the RAF-1 protein kinase activity.

References

Akhtar, S. & Juliano, R.L. (1992): Cellular uptake and intracellular fate of antisense oligonucleotides. Trends Cell Biol 2, 139-144.
App, H., Hazan, R., Ziberstein, A., Ullrich, A., Schlessinger, J. & Rapp, U.R. (1991): Epidermal growth factor (EGF) stimulates associated and kinase activity of Raf-1 with the EGF receptor. Mol. Cell Biol. 11, 913-919.
Baccavin, M., Sabastini, D.M., App, H., Rapp, U.R., & Stanley, E.R. (1990): Colony stimulating factor-1 (CSF-1) stimulates temperature dependent phosphorylation and activation of the RAF-1 protooncogene product. EMBO J. 9, 3649-3657.
Bazan, J.F. (1990): Structural design and molecular evolution of a cytokine receptor superfamily. Proc. Natl. Acad. Sci. USA 87, 6934-6938.
Blackshear, P.J., Haupt, D.M., App, H. & Rapp, U.R. (1990): Insulin activates the Raf-1 protein kinase. J. Biol. Chem. 265, 12131-12134.

Boswell, H.S, Nahreini, T.S., Burgess, G.S., Srivastaua, A., Gabig, T.G., Inhorn, L., Srour, E.F. & Harrington, M.A. (1991): A Ras oncogene imparts growth factor independence to myeloid cells that abnormally regulate protein kinase C. A nonautocrine transformatin pathway. Exp. Hematol. 18, 452-460.

Carroll, M.P., Clark-Lewis, J., Rapp, U.R. & May, W.S. (1990): Interleukin-3 and granulocyte macrophage colony stimulating factor mediate rapid phosphorylation and activation of cytosolic c-raf. J. Biol. Chem. 265, 19812-19817.

Carroll, M.P. & May, W.S. (1992): Identification of the Raf-1 phosphorylation sites: Further characterization of a necessary hematopoietic signal transduction mechanism. Blood 10:(Suppl. 1), 247a, 1992.

Carroll, M.P., Rapp, U.R. & May, W.S. (1992): PKC can directly activate a necessary enzymatic component of hematopoietic proliferation, RAF-1. Exp. Hematol. 20:(Suppl.), 182.

Carroll, M.P., Spivak, J.L., McMahon, M., Weich, N., Rapp, U.R. & May, W.S. (1991): Erythropoietin induces Raf-1 activation and Raf-1 is required for erythropoietin mediated proliferation. J. Biol. Chem., 266, 14964-14969.

Dickson, B., Sprenger, F., Morrison, D. & Hafen, E. (1992): Raf functions downstream of Ras-1 in the sevenless signal transduction pathway. Nature 360, 600-603.

Durunio, V., Nip., L. & Pelech, S.C. (1989): Interleukin-3 stimulates phosphatidylcholine turnover in a mast/megakaryocyte cell line. Biochem. Biophys. Res. Communic. 164, 804-808.

Fields, A.P., Pincus, S.M., Kraft, A.S. & May, W.S. (1989): Interleukin-3 and Bryostatin 1 mediate rapid nuclear envelope protein phosphorylation in growth factor-dependent FCD-P1 hematopoietic cells. A possible role for nuclear protein kinase C. J. Biol. Chem. 264, 21896-21901.

Gewirtz, A.M. & Calabretta, B. (1988): A c-myb antisense oligodeoxyribonucleotide inhibits normal human hematopoiesis in vitro. Science 242, 1303-1306.

Isfort, R.J., Stevens, P., May, W.S., Ihle, J.N. (1988): Interleukin-3 binds to a 140KDa phosphotyrosine-containing cell surface protein. Proc. Natl. Acad. Sci. USA 85, 7982-7986.

Kovacina, K.S., Yonezansa, K., Brautigan, D.C., Tonks, N.K., Rapp, U.R. & North, R.A. (1990): Insulin activates the kinase activity of the Raf-1 protooncoge by increasing its serine phosphorylation. J. Biol. Chem. 265, 12115-12118.

Kraft, A.S., Baker, V.V. & May, W.S. (1987): Bryostatin induces a rapid loss of calcium phospholipid dependetn protein kinase activity from human promyelocytic leukemia cells (HL60). Oncogene 1, 111-118.

May, W.S., Sharkis, S.J., Esa, A.H., Gebbia, V., Kraft, A.S., Pettit, G.R. & Sensenbrenner, L.L. The antineoplastic bryostatins are multipotential stimulators of human hematopoietic progenitor cells. Proc. Natl. Acad. Sci. USA 84, 8483-8487.

May, W.S. & Tyler, P.G. (1987): Phosphorylation of the surface transferrin receptor stimulates receptor internalization in HL60 leukemic cells. J. Biol. Chem. 262, 16710-16718.

Morrison, D.K., Kaplan, D.R., Rapp, U.R. & Roberts, T.M. (1988): Signal transduction from membrane to cytoplasm growth factors and membrane-bound oncogene products increase Raf-1 phosphorylation and associated protein kinase activity. Proc. Natl. Acad. Sci. USA 85, 8855-8859.

Nishimura, M., Kaku, K., Azuno, Y., Okafuji, K., Inoue, Y. & Kanako, T. (1992): Stimulation of phosphoinositol turnover and PKC activation by GM-CSF in HL60 cells. Blood 80, 1045-1051.

Pettit, G.R, Herald, S.I., Double, K.D.L., Arnold, F. & Clardy, J. (1982): Isolation and structure of bryostatin-1. J. Am. Chem. Soc. 104, 6846-6848.

Quelle, F.W. & Wojchowski, D.M. (1991): Proliferative action of erythropoietin is associated with rapid protein tyrosine phosphorylation in responsive BGSut.EP cells. J. Biol. Chem. 266, 609-614.

Rajotte, D., Haddad, P., Haman, A., Cragoe, Jr., E.J. & Huang, T. (1992): Role of protein kinase C and the Na^+/H^+ antiporter in suppressing of apoptosis by GM-CSF and IL-3. J. Biol. Chem. 267, 9980-9987.

Rapp, U.L., Cleveland, J.L., Bonner, T.I. & Storm, S.M. (1988): in The Oncogene Handbook (Reddy, E.P., Skalkan, Amanl Curran, T. eds.) pp 213-253, Elsevier Science Publishers & B.V. Amsterdam.

Sozeri, O., Vollmer, K., Liyanage, M., Firth, D., Kour, G., Mark, III, G.E. & Stabel, S. (1992): Activation of the cRAF protein kinase by protein kinase C phosphorylation. Oncogene 7, 2259-2262.

Spivak, J.L., Fisher, J., Isaacs, M. & Hawkins, W.D. (1992): Protein kinase and phosphatase are involved in erythropoietin signal transduction. Exp. Hematol. 20, 500-504.

Turner, B., Rapp, U., App, H., Greene, M., Doboshi, K. & Reed, J. (1991): Interleukin-2 induces tyrosine phosphorylation and activation of p72-74 Raf-1 kinase in a T-cell line. Proc. Natl. Acad. Sci. USA 88, 1227-1231.

Whetton, A.D., Monk, P.N., Consalvey, S.D., Huany, S.T., Dexter & Downs, C.P. (1988): Interleukin-3 stimulates proliferation via protein kinase C activation without increasing inositol lipid turnover. Proc. Natl. Acad. Sci. USA 85, 3284-3288.

Summary

Hematopoietic growth factors (HGF) such as Erythropoietin (EPO) and Interleukin-3 (IL-3) stimulate net cell growth by binding to specific cell surface receptors. Binding triggers an immediate series of intracellular biochemical and molecular changes, known as transmembrane signaling, that are required to initiate cell growth. Recently, we and others have identified the serine/threonine protein kinases RAF-1 and PKC as crucial components of HGF signaling. We have used antisense oligodeoxyribonucleotides added directly to HGF dependent cells to study the role and mechanism of activation of these signaling components. Through a series of these and other molecular studies our results indicate that the PKCα isoform may perform a potential linker role between the HGF surface receptor and cytoplasmic RAF-1, suggesting that PKC may be the/an upstream activator of RAF-1 (ie a RAF-1 kinase·kinase). With respect to the potential mechanism of PKC activation of RAF-1, our data indicate that EPO or IL-3 mediated ser phosphorylation of RAF-1, likely via PKC, is sufficient to activate RAF-1 protein kinase activity. We have identified the putative PKC mediated RAF-1 ser phosphorylation sites. Finally, our results suggest that activated PKC may also block the apoptosis pathway that is responsible for programmed cell death, perhaps via regulation of bcl2 which is highly expressed in hematopoietic cells. In this manner, net cell growth mediated by HGFs can be envisioned to result from the combined effects of stimulation of a proliferation pathway and inhibition of an apoptosis pathway.

Résumé

Les facteurs de croissance hématopoïétiques tels que l'érythropoïétine (Epo) et l'interleukine-3 (IL-3) stimulent la croissance cellulaire en se liant à des récepteurs de surface spécifiques. La liaison déclenche une série immédiate de changements intracellulaires biochimiques et moléculaires, signaux transmembranaires, nécessaires pour initier la croissance cellulaire. Récemment, nous et d'autres avons identifié la sérine/thréonine protéine kinase RAF-1 et la PKC comme des composants cruciaux pour le signal des facteurs de croissance hématopoïétiques. Nous avons utilisé des oligodéoxyribonucléotides antisenses ajoutés directement à des cellules dépendantes des facteurs de croissance pour étudier le rôle et le mécanisme d'activation de ces signaux. Grâce à ces études moléculaires et à d'autres, nos résultats indiquent que l'isoforme de la PKCα peut jouer un rôle de liaison potentiel entre le récepteur de surface pour le facteur de croissance et le RAF-1 cytoplasmique, ce qui suggère que la PKC pourrait être le ou un activateur en amont de RAF-1 (c.a.d une RAF-1 kinase-kinase). En tenant compte du mécanisme possible de l'activation de RAF-1 par PKC, nos données indiquent que la ser-phosphorylation de RAF-1 par l'Epo ou l'IL-3, vraisemblablement via la PKC, est suffisante pour activer l'activité de la protéine-kinase RAF-1. Nous avons identifié les sites présumés de la ser-phosphorylation de RAF-1 médiée par la PKC. Enfin, nos résultats suggèrent que la PKC activée pourrait aussi bloquer la voie de l'apoptose responsable de la mort cellulaire programmée, peut-être via la régulation de bcl2 qui est fortement exprimé dans les cellules hématopoïétiques. De cette façon, la croissance cellulaire nette médiée par les facteurs de croissance peut être envisagée comme étant le résultat des effets combinés de la stimulation d'une voie de prolifération et de l'inhibition d'une voie d'apoptose.

Developmental regulation of the Raf-1 protein: potential role in erythropoietin responsiveness

Marilyn Sanders, Hsienwie Lu, Frederick Walker and Nicholas Dainiak

Departments of Anatomy, Pediatrics, Medicine and Laboratory Medicine and the Connecticut Cancer Institute, University of Connecticut School of Medicine, 263 Farmington Avenue, Farmington, Connecticut, 06030 USA

ABSTRACT
While fetal-derived erythroid progenitors are more sensitive to the effects of exogenous erythropoietin in culture than adult-derived progenitors (Peschle et al 1981), the mechanism(s) of this enhanced sensitivity are poorly understood. We have shown that antisense oligonucleotides to *raf*-1 inhibit adult bone marrow CFU-E-derived colony formation implying that Raf-1 protein may play an integral role in post-receptor erythropoietin-induced signal transduction (Sanders et al 1992). Here, we show that the message for *raf*-1 is present in CFU-Es purified from fetal, neonatal and adult sources. We also report the presence of a 74 kD protein, consistent with Raf-1, in fetal and adult CFU-Es. These data support our hypothesis that the Raf-1 protein is involved in human erythropoietin-induced signal transduction. The role of Raf-1 in enhanced erythropoietin sensitivity of fetal erythroid progenitors remains to be determined.

KEY WORDS
Raf-1 protein, developmental regulation, erythropoietin

INTRODUCTION
The results of recent studies implicate Raf-1, the protein product of the *raf*-1 proto-oncogene, in post-receptor signal transduction for a variety of growth factors, including insulin (Blackshear et al 1990), platelet-derived growth factor (Morrison et al 1989), epidermal growth factor (App et al 1991), interleukins 2 and 3 (Turner et al 1991, Carroll et al 1990), colony-stimulating factors (Carroll et al 1990) and erythropoietin (Carroll et al 1991). This 74 kD cytoplasmic serine/threonine kinase may participate in a phosphorylation cascade involving several intermediate proteins and ultimately, resulting in induction of nuclear proto-oncogenes e.g. *c-fos* and *c-jun*, whose protein products increase DNA synthesis and thus enhance cellular proliferation. Because Raf-1 protein is implicated in post-receptor erythropoietin-mediated signal transduction for murine erythroid cell lines (Carroll et al 1991), we examined the pattern of *raf*-1 mRNA and protein production by human committed erythroid progenitors and the ability of antisense oligonucleotides to *raf*-1 to inhibit CFU-E-derived colony formation by fetal, neonatal and adult hematopoietic tissue.

MATERIALS AND METHODS
Erythroid progenitor cells were purified from adult bone marrow, neonatal umbilical cord blood or fetal liver. All protocols were reviewed by the Institutional Review Committee with informed consent obtained as required. Mononuclear cells were obtained after Ficoll Paque density gradient centrifugation and cultured in serum-substituted conditions with human

recombinant erythropoietin (rHuEpo) using standard methodology. For studies involving oligonucleotides, the mononuclear cells were incubated overnight with antisense/sense oligomers, shaking 5% CO_2, 37°C. After a 12-14 hour incubation, cell suspensions were supplemented with rHuEpo ± antisense/sense oligomers and cultured in fibrin clots. After 7 days incubation, the cells were removed from fibrin clots, fixed with glutaraldehyde and stained with benzidine/hematoxylin. CFU-E-derived colonies were recognized as clusters of 8-49 nucleated benzidine-staining cells. 18 bp antisense/sense oligonucleotides to *raf*-1 were prepared on an Applied Biosystems DNA synthesizer. All experiments included a control culture with erythropoietin alone as well as test cultures with antisense or sense oligomers (5-15µM) and rHuEpo.

For studies requiring cultures of purified progenitors, mononuclear cells were depleted of accessory cells by monocyte adherence to plastic flasks, soybean agglutination and sheep RBC rosetting. Cells were then incubated with 30% fetal calf serum and rHuEpo in 0.9% methylcellulose in 5% CO_2, 37°C. At 7 days, purified CFU-Es were removed from methylcellulose, washed and incubated in growth factor-free medium prior to use in experiments.

RNA was extracted with guanidinium/phenol/chloroform and levels of *raf*-1 and control actin mRNA were analyzed on Northern blots using [^{32}P]dCTP labeled cDNA probes. Signals were quantified using a Betascope Model 603 blot analyzer.

Protein was identified by labeling intact cells with ^{35}S-methionine after incubation in methionine-free medium. Cells were then lysed and Raf-1 protein immunoprecipitated with NH-7, a Raf-1 antipeptide antisera to residues 638-648 (McGrew et al 1992). Immunoprecipitated samples were loaded on a 9% SDS polyacrylamide gel for resolution by electrophoresis. The signal was visualized using autoradiography, -70°C for 2-5 days.

RESULTS
As shown in Fig. 1, addition of 5-15 µM antisense oligonucleotides to *raf*-1 to adult bone marrow-derived progenitors resulted in up to 78% inhibition of CFU-E-derived colony formation at 7 days, compared to control cultures with rHuEpo alone. Addition of similar amounts of sense oligomers did not affect colony formation. Addition of similar amounts of oligomers to fetal and neonatal-derived progenitors failed to inhibit colony formation.

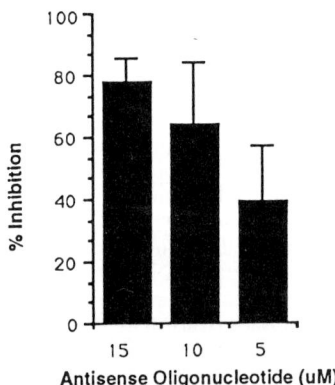

Fig. 1 Effects of antisense oligomers on adult bone marrow erythroid colony formation. Results are the average of two experiments for light density mononuclear cells cultured in serum-substituted medium containing rHuEpo and antisense oligomers to raf-1 at the indicated final concentrations. Values represent % inhibition of colony formation compared to control cultures with rHuEpo alone

Purified CFU-Es from fetal, neonatal and adult sources expressed the message for *raf*-1. Shown in Fig. 2 is an autoradiogram of a typical experiment. The relevant signal is a 3.1 kb band seen in fetal, neonatal and adult CFU-Es. This signal is also visualized in RNA from ECO NIH 3T3 cells, a transformed cell line which overexpresses Raf-1 protein.

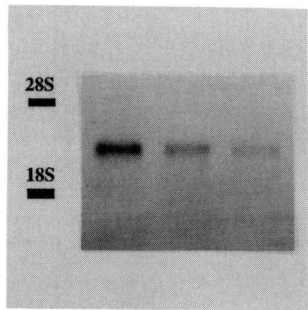

Fig. 2. Autoradiogram of an experiment showing raf-1 mRNA from (left to right) fetal, neonatal and adult-derived CFU-Es.

Initial studies have shown a 74 kD protein, corresponding to Raf-1, in adult bone marrow and fetal-derived CFU-Es which is also seen in ECO NIH 3T3 cells. Shown in Fig. 3 is an autoradiogram of immunoprecipitated proteins showing a 74 kD protein in cell lysates from ECO NIH 3T3 cells (lane 1) and adult-derived CFU-Es (lane 2). Lane 3 shows the effect of addition of excess Raf-1 peptide which blocks visualization of the protein, verifying the specificity of the antibody to Raf-1. Additional studies are underway to clarify whether there is Raf-1-associated phosphorylation and activation upon growth-factor induction.

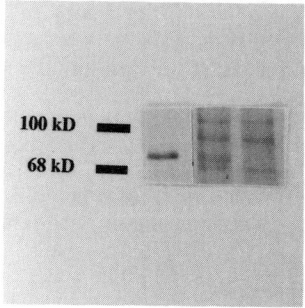

Fig. 3. Autoradiogram of immunoprecipitated proteins showing a 74 kD protein corresponding to Raf-1 from ECO NIH 3T3 cells (lane 1) and adult-derived CFU-Es (lane 2). Lane 3 shows addition of excess Raf peptide to lysates of adult-derived CFU-Es which prevents immunoprecipitation of this 74 kD protein.

DISCUSSION

Enhanced erythropoietin sensitivity seen in fetal erythroid progenitors compared to adult-derived progenitors may be explained either by changes in : 1) erythropoietin receptor number or affinity or 2) a post-receptor signal transduction mechanism. Data from our preliminary antisense oligonucleotide experiments showing dose-dependent inhibition of adult CFU-E-derived colony formation suggests that the Raf-1 protein is integral to erythropoietin-mediated signal transduction. The failure of similar concentrations of oligomers to inhibit fetal and neonatal colony formation suggests that the message for Raf-1 protein may be developmentally regulated and thus increased doses of oligomers required to inhibit colony formation (experiments ongoing in our laboratory).

These results led us to speculate whether the transcription of *raf*-1 mRNA is itself developmentally regulated. Although our initial experiments suggested an increase in fetal compared to adult message for *raf*-1, further experiments suggested differences in CFU-E purity which could explain these findings. Experiments are currently underway to refine our cultures to enable direct comparisons between tissue sources. Should this increase persist in more highly purified CFU-Es, it is possible that increased transcription, decreased message degradation and/or amplified translation of Raf-1 protein may contribute to enhanced erythropoietin sensitivity of fetal CFU-Es.

Carroll et al have demonstrated increased phosphorylation of HCD-57 cells within 1 minute of addition of erythropoietin, a response which peaks by fifteen to thirty minutes and returns to baseline by 90 minutes. We have visualized a 74 kD protein from isolates of fetal and adult CFU-Es which corresponds to the Raf-1 protein, and additional experiments are in progress to document the course of Raf-1 phosphorylation.

In summary, evidence exists to support the role of Raf-1 protein in erythropoietin-mediated signal transduction in human erythroid progenitor cells. Both the message for Raf-1 and the protein itself have been demonstrated in purified human CFU-Es. An understanding of the developmental regulation of Raf-1 protein awaits further experiments designed to quantify relative message and protein levels in fetal, neonatal and adult-derived CFU-Es.

REFERENCES
App H., Hazan R., Zilberstein A., Ulrich A., Schlessinger J. and Rapp U. (1991) Epidermal growth factor (EGF) stimulates association and kinase activity of Raf-1 with the EGF receptor. Mol. Cell. Biol. 11:913-919.
Blackshear P.J., Haupt D.M., App H. and Rapp U.R. (1990) Insulin activates the Raf-1 protein kinase. J Biol. Chem. 265:12131-12134.
Carroll M.P., Clark-Lewis I., Rapp U.R., May W.S. (1990) Interleukin-3 and granulocyte-macrophage colony-stimulating factor mediate rapid phosphorylation and activation of cytosolic c-raf. J Biol. Chem. 265:19812-19817.
Carroll M.P., Spivak J.L., McMahon M., Weich N., Rapp U.R., May W.S. (1991) Erythropoietin induces Raf-1 activation and Raf-1 is required for erythropoietin-mediated proliferation. J Biol. Chem 266:14964-14969.
McGrew B.R., Nichols D.W., Stanton V.P. Jr., Cai H., Whorf R.C., Patel V., Cooper G.M., Laudano A.P. (1992) Phosphorylation occurs in the amino terminus of the Raf-1 protein. Oncogene 7:33-42.
Morrison D.K., Kaplan D.R., Escobedo J.A., Rapp U.R., Roberts T.M. and Williams L.T. (1989) Direct activation of the serine/threonine kinase activity of Raf-1 through tyrosine phosphorylation by the PDGF-ß receptor. Cell 58:649-657.
Pelech S.L. and Sanghera J.S. (1992) MAP kinases: charting the regulatory pathways. Science 257:1355-1356.
Peschle C., Migliaccio A.R., Migliaccio G., Ciccariello R., Lettieri F., Quattrin S., Russo G. and Mastroberardino G. (1981) Identification and characterization of three classes of erythroid progenitors in human fetal liver. Blood 58:565-572.
Sanders M.R., Walker F., Dainiak N. (1992) Erythropoietin responsiveness and Raf-1 mediated signal transduction in serum-depleted cultures of human fetal, neonatal and adult progenitor cells. Pediatr. Res. 31:270A.
Turner B., Rapp U., App H., Greene M., Dobashi K., Reed J. (1991) Interleukin 2 induces tyrosine phosphorylation and activation of p72-74 Raf-1 kinase in a T-cell line. Proc. Natl. Acad. Sci. USA 88:1227-1231.

Inhibition of S phase in chronic myeloid leukemia by antisense oligonucleotides

F.X. Mahon, F. Belloc, A. Rice, J.M. Boiron, P. Bernard, A. Broustet and J. Reiffers

Laboratoire de Greffe de Moelle, URA CNRS 1456, Université de Bordeaux II, Bordeaux, France

To determine the function relevance of the BCR-ABL gene for the proliferation of Chronic Myelogenous Leukemia (CML) cells, we used synthetic antisense 18-mer oligonucleotides (ASO) complementary to the two different junctions (b2a2 or b3a2) of the hybrid mRNA. Eight patients were evaluated with this technique at diagnosis before any treatment, six patients were in chronic phase, one in acceleration phase. Another patient with myelodysplastic syndrome served as control. We used peripheral blood obtained after cytapheresis and mononuclear cells were separated by Ficoll sedimentation. Cells were exposed to appropriated oligomers during 14 hours in liquid medium containing RPMI and low concentration of fetal calf serum ($\leq$ 5 %) to decrease the degradation of oligonucleotides. For each oligomer the final concentration was 14 mM. The cells were pulse labelled with 10 mM of BrdUrd, then reacted with monoclonal antiBrdUrd antibody FITC and stained with propidium Iodide (PI). The bivariate BrdUrd/DNA fluoresence distribution was evaluated.The index of S phase cell cycle (IS) which measured the percentage of S phase cells as well as the fluorescence intensity was calculated (Lacombe F et al,1988). The type of mRNA was determined by reverse transcriptase polymerase chain reaction. For the patient with CML in acceleration as well as for the patient with myelodysplastic syndrome, no modification of cell cycle was observed. For the other patients, the IS demonstrated that AS induced a slowing down of DNA synthesis [IS varied from 13% to 90% (mean= 50+/- 32 % p<0.01)] of the control values for the 6 patients in chronic phase.These results were confirmed by CFU-GM studies. (Mahon FX et al, 1993).

In this study, we were able to inhibit cell proliferation measured by cell cycle analysis in patients with CML in chronic phase.It is interesting to point that the blockage of mRNA even transiently was sufficient to modify the S phase of the cell cycle. The role of BCR-ABL on the cell cycle remains incompletely understood, ABL protein alone is phosphorylated by p^{34cdc2} kinase during the cell cycle and undergoes a cell cycle regulated serine threonine phosphorylation (Kipreos ET& Wang JY,1992:). We showed herein that the chimeric protein seems to be necessary to promote S phase in chronic phase CML patients.

The role of different ASO has been recently investigated in normal marrow as well as in CML marrow cells (Carraciolo D et al, 1989; Rosti V et al.1992). Szczylick et al used specific anti BCR-ABL ASO and obtained a specific inhibition of leukemic blast cells from a patient with myeloid CML blast crisis (Szczylik C et al,1991). In our hands no inhibition was observed for the only patient tested in accelerated phase. However, it is well known that other molecular abnormalities could affect blast cells during the progression of disease to the blastic phase (Ahuja H et al, 1991; Towatari M et al, 1991).We conclude that these techniques are interesting to study the role of BCR-ABL and particulary in cell cycle (S phase).

REFERENCES

Ahuja Het al .(1991): The spectrum of molecular alterations in the evolution of chronic myelocytic leukemia. J Clin Invest. 87, 2042-2049.

Carraciolo D et al. (1989). Lineage specific requirement of cabl function in normal hematopoiesis. Science. 245,1107-1110.

Lacombe F et al.(1988): Evaluation of four methods of DNA distrbution data analysis based on bromodeoxyuridine/DNA bivariate data. Cytometry . 9 , 245-253.

Lugo TG et al.(1990): Tyrosine kinase activity and transformation potency of BCR-ABL oncogene products. Science. 247, 1079-1082.

Mahon FX et al.(1993): Antisense oligomers in chronic myeloid leukemia. Lancet. 341, 566.

Rosti V et al.(1992): C-ABL function in normal and chronic myelogenous leukemia hematopoiesis: in vitro studies with antisense oligomers. Leukemia. 6, 1-7.

Shtivelman E et al. (1985): Fused transcript of abl and bcr genes in chronic myelogenous leukemia. Nature. 315, 550-554.

Szczylik C et al. (1991): Selective inhibition of leukemia cell proliferation by BCR-ABL antisense oligonucleotides. Science. 253, 562-565.

Towatari M et al.(1991): Absence of the human retinoblastoma gene product in the megakaryoblastic crisis of chronic myelogenous leukemia. Blood. 78,2178-2181.

Positive and negative transacting factors and their modification

David E. Fisher

Center for Cancer Research, Massachusetts Institute of Technology, Cambridge, MA 02139, and Dana Farber Cancer Institute, Harvard Medical School, Boston, MA 02115, USA

Gene transcription is a central process in the metabolism of a cell. It largely dictates the expression of all phenotypes, healthy or pathological. Regulation of transcriptional pathways constitutes one of the most highly studied topics in modern molecular biology due to the very broad implications of this regulation for diverse cellular activities. While eukaryotic transcription is comprised of numerous biochemical processes, one fundamental facet is the sequence-specific recognition of DNA by protein transcription factors. An analysis of such protein:DNA recognition will be described in this manuscript as applied to the oncogenic transcription factor Myc. The experimental approach outlined below is applicable to many biomolecular recognition processes. In elucidating critical features of this recognition, these studies have also suggested ways to successfully design high affinity analogs which may serve as blueprints for the design of transcription targeted therapeutics.

Introduction

Oncoproteins are products of genes capable of conferring upon their host cell phenotypic alterations resulting in malignancy. A large number of these factors have been associated with specific human malignancies and have stimulated enormous research efforts which target all the cellular compartments in which oncoproteins are located. These molecular pathogens have been associated with extracellular ligand, receptor, inner plasma membrane, second messenger, and nuclear niches within the cell. One of the largest groups of oncoproteins are the nuclear ones. Within the nucleus, factors have been identified with dominant acting or recessive (tumor suppressive) activities in transformation pathways. Nuclear oncoproteins are likely to play key roles in pathways involving gene transcription, DNA replication, or RNA processing. As such, an understanding of their biochemical interactions may provide information about fundamental cellular metabolic pathways as well as the process of malignant transformation.

Eukaryotic gene transcription occurs through sets of factors which can be roughly categorized as core factors and regulatory factors. The core proteins include TATA binding protein, RNA polymerase II, and associated factors. Regulatory transcription factors bind specific target DNA sequences in promoter or enhancer elements and are capable of transmitting regulatory signals to the core transcriptional machinery for either stimulation or suppression of basal transcription. These factors usually contain distinct DNA binding and activation domains. The signals which they transmit to the core machinery appear to be mediated, at least in part by a separate family of proteins which are capable of interactions with both the regulatory factors and the basal transcription machinery.

Suppression of transcription by regulatory factors occurs, in nature, by a number of mechanisms. The Drosophila Krüppel factor contains a distinct domain with suppressive activity (Licht et al., 1990). Modifications of stimulatory transcription factors may occur (for example by phosphorylation, sequestration, or a requirement for ligand binding) which control the ability of a factor to bind its DNA recognition element. Examples of this are the Max protein which cannot bind DNA upon phosphorylation by casein kinase II (Berberich & Cole, 1992), NFκB which is sequestered away from DNA when bound by its inhibitor IκB (Ghosh & Baltimore, 1990), and steroid hormone receptors which require ligand binding for activity. A third mechanism for transcriptional repression is transdominant suppression. An example is the TFE3 factor which may be alternatively spliced (Roman et al., 1991) to either contain or delete a transactivating domain. Deletion of the transactivating domain while retaining DNA binding ability produces a protein which fails to directly activate transcription. Importantly, such a factor also blocks other proteins from binding the DNA target element by competition.

An understanding of the molecular nature of sequence specific DNA recognition by the DNA binding domain of a biologically important transcription factor sheds light on biological recognition per se as well as approaches to the design of artificial transdominant suppressors. Molecules with enhanced DNA binding affinity may compete well against endogenous transcription factors and provide attractive therapeutic targets. The studies described below examine the DNA recognition properties of the Myc oncoprotein and related transcription factors. Distinct structural features allowed the design of high affinity analogs which bind approximately 35-fold more avidly to DNA than the binding domain of Myc.

Materials and Methods

Myc and TFEB plasmid constructs were in Bluescript-SK containing an oligonucleotide designed for translation initiation (Baldwin et al., 1990). Purified TFEB protein was made as a histidine fusion (Novagen) and purified from BL21 cells by nickel chelate chromatography according to the manufacturer's recommendations (Qiagen Corp.). Circular dichroism spectroscopy was performed using an Aviv 60DS spectrapolarimeter over the wavelength range shown in Figure 3, at 25°C with signal averaging of 10 minutes for each data point. The protein was dialyzed into 10 mM sodium phosphate buffer, pH 7.0 with 100 mM sodium chloride. and used directly for CD analysis (Fisher et al., 1993) in either the absence or presence of a double stranded oligonucleotide containing the CACGTG consensus binding site plus 5 ng per ml poly dI-dC. Electrophoretic mobility shift assays (EMSA) were performed using TFEB wild type or mutants or TFEB/Myc chimeras synthesized by in vitro transcription and translation using rabbit reticulocyte lysate (Fisher et al., 1993). Double stranded ^{32}P-labeled DNA probes were generated by Klenow end-filling of restriction digested inserts from plasmids containing the CACGTG consensus binding site (Carr & Sharp, 1990) as it occurs in the adenovirus major late promoter. Native polyacrylamide gels were run in Tris-Glycine-EDTA buffer system, dried and exposed. Quantitation of binding was determined by counting "shifted" (protein-bound) DNA probe using a Phosphorimager. The linear range for DNA binding was found to be approximately 5 to 10 fold above or below the protein levels used in these experiments. Site directed mutagenesis was performed in accordance with the manufacturer's recommendations (Amersham). Mutants were verified by DNA sequencing.

Results

Basic domain homologies

Myc oncoprotein belongs to a family of transcription factors which contain basic, helix-loop-helix and leucine zipper motifs (b-HLH-ZIP) arranged contiguously. The ~20 amino acid basic domain is likely to contact specific DNA base pairs while the helix-loop-helix and leucine zipper domains specify protein dimerization interactions. There are approximately 10 proteins in this b-HLH-ZIP family which in addition to sharing motif structures are also capable of binding to the identical target DNA sequence, which consists of the palindrome CACGTG. As a starting point to determine

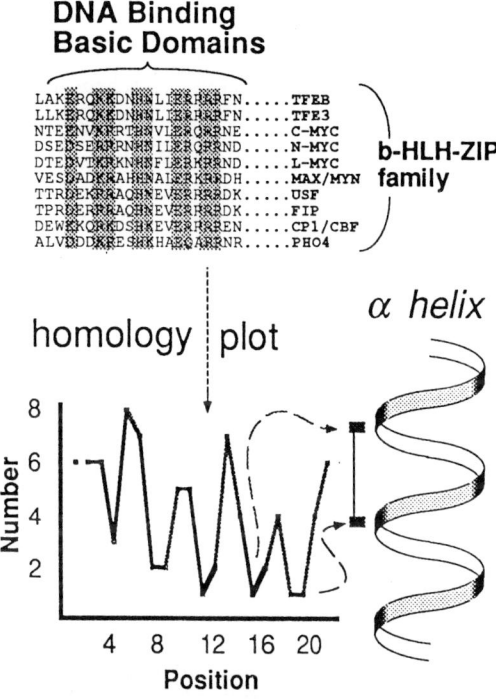

Figure 1. Basic domain amino acid sequence homologies of b-HLH-ZIP proteins. Ten b-HLH-ZIP proteins which all recognize the DNA sequence CACGTG were aligned and their sequence variation plotted as a function of amino acid position. The nadirs on the homology plot represent conserved positions and are spaced on average 3-4 amino acids apart, a spacing reminiscent of consecutive α-helical turns.

which amino acids are important for DNA recognition, the basic domains of these 10 proteins were aligned and sequence homologies were plotted as a function of position (Fig. 1) (Fisher et al., 1991). Conserved positions were detected and were felt likely to represent positions of importance for DNA recognition. The spacing of these conserved positions, every 3-4 amino acids, was suggestive of the spacing for consecutive turns of a protein α-helix. Therefore, on the basis of sequence homologies it appeared likely that these b-HLH-ZIP proteins utilize an α-helical basic domain structure and recognize DNA through interactions on one face of that α-helix.

α-helical structure

A more direct assessment of α-helical structure was carried out using circular dichroism (CD) spectroscopy. CD spectra of α-helices are characterized by absorption minima at 208 and 222 nm. Figure 2 shows that in the absence of DNA some α-helical structure was detected. However upon addition of DNA the amplitude at 208 and 222 nm increased strikingly, suggesting that upon binding to DNA a new region of α-helix was formed in the protein. Since the basic domain is likeliest to be contacting DNA, these results suggest that DNA recognition occurs through an α-helical basic domain structure. In addition, the increase in amplitude upon binding to DNA implies that the basic domain is not α-helical in the absence of DNA, but requires DNA to stabilize its configuration.

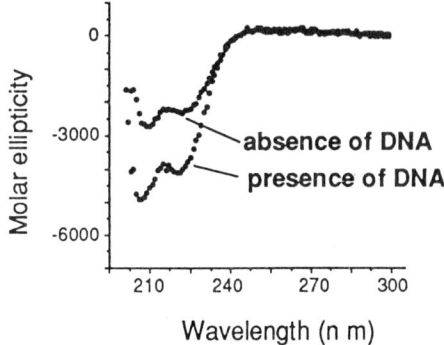

Figure 2. CD spectra of purified TFEB shows DNA induced α helical folding. Purified TFEB (used as a model for Myc because it can bind DNA as a homodimer) was tested either in the absence or presence of CACGTG containing DNA. The increased amplitudes at 208 and 222 nm in the presence of DNA indicate new α-helical structure.

Mutagenesis

To further delineate specific amino acid residues in the basic domain which likely play roles in DNA recognition, exhaustive mutagenesis was carried out utilizing alanine scanning. Alanine substitution was chosen because this amino acid has the highest α-helical propensity of any amino acid and it was desirable to alter amino acid side chains without interfering with the α-helical structure of the basic domain. In addition, at positions felt to be important for DNA recognition, conservative substitutions were then made to assess the charge constraints at these positions. Figure 3 shows a summary of the results of these mutational effects. All of the nonconserved positions tolerated alanine mutations without substantial effect on DNA affinity. Four positions were found to be "critical" for DNA recognition, in that alanine or even conservative mutations greatly diminished DNA binding at those sites. The remaining three positions (the three arginines flanking the most C-terminal "critical" arginine) were more tolerant of conservative mutations while still showing substantial effects upon alanine substitution. Some of these differences were elucidated by testing these mutants for binding to a lower affinity DNA site containing the core sequence CATGTG from the immunoglobulin heavy chain enhancer μE3 site (Fisher et al., 1991). Based on these mutational studies, the critical face of the basic region α-helix was identified and shown to

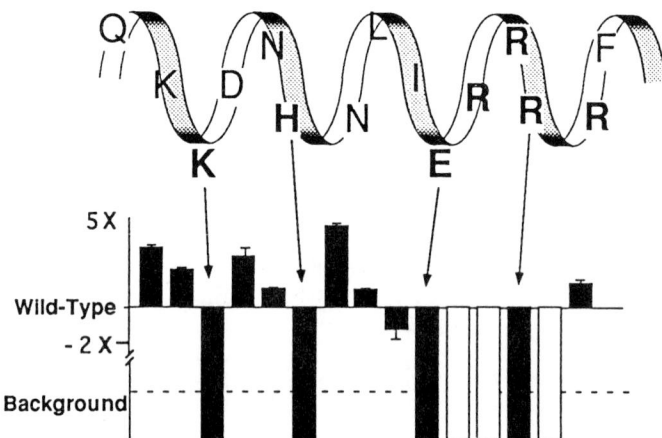

Figure 3. Scanning mutational analysis. TFEB was used as substrate for alanine scanning mutagenesis across the basic domain. The histogram of DNA binding shows positions (black) where alanine mutation had little effect. Four others (indicated by arrows) were sensitive to both alanine and conservative substitutions. The three positions shown with clear boxes were sensitive to alanine mutation, but tolerated conservative mutation (lysine) on either the high affinity or low affinity DNA probes.

correspond to the conserved amino acids seen in the sequence comparisons above (Fig. 1). The three flanking arginines near the most C-terminal "critical" position displayed evidence of contributing to the stability of the protein:DNA complex, although with looser constraints than the four "critical" residues. Since these three amino acids are expected to reside at various positions around one full helical turn of protein, it is reasonable to postulate that they may make arginine:phosphate contacts with the DNA backbone at several positions across the major groove of DNA. There are large potential benefits, energetically, of multiple contiguous DNA contacts over one turn of an unstable α-helical peptide segment, and this group of flanking arginines has been termed a "clamp" motif on that basis.

High affinity analogs

Since the basic domain appears to require a DNA induced stabilization of α-helical folding, attempts were made to increase the α-helicity of the basic domain by systematic introduction of alanine residues at all positions which do not directly contribute to DNA contact. Substitution of multiple alanines in these artificial basic domain constructs would test whether spontaneous α-helicity can be enhanced (even in the absence of DNA) and also would verify the sufficiency of the above identified positions for protein:DNA recognition. The effects of multiple alanine substitutions within the basic domain are shown in Figure 4. Since Myc is unable to form stable homodimers (due to constraints within its HLH-ZIP region) a chimeric protein was generated substituting Myc's basic domain for TFEB's within the background of TFEB's HLH-ZIP region. This construct ("MycBR") bound DNA specifically and served as the basis for comparison of the various analog constructs. Systematic substitution of all nonconserved residues with alanine was seen to enhance DNA affinity both in the case of the basic region of c-Myc (Figure 4) and TFEB (data not shown). Additionally, one position in the basic domain of c-Myc (indicated in Figure 4 by asterisk) was shown to confer enhanced affinity when substituted to lysine from arginine. Interestingly lysine is the naturally occurring amino acid at this position in the b-HLH-ZIP proteins TFEB and TFE3 while most others, including Myc, contain arginine. The composite effects of multiple alanine substitutions and arginine to lysine mutation at the sixth position in the basic region produced net 35-fold enhanced affinity relative to the affinity of c-Myc's basic region. The enhanced affinity was shown to be due to enhanced sequence specific DNA recognition using nonspecific DNA competitors in quantitative analysis.

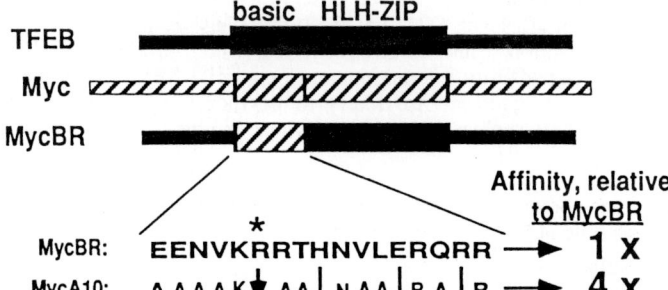

Figure 4. High affinity DNA binding analogs. To test Myc's basic region in a homodimeric context, a chimera (MycBR) was made by substituting its basic region into TFEB, a factor which is capable of homodimerization. Multiple alanines were systematically substituted at positions found not to directly contribute to DNA contact and the critical arginine shown with an asterisk was mutated to lysine. Collectively, these changes produced 35-fold enhanced DNA affinity as compared to Myc-BR.

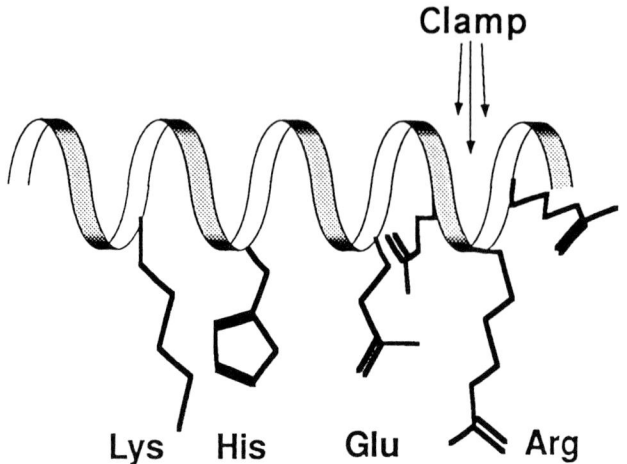

Figure 5. Structural elements of Myc's basic domain which are essential for DNA binding. Key features include DNA induced α-helical folding of the backbone, 4 critical amino acids along one face, and flanking arginines which comprise the "clamp" motif.

To assess whether the enhanced affinity of multiple alanine-containing basic domains was due to higher spontaneous α-helix formation, peptides corresponding to the wild-type TFEB basic domain and the 12-alanine analog (K-A12) were studied. Molar ellipticity at 222 nm was measured by CD, an index of α-helical content. By this assay, less than 10% of the wild type peptide was α-helical in the absence of DNA under conditions in which >50% of the alanine mutant was α-helical (data not shown).

Discussion

Analysis of the basic regions from b-HLH-ZIP proteins capable of binding the hexanucleotide CACGTG has revealed strong evidence for recognition by an α helix of protein which likely fits in the major groove of DNA. CD suggests that the basic domain is intrinsically disordered, but undergoes an α-helical folding transition upon DNA binding. Scanning mutagenesis permitted identification of amino acids likely to make base pair contacts. A "clamp" motif consisting of several basic amino acids which flank a residue involved in base-pair recognition may help stabilize the α-helical folding transition (Fig. 5). The discovery that basic domains of this family of proteins require DNA to induce the α-helical fold parallels a similar observation in the basic domains of basic-leucine zipper proteins (Talanian et al., 1990; O'Neil et al., 1990; Weiss et al., 1990; Patel et al., 1990). The likelihood that this transition results in an energy loss which produces net lower protein:DNA affinity suggests that lower affinity interactions are beneficial for these classes of transcription factors. Lower affinity interactions are likely to be more easily regulated or suppressed and may be desirable for factors such as oncoproteins whose binding to DNA produces major phenotypic change in a cell.

Utilizing information from the studies above, higher affinity analogs were designed by substituting the more helix-prone amino acid alanine at positions in the basic domain felt not to be involved in direct DNA contact. In this way, substitutions which produced basic regions with high (~50%) α-helix in the absence of DNA yielded proteins with substantially enhanced affinity for the specific DNA sequence. Interestingly, the general availability of alanine suggests that optimal high affinity proteins are not physiologically desirable in cells, perhaps due to diminished susceptibility to regulation, as discussed above.

The understanding of basic structural constraints for DNA recognition by factors such as the Myc family of oncoproteins offers clues to the complementarity of protein:DNA recognition surfaces. Additionally, the opportunity to exploit certain predicted structural features (such as the α-helical backbone) in the design of higher affinity analogs permits potential use of this information as a possible blueprint for small molecule design. Such molecular mimics may function as transdominant suppressors of transcription and prove valuable in efforts to understand and treat Myc induced carcinogenesis.

Acknowledgements

The author gratefully acknowledges Dr. Phillip Sharp in whose laboratory most of these experiments were carried out; Drs. R. Sauer and M. Milla for assistance with the CD analyses; L. Parent for technical assistance; and members of the Sharp lab for useful discussions. This work was supported by a Postdoctoral Fellowship for Physicians from the Howard Hughes Medical Institute, by United States Public Health Service grant PO1-CA42063 from the National Institutes of Health and cooperative agreement CDR-8803014 from the National Science Foundation to Phillip Sharp, and partially by National Cancer Institute Cancer Center Support (core) grant P30-CA14051.

References

Baldwin, A.S., LeClair, K.P., Singh, H., & Sharp, P.A. (1990): A large protein containing zinc finger domains binds to related sequence elements in the enhancers of the class I major histocompatibility complex and kappa immunoglobulin genes. Mol. Cell. Biol. 10, 1406-1414.

Berberich, S.J. & Cole, M.D. (1992): Casein kinase II inhibits the DNA-binding activity of Max homodimers but not Myc/Max heterodimers. Genes & Dev. 6, 166-176.

Carr, C.S. & Sharp, P.A. (1990): A helix-loop-helix protein related to immunoglobulin E box-binding proteins. Mol. Cell. Biol. 10, 4384-4388.

Fisher, D.E., Carr, C.S., Parent, L.A., & Sharp, P.A. (1991): TFEB has DNA-binding and oligomerization properties of a unique helix-loop-helix/leucine-zipper family. Genes & Dev. 5, 2342-2352.

Fisher, D.E., Parent, L., & Sharp, P.A. (1993): High affinity DNA-binding Myc analogs: recognition by an α helix. Cell 72, 267-276.

Ghosh, S. & Baltimore, D. (1990): Activation in vitro of NFκB by phosphorylation of its inhibitor IκB. Nature 344, 678-682.

Licht, J.D., Grossel, M.T., Figge, J., & Hansen, U.M. (1990): Drosophila Krüppel protein is a transcriptional repressor. Nature 346, 76-79.

Roman, C., Cohn, L., & Calame, K. (1991): Creation of a transdominant negative form of transcription activator mTFE3 by differential splicing. Science 254, 94-97.

O'Neil, K.T., Hoess, R.H., & DeGrado, W.F. (1990): Design of DNA-binding peptides based on the leucine zipper motif. Science 249, 774-778.

Patel, L., Abate, C., & Curran, T. (1990): Altered protein conformation on DNA binding by Fos and Jun. Nature 347, 572-575.

Talanian, R.V., McKnight, C.J., & Kim, P.S. (1990): Sequence-specific DNA binding by a short peptide dimer. Science 249, 769-771.

Weiss, M.A., Ellenberger, T., Wobbe, C.R., Lee, J.P., Harrison, S.C., & Struhl, K. (1990): Folding transition in the DNA-binding domain of GCN4 on binding to DNA. Nature 347, 575-578.

Résumé

La régulation de l'expression génique est critique pour le développment normal et la différenciation des cellules. Les facteurs de transcription sont des protéines qui se lient aux séquences d'ADN proches du gène exprimé et régulent les mécanismes transcriptionnels de différentes façons afin de moduler les niveaux d'expression. Ces facteurs représentent potentiellement d'importantes cibles thérapeutiques en raison du pouvoir présumé de leurs effets positifs ou négatifs sur la transcription génique. Une séquence de reconnaissance spécifique d'un site cible de liaison d'ADN a été décryptée dans ce travail pour l'oncoprotéine Myc. Myc doit se lier au palindrome CACGTG pour transformer les cellules. La spectroscopie circulaire par dichroisme a démontré que la liaison à l'ADN induit directement la formation d'une nouvelle région de la protéine en hélice alpha. L'analyse mutationelle a révélé des acides aminés particuliers dans une région connue comme étant le " domaine basique" impliqués dans des contacts spécifiques avec l'ADN . Comme ce domaine parait être une hélice alpha instable en l'absence d'ADN, des efforts pour augmenter son hélicité intrinsèque ont été entrepris en remplaçant systématiquement tous les acides aminés non en contact avec l'ADN par de l'alanine, acide aminé ayant la plus haute propension à l'hélice alpha. De cette façon, des analogues ont été générés avec une hélicité substantiellement plus grande et par conséquent une affinité pour l'ADN plus élevée.

Summary

Regulated gene expression is critical to the normal development and differentiation of cells. Transcription factors are proteins which bind DNA sequences near the expressed gene and regulate transcriptional machinery in various ways to modulate expression levels. These factors represent potentially important therapeutic targets due to the anticipated potency of their positive or negative effects on gene transcription. Sequence specific recognition of a target DNA binding site has been deciphered in this study for the DNA binding oncoprotein Myc. Myc binds the palindrome CACGTG and must do so in order to transform cells. Circular dichroism spectroscopy demonstrated that DNA binding directly induces formation of a new region of protein alpha helix. Mutational analysis revealed particular amino acids in a region known as the "basic domain" which likely make specific DNA contacts. Since this domain appears to be an unstable alpha helix in the absence of DNA, efforts to increase its intrinsic helicity were undertaken by systematically replacing all non-DNA-contacting amino acids with alanine, the amino acid with the highest alpha helical prope nsity. In this way, analogs were generated with substantially higher helicity and correspondingly higher DNA affinity.

Genetic analysis of the LIF cytokine during mouse embryogenesis

J. Perreau*, D. Dumenil**, J.L. Escary*, L. Tiret*, F. Conquet*,
Y. Lallemand* and P. Brûlet*

*Unité d'Embryologie Moléculaire, Institut Pasteur, 25, rue du Docteur-Roux, 75724 Paris Cedex 15;
**INSERM U 362, Institut Gustave-Roussy, 94800 Villejuif, France

Cytokines are involved in cell-cell interactions in widely different biological process: from embryogenesis to the immune response and neuronal communication. Deciphering the role of one cytokine is not an easy task because they are biologically active in minute amounts and are highly specific for some cell types. At the same time redundancy, synergy and pleiotropy seem to be the rule among cytokines. We have used new genetic techniques to establish the mode of action of the LIF (Leukaemia Inhibitory Factor) *in vivo*, with a special emphasis on stem cells during embryogenesis.

SPECIFIC MUTAGENESIS IN THE MOUSE

About a hundred mice with an altered gene have now been generated through gene targeting in embryonic stem (ES) cells (Cold Spring Harbor, 1992). First, the desired mutation is introduced into the cloned sequence of a given locus. The mutation can be a point mutation, a few kilobases long deletion, a substitution of the coding domain or a modification of the promoter region. The mutated DNA electroporated into ES cells will integrate into its appropriate genomic location by using the cellular machinery for homologous recombination. Germ line chimerae are then generated by microinjecting recombined ES cells into blastocysts or morulae (Lallemand & Brulet, 1990) . The interbreeding of heterozygous siblings will yield animals homozygous for the desired mutation. Such a scheme allows the detailed analysis of the mutated gene under various genetic backgrounds. A sophisticated genetic analysis is required to understand the complex biological process of mammalian embryogenesis. We expect that many genes differentially regulated during embryogenesis form networks that underlie morphogenesis. A simple null mutation could lead to a complete disruption of a regulatory network, or, on the other hand, the differences between mutant and wild-type phenotypes might be minor because of the possible redundancy of genes controlling embryogenesis.
We therefore have developed a procedure whereby a chosen gene is not only inactivated but also replaced by a functional reporter gene, the *Escherichia coli* Lac Z (Le Mouellic et al., 1992). After genetic recombination, the endogenous promoter of the targeted gene controls the reporter gene, whose expression can be followed *in situ* throughout embryogenesis of the mutant animal. This is specially useful as a means to identify the cells in an homozygous embryo that should have been transcribing the targeted gene. In heterozygous embryos and adults, the pattern of Lac Z staining should reflect the transcription pattern of the gene provided that no genetic regulatory element has been affected by the mutation. Again, this technique is quite universal and various genes can be used as a substituting gene. For instance, we are trying to immortalize defined embryonic cells by substituting a gene transcribed in those cells with one copy of a conditional oncogene.

LIF EARLY TRANSCRIPTION PATTERN

The cytokine LIF, for Leukemia Inhibitory Factor, was identified by Metcalf and collaborators because of its differentiating activity on some leukemic cells. *In vitro,* LIF exhibits a broad spectrum of biological effects on different cell types (Hilton & Gough, 1988), in particular, it influences several stem cell systems, inhibiting differentiation of ES cells (Smith et al., 1988; Williams et al., 1988) and promoting survival and/or proliferation of primitive hematopoietic precursors (Fletcher et al., 1991; Leary et al., 1990) and of primordial germ cells (Matsui et al., 1991). In the adult, LIF is produced by several cell types including alloreactive T cell clones, thymic epithelium, mitogen-treated splenocytes, uterine endometrial glands, and pituitary follicular cells. Because of the reported biological effect of LIF on ES cells *in vitro,* we first wanted to establish that the LIF gene was transcribed *in vivo* during mouse embryogenesis. By a variety of methods, we were able to establish that the LIF gene was transcribed as early as the blastocyst stage (Conquet & Brulet, 1990). Blastocysts are composed of two different cell types (Fig. 1): the trophectoderm which is an epithelium of differentiated cells surrounding a few totipotent cells called the inner cell mass (ICM). Trophectoderm cells are likely to produce the LIF which acts onto the neighboring totipotent ICM cells. After implantation into the uterus, LIF transcription is particularly strong in the extraembryonic part of the implanted embryo. Again, we postulate that the LIF protein synthetized in the extra-embryonic tissue is active onto the pluripotent cells of the primitive ectoderm or epiblast. After gastrulation, no transcript could be detected in the embryo proper, LIF gene transcription is confined to the extra-embryonic tissues and placenta.

LIF producing cells seem to be in close proximity of the likely target cells. During mouse embryogenesis, LIF transcripts are detected when the first undifferentiated cells of the ICM appear. After implantation, the epiblast cells are close to the LIF producing tissue. Between day 6.5 and 7.5 post coïtum, the first mesodermal cells appear in the primitive streak. From fate map studies at the early streak stage (Lawson & al., 1991), they will form the blood islands in the visceral mesoderm after migration to the LIF producing extra-embryonic tissue. In those blood islands are localised the embryonic hemopoietic stem cells. Finally, just posterior to the definitive primitive streak in the extra embryonic mesoderm close to the amniotic fold lies a cluster of alkaline phosphatase positive cells (Ginsburg & al., 1990) . They are likely candidates to be the primordial germ cells, which *in vitro* at least are also LIF target cells.

GAIN OF FUNCTION MUTATION

Our first approach was to overexpress LIF in the embryonic cells of the epiblast *in vivo* and look for a phenotype at gastrulation (Conquet & al., 1992). At this time in development, the embryonic cells should undergo major reorganisation, differentiation and proliferation. LIF is known to have two isoforms (Ratjen & al., 1990) one protein is soluble (LIF/$_D$), whereas an extra cellular matrix bound form (LIF/$_M$) was also reported. The various experimental parameters were established using the ES cells *in vitro* . To drive the expression of LIF cDNAs, we selected the cytomegalovirus promoter CMV. After introduction of the DNA molecules into the ES cells, various clones overexpressing either the diffusible or the membrane bound form of the LIF were selected. At least, for a few passages *in vitro,* those cells could proliferate without feeder cells or added exogenous LIF. The LIF expressing ES cells behave like autocrine cells *in vitro* although we did not prove it formally , no antibodies again the LIF receptor being available. Overexpressing cells were then injected into morulae instead of blastocyst (Lallemand & Brulet, 1990). The main advantage of this technique is that the amount of chimerism achieved is reproducibly very high. Within a litter every manipulated embryo is between 90 to 100% chimeric. After injection of a few LIF overexpressing ES cells into morulae, the manipulated embryos were reimplanted into a foster mother. To analyse LIF effect on the gastrulation process, the foster mother was sacrified a few days later, when the embryos were at the stage 8.0 to 9.0 post coïtum. Two different phenotypes are obtained, embryos that are overexpressing the diffusible form of LIF are essentially normal whereas those overexpressing the LIF/$_M$ did not enter gastrulation. No mesoderm is detectable in what look like essentially a degenerating egg cylinder stage embryos. We do not know how is significant the difference observed between the two isoforms. It could just be a quantitative effect as we could show that the phenotype appeared only with the embryos presenting the highest amount of chimerism.

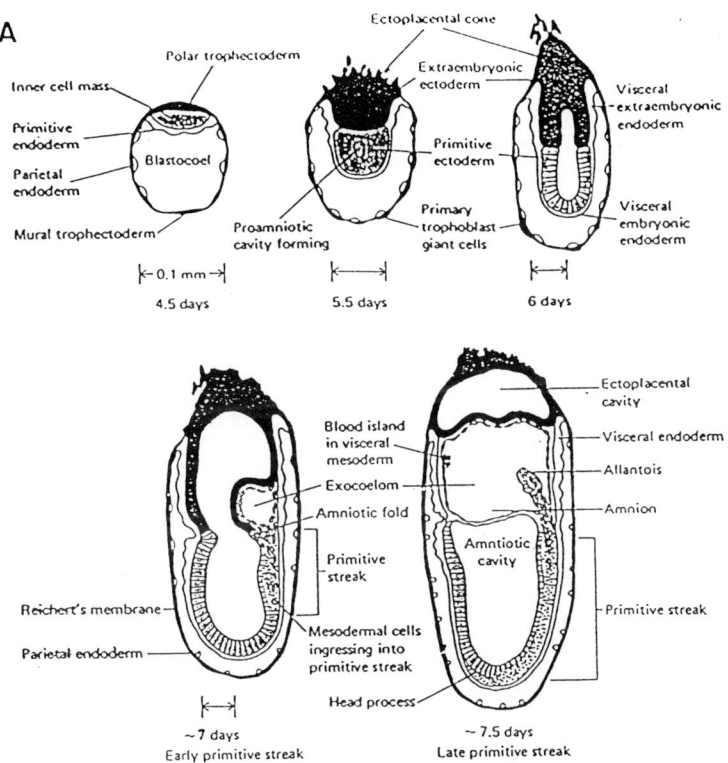

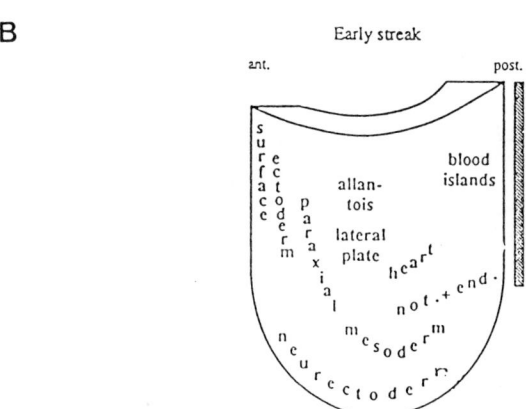

Fig. 1. (A) Schematic representation of early postimplantation development of the mouse embryo.
(In Manipulating the mouse embryo, B. Hogan, F. Costantini and E. Lacy. Cold Spring Harbor Laboratory, p 50).

(B) Cell fate in the early postimplantation embryo.
(In postimplantation development in the mouse, Ciba Foundation Symposium 165, p 15. Ed. John Wiley & Sons).

Nevertheless, under certain experimental conditions, overexpressing LIF in the embryonic ectodermal stem cells interferes with their decision between proliferation, survival and differentiation. In those manipulated embryos, the primitive ectoderm layer seems normal. Implantation apparently proceeds correctly. Only when the ectodermal cells should differentiate into mesoderm, did we detect the first abnormality. LIF overproduction renders the primitive ectodermal stem cells insensitive to the normal mesoderm inductive signals present in the vicinity of the streak at the beginning of gastrulation. LIF could be acting as a negative regulator for the normal differentiation program of the pluripotent embryonic cells. Strong overexpression would lock the stem cells in a state where the normal genetic programme for their differentiation cannot be activated. Finally in these experiments, we did not notice a strongly abnormal size of the embryonic ectoderm. It seems therefore that LIF did not act as a true proliferation factor for the embryonic stem cells.

LOSS OF FUNCTION MUTATION

The LIF gene is a single copy gene in the mouse genome, localized on the short arm of chromosome 11. LIF- mice were generated (Escary & al., 1993) by deleting a 3.3 kbp DNA fragment that contains all the known exons and introns plus a part of the 3' untranslated region of the gene (Fig. 2). This DNA fragment codes for the two reported isoforms of the murine LIF protein.

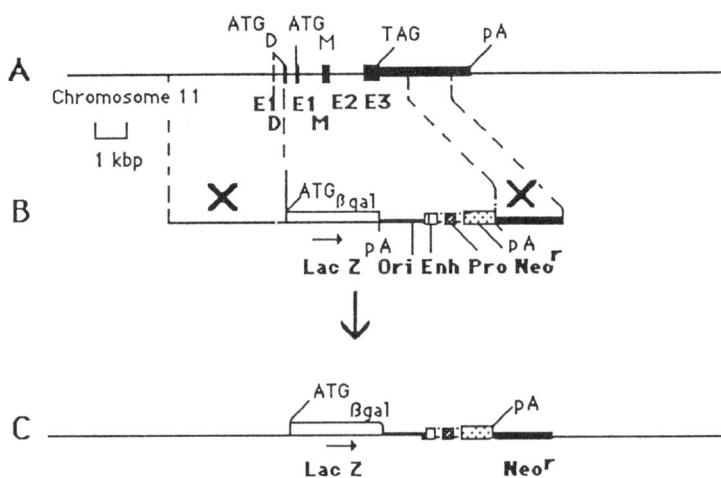

Fig. 2 A) Targeting strategy for substituting the murine LIF coding region with the reporter Lac Z in ES cells. A 3301 bp DNA fragment of the gene including all the known introns and exons of the two forms of LIF (LIF/D and LIF/M) plus 571 bases of the common 3' untranslated region was deleted. After targeting, the most 5' ATG was provided by LacZ and the junction with the LIF diffusible promoter was verified by sequencing. The mutated ES clones were microinjected into C57Bl/6 blastocysts which were implanted into pseudopregnant females. Positions of the exons (E1$_D$, E1$_M$, E2, E3) and initiation codons (ATG$_D$, ATG$_M$) of the two forms of LIF, termination codon (TAG), a 3'-untranslated region (3,2 kb) and the polyadenylation site (pA) are shown. B) Replacement vector containing the cloning vector pGN (Le Mouellic et al., 1990) and two flanking regions of the LIF coding region. The 5' LIF fragment 3853 bp along with the 3' LIF 1422 bp fragment were subcloned into pGN. C) Structure of the targeted LIF locus after recombination.

Analysis of 230 five week old animals obtained by crossing heterozygotes revealed 25% wild type, 61% heterozygous and 14% homozygous animal. Almost half the homozygous mutant animals were missing. Thus, either there is a distortion in the transmission of the LIF- sperm or more likely the mutation has variable penetrance and is lethal in some homozygous embryos. In the present loss of function experiments, surviving LIF- animals however appeared normal at birth and later, and histological analysis of many organs did not disclose abnormalities. Effect of the LIF deficiency on the peripheral nervous system is still being investigated. Initially, LIF- mice were smaller than wild type mice and, to a lesser extent, this was true of heterozygous animals as well (Fig 3).

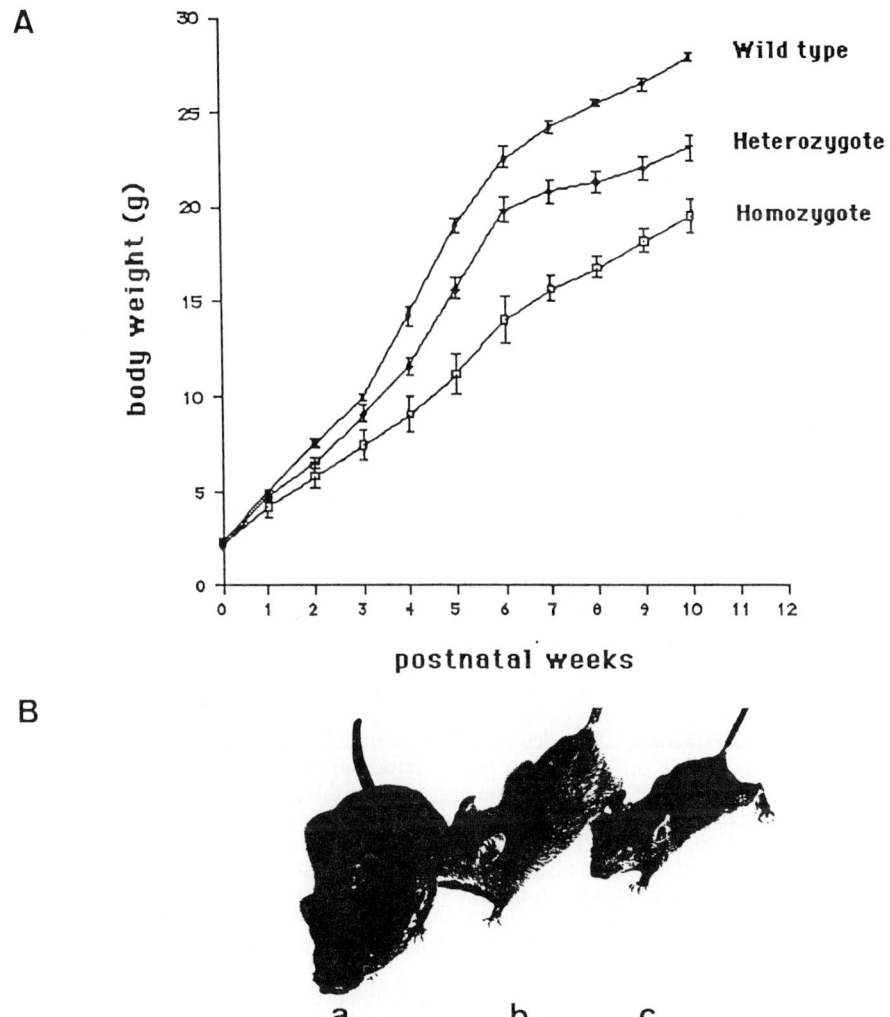

Fig. 3. A) Growth curves of wild type, heterozygous and homozygous littermates derived from five heterozygous intercrosses. The mean weights of ten wild type, twenty heterozygous and five homozygous animals of both sexes are plotted. Verticals bars represent the range in values from pooled animals. The difference was most obvious between the first and second months, when the body weight of deficient mice was only 63% of the wild type. Subsequently, the difference became smaller, stabilizing at 76%. B) Six weeks old mice, wild type (a), heterozygous (b) and homozygous (c) littermates are shown.

While males were fertile, homozygous females produce blastocysts but they fail to implant and do not develop further. To examine the cause, reimplantation experiments were performed in which wild type 3,5 days blastocysts were injected into the uterus of pseudopregnant homozygous, heterozygous and wild type females. The wild type and the heterozygous but not the LIF- females generated normal litters (Table 1).

Table 1. Fertility of wild type, heterozygous, and homozygous females.

	Genotype		
	Wild type (+/+)	Heterozygotes (+/-)	Homozygotes (-/-)
Number of females	10	10	15
Number of plugs for each female	5	5	10
Number of litters for each female	5	5	0
Number of pups	40	38	0
Number of reimplanted females*	10	10	10
Number of litters after reimplantation	7	6	0
Pups born after reimplantation	56	45	0

*Ten wild type blastocysts were reimplanted per female in one uterine horn.

Our results, together with the reported maternal expression of LIF by uterine endometrial glands at the time of blastocyst implantation argue for a role of LIF in this process (Stewart & al., 1992). Our data demonstrate several roles for LIF during embryogenesis: its expression is required in the mother for itself and for correct gestation certainly at blastocyst implantation and it contributes significantly to normal embryonic development. Moreover, LIF is expressed by pituitary follicular cells and LIF expression by uterine endometrial glands can be induced by estrogen.

In addition to reproductive defects, LIF- mice exhibited immunological abnormalities. The effects of LIF deficiency on hemopoeitic differentiation are described in another paper (Dumenil & al., this volume; Escary et al., 1993). For the sake of the argument in this article, we will only mention the mutation effect on the hemopoietic stem cells. We find a dramatic decrease of the pluripotent stem cells CFU-S in the spleen and, to a lesser extent, in the bone marrow. By transplanting LIF- hemopoietic stem cells into lethally irradiated animals, it was also shown that the mutation did not affect the potentialities of the stem cells themselves. A cellular microenvironment producing LIF is necessary to maintain the pool of hemopoietic stem cells. Finally, all the differentiated cells of the hemopoietic and immune system seems to be present.in the proper amount and functional. These deficits were also detected in heterozygous animals but to a lesser extent

To summarize if the microenvironment does not produce LIF, the pool of stem cells is not maintained but their differentiated derivatives are nevertheless produced. We have postulated LIF to be a negative regulator of the normal differentiation program of the pluripotent embryonic stem cells. Our results show that it could have the same role on the hemopoietic stem cells. In the absence of LIF, the stem cells do not proliferate but differentiate. For each stem cell which are potential targets of LIF, a specific differentiation program will be activated depending upon the cell developmental history. In the LIF- animals, we do not yet know the fate of the embryonic hematopoietic stem cells in the blood island. Nor do we know the fate of the pluripotent stem cells of the primitive ectoderm. A study of the LIF- mice embryogenesis is complicated by several factors. Firstly, the homozygous LIF- mothers are sterile. We have, therefore, to study the embryos in an heterozygous LIF- mother which produces LIF.

Secondly LIF, granulocyte colony stimulating factor (G-CSF), interleukin-6 (IL-6), oncostatin-M (ONC-M) and ciliary neurotrofic factor (CNTF) constitute a structurally and functionally related family of cytokines that display redundant activities (Bazan, 1991) Moreover, LIF, ONC-M, CNTF and IL-6 receptors share a common subunit, the IL-6 signal transducer gp 130 (Davis et al., 1991; Gearing et al., 1992; Kishimoto et al., 1992) During mouse embryogenesis, the absence of LIF may be partially compensated for by these other cytokines.

LIF BIOLOGICAL MODE OF ACTION

As already mentionned above, LIF acts *in vitro* on many different cell types either as a differentiation inducer, a proliferation factor or a survival molecule (Hilton & Gough, 1991; Smith & al., 1988); Williams et al., 1988; Fletcher et al., 1991; Leary et al., 1990). On the primitive ectodermal stem cells before gastrulation and problably also on the hematopoietic stem cells, our experimental data can be rationalized by postulating that LIF acts as a negative regulator of the differentiation programme. This differentiation program is specific of the stem cell developmental history. If LIF is overproduced, the stem cell differentiation program is not activated. In absence of LIF, the population of stem cells is not properly maintained. With regard to these stem cells, LIF is not a proliferation or a differentiation factor but rather would select one genetic program, the maintenance of the stemness state, against another genetic program, the start of differentiation.
Two additional points are worth emphasizing. Firstly, LIF producing cells are in close proximity to the target stem cells indicating a short range action. Secondly, in the LIF deficient mice, a strong heterozygous effect is detected with the hematopoietic and immune disorders as well as in growth retardation. The origin of this dominant effect is unclear suggesting however a dosage effect at the gene product level.
Finally, a further complication might arise in our interpretation because the population of stem cells may not be homogeneous. The ICM cells as well as the primitive ectodermal cells can be dissociated and reinjected into a blastocyst to produce a new individual. According to this criteria, those totipotent cells seem to be homogeneous. However it seems unlikely that these ICM and epiblast cells would express the same genetic and epigenetic information. We rather expect some positional information along the cylindrical primitive ectoderm.
The same heterogeneity might also be found *in vitro* in an ES cell population. New embryonic stem cells are produced by letting blastocysts attach onto a culture dish. After a few days, the outgrowth is dissociated and some cells might proliferate. A new ES line will be established if upon reinjection into a host blastocyst, the cell contribute to every embryonic tissues including the germ line. However a much more stringent test of their developmental capacities is to inject one to three ES cells into a morula before the ICM formation (Lallemand & Brulet, 1990). A high amount of chimerism is then reproducibly obtained. But living animals can only be obtained if the ES line has a small number of *in vitro* passages. It is quite common that injected ES cells from a given line will develop into a living chimera when injected into a blastocyst but not when injected into a morula. In fact, we postulate that some developmental information is lost *in vitro* as the number of passages increases. An ES cell line could in fact be a population.of stem cells with closely related developmental states. The longer time spent *in vitro*, the wider the distribution of developmental states would be and therefore a reduced ability for a single cell to colonize a morula. LIF's action on such an heterogeneous population might be more complex than in our above assumption. For instance, LIF could affect only defined subpopulation of stem cells.

ACKNOWLEDGEMENTS

This work was supported by grants from the Association pour la Recherche sur le Cancer, the Association Française contre les Myopathies, the Centre National de la Recherche Scientifique, the Institut National de la Santé et de la Recherche Médicale, the Ligue Nationale Française contre le Cancer, the Ministère de la Recherche et de la Technologie, and the programme du génome humain de la C.E.E.

REFERENCES

Bazan, J.F. (1991): Neuropoietic cytokines in the hematopoietic fold. *Neuron.* Vol. 7, 197-208.

Cold Spring Harbor, Mouse Molecular Genetics Meeting, 26-30 Août 1992.

Conquet, F., and Brûlet, P. (1990): Developmental expression of myeloid leukemia inhitory factor gene in preimplantation blastocysts and in extraembryonic tissue of mouse embryos. *Mol. Cell. Biol.* July 1990, 3801-3805.

Conquet, F., Peyrieras, N., Tiret, L. and Brulet, P. (1992): Inhibited gastrulation in mouse embryos overexpressing the leukemia inhitory factor. *Proc. Natl. Acad. Sci.* Vol. 89, 8195-8199.

Davis, S., Aldrich, T.H., Valenzuela, D.M., Wong, V., Furth, M.E., Squinto, S.P. and Yancopoulos, G.D. (1991) : The receptor for ciliary neurotrophic factor. *Science.* Vol. 253, 59-63.

Escary, J.L., Perreau, J., Dumenil, D., Ezine, S. and Brulet, P. (1993): Maintenance of hematopoietic stem cells and thymocyte stimulation require leukaemia inhibitory factor. *Nature.* in press.

Fletcher, F.A., Moore, K.A., Ashkenazi, M., De Vries, P., Overbeek, P.A., Williams, D.E. and Belmont, J.W. (1991): Leukemia Inhibitory Factor improves survival of retroviral vector-infected hematopoietic stem cells in vitro, allowing efficient long-term expression of vector-encoded human adenosine deaminase. *J. Exp. Med.* 174, 837-845.

Gearing, D.P., Comeau, M.R., Friend, D.J., Gimpel, S.D., Thut, C.J., McGourty, J., Brasher, K.K., King, J.A., Gillis, S., Mosley, B., Ziegler, S.F., and Cosman, D.(1992) : The IL-6 signal transducer, gp 130: An oncostatin M receptor and affinity converter for the LIF receptor. *Science.* Vol. 255, 1434-1437.

Ginsburg, M., Snow, M.H.L. and McLaren, A. (1990) : Primordial germ cells in the mouse embryo during gastrulation. *Development.* 110, 521-528.

Hilton, D.J. and Gough, N.M. (1991) : Leukemia Inhibitory Factor: a biological perspective. *J. of Cell. Bioch.* 46, 21-26.

Kishimoto, T., Akira, S. and Taga, T. (1992) : Interleukin-6 and its receptor: A paradigm for cytokines. *Science.* Vol. 258, 593-596.

Lallemand, Y. and Brulet, P. (1990) : An in situ assessment of the routes and extents of colonisation of the mouse embryo by embryonic stem cells and their descendants.*Development.* 110, 1241-1248.

Lawson, K.A., Meneses, J.J. and Pedersen R.A. (1991) : Clonal analysis of epiblast fate during germ layer formation in the mouse embryo. *Development.* 113, 891-911.

Leary, A.G., Wong, G.G., Clark; S.C., Smith, A.G. and Ogawa, M. (1990) : Leukemia Inhibitory Factor differentiation-inhibiting activity human interleukin for DA cells augments proliferation of human hematopoietic stem cells. *Blood.* 75, 1960-1964.

Le Mouellic, H., Lallemand, Y. and Brulet, P. (1990) : Targeted replacement of the homeobox gene Hox-3.1 by the *Escherichia coli lac Z* in mouse chimeric embryos. *Proc. Natl. Acad. Sci.* Vol. 87, 4712-4716.

Le Mouellic, H., Lallemand, Y. and Brulet, P. (1992) : Homeosis in the mouse induced by a null mutation in the Hox-3.1 gene. *Cell.* Vol 69, 251-264.

Matsui, Y., Toksoz, D., Nishikawa, S., Nishikawa, S.I., Williams, D., Zsebo, K. and Hogan, B.L.M. (1991) : Effect of Steel factor and leukaemia inhibitory factor on murine primordial germ cells in culture. *Nature.* 353, 750-752.

Ratjen, P.D., Toth, S., Willis, A., Heath, J.K. and Smith, A.G. (1990) : Differentiation inhibiting activity is produced in matrix-associated and diffusible forms that are generated by alternative promoter usage. *Cell.* 62, 1105-1114.

Smith, A.G., Heath, J.K., Donaldson, D.D., Wong, G.G., Moreau, J., Sthal, M. and Rogers, D. (1988) : Inhibition of pluripotential embryonic stem cell differentiation by purified polypeptides. *Nature.* Vol. 336, 688-690.

Stewart, C.L., Kaspar, P., Brunet, L.J., Bhatt, H., Gadl, I;, Kontgen, F. and Abbondanzo, S.J. (1992) : Blastocyst implantation depends on maternal expression of leukaemia inhitory factor. *Nature.* 359, 76-79.

Williams, R.L., Hilton, D.J., Pease, S., Willson, T.A., Stewart, C.L., Gearing, D.P., Wagner, E.F., Metcalf, D., Nicola, N.A. and Gough, N.M. (1988) : Myeloid leukaemia inhibitory factor maintains the developmental potential of embryonic stem cells. *Nature.* Vol. 336, 684-687.

Résumé

La cytokine LIF, Leukemia Inhibitory Factor, agit sur de nombreux types cellulaires *in vitro*, en particulier, sur les cellules souches embryonnaires ES. Pour définir son mode d'action sur les cellules souches, une étude *in vivo* par mutagenèse dirigée a été entreprise. Les souris dont le gène LIF a été inactivé présentent de graves troubles du système hématopoïétique. Leur analyse fine nous permet de conclure que la cytokine LIF est nécessaire au maintien du pool de cellules souches hématopoïétiques *in vivo*. Par contre la surexpression du LIF, *in vivo,* inhibe la différentiation des cellules souches de l'ectoderme primitif en mésoderme. En réponse aux signaux de son environnement, une cellule souche prolifère ou se différencie. D'après nos résultats, *in vivo*, les cellules souches auraient tendance à différencier ou à disparaître en l'absence de LIF. Un excès de LIF, par contre, les bloquerait dans leur programme de prolifération.

Hematopoietic disorders in LIF-mutant mice

Dominique Dumenil*, Sophie Ezine**, Jacqueline Perreau***,
Jean-Louis Escary*** and Philippe Brûlet***

*Institut Gustave-Roussy, U 362, 94800 Villejuif, France; ** INSERM U 345, CHU Necker, 75015 Paris Cedex 15, France; ***Unité d'Embryologie Moléculaire de l''Institut Pasteur, URA 1148 CNRS, 75724 Paris Cedex 15, France

ABSTRACT

Analysis of the hematopoietic system in LIF deficient mice (LIF⁻) derived by gene targeting techniques revealed a dramatic decrease in spleen day-12 CFU-S and a significant but lesser decrease in marrow CFU-S. Injection of spleen and marrow cells from LIF⁻ mice into lethally irradiated wild type animals promoted long-term survival of 70% and 86% of the recipients respectively. Committed progenitors (BFU-E and GM-CFC) were also reduced in number in spleens of LIF⁻ mice whereas marrow progenitors and circulating mature cells were normal. Heterozygous animals exhibited an intermediate phenotype, suggesting a dosage effect for the LIF gene product.

In addition, T cell responses to concanavalin A (ConA) or allogeneic stimulation were significantly reduced in the thymus of homozygous and heterozygous animals.

Defects in stem cell number were compensated by exogenous LIF delivered by mini-osmotic pump. LIF is therefore required to promote survival of the normal pool of stem cells (CFU-S) but not their terminal of stem cells (CFU-S) but not for their terminal differentiation. Hematopoiesis was more affected in spleen than in marrow of LIF⁻ mice, suggesting that stem cells interact differently with the splenic and medullary microenvironnement.

Leukemia Inhibitory Factor (LIF) is a cytokine with a pleiotropic activity in several adult and embryonic systems (Hilton, 1992). The gene encoding LIF is cloned (Gearing et al., 1987) and located on chromosome 11 in mouse (Kola et al., 1990) and on chromosome 22 in man (Sutherland et al., 1989) Two transcripts, generated by alternative splicing and diverging throughout exon 1, give rise to two forms of the LIF protein : a matrix associated form (LIF_M) and a diffusible form (LIF_D) (Rathjen et al., 1990). In the hematopoietic system, LIF induces the differentiation of some leukemic cell lines (Tomida et al., 1984 ; Hilton et al., 1988), the proliferation without differentiation of hematopoietic stem cells (Leary et al., 1990), the proliferation of megakaryocyte progenitor cells (Metcalf et al., 1990) and DAI cells (Moreau et al., 1987) and the survival of hematopoietic stem cells infected with a retrovirus (Fletcher et al., 1991). LIF also influences the functional activity of differentiated cells as liver parenchymal cells and autonomic nerves. Furthermore, LIF suppresses Embryonic Stem (ES) cell differentiation (Smith et al., 1988 ; Williams et al., 1988) and sustains the survival and proliferation of primordial Germ Cells (De Felici and Dolci, 1991 ; Matsui et al., 1991). The LIF receptor is a member of the cytokine receptor family which lacks a tyrosine kinase domain and shares structural homologies with oncostatin M (OSM), IL-6 (Interleukin-6) and G-CSF(Granulocyte-Colony Stimulating factor) (Rose and Bruce, 1991). OSM binds the high affinity LIF membrane receptor (Gearing and Bruce, 1992). Furthermore the IL-6 (Gearing et al., 1992).and IL-11 (Yin et al., 1992) signal transducer, a membrane glycoprotein named gp130, is an OMS receptor and converts the low affinity LIF receptor into the high affinity one The fact that these factors interact with similar receptors and exhibit a plurifunctionality on the same cell populations may account for their redundancy.

In order to clarify the role of LIF in vivo, mutant mice were generated in which the LIF gene was inactivated by homologous recombination. The hematopoiesis and immunologic responses of these LIF⁻ mice were analysed.

MATERIAL AND METHODS

a) Hematopoietic progenitors

Bone marrow and spleen cells were dispersed and suspended in Iscove's Modified Dulbecco's medium (IMDM). CFU-S numbers were determined by Till and Mc Culloch's technique (Till & Mc Culloch, 1961). 10^5 bone marrow and 10^6 spleen cells were injected into lethally irradiated mice and macroscopic spleen colonies were counted 12 days after injection. The number of clonogenic committed progenitor cells (GM-CFC and BFU-E) was determined in semi-solid methyl cellulose culture as described by Worton et al., 1969. Briefly, 5×10^4 and 5×10^5 bone marrow and spleen cells, respectively, were used to determine GM-CFC numbers and 10^5 or 10^6 bone marrow and spleen cells, respectively for BFU-E. The culture mixture is composed of methyl cellulose (0,0 %), 20 % fetal calf serum and 20 % growth factors (lung and heart conditioned medium for GM-CFC and Wehi-3B conditioned medium plus Erythropoïetin (1 U/ml), for BFU-E). Cultures were plated in quadriplicate and counted at day 7 for GM-CFC and day 10 for BFU-E.

b) Hematopoietic reconstitution

Lethally irradiated mice (9.5 Gy) were grafted with cells from bone marrow or spleen of homozygous LIF mutant or wild type mice. The number of nucleated cells injected was adjusted so that grafts contained approximately the same number of CFU-S (arbitrarily : 16.0±1.0). Fifteen recipient mice were used per group. 38 days after the graft, 3 mice were sacrified in each group, the hematopoietic progenitor content (CFU-S, BFU-E and GM-CFC) of bone marrow and spleen was determined.

c) Effects of exogenous LIF delivery

Exogenous LIF was delivered in mutant mice by mini-osmotic pump (ALZET, model 2001) delivering rLIF at a rate of 10^3 IU per hour. The pumps were implanted into the peritoneum of 3 LIF⁻ and 3 wild type mice. After 7 days of LIF treatment, the number of hemopoietic progenitors was determined.

d) Concavalin A and Allogeneic stimulations

Thymocytes (4×10^5 per well) were cultured in 0.2 ml IMDM/10 % FCS in 96-well flat-bottom trays together with ConA (3g.ml^{-1}) or 10^6 irradiated (3.000 rads) splenocytes from C$_3$H/HeJ$^{(H-2k)}$ mice. They were pulsed with 1Ci (^{3}H) thymidine on day 2 for ConA stimulation or on day 3 for allogenic stimulation and collected 4 hours later.

RESULTS

a) Hemopoietic disorders in bone marrow and spleen of LIF⁻ mice

A dramatic decrease in all stem cell compartments tested is observed in the spleen of LIF⁻mice. Pluripotent stem cells (CFU-S) were reduced by 88 %, committed erythroid (BFU-E) progenitors by 85 % and granulo-macrophagic (GM-CFC) progenitors by 73 % as compared to wild type controls (Table 1A). The number of progenitors in the spleen of heterozygous mice was intermediate between those of wild type and homozygous mutant animals indicating a dosage effect for the LIF gene product. In the bone marrow of homozygotes, the CFU-S number was also significantly decreased but not to the same extent (59 % reduction, p<0.01 (Table 1B)). GM-CFC and BFU-E compartments in bone marrow were not affected. The total nucleated cell number in the spleen ($177.0 \pm 10 \times 10^6$) and the bone marrow ($27.6 \pm 1.9 \times 10^6$ /leg), the numbers of circulating mature red and white blood cells, and in particular blood platelet level ($1040 \pm 145 \times 10^6$/ml), were normal in LIF⁻ mice.

Table 1 : Hematopoietic progenitors in LIF⁻ mice.

A : Number of spleen progenitors in LIF⁻ mice

MICE	CFU-S	BFU-E	GM-CFC
+/+	4,293 ± 675*	10,272 ± 2 021	29,360 ± 5,903
	(7)**	(7)	(5)
+/-	1,279 ± 528 °	3,358 ± 891°	8,380 ± 2,289°
	(6)	(6)	(5)
-/-	456 ± 138°°	1,699± 357°°	7,756 ± 2,427°°
	(10)	(10)	(9)

B : Number of bone marrow progenitors in LIF⁻ mice

MICE	CFU-S	BFU-E	GM-CFC
+/+	4,885 ± 1 108*	3,169 ± 601	55,320 ± 13,261
	(7)**	(7)	(5)
+/-	3,470 ± 990	4,606 ± 1 289	71,320 ± 16,744
	(5)	(5)	(5)
-/-	2,020 ± 418°°	2,820 ± 629	54,944 ± 5,536
	(10)	(10)	(9)

* Mean value ± SE, ** (n) number of animals tested, ° Significantly different from +/+ mice $p \leq 0.05$ (t test), °° $p \leq 0.01$

b) Stem cells from bone marrow and spleen of LIF⁻ mice are able to reconstitute irradiated mice

Irradiated mice were injected with different numbers of bone marrow and spleen cells of wild type and mutant mice. We compared the survival of mice grafted with an identical number of CFU-S (16.0 ± 1.0) during four months. Marrow or spleen cells from wild type mice enabled survival of 80 % and 93 % of the animals respectively. When bone marrow or spleen cells from LIF⁻ mice were injected into irradiated mice, respectively 86 and 70 % of mice survived. At thirty eight days after the graft, bone marrow or spleen cells from LIF⁻ mice have led to a complete hematologic reconstitution (Table 2). Slightly lower numbers of CFU-S and BFU-E were present in the spleens of animals reconstituted with LIF⁻ bone marrow and spleen cells as compared with those reconstituted with wild type cells. The differences, however, were not statistically significant and the high proportion of mice surviving 4 months after receiving either bone marrow or spleen cells from LIF⁻ mice argue against any gross abnormalities in the capacity of LIF⁻ stem cells to reconstitute a functional hematopoietic system. The proportion of donor-type reconstitution was over 75 % in the 12 animals grafted in sex-mismatched condition as assessed on splenic and marrow DNAs with a Y chromosome specific probe (not shown).

c) Exogenous LIF delivery compensates for the consequences of the mutation

After 7 days of LIF infusion (10^3 IU per hour) (Fig 1), two mice exhibited normal numbers of bone marrow (88 % and 120 % of the control CFU-S value) and spleen progenitors (90 % and 104 % of CFU-S, 50 % and 70 % of BFU-E, 130 % and 150 % of GM-CFC control values). One mouse showed partial recovery in the spleen : CFU-S compartment (53 % of the control value) and BFU-E compartment (10 % of the control value) were not completely restored.

Table 2 : Hematopoietic progenitors in the bone marrow (A) and in the spleen (B) of irradiated recipient mice 38 days after the graft of wild type (+/+) or mutant (-/-) bone marrow (BM) or spleen (Sp) cells

A : Bone marrow

GRAFT	GM-CFC	BFU-E	CFU-S
BM +/+	25,385 ± 7,188	3,038 ± 860	3,447 ± 1,383
BM -/-	31,652 ± 2 003	3,254 ± 1,087	2,782 ± 467
Sp +/+	31,108 ± 6,161	2,641 ± 759	3,376 ± 1,000
Sp -/-	49,760 ± 14,314	6,041 ± 1,941	3,385 ± 982

B : Spleen

GRAFT	GM-CFC	BFU-E	CFU-S
BM +/+	20,747 ± 14,773	28,779 ± 19,767	5,977 ± 2,508
BM -/-	15,436 ± 2,390	14,916 ± 3,831	5,256 ± 986
Sp +/+	3,652 ± 1,600	12,806 ± 1,154	3,438 ± 434
Sp -/-	20,008 ± 10,177	5,064 ± 1,505	1,921 ± 864

d) Thymic disorders in LIF⁻ mice

LIF⁻ mice had normal numbers of mature lymphoid cells but exhibited abnormal thymic T cell responses to challenge. B cell subsets in the bone marrow, spleen and peritoneum were normal, as were IgM, IgG and IgA levels (not shown). In the thymus, the average total cell number was similar in normal and deficient mice (67.7×10^6 (n = 11) versus 49.7×10^6 (n = 12)). Furthermore, two and three-color flow cytometry did not reveal major differences in thymocyte subsets detected with anti-CD4, CD8 and CD3 monoclonal antibodies (not shown). Nevertheless, thymic T cell response to ConA or allogeneic

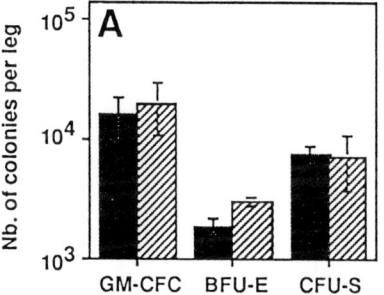
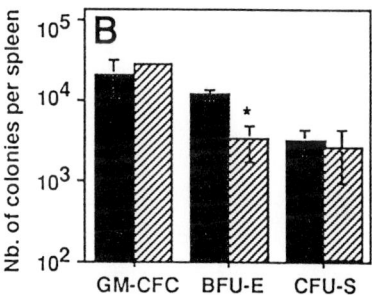

Fig 1 : Effect of exogenous LIF. Mini-osmotic pumps were implanted in 3 wild-type (closed bar) or in 3 LIF⁻ mice (hatched bar). Progenitors were assessed in the bone marrow (A) or in the spleen (B) of treated animals after 7 days of infusion.* Significantly different from +/+, P≤ 0.02.

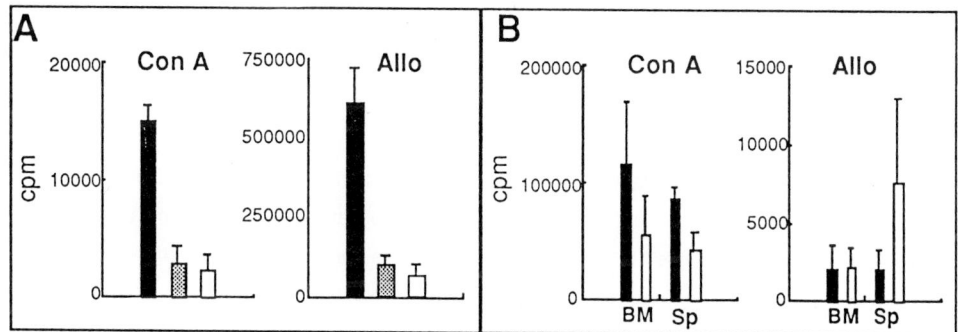

Fig 2 : Thymic T cell response after concavalin A (con A) or allogeneic stimulation.

A : Four animals of each genotype (+/+ : closed bars, +/- : hatched bars, and -/- open bars) were tested.
B : T cell response from thymus of letally irradiated mice reconstituted with bone marrow (BM) or spleen (Sp) from wild type (closed bars) or LIF⁻ mice (open bars) were represented.

stimulation was dramatically reduced in heterozygous and homozygous 6-8 weeks old mice (Fig. 2A). In contrast peripheral mature T cells of the spleen and lymph nodes were normal as were T cell responses to ConA or allogeneic stimulation (not shown). In the reconstitution experiments described above, thymic T cell responses from the recipient animals (wild type environment) were normal whether the genotype of the grafted marrow or spleen cells was wild type or LIF⁻ (Fig. 2B).

d) Thymic disorders in LIF⁻ mice

LIF⁻ mice had normal numbers of mature lymphoid cells but exhibited abnormal thymic T cell responses to challenge. B cell subsets in the bone marrow, spleen and peritoneum were normal, as were IgM, IgG and IgA levels (not shown). In the thymus, the average total cell number was similar in normal and

DISCUSSION

It has recenlty been shown that females lacking a functional LIF gene are fertile but their blastocysts fail to implant and do not develop (Stewart et al., 1992, Perreau et al. this issue). Here, we describe other deficiencies in hemopoietic and thymic tissues of LIF⁻ mice.

The T cell response to mitogenic stimulation is altered in the thymus and a dramatic decrease in the number of the spleen hemopoietic progenitors is observed in LIF deficient mice. The bone marrow CFU-S pool is also reduced but to a lesser extent. However, the spleen and bone marrow stem cells are able to reconstitute hemopoiesis of irradiated mice. Thirty eight days after the graft of splenic or bone marrow cells from LIF⁻mice, the recipient mice have the same number of hemopoietic progenitors in their spleen and bone marrow than recipient mice grafted with normal bone marrow or spleen cells. In addition, the response to stimulation of T cells in the thymus of these reconstituted animals is normal. Furthermore, exogenous LIF delivery in LIF⁻mice compensates for the consequences of the mutation. After 7 days of treatment, the numbers of hemopoietic progenitors return to normal values in the spleen.

All these results demonstrate that LIF is necessary for both stem cell maintenance and T cell function but the LIF deficiency does not seem to alter the stem cell potentialities. However stem cells of LIF⁻ mice are still functional. When these LIF⁻ stem cells are placed in a normal microenvironment or when LIF is provided exogenously, they are still able to express their potentialities. LIF does not seem to play a crucial role in hemopoietic differentiation since, in LIF⁻ mice, terminal differentiation which occurs from a reduced number of stem cells, is normal. However, we cannot exclude that this LIF function is compensated for in LIF⁻ mice, by another member of the LIF cytokine family (Oncostatin-M, IL6, Ciliary neutrophic factor,CNTF, G-CSF).

To account for the specificity of LIF activity on a given organ, at a given time, one has to postulate that LIF is produced locally and has an exclusive in situ activity. This hypothesis, already expressed by Metcalf (1991), is in agreement with our results on the role of LIF secreted by marrow stromal and thymic endothelial cells and with the fact that LIF exists in a membrane-associated form.

These results suggest that LIF may be essential to maintain a normal stem cell pool (CFU-S), but is not absolutely required for their terminal differentiation as shown by normal numbers of circulating myeloid and lymphoid cells in LIF⁻ mice. LIF deficiency selectively altered spleen hemopoiesis as compared to marrow hemopoiesis which suggests a differential interaction of stem cells with the splenic and medullary microenvironment producing LIF.

Acknowledgments :

We thank C. Lacout, A. Grandien and A.M. Joré for technical assistance, N. Blumenfeld for critically reading the manuscript and V. Payen for typing the paper. This work was supported by grant from Recherche et Partage.

REFERENCES

De Felici M, Dolci S: Leukemia inhibitory factor sustains the survival of mouse primordial germ cells cultured on TM4 feeder layers. Dev Biol 147:281, 1991

Fletcher FA, Moore KA, Ashkenazi M, De Vries P, Overbeck PA, Williams DE, Belmont JW: Murine leukemia inhibitory factor enhances retroviral-vector infection efficiency of hematopoietic progenitors. J Exp Med 174:837, 1991

Gearing DP, Gough NM, King JA, Hilton DJ, Nicola NA, Simpson RJ, Nice EC, Kelson A, Metcalf D: Molecular cloning and expression of cDNA encoding a murine myeloid leukemia inhibitory factor (LIF). Embo J. 6:3995, 1987

Gearing DP, Bruce AG: Oncostatin M binds the high-affinity leukemia inhibitory factor receptor. New Biol. 4:61, 1992

Gearing DP, Comeau MR, Friend DJ, Gimpel SD, Thut CJ, McGourty J, Brasher KK, King JA, Gillis S, Mosley B, Ziegler SF, Cosman D: The IL-6 signal transducer, gP130 : an oncostatin M receptor and affinity converter for the LIF receptor. Science 255:1434, 1992

Hilton DJ, Nicola NA, Metcalf D: Purification of a murine leukemia inhibitory factor from krebs ascites cells. Anal Biochem 173:359, 1988

Hilton DJ: LIF : lots of interesting functions. TIBS 17:72, 1992

Kola I, Davey A, Gough NM: Localization of the murine leukemia inhibitory factor gene near the centromere on chromosome 11. Growth Factors 2:235, 1990

Leary AG, Wong GG, Clark SC, Smith AG, Ogawa M: Leukemia inhibitory factor differentiation inhibitory activity/human interleukin for DA cells augments proliferation of human hematopoietic stem cells. Blood 75:1960, 1990

Matsui Y, Toksoz D, Nishikawa SL, Williams D, Zsebo K, Hogan BLM: Effect of steel factor and leukaemia inhibitory facter on murine primordial germ cells in culture. Nature 353:750, 1991

Metcalf D, Gearing DP: A fatal syndrome in mice engrafted with cells producing high levels of leukemia inhibitory factor. Proc Nat Acad Sci USA 86:5948, 1986

Metcalf D: The leukemia inhibitory factor (LIF). J Cell Cloning 9:95, 1991

Moreau JF, Bonneville M, Godard A, Gascan H, Gruart V, Moore MA, Soulillou JP: Characterization of a factor produced by human T cell clones exhibiting eosinophil-activating and burst-promoting activities. J. Immunol. 138:3844, 1987

Ratjen PD, Toth S, Willis A, Heath JK, Smith AG: Differentiation inhibiting activity is produced in matrix-associated and diffusible forms that are generated by alternate promoter usage. Cell 62:1105, 1990

Rose TM, Bruce G: Oncostatin M is a member of a cytokine family which includes leukemia inhibitory factor, granulocyte colony-stimulating factor and interleukin-6. Proc Nat Acad Sci USA 88:8641, 1991

Smith AG, Heath JK, Donaldson DD, Wong GG, Moreau J, Sthal M, Rogers D: Inhibition of pluripotential embryonic stem cell differentiation by purified polypeptides. Nature 336:688, 1988

Stewart CL, Kaspar P, Brunet LG, Bhatt H, Gadi I, Köntgen F, Abbondanzo SJ: Blastocyst implantation depends on maternal expression of leukaemia inhibitory factor. Nature 359:76, 1992

Sutherland GR, Baker E, Hyland VI, Callen DF, Stahl J, Gough NM: The gene for human leukemia inhibitory factor (LIF) maps to 22q12. Leukemia 3:9, 1989

Till JE, McCulloch EA: A direct measurement of the radiation sensitivity of normal mouse bone marrow cells. Radiat. Res. 14:213, 1961

Tomida M, Yamamoto-Yamaguchi Y, Hozumi M: Purification of a factor inducing differentiation of mouse myeloid leukemic M1 cells from conditioned medium of mouse fibroblast L929 cells. J Biol Chem 259:10978, 1984

Williams RL, Hilton DJ, Pease S, Wilson TA, Stewart CL, Gearing DP, Wagner EF, Metcalf D, Nicola NA, Gough NM: Myeloid leukemia inhibitory factor (LIF) maintains the developmental potential of embryonic stem cells. Nature 336:684, 1988

Worton RG, McCulloch EA, Till JE: Physical separation of hemopoietic stem cells from cells forming colonies in culture. J Cell Physiol 74:171, 1969

Yin T, Taga T, Tsang MLS, Yasukawa K, Kishimoto T, Yang YC: Interleukin (IL)-6 signal transducer, GP130, is involved in IL-11 mediated signal transduction. Blood 80: 151, 1992.

Résumé

L'analyse du tissu hématopoïétique de souris rendues deficiente en Lif (Leukemia inhibitory factor) par recombinaison homologue a montré que le nombre de CFU-S 12 jours, dans la rate, est dramatiquement réduit. Le nombre de CFU-S médullaire est également diminué mais à un moindre degré. Les cellules de la moelle et de la rate des souris Lif⁻ sont capables d'induire la survie à long terme d'animaux irradiés à dose létale suggérant la présence dans ces populations de cellules souches primitives ayant conservé leurs potentialités de reconstitution du tissu hématopoïétique.

Le nombre des progéniteurs (BFU-E et GM-CFC) est également diminué dans la rate des animaux Lif⁻ alors qu'il est normal dans la moelle de ces animaux. Le nombre de cellules du sang (globules rouges, blancs et plaquettes) des souris Lif⁻ est équivalent à celui des souris sauvages. Les souris hétérozygotes pour la mutation ont un phénotype intermédiaire entre celui des homozygotes et des souris sauvages ce qui suggère un effet dose du produit du gène Lif.

De plus, la réponse des lymphocytes T du thymus des souris Lif⁻ à une stimulation par la concavalin A ou à une stimulation allogénique est significativement réduite comparée à la réponse des lymphocytes T d'une souris normale.

L'implantation de mini-pompe délivrant du Lif dans le péritonéum des souris. LIf⁻ corrige les déficites hématologiques observés dans ces souris. Lif est donc nécessaire pour le maintien du pool des CFU-S et ne semble pas indispensable pour leur différenciation. L'hématopoïèse splénique est plus perturbée que l'hématopoïèse médullaire, ceci suggère que les cellules souches interagissent différemment avec les cellules du microenvironnement médullaire et splénique qui produisent Lif.

Inactivation of the Stem Cell Inhibitor (SCI) gene by homologous recombination

Don Cook, Suzanne Kirby, Tom Coffman*, Mark Plumb**, Ian Pragnell** and Oliver Smithies

*Departments of Pathology and Medicine, University of North Carolina, Chapel Hill, NC, **CRC Beatson Laboratories, Glasgow, and *Department of Medicine, Duke University, Durham, NC, USA*

Hematopoietic stem cells (HSC) are largely maintained in a nonproliferative state; however, when challenged by a physiologic stress or toxic insult, HSC are recruited into cycle. These baseline homeostasis and proliferative responses to stress are controlled by both positive and negative regulators. One negative regulator that has been shown to affect HSC is stem cell inhibitor (SCI) or macrophage inflammatory protein-1-alpha. SCI is a small, heparin-binding, aggregating protein, normally produced by stimulated macrophages and spleen cells. Its bioactivities include induction of local inflammation and some granulocyte functions and synergistic activity for myeloid precursors (Wolpe et al., 1988; Broxmeyer, et al., 1990, Graham et al., 1992). Recently it also been shown to exhibit reversible, dose-dependent, inhibition of DNA synthesis and cycling of CFU-S and CFU-A (Graham et al., 1990, Dunlop et al., 1992).

In order to better define the role of SCI in vivo we have used homologous recombination to disrupt the SCI locus in embryonic stem (ES) cells. We initially created a targetting construct using DNA from a DBA liver library; however, we failed to demonstrate any correctly targetted clones in over 1500 colonies screened. Since it has been demonstrated that isogenic DNA can increase targetting efficiency, we created a second construct using DNA from a Sau3A library derived from embryonic stem (ES) cell line (E14TG2a). This construct (Fig. 1) contains 10 kb of homology to the endogenous locus and includes two selectable markers (Fig. 1). We have replaced the promoter, exon 1, and part of exon 2 with the neomycin resistance (neo) gene to allow for G418 selection of cells incorporating the foreign DNA. In addition, an HSV thymidine kinase gene was included downstream of the 3' homology to allow for negative selection of random integrants with ganciclovir. Doubly-resistant ES cell clones were screened for the desired modification by PCR and Southern blot hybridization. On average, one in 20 clones surviving selection was correctly targetted, representing one in a million electroporated cells. We are presently injecting targetted lines into blastocysts to obtain chimeric mice with tissues derived from modified ES cells. Germline transmission from chimeric animals is expected to generate animals heterozygous for SCI inactivation.

We anticipate that bone marrow stroma from these animals may allow better proliferation of HSC in vitro. Since at least 6 months are required to obtain homozygotes from the heterozygotes, we have begun to employ two other strategies to study homozygous SCI deficiency in vitro, using LTBMC and/or LTC-IC systems. First, we will use the heterozygous ES cells to select for events that result in cells with both SCI loci disrupted by increasing the G418 concentration in the ES cell growth medium (Mortensen et al., 1992). Second, given the high frequency of targetting, we will target the unmodified allele in the heterozygous cells with a construct containing the hypoxanthine guanine phosphoribosyl transferase (HPRT) gene in place of the neo gene. Doubly-targetted ES cells derived by either method can then be used to create chimeric animals. Bone marrow obtained from these chimeras can also be cultured in the presence of G418 to directly yield SCI-null stromal cells which can be assayed for HSC growth support function.

Investigations of the function of SCI will include characterization of the hetero- and homozygous phenotypes in mice and potentially breeding of SCI-deficient mice to TGF-beta-deficient mice to determine whether any growth advantage is conferred when two inhibitors of HSC proliferation are absent. Such a growth advantage induced by any method could facilitate the use of bone marrow as a source for gene therapy and would be a necessary step for gene targetting of HSC directly, given present levels of targetting efficiency.

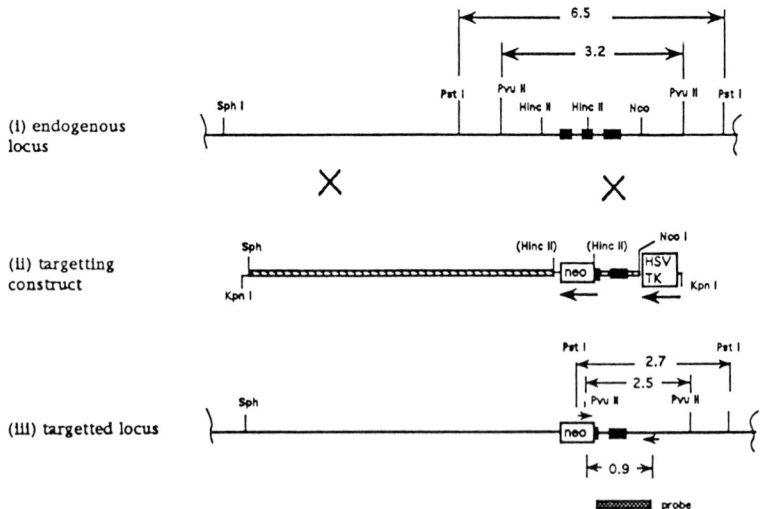

Figure 1. The endogenous SCI locus (i) and the *Kpn*-linearized targetting construct (ii) are shown. Exons of the SCI gene are represented by solid black rectangles, and the direction of transcription of the neo and HSV TK genes is shown by the large arrows. Hatched regions represent DNA derived from the SCI locus. The structure of the targetted SCI locus is shown (iii), together with the novel *Pvu* II and *Pst* I fragments generated by the recombination event. Small arrow heads represent the primers for the neo gene and the 3' SCI region used for PCR amplification of the targetted locus.

REFERENCES

Broxmeyer, H.E., et al., (1990): Enhancing and Suppressing Effects of Recombinant Murine Macrophage Inflammatory Proteins on Colony Formation In Vitro by Bone Marrow Myeloid Progenitor Cells. *Blood* 76(6):1110-11.

Dunlop, D.J., et al. (1992): Demonstration of Stem Cell Inhibition and Myeloprotective Effects of SCI/rhMIP1a In Vivo. *Blood* 79(9):2221-2225.

Graham, G.J. et al. (1990): Identification and Characterization of an Inhibitor of Haemopoietic Stem Cell Proliferation. *Nature* 344:442-444.

Graham, G.J. et al. (1992): Purification and Biochemical Characterization of Human and Murine Stem Cell Inhibitors (SCI). *Growth Factors* 7:151-160.

Mortensen, R.M. et al. (1992): Production of Homozygous Mutant ES Cells with a Single Targeting Construct. *Mol. Cell. Biol.* 12(5):2391-2395.

Wolpe, S.D. (1988): Macrophages Secrete a Novel Heparin-Binding Protein with Inflammatory and Neutrophil Chemokinetic Properties. *J. Exp. Med.* 167:570-581.

V. Inhibition of hematopoiesis by leukemic cells

V. *Inhibition de l'hématopoïèse par les cellules leucémiques*

Role of cytokines in growth control of acute lymphoblastic leukemia cell lines

Amos Cohen, Tom Grunberger, Wilma Vanek and Melvin H. Freedman

Divisions of Immunology/Cancer Research, and Hematology/Oncology, The Hospital for Sick Children, 555 University Ave., Toronto, Ontario, Canada, M5G 1X8

ABSTRACT

Five cell lines were established from patients with pre-B ALL and their growth characteristics were analyzed in detail. The cell lines express phenotypic markers characteristic of freshly isolated pre-B ALL cells. All the cell lines express the mRNA for several hematopoietic growth factor receptors as well as receptors for various cytokines. Several cytokines such as interleukin-1 (IL-1), IL-3, granulocyte-macrophage colony stimulating factor (GM-CSF), stem cell factor (SCF) and IL-7 are able to stimulate the growth of cell lines that express the corresponding receptors. Other cytokines, such as tumor necrosis factor-α (TNF-α), interferon-γ (IFN-γ) and IL-6 are negative regulators of cell lines that express their receptors. In three of the cell lines we were able to demonstrate the existence of IL-1 or GM-CSF growth stimulatory autocrine cycles.

INTRODUCTION

Hematopoietic cells are derived from limited numbers of self-renewing pluripotent stem cells that generate progenitor cells committed to proceed along one of the maturation pathways. To regulate the production rate of blood cells of specific lineages, a tight control of the process of cell renewal, commitment, and differentiation is required. The hematopoietic growth factors and cytokines play a critical in these processes.

Acute lymphoblastic leukemia (ALL) is characterized by an arrest of differentiation and the subsequent proliferation of undifferentiated leukemic cells in the bone marrow. Since the proliferation of normal cells of hemopoietic origin is controlled by growth factors, it has been suggested that leukemias may arise by dysregulation of these pathways (Greaves 1986). Leukemic cells may become independent of exogenous growth factors which normally control proliferation and differentiation by acquiring the ability to synthesize and respond to endogenous growth factors.

Autocrine growth, whereby a cell constitutively produces its own growth factor, has been demonstrated with various malignancies(Sporn 1985) including leukemia (Cohen et al. 1991). Several studies have shown that interleukin-3 (IL-3) or granulocyte macrophage-colony stimulating factor (GM-CSF) are involved in autocrine pathways regulating the proliferation of murine leukemic cells (Duhrsen U 1988; Schrader J. W 1983). However, despite efforts by different groups, the mechanisms involved in the growth control of ALL cells are still unclear. The ability of ALL cells to form colonies in semisolid media containing various growth factors is still very limited. For this reason we have concentrated our efforts to establish cell lines from primary ALL cells that would allow us to investigate the role of cytokines in positive and negative regulation of leukemic cell growth.

MATERIALS AND METHODS

ALL cell lines:

Leukemic cells were obtained from bone marrow aspirates 5 children diagnosed with pre-B ALL. Leukemic cells were isolated from bone marrow by Percoll (Pharmacia, Piscataway, N.J.) density gradient centrifugation (1.077 g/ml, 400g, 4°C) and cultured with weekly medium changes in 25-cm^2 tissue culture flasks (Corning Glass Works Corning, N.Y.) in α-modified minimum essential medium (MEM, GIBCO, Grand Island, N.Y.) containing l0% heat inactivated fetal bovine serum (FBS; Flow, McLean, VA.) Medium was changed weekly and cultures were incubated at 37°C in a humidified incubator with 5% CO_2 and 95% ambient air. After two months the cells grew as a single cell suspension with a doubling time of forty to fifty hours. The cell lines were negative for mycoplasma and did not carry Epstein-Barr viral sequences.

Cell lines were established by incubating leukemic cells (10^6/ml) obtained from bone marrow aspirates in alpha-modified minimal essential medium (MEM) containing 10% heat inactivated FCS. After eight to twelve weeks, the cells were cloned in semisolid methylcellulose and single colonies were isolated and expanded in liquid culture medium. Cell lines established this way resembled the donors' leukemic cells. For instance the karyotype of the B1cell line showed the (4;11) (q21;q23) in all metaphases. In addition, other chromosomal abnormalities include trisomy 6, der(1)t(1;8) (p36;q13), der(10)t(1;10)(q11;p15), were consistently observed in all metaphases of this cell line. Similarly, the W1 cell showed monosomy 7 that was also present in the leukemic sample at diagnosis. In addition there was a general agreement between cell surface markers present in leukemic cells at diagnosis and those found on the corresponding established cell lines. These observations led us to believe that may offer a representative in vitro model for the study of growth regulation of ALL.

Colony Growth Assays

Colony growth assays were performed in semi-solid medium. Leukemic cells, at the indicated density, were plated in 35mm Lux suspension dishes (Nunc, Inc., Naperville, IL.) in αMEM medium containing 10% FBS (Flow-Select Fetal Bovine Serum, Mclean, VA.) in methyl cellulose (0.9%, in Methocel, Dow Chemical Co., Midland, Michigan). Duplicate culture dishes were incubated at 37°C with 5% CO_2 in air in a humidified atmosphere. Colonies (≥ 20 cells) were counted after eight days using an inverted microscope.

Quantitation of mRNA of Cytokines and Cytokine Receptors by the Polymerase Chain Reaction

The quantitation of specific mRNA species by the polymerase chain reaction (PCR) was performed using synthetic DNA as an internal standard as previously described (Cohen, Grunberger et al. 1991). The technique involves co-amplification of a target cDNA (produced from the corresponding mRNA by reverse transcription) and of the internal standard. The target cDNA and the internal standard utilize the same primer sequences but yield PCR products of different sizes which can be separated by gel electrophoresis. Total cellular RNA was prepared according to the methods of Chomczynski and Sacchi (Chomczynski and Sacchi 1987). RNA concentrations were measured spectrophotometricaly (Beckman Instruments DU-40; Fullerton, CA.). The integrity of the RNA was confirmed by electrophoresis under denaturing conditions on a 1% agarose gel.

cDNA was prepared by reverse transcription at 37°C for 60 minutes in a 20 µl reaction mixture containing 2 µg of total cellular RNA, 1 mM dithiothreitol, 0.5 mM dNTP, 0.1µmol of a specific oligonucleotide primer and 100 units of recombinant MMLV reverse transcriptase (Bethesda Research Laboratories). Internal standard DNA (0.1 - 3.0 pg) and cDNA (produced from 100-300 ng total cell RNA) were amplified together with 1 unit of *Thermus aquaticus* DNA polymerase (Taq polymerase, Perkin-Elmer/Cetus) and 50 mM dNTP, 0.1 mM each of the 5' and 3' primers (1 x 10^6 cpm of γ^{32}P-end labeled primer was added where indicated) in a total volume of 50 µl. The mixture was overlaid with mineral oil and then amplified using a thermal cycler (Ericomp; San Diego, CA) over a number of cycles (indicated in text) denaturing at 95°C for 30 seconds, primer annealing at 54°C for 30 seconds, and extending at 72°C for 1 minute. Oligonucleotides were end-labeled with [γ-^{32}P] ATP and separated from unincorporated nucleotide with a Sephadex G-50 column. Ten microliters of PCR product were electrophoresed using 8% polyacrylamide gels in Tris borate/EDTA buffer. Gels were stained with ethidium bromide and photographed. When radiolabelled oligonucleotides were incorporated into the amplified product, the gels were dried and autoradiographed at -70°C. For quantitative analysis the appropriate bands were excised from the gel and radioactivity determined by scintillation counting. The amount of radioactivity recovered from the excised internal standard gel bands was plotted against the corresponding internal standard DNA concentration and the amount of cDNA present in each sample calculated from the standard curve.

Cytokines, Antibodies and ELISA

Human recombinant IL-1α (specific activity 3×10^8 units/mg) was a gift from Hoffmann-La Roche, (Nutley, NJ); recombinant human IL-2 (3×10^6 units/mg protein) was purchased from Genzyme Co. (Boston, MA); human recombinant IL-3 (6×10^6 units/mg), human recombinant IL-4 (1.2×10^4 units/mg), human recombinant IL-5, (1.2×10^3 units/mg), human recombinant IL-6, (4×10^6 units/mg), human recombinant IL-7, (3×10^3 units/mg), human recombinant GM-CSF, (10.6×10^6 units/mg), and human recombinant G-CSF (3×10^4 units/ml), were a gift from Genetics Institute (Cambridge, MA); recombinant human TNF-α (5.6×10^7 units/ml) was a gift from Genentech, Inc. South San Francisco CA); purified rabbit anti-human IL-1β antibody was purchased from Genzyme (Boston MA); γIFN (1×10^7 units/mg) was a gift from Biogen (Cambridge MA). Antibodies used in flow cytometry experiments were purchased from Coulter (Hialeah FL) as rhodamine conjugates for B4-IgG and T3-IgG and FITC conjugates for My-9-IgG, My4-IgG and I3-IgG. An ELISA specific for IL-1β was obtained from Cistron, Pine Brook, NJ. Sensitivity of the assay was shown to be 4 pg/ml. IL-6 receptor assay was performed using phycoerythrin-labeled IL-6 purchased from R&D Systems (Minneapolis, MN).

RESULTS AND DISCUSSION

The differentiation phenotype of the five ALL lines is summarized in Table 1. All five cell lines exhibited phenotypic markers characteristic of pre-B ALL such as high levels of CD19 and HLA-DR. However, there are some differences among the cell lines with regard to the degree of maturation along the B cell lineage. The A-1, B-1 and G-2 cell lines do not express the common ALL antigen, CD10, consistent with somewhat immature pre-B phenotype compared to the other two cell lines C-1 and W-1 that are CD10 positive (Zola 1987). The B-1 cell line is an immature biphenotypic pre-B/myeloid cell line (characteristic of ALL with 4;11 translocation) which expresses the stem cell marker CD34 and can be induced to differentiate along the myeloid lineage by IL-6 (Cohen 1992; Cohen et al. 1991).

To assess the possible role of growth factors in the regulation of the growth of these ALL cell lines we first examined the possibility that some of the lines promote their own growth via an autocrine cycle. The growth characteristics of some of the cell line are consistent with autocrine growth, for example, cell density-dependent growth regulation. In a colony assay, cell growth, as a function of density of the initial cells seeded, was slow below 1×10^4 cells/ml and became linear above this cell density. No growth was observed at or below a cell density of 2.5×10^3 cells/ml.

TABLE 1

Phenotypic Analysis of ALL Cell Lines

Percent of antigen expressing cells.

Cell line	HLA-DR	CD10	CD19	CD20	CD34	CD44.
A-1	100	0.4	98.7	0	nd	100
B-1	92.0	0.3	99.4	0	13.0	99.8
C-1	98.8	91.3	99.6	0.3	nd	100
G-2	96.9	0.6	94.4	10.0	nd	100
W-1	90.0	23.0	90.0	8.0	nd	100

This pattern of density dependent cell growth is consistent with the secretion of an autocrine factor by B1 cells which stimulates their growth. According to this hypothesis the threshold concentration of the growth factor required to stimulate its own receptor is reached at cell densities above 2.5×10^3 cells/ml. To test this hypothesis further we carried out experiments in which the growth of the ALL cells, seeded at low density (5×10^3 cells/ml), were grown in the presence of supernatant from high density serum free suspension cultures of the corresponding cell lines (Figure 1). The supernatants from all stimulate their own growth providing evidence for the presence of autocrine growth factors in these cell lines. To identify the growth factor involved in growth regulation in these cell lines we first examined the expression of known cytokines genes and their receptors in the ALL lines. For that purpose we employed quantitative RT-PCR assays that utilize synthetic internal standard for measurements of cytokine and cytokine receptor mRNA levels in these cell lines (Table 2).

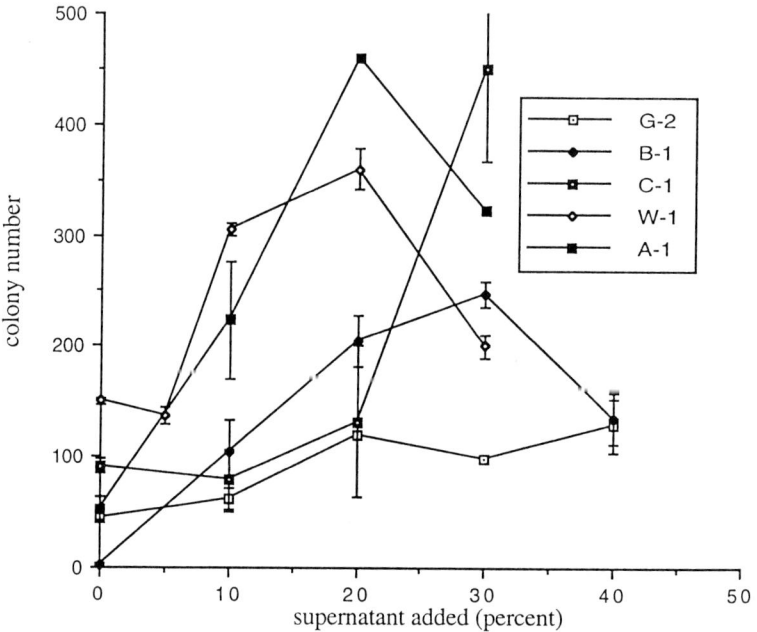

Figure 1. *Effect of cell supernatant on the clonogenic growth of ALL cell lines. A-1, B-1, C-1, G-2 and W-1 cell lines were seeded in semisolid medium at low cell density (3×10^3 cells/ml) and various concentrations of the corresponding supernatants from high density (1×10^6 cells/ml) cultures were added as indicated.*

Examination of the expression of the mRNAs of cytokine and their receptors in the five ALL cell lines demonstrates a common pattern of expression with some significant differences possibly reflecting the phenotypic heterogeneity of individual pre-B ALL cases. In general, the expression of mRNA of the various cytokine receptors was common in these cell line while the corresponding cytokine mRNA was rarely expressed. For instance the mRNA of the receptors for IL-4, IL-6, GM-CSF, INF-γ, and TNF was expressed in all five cell lines and most cell lines also expressed the mRNA of the receptors for IL-1, IL-3, IL-7, SCF and erytherpoietin. In a few cases individual cell lines expressed both the cytokine and its receptor raising the possibility of a functional autocrine cycle that promotes the growth of the leukemic cells.

TABLE 2

Expression of Cytokine mRNA in ALL Cell Lines

Cytokine mRNA(C) cytokine receptor mRNA(R)

line	IL-1	IL-2	IL-3	IL-4	IL-5	IL-6	IL-7	GM-CSF	IFN-γ	TNF	SCF	EPO
A-1	-/-	-/-	C/R	-/R	-/-	-/R	-/-	-/R	-/R	-/R	-/R	-/R
B-1	C/R	-/-	-/-	-/R	-/-	-/R	-/R	-/R	-/R	-/R	-/R	-/-
C-1	-/R	-/-	C/R	-/R	-/-	-/R	-/-	-/R	-/R	-/R	-/R	-/R
G-2	-/-	-/-	-/R	-/R	-/-	-/R	-/R	C/R	-/R	C/R	-/-	-/R
W-1	C/R	-/-	-/R	-/R	-/-	-/R	-/R	-/R	-/R	-/R	-/R	-/R

The profile of cytokine receptor expression provide the basis for functional analysis of specific cytokines on the growth of the leukemic cell lines. The analysis of the effect of various cytokines on the growth of leukemic cells was performed using clonogenic assays as depicted in Figure 2. In general there is good correlation between the expression of specific cytokine receptor and positive or negative growth regulation of the corresponding growth factor. The response to several cytokines is uniform across all five ALL cell lines while other cytokines act specifically on individual cell lines reflecting both the commonalty and subtle differences of pre-B ALL. Thus both TNF-α and IFN-γ completely inhibit colony formation of each one of the ALL lines, suggesting a potential clinical use for these cytokines in this disease.

On the other hand, both IL-3 and GM-CSF increase the number of colonies in four of the ALL cell lines by two to five fold. There was absolute correlation between the response of cell lines to IL-3 and GM-CSF reflecting the use of a common signal transducing β subunit by the two receptors (Linnekin and Farrar 1990). The expression of IL-3 and GM-CSF in the ALL cell lines and the effects of their ligands in promoting the growth of these lines are consistant with the reports of the effects of these cytokines on freshly isolated pre-B ALL cells (Freedman M. H 1993; Wormann B 1989). The universal stimulation of ALL cells by GM-CSF may be of clinical importance in view of the increasing use of this cytokine in leukemic patients with chemotherapy-induced neutropenia and following bone marrow transplantation and thus the risk of stimulating residual leukemic cells in these patients. The only cell line that failed to respond to IL-3 and GM-CSF was B-1 which stands out in this group due to its distinct biphenotypic pre-B/myeloid features (Cohen et al. 1991).

All the cell lines were stimulated by SCF however, the degree of stimulation varied between 20% in G-2 cells to a maximum of 300% in B-1 cells. Similar growth stimulatory effects of SCF have been reported on murine pro-B cells (McNiece 1991). The growth of two other cell lines (C-1 and G-2) was significantly stimulated by IL-7 consistent with the reported response of some cases of pre-B ALL to this cytokine (Skonberg 1991). Surprisingly, erythropoietin (EPO) stimulated the growth of all cell lines although four of the responses were to a limited degree. In view of these results the potential growth regulatory effect of erythropoietin in freshly isolated ALL cells as well in normal B cell development have to be examined.

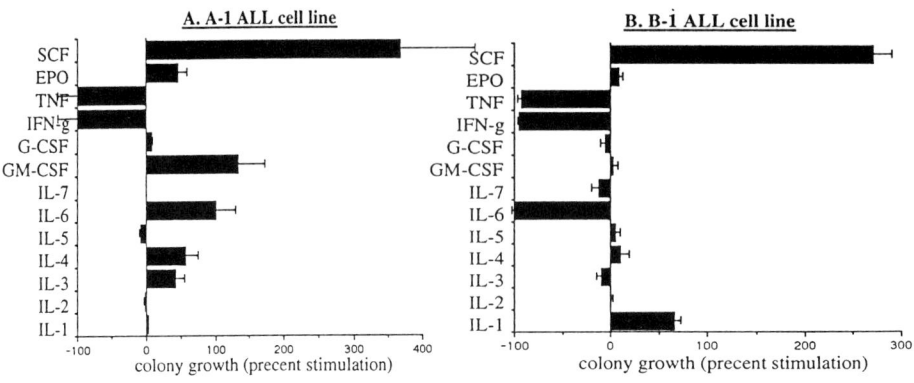
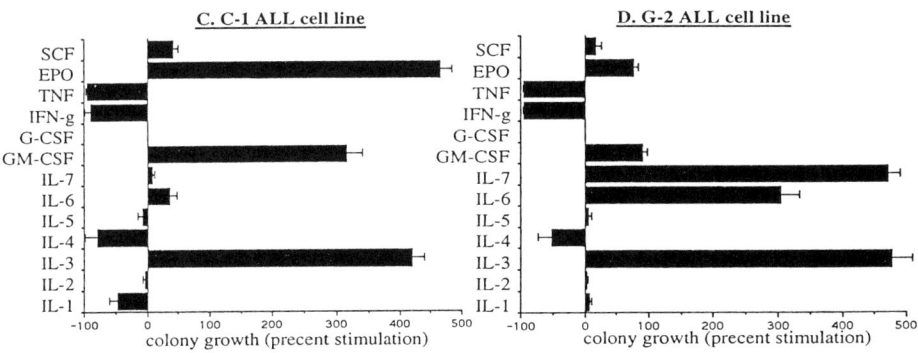
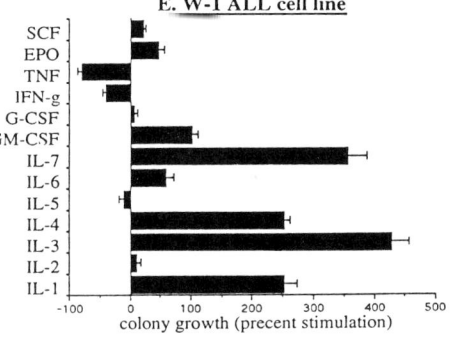

Figure 2. Effect of cytokines on colony growth in ALL cell lines. A-1 (A), B-1 (B), C-1 (C), G-2 (D) and W-1 (E) were seeded in semi solid medium (10,000-50,000 cells/ml) in the presence or absence of IL-1 (100 u/ml), IL-2 (100 u/ml), IL-3 (40u/ml), IL-4 (100u/ml), IL-5 (100u/ml), IL-6 (40u/ml), IL-7 (100 u/ml), GM-CSF (5 u/ml), TNF (100 u/ml), SCF (25 ng/ml) and EPO (5 ng/ml). The number of colonies of 20 cells or more was counted after 10-14 days.

In a few cases we have observed the presence of the mRNAs of both individual cytokine and its receptor on the surface a particular cell line. We have also shown that some of these cytokines when added exogenously can further stimulate the growth of particular cell lines supporting the possibility of autocrine cycles operating in these cell lines. For instance B-1 and W-1 cell lines express both IL-1β and its receptors and thus their growth may be promoted by an IL-1 mediated autocrine cycle. We have conducted a set of experiments to test this hypothesis, we have shown that specific antibodies against IL-1β are able to interrupt the autocrine cycle and inhibit the growth of B-1 (Cohen et al. 1991) and W-1 (data not shown) cells. Similar experiments were performed with G-2 cells demonstrating the operation of a GM-CSF-mediated autocrine cycle in this cell line (Freedman M. H 1993). In many cases we have found the expression of various cytokines and their receptors in freshly isolated leukemic cells from ALL patients consistent with the existent of autocrine cycles contributing to the malignant process.

REFERENCES

Chomczynski, P. and N. Sacchi. (1987). "Single-step method of RNA isolation by acid guanidinium thiocyanate-phenol chloroform extraction." *Anal. Biochem.* 162: 156.

Cohen, A., Petsche, D., Grunberger, T. and Freedman, M. (1992). "Interleukin-6 induces myeloid differentiation of a human biphenotypic cell line." *Lrukemia Res.* 16: 751-760.

Cohen, A., T. Grunberger, W. Vanek, I. D. Dube, P. J. Doherty, M. Letarte, C. Roifman and M. H. Freedman. (1991). "Constitutive expression and role in growth regulation of interleukin-1 and multiple cytokine receptors in a biphenotypic leukemic cell line." *Blood.* 78(1): 94-102.

Duhrsen U, M. D. (1988). "A model system for leukemic transformation of immortalized hemopoietic cells in irradiated recipient mice." *Leukemia.* 2: 329-335.

Freedman M. H, G. T., Correa, P, Axelrad A. A, Dube, I. D, and Cohen A. (1993). "Autocrine and paracrine growth control by GM-CSF of acute lymphoblastic leukemia cells." *Blood.* in press:

Greaves, M. F. (1986). "Differentiation-linked leukemogenesis in lymphocytes." *Science.* 234(4777): 697-704.

Linnekin, D. and W. L. Farrar. (1990). "Signal transduction of human interleukin 3 and granulocyte-macrophage colony-stimulating factor through serine and tyrosine phosphorylation." *Biochem J.* 271(2): 317-24.

McNiece, I. K., Langely, K. E., Zsebo, K. M. (1991). "The role of recombinant stem cell factor in early B cell development. Synergistic interaction with IL-7." *J Immunol.* 146: 3785-3790.

Schrader J. W, C. R. M. (1983). "Autogenous production of a hemopoietic growth factor, persisting-cellstimulatory factor, as a mechanism for transformation of bone-marrow derived cells." *Proc. Natl. Acad. Sci. USA.* 80: 6892-6897.

Skonberg, C., Erikstein, B. K., Smeland, E. B. Lie, S. O., Funderud, S., Beiske, K., Blomhoff H. K. (1991). "Interleukin-7 differentiates a subgroup of acute lymphoblastic leukemias." *Blood.* 77(2445-2451):

Sporn, M. B., Roberts, A. B.,. (1985). "Autocrine growth factors and cancer." *Nature.* 313: 745-749.

Wormann B, G. T. G., Mufsou R. A, Le Bien T. W. (1989). "Proliferation effect of interleukin-3 on normal and leukemic human B cell precursors." *Leukemia.* 3: 399-405.

Zola, H. (1987). "The surface antigens of human B lymphocytes." *Immunol. Today.* 8: 308-314.

Résumé

Les caractéristiques en terme de croissance cellulaire de 5 lignées cellulaires établies à partir de patients atteints de LAL pré-B ont été analysées en détail. Ces lignées cellulaires expriment des marqueurs phénotypiques identiques aux cellules de LAL pré-B fraichement isolées. Toutes les lignées expriment l'ARNm de plusieurs récepteurs de facteurs de croissance hématopoïètiques ainsi que les récepteurs de diverses cytokines. Plusieurs cytokines telles que l'interleukine 1 (IL-1), l'IL-3, le granulocyte-macrophage colony stimulating factor (GM-CSF), le stem cell factor (SCF) et l'IL-7 sont capables de stimuler la croissance de ces lignées lorsque celles-ci expriment les récepteurs correspondants. D'autres cytokines, telles que le tumor necrosis factor-α (TNFα), l'inteféron-γ (IFN γ) et l'IL-6, apparaissent comme des régulateurs négatifs pour les lignées qui expriment leurs récepteurs. Dans trois de ces lignées, nous avons été capables de démontrer l'existence pour l'IL1 et le GM-CSF de boucles autocrines de croissance.

Effect of interferons on normal and leukemic hematopoiesis

Carmelo Carlo-Stella, Lina Mangoni and Vittorio Rizzoli

Department of Hematology, Bone Marrow Transplantation Unit, University of Parma, Parma, Italy

Summary. Interferons (IFNs) are a heterogeneous family of proteins able to modulate the immune system, affect cell differentiation, enhance protection against viral infection, and exert antiproliferative activities. The availability of clonogenic assays for the in vitro growth of pluripotent and lineage-restricted hematopoietic progenitor cells provided the basis for studying the effects of recombinant IFNs on the hematopoietic system. In this paper, the inhibitory effects of IFNs at the level of normal and leukemic progenitors as well as their mechanisms of action will be reviewed. Data are presented showing that the antitumor effects of IFNs occur not only as a consequence of their antiproliferative effect on progenitor cells, but also involve different mechanisms, such as induction of cell differentiation with reduction of the self-renewal of the neoplastic clone and augmentation of host effector mechanisms.

The definition of "biological response modifier" (BRM) includes agents or approaches that alter interactions between tumors and host toward therapeutic advantage, primarily through modification of the host response to tumors (Mihich, 1986). A common denominator in the action of most BRMs is the presence of a specific receptor on the surface of the target cell that may be represented by the tumor cell, which is directly affected by the BRM, or it may be the host cell that is triggered to exert antitumor activity either directly or through a cascade of effects (Mihich, 1986). Typical BRMs include the interferons (IFNs), a heterogeneous family of proteins able to modulate the immune system, affect cell differentiation, enhance protection against viral infection, and exert antiproliferative activities (Kirkwood & Ernstoff, 1984; Paulnock & Borden, 1985; Burke, 1986; Borden & Ball, 1981). IFNs are now divided into three major classes depending on the antigenic specificity detected by monoclonal neutralising antibodies (Committee on Nomenclature, 1980): IFN-alpha, IFN-beta, and IFN-gamma. IFN-alpha and IFN-beta, collectively termed type I or viral IFN are structurally related and acid-stable molecules. IFN-gamma, type II or immune IFN, is a structurally distinct, acid-labile glycoprotein.

INTERFERON PRODUCTION

IFNs-alpha, a family of at least 15 highly homologous species, are produced by a variety of cell types including monocytes, null cells (non-B, non-T lymphocytes) and B-cell lines. IFN-beta is mainly produced by fibroblasts and epithelial cells. IFN-alpha and IFN-beta are produced in response to the same inducers, being represented by viruses and a wide variety of natural and synthetic non-viral agents (Morris & Burke, 1983). IFN-gamma is produced by T cells in response to mitogens or specific antigens, as well as by natural killer cells (Perussia et al., 1980). IFN-gamma production by T-lymphocytes is dependent upon the secretion of interleukin-1 (IL-1) by accessory adherent cells and of interleukin-2 (IL-2) produced mostly by helper T cells (Farrar et al., 1981). The production of IFN-gamma by NK cells in response to IL-2 might be an important amplification mechanism in vivo; i.e., the IL-2 produced by T cells in response to an antigenic stimulus may recruit NK cells into the IFN-gamma-producing population (Trinchieri & Perussia, 1985).

THE INTERFERON MOLECULES AND THEIR GENES

The genes for IFNs have been identified and their cloned complementary DNA sequences expressed (Pestka, 1981). At least 18 immunologically related, highly homologous (70-90% homology) gene products for the human IFN-alpha gene family and 2-5 separate genes for IFN-beta have been identified. IFN-alpha and IFN-beta genes code for a protein of 166 amino acids with an additional 20 amino acid secretory peptide present on the amino-terminal end. The IFN-alpha genes have been localised on human chromosome 9. The IFN-beta genes are located on chromosomes 2 and 5. A single gene coding for human IFN-gamma has been identified on chromosome 12. The IFN-gamma gene codes for a protein of 166 amino acids, but the 23 amino acids at the N-terminal end constitute a signal sequence that is cleaved during biosynthesis of the protein. Only very little sequence homology to type I IFNs can be found.

INTERFERON RECEPTORS

The interaction of IFNs with susceptible cells is mediated by binding to specific high-affinity cell membrane receptors, a step which is a prerequisite, but of itself insufficient, for the subsequent biological effects of IFNs (Zoon & Arnheiter, 1984). Both IFN-alpha and IFN-beta bind to the same high-affinity species-specific receptors (type I receptors) coded by a gene on chromosome 21, whereas IFN-gamma acts through type II IFN receptors (Orchansky et al., 1984). The two receptor systems appear to be independent (Branca & Baglioni, 1981). Cellular processes subsequent to the binding of IFN show general similarities with many other polypeptide hormone-receptor systems. At 37°C IFN-receptor complexes are rapidly internalised (Branca et al., 1982). Reappearance of receptor activity occurs after de novo protein synthesis and re-assembling of new receptor proteins. Recycling of internalised receptors after the dissociation of bound ligands has also been shown (Branca & Baglioni, 1982).

MECHANISM OF ACTION OF INTERFERONS

The signal subsequent to IFN-receptor binding is transduced to the nucleus. A number of genes are newly transcribed into mRNA and, finally, a number of proteins that are not found in untreated cells are synthesised. Both IFN-alpha and IFN-gamma induce the synthesis of several common polypeptides, but that IFN-gamma also induces the synthesis of 12 distinct polypeptides (Weil et al., 1983). Although IFN-alpha and IFN-gamma share some common mechanisms of action, the strong synergism described between type I and type II IFNs suggests that the molecular pathways induced by the two types of IFN are not completely identical. Two enzymes (2'-5'-oligoadenylate synthetase and a protein kinase) have been identified which are relevant to the mechanisms of IFN antiviral and antiproliferative activity (Senn, 1984). A protein kinase able to phosphorylate and thereby inactivate the peptide elongation initiation factor (eIF2) involved in beginning protein synthesis, and a phosphodiesterase pathway represent mechanisms of protein synthesis inhibition which are parallel to, but independent of, 2'-5'-oligoadenylate synthetase (Lebleu & Content, 1982).

INTERFERONS AS AGENTS OF CELLULAR DIFFERENTIATION INDUCTION

IFN-alpha and IFN-beta at concentrations of 400-1000 U/ml have a marked antiproliferative effect and promote differentiation induction of monoblast cell line U937

(Hattori et al., 1983), but not of HL-60 (Ball et al., 1984), towards the monocytic pathway. IFN-alpha and IFN-beta at concentrations of 500-1000 U/ml enhance the differentiation of HL-60 seen with retinoic acid. IFN-alpha also restores the inducibility of HL-60 subclones which are resistant to the differentiative effect of retinoic acid or dimethylsulfoxide (Grant et al., 1985). IFN-gamma preparations induce monocytic differentiation in both HL-60 and U937 cell lines (Perussia et al., 1983), as measured by morphological, enzymatic, immunological and functional parameters. Several recent papers have emphasised the differentiation-inducing activity exerted by rIFN-gamma in association with retinoic acid or 1-alpha,25-dihydroxyvitamin D_3 in continuous cell line models (Gullberg et al., 1985; Ball et al., 1986). Used as a single agent, rIFN-gamma exerts differentiation-inducing activity on fresh blast cells from patients with acute myeloid leukemia (AML) and myelodysplastic syndromes (MDS) (Perussia et al., 1983; Matsui et al., 1985; Carlo-Stella et al., 1988). This activity - particularly evident as receptor enhancement for the complement fragment C3bi - is exerted by rIFN-gamma at immunologic, morphologic, enzymatic and functional levels and is consistent with monocytic maturation. Cytogenetic evidence suggests that differentiation induction events are not due to the expansion of residual nonleukemic precursor cells and strongly supports the leukemic origin of both the blasts and the more differentiated cells (Carlo-Stella et al., 1988). Both Type I and Type II IFNs induce or enhance the expression of class I MHC antigens on various cell types, although IFN-gamma is somewhat more efficient (Gresser, 1984). In addition, IFN-gamma is a very potent inducer of class II MHC antigens, while type I IFNs induce low levels of class II antigens and only at doses several orders of magnitude higher than those of IFN-gamma (Steeg et al., 1982). Induction of MHC class II antigens on the U937 cell line by rIFN-gamma is associated with the increased sensitivity of these cells to other inhibitory factors, such as acidic isoferritins and lactoferrin (Broxmeyer et al., 1986), and this activity is counteracted by prostaglandin E (Piacibello et al., 1986).

INTERFERONS AS INHIBITORS OF CELL PROLIFERATION

Natural and recombinant IFNs inhibit the growth both of normal and neoplastic cells (Taylor-Papadimitriou, 1980). IFNs slow or stop cell proliferation by extending all phases of the cell cycle (Balkwill & Taylor-Papadimitriou, 1978). The availability of clonogenic assays for the in vitro growth of pluripotent (CFU-GEMM, colony-forming unit-granulocyte, erythrocyte, macrophage, megakaryocyte) and lineage-restricted (BFU-E, burst-forming unit-erythroid; CFU-E, colony-forming unit-erythroid; CFU-GM, colony-forming unit-granulocyte-macrophage; CFU-Mk, colony-forming unit-megakaryocyte) hematopoietic progenitor cells provided the basis for studying the effects of IFNs on the hematopoietic system.

Continuous exposure of bone marrow cells, throughout the entire culture period (7-21 days), to natural alpha, beta, and gamma IFNs suppresses in a dose-dependent manner the growth of human pluripotent hematopoietic progenitor cells as well as that of the lineage-restricted ones (Neumann & Fauser, 1982; Broxmeyer et al., 1983). However, natural IFN preparations were probably contaminated with a variety of biologically active molecules that exhibit properties similar or antagonistic to those of IFN (Degliantoni et al., 1985). Studies performed with highly purified recombinant-DNA-derived IFNs and monoclonal antibodies against rIFNs have confirmed that continuous exposure of bone marrow cells to rIFNs results in a marked antiproliferative activity on CFU-GEMM, BFU-E, CFU-E, and CFU-GM. Recombinant IFNs exert a potent antiproliferative effect on in vitro megakaryocytic colony (CFU-Mk) (Ganser et al., 1987). The degree of CFU-Mk growth inhibition is essentially the same for both rIFN-alpha and rIFN-gamma. The dose dependence of CFU-Mk suppression is similar to that observed for erythroid and granulocyte-macrophage progenitor cells. Short-term exposure of bone marrow cells to rIFNs before culture fails to cause a significant reduction in CFU-GEMM and CFU-Mk growth, even when 10^4 U/ml of rIFNs are used. The demonstration of an antiproliferative effect by continuous exposure reflects a cytostatic activity, whereas antiproliferative effects observed after short-term exposure suggest a cytotoxic mechanism. The inhibitory activity of rIFNs is neutralised by preincubating each rIFN prior to culture with its respective monoclonal antibody, thus demonstrating that the inhibitory action is due to a specific rIFN preparation. The inhibition exerted by both rIFN-alpha and rIFN-gamma can be abrogated by the addition of neutralising monoclonal antibodies as late as one hour after the onset of incubation; this process is no longer reversible after four hours and no further neutralisation can be obtained after 24 and 96 hours of culture (Ganser et al., 1987). Small amounts of rIFN-gamma have been shown to act synergistically with rIFN-alpha in suppressing the growth of normal human bone marrow CFU-GEMM, BFU-E, and CFU-GM (Broxmeyer et al., 1985).

Two general mechanisms could account for the responses observed in a clonogenic assay: a direct effect on the hematopoietic progenitor cells or an indirect action mediated by stimulation of host cytotoxic accessory cells (cytotoxic T lymphocytes, NK cells, monocytes). The role of accessory cells (T-lymphocytes and monocytes) in the mediation of IFN suppression of hematopoietic progenitors has been extensively analysed. Ganser et. al. (1987) indicated that the mechanisms underlying the suppressive effect of the two types of rIFNs on CFU-GEMM and CFU-Mk are different. The inhibitory effect of rIFN-gamma, but not alpha, was significantly reduced after depletion of adherent and/or T-cells. Recombinant IFN-alpha acts directly on the progenitor cells, whereas the influence of rIFN-gamma is largely mediated by accessory cells.

Highly purified monocyte preparations, but not resting T-lymphocytes, release granulocyte-monocyte stimulating factors upon stimulation with low doses of IFN-gamma (0.01-10 U/ml). Recombinant IFN-gamma can enhance the release of factors from peripheral blood T lymphocytes that stimulate CFU-GEMM, BFU-E, and CFU-GM growth when the T-cells are stimulated with phytohemagglutinin; this activity is exerted by those T-cells with a helper/inducer phenotype (Piacibello et al., 1985; Herrman et al., 1986). In vitro and in vivo competition between rIFNs and growth stimulating factors may have physiologic and pathologic relevance. Mamus et al. (1986) failed to prevent or reverse the inhibitory activity of rIFN-alpha and rIFN-gamma on BFU-E growth in culture by adding increasing doses of natural erythropoietin. In contrast, rIFN-gamma-induced inhibition of CFU-E has been shown to be corrected by recombinant erythropoietin (Means & Krantz, 1991). Reductions in the concentrations of fetal calf serum (FCS) or human placental conditioned medium (HPCM), both of which contain positive growth factors required for in vitro colony formation, increase the inhibitory effect of rIFNs on CFU-GM, whereas higher concentrations of FCS or HPCM may partially overcome the inhibitory effect of rIFNs (Raefsky et al., 1985).

Taetle et al. (1980) showed that IFN-beta is equally inhibitory for blast progenitor cells and normal bone marrow CFU-GM. The self-renewal capacity of blast progenitors - as detected in secondary cultures - was inhibited, and this effect was specific since no reduction in secondary plating efficiency was seen with adriamycin, a known cytostatic drug that can suppress primary colony formation. Freedman et al. (1985) tested the effect of various rIFNs-alpha on blast progenitor cells and on their capacity for self-renewal, and confirmed that continuous exposure to rIFNs-alpha suppresses blast colony growth and the self-renewal capacity of blast progenitor cells in a dose-dependent fashion.

The in vitro sensitivity to rIFN-alpha and rIFN-gamma of circulating CFU-GEMM, CFU-Mk, BFU-E and CFU-GM from patients with myelofibrosis with myeloid metaplasia is similar to that of normal progenitors, and rIFN concentrations ranging from 16 to 187 U/ml cause 50% inhibition of colony formation (Carlo-Stella et al., 1987). Inhibitory activity has been reported for natural leukocyte and fibroblast IFN preparations on chronic myelogenous leukemia (CML) progenitor cells, showing that the susceptibility of CML-derived granulocyte-macrophage progenitor cells to the suppressive effect of IFNs is similar to that seen in normal subjects. The effects of recombinant IFNs on CML marrow were extensively investigated. When added to chronic myelogenous leukemia marrow cells both rIFN-alpha and rIFN-gamma significantly reduced colony formation, with 50% inhibition occurring at 71 and 186 U/ml for CFU-GEMM, 40 and 152 U/ml for CFU-Mk, 222 and 1,458 U/ml for BFU-E, and 119 and 442 U/ml for CFU-GM, respectively (Carlo-Stella et al., 1988). A small amount of rIFN-gamma (5 U/ml) acted synergistically with increasing doses of rIFN-alpha. This synergy was evident even when rIFN-gamma was added 72 hrs after the initiation of cultures but was completely lost when the target cells were depleted of accessory cells (Carlo-Stella et al., 1988). When a low dose of rIFN-alpha (5 U/ml) was added to rIFN-gamma the 50% inhibitory concentration values were decreased up to 10-fold (Carlo-Stella et al., 1988). Addition of recombinant IFN-gamma to cultures of bone marrow-derived CML progenitor cells might increase the proportion of Ph-negative colonies (McGlave at al., 1987).

CML has been used as an optimal model to investigate the complex regulatory effects of IFNs on the hematopoietic system. Deregulation of the mechanism that normally maintains the majority of primitive hematopoietic cells in a quiescent state is a characteristics of CML. Several mechanisms have been proposed to explain an abnormal control of neoplastic progenitor cell cycling in CML patients. These include defective adhesive interactions of primitive Ph-positive cells with stromal cells or their products, autocrine or paracrine growth factor production, the possibility that primitive Ph-positive progenitors may have an altered responsiveness to either positive or negative growth factors produced by stromal cells. The defective capability of CML cells to adhere to preformed allogeneic marrow stroma is supported by the deficiency of a phosphatidylinositol-linked cell adhesion molecule (Gordon et al., 1991). Recent data by Osterholz et al. (1991) and Dowding et al. (1991) show that IFN-alpha enhances the capacity of marrow stromal cells to bind CML progenitor cells. It has also been shown that CML progenitor cells are deficient in their expression of the cytoadhesion molecule lymphocyte function antigen-3 (LFA-3) and that IFN-alpha restores the deficient expression of this molecule (Upadhyaya et al., 1991).

CONCLUSIONS

IFNs have the characteristics of classical biological response modifiers in that they modify both host and neoplastic cells. The antitumor effect of IFNs occurs as a consequence of at least three different mechanisms: (i) induction of cell differentiation with reduction of neoplastic clone self-renewal; (ii) antiproliferative effect on progenitor cells, and (iii) augmentation of host effector mechanisms. It is likely that IFN-gamma - more than IFN-alpha - is able to modulate all these functions, although less is known concerning the clinical efficacy of immune IFN. Probably the relevance of each mechanism varies from patient to patient depending on the immune status of the host and the characteristics of the tumor. In vivo interrelationships with other cytokines, endogenous levels of hematopoietic colony stimulating factors, dose, scheduling, and route of administration of a certain IFN subtype will all be important factors in determining the nature of the clinical response. Application of preclinical models to the investigation of open problems such as synergy between rIFNs and other cytokines will offer new possibilities for the therapeutic use of rIFNs in cancer patients.

REFERENCES

Balkwill, F., & Taylor-Papadimitriou, J. (1978): Interferon affects both G1 and S + G2 in cells stimulated from quiescence to growth. *Nature* 274, 798-800.

Ball, E.D., Guyre, P.M., Shen, L., Glynn, J.M., Maliszewski, C.R., Baker, P.E., & Fanger, M.W. (1984): Gamma interferon induces monocytoid differentiation in the HL-60 cell line. *J. Clin. Invest.* 73, 1972-1077.

Ball, E.D., Howell, A.L., & Shen, L. (1986): Gamma interferon and 1,25 dihydroxyvitamin D_3 cooperate in the induction of monocytoid differentiation but not in the functional activation of the HL-60 promyelocytic leukemia cell line. *Exp. Hematol.* 14, 998-1005.

Borden, E.C., & Ball L.A. (1981): Interferons: biochemical, cell growth inhibitory, and immunological effects. *Prog. Hematol.* XII, 299-339.

Branca, A., & Baglioni, C. (1981): Evidence that type I and II IFNs have different receptors. *Nature* 294, 768-770.

Branca, A.A., & Baglioni, C. (1982): Down-regulation of the interferon receptor. *J. Biol. Chem.* 257, 13197-13200.

Branca, A.A., Faltynek, C.R., D'Alessandro, S.B., & Baglioni, C. (1982): Interaction of interferon with cellular receptors: internalization and degradation of cell-bound interferon. *J. Biol. Chem.* 257, 13291-13296.

Broxmeyer, H.E., Cooper, S., Rubin, B.Y., & Taylor, M.W. (1985): The synergistic influence of human interferon-gamma and interferon-alpha on suppression of hematopoietic progenitor cells is additive with the enhanced sensitivity of these cells to inhibition by interferons at low oxygen tension in vitro. *J. Immunol.* 135, 2502-2506.

Broxmeyer, H.E., Lu, L., Platzer, E., Feit, C., Juliano, L., & Rubin, B.Y. (1983): Comparative analysis of the influences of human gamma, alpha, and beta interferons on human multipotential (CFU-GEMM), erythroid (BFU-E) and granulocyte-macrophage (CFU-GM) progenitor cells. *J. Immunol.* 131, 1300-1305.

Broxmeyer, H.E., Piacibello, W., Juliano, L., Platzer, E., & Berman, E. (1986): Gamma interferon induces colony-forming cells of the human monoblast cell line U937 to respond to inhibition by lactoferrin, transferrin, and acidic isoferritins. *Exp. Hematol.* 14, 35-43.

Burke, D.C. (1986): Interferon and cell differentiation. *Br. J. Cancer* 53, 301-306.

Carlo-Stella, C., Cazzola, M., Ganser, A., Barosi, G., Dezza, L., Meloni, F., Pedrazzoli, P., Hoelzer, D., & Ascari, E. (1987): Effects of recombinant alpha and gamma interferons on the in vitro growth of circulating hemopoietic progenitor cells (CFU-GEMM, CFU-Mk, BFU-E, and CFU-GM) from patients with myelofibrosis with myeloid metaplasia. *Blood* 70, 1014-1019.

Carlo-Stella, C., Cazzola, M., Ganser, A., Bergamaschi, G., Pedrazzoli, P., Hoelzer, D., & Ascari, E. (1988): Synergistic antiproliferative effect of recombinant interferon gamma with recombinant interferon alpha on chronic myelogenous leukemia hematopoietic progenitor cells (CFU-GEMM, CFU-Mk, BFU-E, CFU-GM). *Blood* 72, 1293-1299.

Carlo-Stella, C., Cazzola, M., Ganser, A., Meloni, F., Pedrazzoli, P., Bernasconi, P., Invernizzi, R., Hoelzer, D., & Ascari, E. (1988): Recombinant interferon gamma induces in vitro monocytic differentiation of blast cells from patients with acute non lymphocytic leukemia and myelodysplastic syndromes. *Leukemia* 2, 55-59.

Committee on Nomenclature. (1980): Interferon nomenclature. *Nature* 286, 110.

Degliantoni, G., Murphy, M., Kobayashi, M., Francis, M.K., Perussia, B., & Trinchieri, G. (1985): Natural killer (NK) cell-derived hematopoietic colony-inhibiting activity and NK cytotoxic factor. Relationship with tumor necrosis factor and synergism with immune interferon. *J. Exp. Med.* 162, 1512-1530.

Dowding, C., Guo, A.P., Osterholz, J., Siczkowski, M., Goldman, J., & Gordon, M. (1991): Interferon-alpha overrides the deficient adhesion on chronic myeloid leukemia primitive progenitor cells to bone marrow stromal cells. *Blood* 78, 499-505.

Farrar, V.L., Johnson, H.M., & Farrar, J.J. (1981): Regulation of the production of immune interferon and cytotoxic T lymphocytes by interleukin-2. *J. Immunol.* 126, 1120-1125.

Freedman, M.H., Fish, E., Estrov, Z., Grunberger, T., & Williams, B.R.G. (1985): Growth inhibitory effect of gene-cloned interferons on human myeloblast colonies. *Exp. Hematol.* 13, 932-936.

Ganser, A., Carlo-Stella, C., Völkers, B., Greher, J., & Hoelzer, D. (1987): Effect of recombinant interferons alpha and gamma on human bone marrow-derived megakaryocytic progenitor cells. *Blood* 70, 1173-1179.

Gordon, M.Y., Atkinson, J., Clarke, D., Dowding, C.R., Goldman, J.M., Grimsley, P.G., Siczkowski, M., & Greaves, M.F. (1991): Deficiency of a phosphatidylinositol-anchored cell adhesion molecule influences haemopoietic progenitor binding to marrow stroma in chronic myeloid leukemia. *Leukemia* 8, 693-698.

Grant, S., Bhalla, K., Weinstein, I.B., Pestka, S., Mileno, M.D., & Fisher, P.B. (1985): Recombinant human interferon sensitizes resistant myeloid leukemic cells to induction of terminal differentiation. *Biochem. Biophys. Res. Commun.* 130, 379-388.

Gresser, I. (1984): The effect of interferon on the expression of surface antigens. In *Interferons and the Immune System*, ed. J. Vilcek, & E. De Maeyer, pp. 113-132. Amsterdam: Elsevier.

Gullberg, U., Nilsson, E., Einhorn, S., & Olsson, I. (1985): Combinations of interferon-gamma and retinoic acid or 1-alpha,25-dihydroxycholecalciferol induce differentiation of the human monoblast leukemia cell line U-937. *Exp. Hematol.* 13, 675-679.

Hattori, T., Pack, M., Bougnoux, P., Chang, Z.L., & Hoffman, T. (1983): Interferon-induced differentiation of U937 cells. Comparison with other agents that promote differentiation of human myeloid or monocyte-like cell lines. *J. Clin. Invest.* 72, 237-244.

Herrman, F., Cannistra, S.A., & Griffin, J.D. (1986): T-cell monocyte interactions in the production of humoral factors regulating human granulopoiesis in vitro. *J. Immunol.* 136, 2856-2861.

Kirkwood, J.M., & Ernstoff, M.S. (1984): Interferons in the treatment of human cancer. *J. Clin. Oncol.* 2, 336-352.

Lebleu, B., & Content, J. (1982): Mechanisms of interferon action: biochemical and genetic approaches. In *Interferon 4*, ed I. Gresser, pp. 48-94. London: Academic Press.

Mamus, S.W., Oken, M.M., & Zanjani, E.D. (1986): Suppression of normal human erythropoiesis by human recombinant DNA-produced Alpha-2-interferon in vitro. *Exp. Hematol.* 14, 1015-1022.

Matsui, T., Takahashi, R., Mihara, K., Nakagawa, T., Koizumi, T., Nakao, Y., Sugiyama, T., & Fujita, T. (1985): Cooperative regulation of c-myc expression in differentiation of human promyelocytic leukemia induced by recombinant gamma-interferon and 1,25 dihydroxyvitamin D_3. *Cancer Res.* 45, 4366-4371.

McGlave, P., Mamus, S., Vilen, B., & Dewald, G. (1987): Effect of recombinant gamma interferon on chronic myelogenous leukemia bone marrow progenitors. *Exp. Hematol.* 15, 331-335.

Means, R.T., & Krantz, S.B. (1991): Inhibition of human erythroid colony-forming untis by gamma interferon can be corrected by recombinant human erythropoietin. *Blood* 78, 2564-2567.

Mihich, E. (1986): Future perspectives for biological response modifiers: a viewpoint. *Semin. Oncol.* 13, 234-254.

Morris, A.G., & Burke, D.C. (1983): The variety of interferons - molecular types, modes of induction and biological effects. In *Recent Advances in Clinical Immunology*, ed. R.A. Thompson & N.R. Rose, pp. 37-49. Edinburgh: Churchill Livingstone.

Neumann, H.A, & Fauser, A.A. (1982): Effect of interferon on pluripotent hemopoietic progenitors (CFU-GEMM) derived from human bone marrow. *Exp. Hematol.* 10, 587-590.

Orchansky, P., Novick, D., Fisher, D.G., & Rubinstein, M. (1984): Type I and type II interferon receptors. *J. Interferon Res.* 4, 275-282.

Osterholz, J., Dowding, C., Guo, A.P., Siczkowski, M., & Goldman, J.M. (1991): Interferon-alpha alters the distribution of CFU-GM between the adherent and nonadherent compartments in long-term cultures of chronic myeloid leukemia marrow. *Exp. Hematol.* 19, 326-331.

Paulnock, D.M., & Borden, E.C. (1985): Modulation of immune functions by interferons. In: *Immunity to Cancer*, pp. 545-559. New York: Academic Press.

Perussia, B., Mangoni, L., Engers, H.D., & Trinchieri, G. (1980): Interferon production by human and murine lymhpocytes in response to alloantigens. *J. Immunol.* 25, 1589-1595.

Perussia, B., Dayton, E.T., Fanning, V., Thiagarajan, P., Hoxie, J., & Trinchieri, G. (1983): Immune interferon and leukocyte-conditioned medium induce normal and leukemic myeloid cells to differentiate along the monocytic pathway. *J. Exp. Med.* 158, 2058-2080.

Perussia, B., Dayton, E.T., Lazarus, R., Fanning, V., & Trinchieri, G. (1983): Immune interferon induces the receptor for monomeric IgG1 on human monocytic and myeloid cells. *J. Exp. Med.* 158, 1092-1113.

Pestka, S. (1981): The human interferons - From protein purification and sequence to cloning and expression in bacteria: before, between, and beyond. *Arch. Biochem. Biophys.* 221, 1-37.

Piacibello, W., Lu, L., Wachter, M., Rubin, B., & Broxmeyer, H.E. (1985): Release of granulocyte-macrophage colony stimulating factors from major histocompatibility complex class II antigen-positive monocytes is enhanced by human gamma interferon. *Blood* 66, 1343-1351.

Piacibello, W., Rubin, B.Y., & Broxmeyer, H.E. (1986): Prostaglandin E counteracts the gamma interferon induction on U937 cells and induction of responsiveness of U937 colony-forming cells to suppression by lactoferrin, transferrin, acidic isoferritins, and prostaglandin E. *Exp. Hematol.* 14, 44-50.

Raefsky, E.L., Platanias, L.C., Zoumbos, N.C., & Young, N.S. (1985): Studies of interferon as a regulator of hemopoietic cell proliferation. *J. Immunol.* 135, 2507-2512.

Senn, C.C. (1984): Biochemical pathways in interferon action. *Pharmacol. Ther.* 24, 235-257.

Steeg, P.S., Moore, R.N., Johnson, H.M., & Oppenheim, J.S. (1982): Regulation of murine macrophage Ia antigen expression by a lymphokine with immune interferon. *J. Exp. Med.* 156, 1780-1793.

Taetle, R., Buick, R.N., & McCulloch, E.A. (1980): Effect of interferon on colony formation in culture by blast cell progenitors in acute myeloblastic leukemia. *Blood* 56, 549-552.

Taylor-Papadimitriou, J. (1980): Effects of interferons on cell growth and function. *Interferon* 2, 13-46.

Trinchieri, G., & Perussia, B. (1985): Immune interferon: a pleiotropic lymphokine with multiple effects. *Immunolgy Today* 6, 131-136.

Upadhyaya, G., Guba, S.C., Sih, S.A., Feinberg, A.P., Talpaz, M., Kantarijian, H.M., Deisseroth, A.B., & Emerson, S.G. (1991): Interferon-alpha restores the deficient expression of the cytoadhesion molecule lymphocyte function antigen-3 by chronic myelogenous leukemia progenitor cells. *J. Clin. Invest.* 88, 2131-2136.

Weil, J., Epstein, C.J., & Epstein, L.B. (1983): A unique set of polypeptides is induced by gamma interferon in addition to those induced in common with alpha and beta interferons. *Nature* 301, 437-439.

Zoon, K.C., & Arnheiter, H. (1984): Studies of the interferon receptors. *Pharmac. Ther.* 24, 259-278.

ACKNOWLEDGMENTS

This work was supported in part by grants from Consiglio Nazionale delle Ricerche (CNR, PFO contract no. 92.02254), MURST (40% - 60%), and Associazione Italiana per la Ricerca sul Cancro (AIRC).

Résumé

Les interférons appartiennent à une famille hétérogène de protéines capables de moduler le système immunitaire, d'affecter la différenciation cellulaire, d'accroitre la protection contre les infections virales, et d'exercer des activités anti-prolifératives. L'existence de tests clonogéniques pour les progéniteurs hématopoïetiques pluripotents et engagés permet de servir de base afin d'étudier les effets des interférons recombinants sur le système hématopoïètique. Dans cet article, les effets inhibiteurs des interférons sur les progéniteurs normaux et leucémiques ainsi que leurs mécanismes d'action sont passés en revue. Les données présentées montrent que les effets antitumoraux des interférons ne sont pas uniquement la conséquence de leurs effets antiprolifératifs sur les cellules progénitrices, mais également dus aux d'autres mécanismes tels que l'induction de la différencaition cellulaire, la réduction de l'autorenouvellement du clone néoplasique et l'augmentation des mécanismes de défense de l'hôte.

The status of haematopoietic stem cell proliferation regulators in murine myeloid leukaemia

Andrew Riches and Simon Robinson

School of Biological and Medical Sciences, University of St. Andrews, Scotland, UK

Leukaemic patients usually present clinically with anaemia, thrombocytopenia and often with recurrent infections. These symptoms are all indicative of an apparent suppression of 'normal' haematopoiesis by the leukaemic cells. This phenomenon is proposed to be due to the production of leukaemia-associated factor(s), which act to suppress the proliferation of 'normal' haematopoietic tissue, but to which the leukaemic cells are themselves insensitive. Direct-acting, haematopoietic stem cell-specific, proliferation regulators modify the proportion of the haematopoietic stem cell population in S-phase. A stimulator of haematopoietic stem cell proliferation, present in extracts of "regenerating", or foetal haematopoietic tissues, acts directly to increase the proportion of the haematopoietic stem cell population in S-phase; while an inhibitor of haematopoietic stem cell proliferation, present in extracts of normal haematopoietic tissues and subsequently characterized as MIP1α, acts directly to reduce the proportion of the haematopoietic stem cell population in S-phase.

The levels of these endogenous stem cell proliferation regulators were investigated in a number of X-irradiation-induced, serially-passaged, murine, myeloid leukaemias (Hepburn *et al.*, 1987) using an *in vitro* assay of a primitive murine haematopoietic precursor characterised by a high proliferative potential (high proliferative potential colony-forming cell, HPP-CFC) and behavioural and regulatory properties similar to those reported for the *in vivo* spleen colony-forming unit (CFU-S) and considered a component of the haematopoietic stem cell compartment (Robinson & Riches, 1991).

No evidence of either the haematopoietic stem cell proliferation stimulator, or inhibitor, was detected in extracts of overtly leukaemic, murine bone marrow. In addition, no evidence of a either a direct-acting, leukaemia-associated proliferation inhibitor, or of any leukaemia-associated toxicity was observed. However, while extracts of overtly leukaemic, murine bone marrow did not 'block' the *action* of the haematopoietic stem cell proliferation regulators, extracts did act to compromise their *production*. Interestingly, a factor produced by the haematopoietic stem cell population - 'stem cell feedback factor' (SCFF) (Lord, 1986), is also reported to regulate the production of the stem cell proliferation regulators. Production of the stem cell proliferation stimulator is 'blocked' in the presence of SCFF thereby indirectly regulating the proportion of the haematopoietic stem cell population in S-phase. A subversion of this existing regulatory mechanism by the leukaemic population could be envisaged.

It is thus proposed that the apparent inhibition of 'normal' haematopoiesis by leukaemic cells may be due, in part, to a 'blocking' of the production of the haematopoietic stem cell proliferation regulators. In this respect, 'normal' haematopoiesis would be unable to respond to the haematopoietic 'stress' induced by the leukaemia, with the subsequent deterioration of peripheral blood quality and clinical presentation of the disease.

Supported by the *Leukaemia Research Fund*

References

Hepburn, M., Doherty, I., Briscoe, C. *et al.* (1987): Transplantation and morphological studies of primary and passaged radiation-induced myeloid leukaemias. *Leuk. Res.* 11, 1001-1009.

Lord, B. (1986): Interactions of regulatory factors in the control of haemopoietic stem cell proliferation. In *Biological Regulation of Cell Proliferation. Serono Symposium* 34, ed. R.Baserga, P.Foa, D.Metcalf & E.Polli, 167-177. New York: Raven Press.

Robinson, S. & Riches, A. (1991): Haematopoietic stem cell proliferation regulators investigated using an *in vitro* assay. *J. Anat.* 174, 153-162.

Negative regulation of human myelopoiesis by a Burkitt lymphoblastoid cell line. A new class of inhibitor(s)?

Elisabeth Sumereau-Dassin, Dina Fardoun, Pascal Peulvé, Gorette Teixeira-Lebrun and Jean-Pierre Vannier

Department of Research on Hematopoietic Stem Cells and Extracellular Matrix Molecules, School of Medicine, BP 97, 76803 St-Etienne-du-Rouvray Cedex, France

In this study, Raji cell line has been treated twice with BM Cyclin then maintained with 10% foetal calf serum (FCS) in Iscove Modified Dulbecco's Medium (IMDM), at 37°C and 5% CO2. The effect of four concentrations of Raji cell free supernatants (SCRaji) on bone marrow progenitors have been tested in semi-solid cultures (Vannier et al., 1980). Aliquots of SCRaji were added on day 0 to $2 \times 10^{(5)}$ cells in IMDM containing 20% CSA-Pl, 20% SVF and 0.9% methylcellulose. Colony-forming units, granulocyte - monocyte (CFU-GM) were scored 12 days later. n = 12. *$\alpha < 0.025$; **$\alpha < 0.01$.

SCRaji concentrations	Number of CFU-CM / $2 \times 10^{(5)}$ cells
0	117 ± 47
0.001	109 ± 35
0.01	45 ± 19*
0.1	4 ± 2**
1	1 ± 1**

On the other hand, results shown that SCRaji acts on immature bone marrow progenitors; when SCRaji were added to semi-solid cultures with a delay of 24 hours, 35.2 ± 35.2 % of CFU-GM were present on cultures which did not differ significantly from controls. SCRaji are not cytotoxic as they did not inhibit U 937 cell growth. They are not prostaglandin-like as $2 \times 10^{(-6)}$ M indomethacin (Kurland et al., 1978) did not modified results. The negative regulation was complete which differs from the inhibitory effect observed with Interferon gamma (mw:17.1 Kd) (Raefsky et al, 1985), Tumor Necrosis Factor alpha (mw: 17 Kd) (Aglietta, 1987) and Interleukine-2 (mw: 15.5 Kd) (Nasr et al, 1989).Using a dialysis membrane with a molecular cutoff of 6,000 - 8,000, preliminary studies have shown that the apparent molecular weight of SCRaji inhibitor(s) was inferior to 6 - 8 Kd. So, the SCRaji inhibition seems to be due to a new class of inhibitor(s). This work was supported by ARC and CDLCC (Seine Maritime, Eure).

REFERENCES

Aglietta, M. (1987): Growth factors and growth inhibitors: their interaction in the regulation of myelopoiesis. In The inhibitors of hematopoiesis. Colloque INSERM vol. 162, eds A. Najman, M. Guigon, N.C. Gorin and J.Y. Mary, pp 11-19.

Kurland, J.I., Broxmeyer, H.E., Pelus, L.M., Brockman, R.S. and Moore, M.A.S. (1978): Role of monocyte-macrophage derived colony stimulating factor and prostaglandin E in the positive and negative feed-back control of myeloid stem cell proliferation. Blood 61: 250-256.

Nasr, S. McKolanis, J., Pais, R. , Findley, H., Hnath, R., Waldrep, K. and Ragab, A.H. (1989): A phase I study of interleukin-2 in children with cancer and evaluation of clinical and immunologic status during therapy. Cancer 64: 783-788.

Raefsky, E.L., Platanias, L.C., Zoumbos, N.C., Young, N.S. (1985): Studies of interferon as a regulator of hematopoietic cell proliferation. J. Immunol. 135, 2507-2512.

Vannier, J.P., Monconduit, M. and Piguet H. (1980): Comparaison entre deux gradients de densité pour séparer les CFC. Biomedicine 33, 236-239.

Excessive production of TGF-β by bone marrow stromal cells in B-cell chronic lymphocytic leukemia (B-CLL) inhibits growth of hematopoietic precursors and IL-6 production by the stromal cells

P. Stryckmans, L. Lagneaux, C. Dorval, D. Bron, E. Bosmans and A. Delforge

Institut Jules-Bordet, Brussels and Eurogenetics, Tessenderlo, Belgium

Previous studies have revealed a decreased production of IL-6 by LPS-stimulated B-CLL mononuclear cells compared to normal subjects. To determine whether marrow-derived stromal cells of B-CLL patients would show a similar decrease in IL-6 production, we have examined the production of CSF's (IL-6, G-CSF and GM-CSF) by the adherent cell population of bone marrow derived from five patients (median age 70 years : 51 - 81) with B-CLL ($CD5^+$ and $CD19^+$), in Rai stage 0, I or II, never treated.

The production of CSF's was measured by ELISA in the supernatant after stimulation of stromal cells for 24 hours by LPS or IL-1 and compared to that of stromal cells from 8 healthy subjects. The mean IL-6 productions by LPS or IL-1 stimulated stromal cells were respectively 123 ± 47 (mean $\pm$ SEM) or 87 ± 20 ng/ml for the controls compared to 36 ± 5 or 40 ± 7 ng/ml for the five B-CLL (p respectively < 0.004 and < 0,04).

No difference was seen for the LPS or IL-1 stimulated GM-CSF production : respectively 119 ± 13 or 100 ± 12 pg/ml for controls and 132 ± 31 or 114 ± 33 pg/ml for B-CLL. No difference was seen either for G-CSF production stimulated by LPS : 5.5 ± 1.3 ng/ml for controls versus 3.8 ± 0.6 ng/ml for B-CLL.

Using the blast-colony forming assay (Bl-CFC) and the classical CFU-GM assay, we have found that : (1) marrow stromal cells of B-CLL were able to support only 25 % of the Bl-CFC growth supported by normal marrow stromal cells, (2) this anomaly was partially corrected (from 25 % to 53 %)by the addition of exogenous IL-6, (3) the colony stimulating activity (CSA) (tested on CFU-GM) of the conditioned medium (CM) of B-CLL stromal cells was also decreased when compared to that of normal controls (32 ± 5 % versus 100 ± 9 %) (p<0.0001), (4) addition of 10 % B-CLL CM produced significant inhibition of normal CFU-GM growth in the presence of human placental conditioned medium as CSA source (44 ± 4 CFU-GM versus 83 CFU-GM, p<0.0001), (5) this inhibition could be abrogated by addition of anti-TGF-β neutralizing antibody (6) this antibody corrected also completely the Bl-CFC growth on B-CLL stromal cells (73 ± 14 versus 38 ± 4 % of normal growth) and the IL-6 production by B-CLL stromal cells (104 ± 23 ng/ml versus 45 ± 11 ng/ml), (7) TGF-β production by LPS-stimulated marrow stromal cells (tested by CCL-64 assay) was significantly increased in B-CLL compared to normal (respectively 53 ± 10 versus 15 ± 4 ng/ml, p < 0.03).

In conclusion, in B-CLL, decreased IL-6 production by stromal cells and inhibiting activity on hematopoietic precursors are attributable to increased production of TGF-β. We hypothesize that TGF-β generated by B-CLL marrow stromal cells could play a crucial pathogenic role in some manifestations of this disease (marrow failure and decreased immunoglobulin production, in advanced stages of the disease).

VI. Hematopoiesis and viruses

VI. *Hématopoïèse et virus*

Mechanisms of aplastic anemia in humans

E. Gluckman*,*** and G. Socié**

*Bone Marrow Transplant Unit and Laboratory of Bone Marrow Biology, University Paris 7; **Unité de Recherche sur la Biologie des Cellules Souches (LIRB-CEA/DSV.DPTE); ***Institut d'Hématologie, Hôpital Saint-Louis, Paris, France

Aplastic anemia (AA) is a rare disease characterized by pancytopenia with decreased bone marrow cellularity. It must be diferentiated from pancytopenia associated with malignant diseases : leukemia, myelodysplasia (MDS), malignant metastasis or myelofibrosis. The distinction between AA and MDS is not always easy; it is based on cytology or pathology of bone marrow samples, cytogenetics and may be on studies of clonality.
The etiology is almost always unknown, but some causes, toxic, viral or constitutional are sometimes recognised. Among the viral causes, post hepatitis anemia has been individualised clinically but there is evidence that none of the known viruses involved in human hepatitis are responsible of the disease (Pol et al., 1990). Constitutional aplastic anemias are rare. Among them, Fanconi anemia has been individualised because it can be diagnosed on the increased spontaneous chromosome breaks with an abnormal sensitivity to clastogenic agents in peripheral blood lymphocytes. At least 4 complementation groups of this disease have been recently described (Levine, 1993), and in one of these complementation group (group C) a gene has been cloned and sequenced (FACC) (Strathdee et al., 1992). Other constitutional disorders such as Dyskeratosis Congenita, Shwachmann syndrome, Seckel syndrome have been described but nothing is known on the mechanism of aplasia.
Several theories have been discussed, but few have been proven because of the lack of good in vitro methodology to study this complex syndrome; they are a microenvironment defect, a stem cell defect at the level of the multipotent progenitor cell or of more differentiated stem cells, a defect of production of a growth factor, the production of an inhibitor by direct cell to cell toxicity (autoimmune mechanism) or by the production of a soluble factor, inhibitor or autoantibody, or an intrisic disorder related to a malignant preleukemic disease. These theories will be discussed in this article.

MICROENVIRONMENT IN APLASTIC ANEMIA

There is currently no evidence that aplastic anemia is related to a direct microenvironmental defect, this is based on the fact that most allogeneic bone marrow transplants are not rejected and it is known that, except macrophages and lymphocytes, the other components of the microenvironment remain of recipient origin after allogeneic bone marrow transplantation (Simmons et al., 1987). On the other hand, coculture studies have shown that, in long term culture, the microenvironment can develop in aplastic marrow and can sustain the growth of progenitors from normal marrow (Marsh et al., 1990, 1991). This is only indirect evidence and there is no proof that an imbalance of growth factors production from the microenvironment is not involved in some cases of aplastic anemia.

STEM CELL DEFECT

A direct quantitative destruction of hematopoietic stem cells has been clearly proven in monozygotic twin transplants where it has been shown that, the infusion of bone marrow cells from an healthy monozygotic twin, in an aplastic patient, without any preparative treatment, can lead in half of the cases to a complete hematological reconstitution (Pei Lu, 1981). In the other half, the reconstitution can be obtained only after a myelo-immunoablative treatment. Thus, the twin transplant results demonstrates the heterogeneity of the mechanisms involved in the pathogenesis of aplastic anemia. The first situation suggests a stem cell defect, while the fact that the other 50% of patients require pretransplant immunosuppression to achieve durable hematologic reconstitution indicates that an immune mechanism contributes to the disease pathology.
Long term bone marrow cultures and limiting dilution analysis have shown, in the majority of the cases, that there was a profund decrease of multipotent clonogenic cells in aplastic patients (Marsh et al., 1990, 1991), even when there was a response to treatments, showing that despite an improvement of the peripheral blood cell counts, the hematopoietic stem cell pool remains profundly depressed (De Planque et al., 1989). It is not known if there are normal residual stem cells able of expansion and differenciation which can regenerate entirely the stem cell pool.

GROWTH FACTOR PRODUCTION OR RECEPTORS DEFECT

It is well known that the production of growth factors is increased in response to pancytopenia, The first growth factors described in the past were isolated from the urine of aplastic patients. Till now, the most potent stimulating factor for megakariocytopoiesis is isolated from aplastic urines and has not yet been cloned. For this reason a growth factor defect seemed unlikely until the Steel and W/Wv mice defects were elucidated. It was shown in these congenitally anemic mice that there was a mutation of *kit* molecule (W/Wv mice) or of its ligand steel factor (Sl/Sl mice) which explained the malformations and the hematological abnormalities : Steel mice could not sustain normal hematopoiesis even after bone marrow

transplantation because of the defect of the receptor while W/Wv mice did not produce c.*kit* molecule and were able to have a complete hematological reconstitution after hematopoietic transplantation from normal congeneic mice (rev. in Witte, 1990). As c *kit* and c.*kit* ligand (KL) have been cloned in human , it has been easy to look for a human analog of steel mice defect. There is currently no evidence of a major mutation of c-kit or KL in human aplastic anemia (Stark et al., 1993). Other mutations of other growth factors or hematopoietic regulators have been studied but these studies are stil very incomplete and it is very difficult to rule out the possibility of a dysregulation of an hematopoietic regulator. In Fanconi anemia, it seems that there is an imbalance of TNF alpha and IL-6 production. It has also been shown that the incubation of Fanconi anemia cells with IL-6 corrects partially the chromosomal fragility (Rosselli et al., 1993). By studying the IL-6 m.RNA, obtained from long term bone marrow culture in aplastic anemia patients , it was shown that the IL-6 m.RNA production was normal except in some Fanconi anemia patients were the transcripts of IL6 and to a lesser degree of GM-CSF was reduced (Stark et al., 1993). Recently, serum level (Wodnar-Filipowicz et al., 1993), and KL effects on the in vitro growth of bone marrow cells from aplastic anemia patients have been studied (Alter et al., 1992), (Wodnar-Filipowicz et al., 1992). Both studies suggest that KL, either alone or in combination with other factors, may be of potential value in treatment of aplastic anemia. In cultures of progenitors with various growth factors, an increase of colony growth has been observed in some cases but never as in normal marrow.
Of the others early-acting cytokines, IL-1 has been extensively studied. Monocyte IL-1 production in vitro has been shown to be decreased as measured by bioassay (Childs et al., 1991) and immunoblotting (Nakao et al., 1989).The implication of IL-1 hypoproduction for the pathogenesis of aplastic anemia remains to be elucidated. Demonstration of differentiation disorders in the monocyte macrophage lineage in AA patients (Andreesen et al., 1989) and the failure of response to exogenous IL-1 suggest that the IL-1 deficiency is a secondary abnormality (Laver et al., 1988). Finally, the expression of stem cell inhibitor (MIP 1 alpha) gene in patients with bone marrow failure was shown to be dysregulated, a finding that may possibly contribute to disease progression (Maciejewski et al., 1992).
Attempts of treatment of aplastic anemia with growth factors gives variable results; G or GM-CSF often increases granulocyte production but it has no effect on the other lineages and usually aplasia relapses if the treatment is discontinued. Preliminary phase I/II studies with IL-1, IL-3 or IL-6 have been perofrmed. Number of traeted patients are still too small to allow definitive conclusions on their efficacies (rev in, Gillio & Gabrilove, 1993 and Smith, 1990).

AUTOIMMUNITY.

The autoimmune origin of aplastic anemia has been suspected after the observation that an immunosuppressive treatment can improve hematopoiesis in aplastic patients (rev in Bessen., 1991). There are cases of autologous recovery in patients receiving allogeneic bone marrow transplantation after immunosuppressive conditioning (Weitzel et al., 1988). The second piece of evidence came from the observation that treatment with horse or rabbit anti lymphocyte globulin can cure patients with severe aplastic anemia (rev in Nissen., 1991). More recently, treatment with cyclosporin-A has been shown to have the same beneficial effect, either alone or in association with antilymphocyte globulin, with a complete or a partial response rate of 50% (Gluckman et al., 1992)(Frickhofen et al., 1992).

The mechanism of action of these agents on aplastic anemia is not known, either they act directly on a population of lymphocytes which are suppressive for hematopoietic stem cells or they act by inhibiting the production of an inhibitory factor of hematopoiesis such as interféron gamma, TGF-beta, TNF- alpha or MIP-1 alpha (Rameshwar & Gascon, 1992, Bonnefoy-Berard et al., 1992)...

In vitro, there is data indicating that T cells and their products are involved in the suppression of hematopoietic stem cells (Nissen et al., 1980). Although some studies have noted increased numbers of circulating CD8+DR+ and CD8+CD25+ T cells (Platanias et al., 1987), neither abnormality in the absolute number of circulating T cells nor T cell subsets have been found consistently in aplastic anemia (Torok-Storb et al., 1985)(Zoumbos et al., 1985). Three lymphokines, Interferon-gamma, tumor necrosis factor-alpha and interleukin-2 have been found to be overproduced by AA T cells (Zoumbos et al., 1985). Although , only Interferon gamma has been identified as a mediator of hematopoietic suppression in AA patients, it is unclear if it can induce directly bone marrow failure. More recent studies, suggests that Interferon gamma acts synergiotically with either lymphotoxins produced by T lymphocytes or TNF alpha produced by both T lymphocytes or natural killer cells to suppress day 14 CFU-GM (Broxmeyer et al., 1988).

CLONAL PRELEUKEMIC DISORDER.

Clinical observations on patients treated with antilymphocyte globulin (ALG) show that, in fact, despite a good initial hematological response, these patients never recover a normal hematopoietic stem cell function (Tichelli et al., 1988, 1992) (De Planque., 1988, 1989). It has been shown that at least 30% of paients treated with ALG develop Paroxysmal nocturnal hemoglobinuria. This disease that is related to a clonal abnormality of the PIG-A gene product thats lead to a decreased expression of the Glycosyl- Phosphatidylinositol linked anchor molecules (Takeda et al., 1993). In Fanconi anemia, the risk of developing leukemia or myelodysplasia is increased compared to the normal population with a frequency of 10 to 20% in patients who do not die of aplasia (Auerbach et al., 1991). In the European bone Marrow Transplantation registry, the actuarial risk of developing PNH in 223 patients with severe idiopathic aplastic anemia surviving more than 2 years after treatment with ALG was 20% (De Planque et al., 1989). More recently, the risk of leukemia or myelodysplasia was reevaluated in 860 patients manly treated with ALG. It was shown that the standardized incidence ratio of these complicaltions, as compared to the general population, was 85 (p<0.001) (Socié et al., 1992).

In vitro, cytological findings are very difficult to analyse and, very often, it is impossible to differentiate hypoplasia related to aplasia or myelodysplasia (Tichelli et al., 1992). Cytogenetic analysis of the bone marrow can show clonal abnormalities, in 4 to 8% of otherwise typical idiopathic aplastic anemia. Most often the 7q- abnormality is observed at diagnosis (Appelbaum et al., 1987). The incidence and the frequency of such cytogenetic abnormalities are increased in patients with Fanconi anemia (Auerbach., 1991). The significance of clonal cytogenetic abnormalities is not well known. It has been observed, in Fanconi anemia, that these abnormalities can persist for long periods of time without evidence of leukemic transformation. More recently, clonality studies have been performed using Southern blot analysis of X-linked RFLP.(Van Kamp et al., 1991) (Josten et al., 1991). Results have been contradictory because of the difficulty of interpreting the results in the absence of comparisons of the hematopoietic cells profile with that of other somatic tissues of the same individual and from normal control population.

In conclusion, these studies show the complexity of the mechanisms of bone marrow suppression in human aplastic anemia. An unique mechanism has been ruled out. The hypothesis that the reduction of the stem cell pool would give a selective advantage to abnormal cells is one of these hypothesis. However, after autologous reconstitution following rejection of allogeneic bone marrow transplantation in patients with aplastic anemia, the incidence of leukemia is very low showing the absence of growth advantage of clonal (leukemic ?) cells in the period of regeneration of aplasia. It also remains clear that apalstic anemia cannot be explained any more by the hypothesis that it is related to a simple quantitative reduction of the stem cell pool by an unknown damage viral or toxic. More experimental work as well as development of animal models such as the SCID HU mice or the in utero stem cell transplant model in sheep might help in the understanding of this severe disease.

References.

Alter B.P. et al. (1992) : Effect of stem cell factor on in vitro erythropoiesis in patients with bone marrow failure syndromes. *Blood.* 80, 3000-3008.

Andreesen R. et al. (1989) : Defective monocyte-to-macrophage maturation in patients with aplastic anemia. *Blood.* 74, 2150-2156.

Appelbaum F.R. et al. (1987) : Clonal cytogenetic abnormalities in patients with otherwise typical aplastic anemia. *Exp Hematol.* 15, 1134-1139.

Auerbach A. et al. (1991) : Leukemia and preleukemia in Fanconi anemia patients: A review of the literature and report of the International Fanconi Anemia Registry. *Cancer. Genet. Cytogenet.* 51, 1-12.

Bonnefoy-Berard N. et al. (1992) : Antiproliferative effect of antilymphocyte globulins on B cells and B-cell lines. *Blood.* 79, 2164-2170.

Broxmeyer H. et al. (1988) : Comparative analysis of the influences of human gamma, alpha and beta interferons on human multipotential (CFU-GEMM) erythoid (BFU-E) and granulocytemacrophage (CFU-GM) progenitor cells. *J Immunol.* 131, 1300-1305.

Childs B. et al. (1991) : Hypoproduction of Interleukin 1 Beta by activated monocytes is specific for severe aplastic anemia. *Exp. Hematol.* 19, 525.

De Planque M. et al. (1989) : Long-term follow-up of severe aplastic anaemia patients treated with antithymocyte globulin. *Br J. Haematol.* 73, 121-126.

De Planque. M. et al. (1988) : Evolution of acquired severe aplastic anaemia to myelodysplasia and subsequent leukaemia in adults. *Br. J. Haematol.* 70, 55-62.

Frickhofen N. et al. (1992) : Treatment of aplastic anemia with antilymphocyte globulin and methylprednisone with or without cyclosporine. *N. Engl. J. Med.* 324, 1297-1304.

Gillio A. P. & Gabrilove J. L. (1993) : Cytokine treatment of inherited bone marrow failure syndromes. *Blood.* 81, 1669-1674.

Gluckman E. et al. (1992). Multicenter randomized study comparing cyclosporine-A alone and antithymocyte globulin with prednisone for treatment of severe aplastic anemia. *Blood.* 79, 2540-2546.

Josten K. M. et al. (1991). Acquired aplastic anemia and paroxysmal nocturnal hemoglobinuria : studies on clonality. *Blood.* 78, 3162-3167.

Laver J. et al. (1988). In vitro interferon-gamma production by cultured T-cells in severe aplastic anaemia : correlation with granulomonopoietic inhibition in patients who respond to anti-thymocyte globulin. *Br. J. Haematol.* 69, 545-550.

Levine A. S. (1993). Workshop on molecular, cellular, and clinical aspects of Fanconi anemia.
Exp. Hematol. 21, 703-726.

Maciejewski J. P. et al. (1992). Expression of stem cell inhibitor gene in patients with bone marrow failure.
Exp. Hematol. 20, 1112-1117.

Marsh J. C. W. et al. (1990). The hematopoietic defect in aplastic anemia assessed by long-term marrow culture.
Blood. 76, 1748-1757.

Marsh J. C. W. et al. (1991). In vitro assessment of marrow stem cell and stromal cell function in aplastic anaemia.
Br. J. Haematol. 78, 258-267.

Nakao S. et al. (1989). Decreased interleukin-1 production in aplastic anemia. *Br. J. Haematol.* 80, 106-110.

Nissen C. et al. (1980). Peripheral blood cells from patients with aplastic anaemia in partial remission suppress growth of their own bone marrow precursors in culture.
Br. J. Haematol. 45, 233-243.

Nissen C (1991). The pathophysiology of aplastic anemia.
Seminars Hematol. 28, 313-318.

Pei Lu D (1981). Syngeneic bone marrow transplantation for treatment of aplastic anaemia : report of a case and review of the literature. *Exp. Hematol.* 9, 257-263.

Platanias L. et al. (1987). Lymphocyte phenotype and lymphokines following anti-thymocyte globulin therapy in patients with aplastic anaemia. *Br. J. Haematol.* 66, 437-443.

Pol S. et al. (1990). Is hepatitis C virus involved in hepatitis-associated aplastic anemia ?
Ann. Intern. Med. 113, 435-437

Rameshwar P. & Gascon P. (1992). Release of interleukin-1 and interleukin-6 from human monocytes by antithymocyte globulin : requirement for de novo synthesis. *Blood.* 80, 2531-2538.

Rosselli F. et al. (1993). In vitro and in vivo anomalies in cytokine production in Fanconi anemia: Interleukin 6 and tumor necrosis factor alpha activities. *Exp. Hematol.* 21, 723.

Simmmons P. J. et al. (1987). Host origin of marrow stromal cells following allogeneic bone marrow transplantation.
Nature. 328, 429-432.

Smith D.H. (1990). Use of hematopoietic growth factors for treatment of aplastic anemia. *Am. J. Ped. Hematol. Oncol.* 12, 425-433.

Socié G. et al. (1992). Malignancies occuring after treament for aplastic anemia : A survey on 1680 patients conducted by the European Group for Bone Marrow Transplantation. *Blood.* 80 , 169a.

Stark R. et al. (1993). The expression of cytokine and cytokine receptor genes in long-term bone marrow culture in congenital and acquired bone marrow hypoplasia. *Br. J. Haematol.* 83, 560-566

Strathdee G. A. et al. (1992). Cloning of cDNAs for Fanconi's anaemia by functional complementation. *Nature.* 356, 763-767.

Takeda J. et al. (1993). Deficiency of GPI anchor caused by a somatic mutation of the PIG-A gene in paroxysmal nocturnal hemoglobinuria. *Cell.* 73, 703-711.

Tichelli A. et al. (1988). Late haematological complications in severe aplastic anaemia. *Br. J. Haematol.* 69, 413-418.

Tichelli A. et al. (1992). Morphology in patients with severe aplastic anemia treated with antilymphocyte globulin. *Blood.* 80, 337-345.

Torok-Storb B. et al. (1985). Subsets of patients with aplastic anemia identified by flow microfluorometry. *N. Engl. J. Med.* 312, 1015-1022.

Van Kamp H. et al. (1991). Clonal hematopoiesis in patients with acquired aplastic anemia. *Blood.* 78, 3209-3214.

Weitzel J. et al. (1988). Use of a hypervarible minisatellite DNA probe (33.15) for evaluating engraftment two or more years after bone marrow transplantation for apalstic anaemia. *Br. J. Haematol.* 70, 91-97.

Witte O. N. (1990). Steel locus defines new multipotent groth factor. *Cell.* 63, 5-6.

Wodnar-Filipowicz A. et al. (1992). Stem cell factor stimulates the in vitro growth of bone marrow cells from aplastic anemia patients. *Blood.* 79, 3196-3202.

Wodnar-Filipowicz A. et al. (1993). Levels of soluble stem cell factor in serum of patients with aplastic anemia. *Blood, in press.*

Zoumbos N. C. et al. (1985). Circulated activated suppressor T lymphocytes in aplastic anemia. *N. Engl. J. Med.* 312, 257-265.

Summary

Aplastic anemia is a rare disease characterized by pancytopenia with decreased marrow cellularity. It must be distinguished from pancytopenia associated with malignant diseases. The mechanism is difficult to elucidate. The principal theories are : a microenvironment defect, a stem cell defect at the level of the multipotent progenitor cell or of more differentiated cells, a defect of production of a growth factor, the production of an inhibitor or an intrinsic disorder related to malignant preleukemia disorder.

Résumé

L'aplasie médullaire est une maladie rare caractérisée par une pancytopénie avec diminution du pool des cellules souches hématopoïétiques.

Elle doit être distinguée, bien que ceci ne soit pas toujours facile, de l'insuffisance médullaire liée à une prolifération maligne, car les formes de passage d'une maladie à l'autre sont fréquentes.

Le mécanisme est souvent difficile à élucider. Les principales théories sont :

Un défaut du microenvironnement, un défaut des cellules souches hématopoïétiques à un stade plus ou moins avancé de différentiation, un défaut de production d'un facteur de croissance, la production d'un inhibiteur de l'hématopoïèse ou un désordre intrinsèque lié à une maladie pré leucémique.

The B19 parvovirus receptor: implications for disease pathogenesis

Neal S. Young and Kevin E. Brown

Hematology Branch, National Heart, Lung and Blood Institute, 9000 Rockville Pike, Bethesda, MD 20892, USA

The Parvovirus family is composed of small, single stranded DNA viruses. Parvoviruses cause a variety of different animal diseases, including feline panleukopenia, canine myocarditis, Aleutian mink disease, and fetal wastage in pigs (for recent comprehenisve review see (Young et al., 1993)). B19 parvovirus, the only member of the *Parvoviridae* pathogenic in humans, is the agent responsible for the common childhood rash illness called fifth disease, a polyarthralgia/arthritis syndrome in adults, hydrops fetalis after intrauterine infection, transient aplastic crisis of hemolytic disease, and pure red cell aplasia due to persistent infection in the immunocompromised host. B19 has also been associated in case reports with idiopathic thrombocytopenic purpura, Henoch-Schonlein purpura, agranulocytosis, pancytopenia, as well as post-viral neuropathic syndromes.

B19 parvovirus shares some common structural features with other parvoviruses, including a nonenveloped icosahedral capsid structure, single stranded DNA genome of about 5 kb with terminal iterative repeat sequences required for replication through double stranded DNA intermediates, and an organization that places the coding sequence for a single nonstructural protein gene 5' to overlapping capsid protein genes at the 3' side of the genome. In several important respects, B19 parvovirus differs from the other *Parvoviridae*. At the molecular level, transcription is initiated from a single promoter, and mRNA quantity is regulated by splicing and termination events. A striking feature of B19 biology is the extraordinary tropism of the virus for erythroid precursor cells, and the virus has only been propagated in human bone marrow and peripheral blood, fetal liver, and two leukemic cell lines in the presence of erythropoietin.

The mechanisms responsible for tissue tropism among the parvoviruses vary. For canine and feline parvovirus, which are extremely close relatives, species specificity is determined by virus binding at the cell surface (Parrish, 1990). For the strains of minute virus of mice which preferentially propagate in either fibroblasts or lymphocytes, intracellular events determine specificity, as entry into nonpermissive cells is normal but DNA replication proceeds to single stranded progeny only in permissive cells (Spalholz & Tattersall, 1983).

We recently have determined that B19 parvovirus tropism for erythroid cells requires two conditions: 1) virus binding to a receptor present on erythroid progenitor cells (and a few other cell types) and 2) intranuclear factors necessary for full length transcription. We here summarize the evidence for these conclusions and the implications for disease pathogenesis of these dual constraints.

IDENTIFICATION OF THE HUMAN PARVOVIRUS RECEPTOR

Hemagglutination assay

Hemagglutination was exploited to characterize the cellular receptor. Hemagglutination is a common property of viruses and can be used to measure virus concentration simply, by observation of the physical state of red blood cells in individual microtiter wells containing increasing dilutions of a test sample. For B19 parvovirus, for which culture systems are impractical at generating large quantities of infectious virions, empty capsids have been used as agglutinins. These capsids were produced in an insect cell system by self-assembly after coinfection with recombinant baculoviruses containing the major (VP2) and minor (VP1) structural proteins or infection with baculovirus encoding only the major protein (Kajigaya et al., 1991); both capsid types agglutinated human erythrocytes (Brown & Cohen, 1992).

The relevance of the hemagglutination assay to B19 parvovirus biology was suggested by the observation that agglutination, or more easily inibition of agglutination by membrane preparations, correlated with the known tissue tropism of the virus. Capsid agglutination of human erythrocytes was inhibited by preparations derived from the erythroid fraction of bone marrow (glycophorin bearing cells); the UT-7 cell line, a megakaryocytoblastoid cell line semipermissive for virus propagation; but not by other leukemic cell lines nonpermissive for the virus.

As a first approximation of the biochemical character of the receptor, red cells were treated with chemical agents that can selectively perturb their protein, lipid, or carbohydrate components, or the red cell membranes were subjected to a variety of chemical extraction procedures to fractionate these constituents. Surprisingly, treatment of erythrocytes with trypsin increased hemagglutination, and only periodation, which affects sugar residues of glycosylated molecules, significantly decreased the hemagglutination titer. On fractionation of erythrocyte membranes, protein preparations were without inhibitory activity wherease lipid fractions were inhibitory. Methods that concentrated neutral lipids produced the most potent inhibitors. Mixtures of commercially available purified neutral sphingolipids duplicated this inhibitory activity, and the most potent single molecule was globoside.

Binding assay

Direct binding of virus to globoside was demonstrated using thin layer chromatography. Capsids bound to globoside and also the closely related Forsmann antigen, but no specific binding was observed for ceramide mono-, di-, and trihexoses.

Globoside is the chemical equivalent of erythrocyte P antigen, a member of the P erythrocyte system (Fig. 1). P antigen is a tetrahexose covalently bound to a

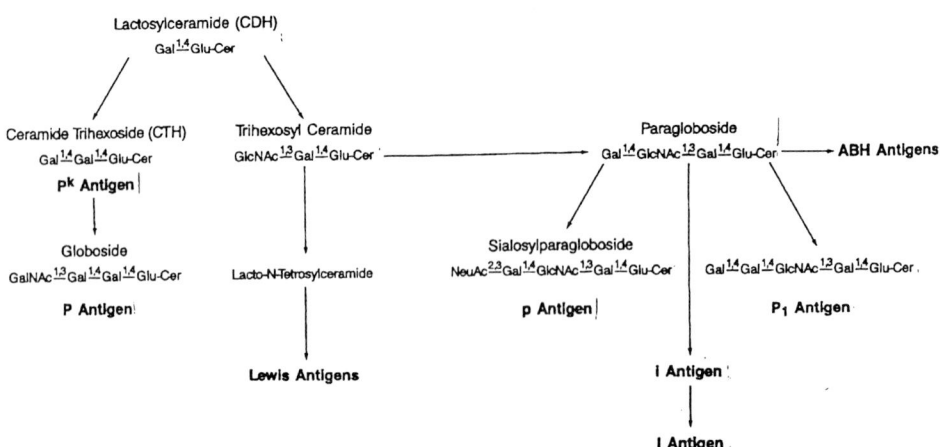

Fig. 1. Synthetic pathways for P antigen system.

ceramide backbone and synthesied from monohexose and trihexose precursors. Expression of P antigen is almost universal, and those rare individuals who fail to express P antigen are presumed to be deficient in the genes for those enzymes required for addition of the appropriate terminal sugars. When examined in the hemagglutination assay, red cells from p and p^k individuals failed to bind capsids, consistent with the absence of P antigen.
Absence of binding and biological activity (see below) for B19 parvovirus of other ceramides indicated that specifity for the virus lay in the carbohydrate sequence rather than lipid structure.

INFECTIVITY STUDIES

To establish the biological relevance of these hemagglutination data to a functional receptor activity, two experiments were performed, both based on the infectivity of B19 parvovirus for marrow cells at the CFU-E or late erythroid progenitor cell stage. In the first, excess of globoside or other similar neutral sphingolipids was added in an effort to compete for cellular binding sites. At 100 µg/plate and appropriate virus concentrations, globoside and Forsmann antigen significantly inhibited virus infection and permitted erythroid colony formation. A dose response relationship of protection relative to globoside addition was also demonstrated, and at the limits of solubility of the lipid component, about 500 µg/plate, almost complete protection of colony formation was achieved. In complementary experiments, monoclonal antibodies to P antigen or to other members of the p antigen system were added to bone marrow cells prior to addition of virus; anti-P but not other antibody preparations protected target cells from the cytopathic effect of B19 parvovirus. These results indicated that globoside operated as a function cellular receptor for B19 parvovirus.

SUSCEPTIBILITY TO PARVOVIRUS INFECTION DEPENDENT ON RECEPTOR EXPRESSION

As a glycolipid, direct transfection of a gene for the parvovirus receptor was not feasible, and because the enzymes responsible for synthesis of globoside have either not been identified or cloned, transfection of the responsible genes also was not possible. However, we took advantage of the importance of the P blood group system in transfusion method to obtain blood and ultimately bone marrow from Tja- individuals, who lack P antigen. Tja- is an important blood bank phenotype, as such individuals are vulnerable to massive hemolysis if transfused with P antigen-positive erythrocytes.

The normal adult population has about a 50% seropositivity rate for antiparvovirus IgG. However, in both retrospective and prospective serologic studies, no Tja- donors showed presence of IgG to B19. Two Tja- individuals donated bone marrow for in vitro infection studies. At concentrations of added parvovirus several orders of magnitude higher than required for ablation of erythroid colony formation in concurrently tested controls, CFU-E-derived colonies developed normally from these p donors. Using immunofluorescence for capsid antigen, in situ hybridization for nonstructural protein DNA and RNA, dot blot and Southern analysis for B19 DNA, there was no evidence of infection by parvovirus of Tja- donor bone marrow.

REPLICATION IS LIMITED BY INTRACELLULAR FACTORS

The transcription map of B19 parvovirus differs from those of other *Parvoviridae*, most importantly in the use of a single left sided promoter. The pattern of RNA species produced in permissive and nonpermissive cells is distinctive: in bone marrow, transcription is full length, and the major RNA species represent capsid protein genes; in nonpermissive cells, the left sided nonstructural protein gene transcript predominates. This difference is the result of a functional block in transcription in nonpermissive cells (Liu et al., 1992). Partial transcription of the genome with production only of the nonstructural protein was predicted to lead to cell death in the absence of virus propagation (Ozawa et al., 1988) and may be the mechanism responsible for thrombocytopenia in transient aplastic crisis (Srivastava et al., 1990). Our preliminary results using monkey cells that bear P antigen suggest that CFU-E-derived colony formation can be nearly completely inhibited, but this striking cytopathic effect is accompanied only by in situ hybridization evidence of partial parvovirus transcription.

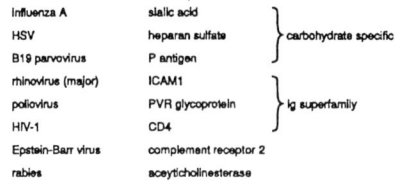

Table

IMPLICATIONS OF P ANTIGEN AS THE PARVOVIRUS CELLULAR RECEPTOR

Blood group antigens as pathogen receptors
P antigen represents the eighth human cellular receptor identified and the first receptor of the Parvovirus family to be characterized (Table). Globoside is the first cellular receptor for a virus of glycolipid structure. However, glycolipids serve as attachment sites for many bacteria, including other members of the P erythrocyte antigen family can serve as attachment sites for certain E. coli strains (Källenius et al., 1980). P antigen is the first distinctive blood group antigen shown to function as a receptor for a virus. Duffy antigen recently has been identified as the receptor for the malaria species Plasmodium ovalum (Hadley & Miller, 1992). For bacterial, protozoal, and now viral infection, genetic absence of the cellular receptor confers protection against infection. From the evolutionary perspective, blood group heterogeneity may be driven by the selective advantage offered by the high degree of resistance some variants confer to those pathogens adapted to the strategy of red cell binding as a part of their life cycle in humans.

Pathogenesis of parvovirus diseases
The distribution of P antigen on human cells offers an explanation for certain aspects of B19 disease (Fig. 2). P on placental cells may serve for intrauterine transmission from the infected mother. P on endothelial cells may help explain the rash and joint symptoms of fifth disease. Forsmann antigen on nasopharyngeal cells that are latently infected may provide an explanation for the initial route of infection.

Knowledge of the receptor also suggests other possible parvovirus illnesses. P is present on megakaryocytes, and thrombocytopenia in association with erythroid aplasia or as a solitary hematologic finding likely is the result of either direct infection of these cells (with cytotoxicity but without viral replication) or by an antibody mechanism in the absence of virus entry. The Donath-Landsteiner antibody of paroxysmal cold hemoglobinuria, a severe immune hemolytic anemia of childhood

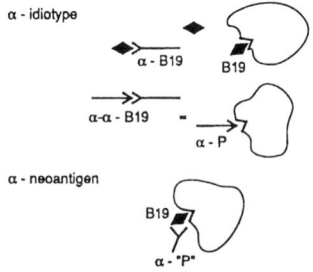

Fig. 2. Possible pathophysiologic mechanisms for paroxysmal cold hemoglobinuria.

described as post-viral, may occur following B19 infection as a result of neoantigen formation after virus binding to P or from anti-idiotype formation (Fig. 3). Fetal or childhood myocarditis could result from direct infection of P antigen-bearing myocytes.

Practical uses
Binding of virus to the tetrahexose fixed to a solid substrate should prove useful in quantitating viral particles and especially in purification of parvoviruses or removal of this agent from blood products. In animals, competition between synthetic carbohydrates and receptor may be helpful in treatment of diseased animals. Finally, parvoviruses can be used to transduce genetic information for gene therapy. Knowledge of their cellular receptor should be useful in selection of target cells or modification of the target cell surface, chemically or enzymatically, for virus entry.

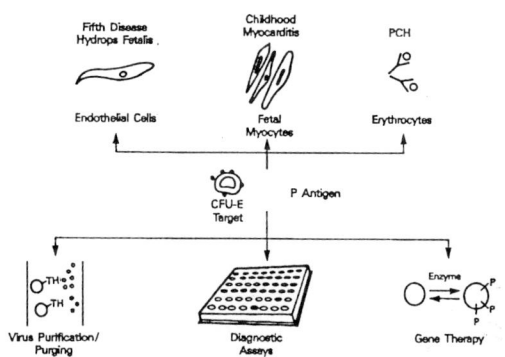

Fig.3. Implications of P antigen as the B19 receptor

REFERENCES

Brown, K.E. & Cohen, B.J. (1992): Haemagglutination by parvovirus B19. *J. Gen. Virol.* 73, 2147-2149.
Hadley, T.J. & Miller, L.H. (1992): Red cell antigens as receptors for malaria parasites. In *Protein Blood Group Antigens of the Human Red Cell Structure, Function, and Clinical Significance*, eds. P.C. Agre& J.P. Cartron, pp. 228-245. Baltimore, MD: Johns Hopkins Univ. Press.
Kajigaya, S., Fujii, H., Field, A.M., Rosenfeld, S., Anderson, L.J., Shimada, T. & Young, N.S. (1991): Self-assembled B19 parvovirus capsids, produced in a baculovirus system, are antigenically and immunogenically similar to native virions. *Proc. Natl. Acad. Sci. USA* 88, 4646-4650.
Källenius, G., Möllby, R., Svenson, S.B., Winberg, J., Lundblad, A., Svensson, S. & Cedergren, B. (1980): The Pk antigen as receptor for the haemagglutinin of pyelonephritic escherichia coli. *FEMS Microbiology Letters* 7, 297-302.
Liu, J., Green, S., Shimada, T. & Young, N.S. (1992): A block in full-length transcript maturation in cells nonpermissive for B19 parvovirus. *J. Virol.* 66, 4686-4692.
Ozawa, K., Ayub, J., Kajigaya, S., Shimada, T. & Young, N.S. (1988): The gene encoding the nonstructural protein of B19 (human) parvovirus may be lethal in transfected cells. *J. Virol.* 62, 2884-2889.
Parrish, C.R. (1990): Emergence, natural history, and variation of canine, mink, and feline parvoviruses. *Adv. Virus Res.* 38, 403-450.
Spalholz, B.A. & Tattersall, P. (1983): Interaction of minute virus of mice with differentiated cells: Strain-dependent target cell specificity is mediated by intracellular factors. *J. Virol.* 46, 937-943.
Srivastava, A., Bruno, E., Briddell, R., Cooper, R., Srivastava, C., van Besien, K. & Hoffman, R. (1990): Parvovirus B19-induced perturbation of human megakaryocytopoiesis in vitro. *Blood* 76, 1997-2004.
Young, N.S., Yoshida, M. & Sugden, W. (1993): Viral pathogenesis of hematologic disorders. In *Molecular Basis of Blood Diseases*, eds. G. Stamatoyanaopoulos, A.W. Nienhuis, P. Leder, P. Mjerus& H. Varmus, Philadelphia: W. B. Saunders.

Summary

B19 parvovirus is tropic for human erythroid progenitor cells. Acute infection causes transient aplastic crisis and fifth disease. Chronic infection results in pure red cell aplasia. In utero infection can produce hydrops fetalis or congenital red cell aplasia. We have identified the viral receptor on cells as erythrocyte P antigen, or globoside. In vitro, infection of parvovirus can be blocked by competition with excess soluble globoside or by antibody to P antigen. Rare humans of the p phenotype cannot be infected by parvovirus, and their bone marrow cells are susceptible to its cytotoxic effects. First identification of cell receptor for parvovirus has important implications for the pathophysiology of human parvovirus diseases.

Résumé

Le parvovirus B19 a un tropisme pour les progéniteurs érythroides humains. L'infection aigue entraine une crise d'aplasie transitoire et la 5ème maladie. L'infection chronique résulte en une aplasie rouge pure. L'infection in utero peut produire une hydrocéphalie ou une aplasie rouge congénitale. Nous avons identifié le récepteur du virus sur les cellules comme étant l'antigène érythrocytaire P ou globoside. In vitro, l'infection par le parvovirus peut être bloquée par compétition avec un excès de globoside soluble ou d'anticorps anti-antigène P. Les rares sujets présentant le phénotype p ne peuvent être infectés par le parvovirus et leurs cellules médullaires sont sensibles aux effets cytotoxiques. La première identification du récepteur cellulaire du parvovirus a d'importantes implications pour la physiopathologie des maladies humaines à parvovirus.

In vivo haemopoietic effects of the parvovirus MVMi in newborn mice

Jose C. Segovia*, Jose M. Almendral*,** and Juan A. Bueren*

*Departmento de Biología Molecular y Celular, CIEMAT, Av. Complutense 22, 28040 Madrid, Spain;
**Centro de Biología Molecular, Universidad Autónoma de Madrid, Cantoblanco, 28049 Madrid, Spain

Parvovirus are small, single-stranded DNA viruses that require for their replication function(s) expressed in proliferating cells at specific differentiated states (Cotmore & Tattersall, 1987). In a previous study, we shown that the immunosupressive strain of the parvovirus minute virus of mice (MVMi) is highly cytotoxic *in vitro* for the committed (CFU-GM, BFU-E) and pluripotent (CFU-S_{12d}) haemopoietic progenitors of the mouse (Segovia et al., 1991; Bueren et al., 1991). MVMi inhibits the colony forming ability of these precursors and yields infectious virus in myeloid cultures. In the present work, we have investigated the potential haemopoietic failures mediated by *in vivo* naturally inoculated mice.

Newborn Balb/c mice (less than 20 hours) were infected intranasally with 10^6 pfu of purified MVMi. At different days post-infection (dpi) the animals were sacrificed and viral DNA replication and production of infectious virus was monitored in the bone marrow and the spleen. In addition, the number of femoral and splenic CFU-GM and BFU-E precursors, together with the titer of antibodies against MVMi in serum was determined.

We have observed an increase of MVMi replication and infectious virus production since the 2nd dpi in all organs studied, with a peak at the 6 dpi. Afterwards, the viral titers in the different organs decreased proportionally with the raise of anti-MVM antibodies in the serum of the animals (fig. 1).

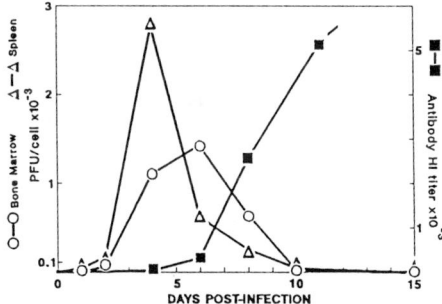

Fig. 1. Virological events in newborn Balb/c mice intranasally inoculated by MVMi.

A decrease of CFU-GM and BFU-E progenitors was observed 8-12 dpi with respect to age-matched noninfected animals, that was more evident in animals that manifested pathological affectation. Animals that survived the infection (10-15%) showed no signs of any haemopoietic dysfunction at 20 dpi.

These results reveal that the parvovirus MVMi is capable of inducing certain *in vivo* haemopoietic crisis manifested by a reduction in the femoral and splenic levels of commited progenitors, occurring with some delay in respect to the replication peak of virus in the organs. The implication of the haemopoietic failure in the death of the animals as well as the direct or indirect mechanisms underlying the progenitors depletion are currently under study.

REFERENCES

Bueren, J.A., Segovia, J.C. and Almendral, J.M. (1991): Cytotoxic infection of hematopoietic stem and committed progenitor cells by the parvovirus Minute virus of Mice. *Ann. New York Acad. Sci.* 628, 262-272.

Cotmore, S.F. and Tattersall, P. (1987): The autonomously replicating parvoviruses of vertebrates. *Adv. Virus Res.* 33, 91-174.

Segovia, J.C., Real, A., Bueren, J.A. and Almendral, J.M. (1991): *In vitro* myelosuppressive effects of the parvovirus Minute Virus of Mice (MVMi) on hematopoietic stem and committed progenitor cells. *Blood* 77(5), 980-988.

Cytomegalovirus mediated cytopenias

Beverly Torok-Storb, Bettina Fries and David Myerson

Fred Hutchinson Cancer Research Center and University of Washington School of Medicine, Seattle, Washington, USA

SUMMARY

The findings presented in this report together with previous observations suggest that different strains of CMV are capable of infecting different target cells and that variations among patients in the specific cell types infected by CMV may contribute significantly to the clinical outcome. These observations indicate that defining exactly how CMV interferes with marrow function requires: 1) determining the specific cell types infected in cases of myelosuppression, 2) characterizing the CMV strain responsible for the infection, and 3) determining how it interferes with cell function. Such information would be imperative for designing strategies to overcome myelosuppression and may eventually shed some light on tropisms and pathogenicity that may be genetically encoded among various CMV strains.

INTRODUCTION

Prior to the use of prophylactic ganciclovir, cytomegalovirus (CMV) infection was a major cause of morbidity and mortality in bone marrow transplant recipients (Goodrich et al., 1993). Approximately 40-70 percent of patients who were either CMV seropositive prior to transplant or who received marrow from a CMV seropositive donor developed CMV infection (Meyers et al., 1986). The most common and severe manifestation of this infection was interstitial pneumonia (Emmanuel et al., 1988; Reed et al., 1988). Less frequent was disseminated disease with the detection of virus in other organs. The actual extent to which CMV contributed to poor marrow function remains controversial.

Previous studies have associated CMV infection with delayed platelet engraftment (Verdonck et al., 1985), phenotypically abnormal peripheral blood leukocytes (Apperley & Goldman, 1988), and graft failure (Bolger et al., 1986). Specifically, Apperley and Goldman reported on a series of 48 allogeneic bone marrow transplant recipients in which 11 of 21 infected patients had thrombocytopenia, 8 demonstrated slow platelet engraftment, 6 had neutropenia, and 4 had graft failure. Other studies have supported the observation that poor marrow function can be associated with CMV (Mutter et al., 1988). However, given the extremely complicated posttransplant course of allogeneic patients, it has been difficult to establish a cause and effect relationship between CMV disease and marrow function. Other complications such as graft-versus-host disease and the treatment thereof, which occur during the same time course, as well as antiviral therapy, can contribute to poor marrow function. The fact that the association of poor marrow function with CMV is more common in allogeneic as opposed to syngeneic or autologous transplant recipients also suggests that other factors besides CMV itself may be responsible for the majority of cases of marrow suppression (Wingard et al., 1988). Reusser et al. (1990) reported a retrospective analysis of 159 patients receiving autologous transplants in which the observation of delayed neutrophil or platelet engraftment could not be associated with CMV infection. A separate study, however, did suggest that platelet recovery was significantly slower in CMV-infected recipients compared to noninfected patients (Verdonck, 1991).

CMV AND MARROW FUNCTION

Given that CMV does contribute in some cases to poor marrow function, several mechanisms acting alone or in combination could mediate this marrow suppression. Hematopoietic progenitor cells or accessory cells such as monocytes and T lymphocytes hypothesized to provide growth factors may be targets for infection (Mackintosh et al., 1993; Rakusan et al., 1989; Sing & Ruscetti, 1990; Rice et al., 1984). Infection of these cells could either directly alter their function or result in cell death. Alternatively, infection of these cells could target them for destruction by the immune response. Since the maintenance of primitive hematopoietic progenitor cells is crucially dependent on their interaction with stromal cells of the marrow microenvironment, the perturbation of stromal cell function by CMV could also result in myelosuppression (Apperley et al., 1989; Simmons et al., 1990). This perturbation could be a direct result of the infection of stromal cells, resulting in altered cell function, or infected cells could be targets for an immune response (Busch et al., 1991).

In previous studies we have shown that, although all CMV isolates inoculated into long-term marrow cultures (LTMC) manage to suppress myelopoiesis, the mechanism of suppression was different depending on the target cell infected. In these studies, various isolates of CMV were added to the LTMC established from normal CMV seronegative donors (Simmons et al., 1990). In this system, myelosuppression was manifested by a dramatic reduction in the number of myeloid cells produced and released in the culture supernatant. Analysis of these LTMC indicated that this suppression could, in some cases, be attributed to direct infection of myeloid cells. This observation was supported, first, by detection of viral antigen in myeloid cells present in the cultures, and second, by the finding that a proportion of clinical isolates inhibited the growth of isolated progenitor cells directly. In other cases, myelosuppression observed in LTMC appeared to be related to the infection of stromal cells. In some cases where stromal cells were shown to be infected as evidenced by the presence of viral antigens in these cells, there was a reduction in the amount of detectable mRNA for G-CSF. The 20 isolates studied in our original series yielded four distinct patterns of reactivity when studied in LTMC, as shown in Table 1.

Our study suggested that direct infection of either myeloid progenitors or critical components of the microenvironment resulted in myelosuppression either by a direct lysis of the target cells or by interfering with their function. Other reports suggest that the inhibition of hematopoiesis in virus-inoculated cultures requires the presence of monocytes and/or lymphocytes, presumably due to the response of these effector cells to the infected targets. In studies by Mackintosh et al. (1993), this immune response was presumably cell mediated since no detectable soluble factors capable of inhibiting colony growth were found. Other investigators argue that soluble factors that are elaborated in vivo as part of the infectious process, particularly gamma interferon and tumor necrosis factor, which have inhibitory effects on progenitor cells in vitro, may in fact

Table 1. Four patterns of activity of 20 CMV strains in long-term marrow cultures.

	1	2	3	4
Suppression of granulocyte production	Yes	Yes	Yes	Yes
Direct infection of myeloid progenitors	No	Yes	No	No
Direct infection of fibroblasts	Yes	No	Yes	Yes
Direct infection of endothelial cells	No	No	Yes	No
Direct infection of adipocytes	Yes	No	Yes	Yes
Direct infection of macrophages	No	Yes	Yes	No
Changes in detectable cytokine mRNA	Yes	No	Yes	Yes

be responsible for the marrow suppression observed in CMV-infected patients (Duncombe et al., 1990). However, the reported effects of CMV infection on cytokine production are inconsistent (Rodgers et al., 1985; Kapasi et al., 1988; Albrecht et al., 1989; Moses & Garnett, 1990). This may be due in part to the fact that various isolates of CMV used in these experiments could have different capabilities of infecting cells and altering cellular gene expression.

Given the panoply of cells that comprise the hematopoietic environment, all of which may be targets of various strains of CMV, there must be multiple ways in which viral infection can interfere with hematopoiesis. By dissecting the hematopoietic microenvironment and determining what function the various cellular components provide, it may become possible to define how infection of a specific cell type results in myelosuppression. However, the observations made in the LTMC system must be correlated with in vivo observations by demonstrating the presence of virus in comparable cell populations isolated from infected patients.

Historically, isolating sufficient numbers of distinct cell types required for this type of evaluation has proven to be a formidable challenge. However, with the application of polymerase chain reaction (PCR) to identify the presence of viral genome in small numbers of cells, it has become possible to address this issue. Our recent efforts in this regard have enabled us to detect the presence of CMV in various subpopulations of marrow obtained from healthy seropositive marrow donors (Torok-Storb et al., 1992). This observation suggests, first, that our initial in vitro studies which indicated various target cells for different clinical isolates, may actually have in vivo relevance. Second, it suggests that it may be possible in some cases to identify CMV-containing cells in donor marrow and, depending on their role in engraftment, to selectively eliminate them. This is of particular importance since the incidence of CMV is significantly higher in seronegative patients receiving marrow from seropositive donors than in those receiving marrow from seronegative donors, although in both cases patients receive screened blood products (Meyers et al., 1986).

CMV AND LYMPHOCYTOPENIA

Studies designed to identify CMV in distinct cell types isolated from patients with neutropenia are more challenging, primarily because of the limited number of cells available for study. In an initial pilot study of infected patients, peripheral blood was obtained at the time of first causes (n=114), the average slopes were positive, being 0.020±0.054 and 0.012±0.074, respectively. In contrast, the average slope was negative, -0.016±0.077 among those who died of disseminated CMV (n=47). The last group was statistically significantly different from the other two groups ($p < .001$). A multivariate analysis of other risk factors indicated that the difference in lymphocyte counts could not be attributed to GVHD, steroids, or transplant type (Fries et al., 1992). Whether patients who had declining lymphocyte counts and subsequently died of CMV also had CMV in their lymphocytes is not known, nor is it possible to make this determination retrospectively. However, since CMV isolates from all these patients have been stored, it will be possible to characterize them should any informative criteria be forthcoming.

ACKNOWLEDGMENTS

This work was supported in part by grants CA18029, CA18221, DK34431, and HL36444 from the National Institutes of Health, DHHS. Dr. Fries was supported by Deutsche Forschungs Gesellschaft (DFG). The manuscript was prepared by Bonnie Larson.

REFERENCES

Albrecht, T., Boldogh, I., Fons, M., Lee, C.H., AbuBakar, S., Russell, J.M. & Au, W.W. (1989): Cell-activation responses to cytomegalovirus infection relationship to the phasing of CMV replication and to the induction of cellular damage. Sub-Cellular Biochemistry 15, 157.

Apperley, J.F. & Goldman, J.M. (1988): Cytomegalovirus: biology, clinical features and methods for diagnosis. Bone Marrow Transplantation 3, 253.

Apperley, J.F., Dowding, C., Hibbin, J., Buiter, J., Matutes, E., Sissons, P.J., Gordon, M. & Goldman, J.M. (1989): The effect of cytomegalovirus on hemopoiesis: in vitro evidence for selective infection of marrow stromal cells. Exp. Hematol. 17, 38.

Bolger, G.B., Sullivan, K.M., Storb, R., Witherspoon, R.P., Weiden, P.L., Stewart, P., Sanders, J., Meyers, J.D., Martin, P.J., Doney, K.C., Deeg, H.J., Clift, R.A., Buckner, C.D., Appelbaum, F.R. & Thomas, E.D. (1986): Second marrow infusion for poor graft function after allogeneic marrow transplantation. Bone Marrow Transplantation 1, 21.

Busch, F.W., Mutter, W., Koszinowski, U.H. & Reddehase, M.J. (1991): Rescue of myeloid lineage-committed preprogenitor cells from cytomegalovirus-infected bone marrow stroma. J. Virol. 65, 981.

Duncombe, A.S., Meager, A., Prentice, H.G., Grundy, J.E., Heslop, H.E., Hoffbrand, A.V. & Brenner, M.K. (1990): Gamma-interferon and tumor necrosis factor production after bone marrow transplantation is augmented by exposure to marrow fibroblasts infected with cytomegalovirus. Blood 76, 1046.

Emmanuel, D., Cunningham, I., Jules-Elysee, K., Brochstein, J.A., Kernan, N.A., Laver, J., Stover, D., White, D.A., Fels, A., Polsky, B., Castro-Malaspina, H., Peppard, J.R., Bartus, P., Hammerling, U. & O'Reilly, R.J. (1988): Cytomegalovirus pneumonia after bone marrow transplantation successfully treated with the combination of ganciclovir and high-dose intravenous immune globulin. Ann. Intern. Med. 109, 777.

Fries, B.C., Khaira, D., Pepe, M., Myers, J. & Torok-Storb, B. (1992): Occurrence of lymphocytopenia in cytomegalovirus (CMV) infected patients predicts fatal outcome. Exp. Hematol. 20, 830a.

Goodrich, J.M., Bowden, R.A., Fisher, L., Keller, C., Schoch, G. & Meyers, J.D. Prevention of cytomegalovirus disease after allogeneic marrow transplant by ganciclovir prophylaxis. Ann. Intern. Med. (in press)

Jiwa, N.M., van Gemert, G.W., Raap, A.K., van de Rijke, F.M., Mulder, A., Lens, P.F., Salimans, M.M., Zwaan, F.E., van Dorp, W. & van der Ploeg, M. (1989): Rapid detection of human cytomegalovirus DNA in peripheral blood leukocytes of viremic transplant recipients by the polymerase chain reaction. Transplantation 48, 72.

Kapasi, K. & Rice, G.P.A. (1988): Cytomegalovirus infection of peripheral blood mononuclear cells, effects on interleukin-1 and -2 production and responsiveness. Virology 62, 3603.

Khaira, D., Myerson, D., Lingenfelter, P.A., Simmons, P. & Torok-Storb B. (1991): CMV DNA in mononuclear cells and concurrent lymphopenia is associated with fatal CMV infection following allogeneic marrow transplantation. Blood 76(Suppl), 548a.

Mackintosh, F.R., Adlish, J., Hall, S.W., St. Jeor, S., Smith, E., Tavassoli, M. & Zanjani, E.D. (1993): Suppression of normal human hematopoiesis by cytomegalovirus in vitro. Exp. Hematol. 21, 243.

Meyers, J.D., Flournoy, N. & Thomas, E.D. (1986): Risk factors for cytomegalovirus infection after human marrow transplantation. J. Infect. Dis. 153, 478.

Moses, A.V. & Garnett, H.M. (1990): The effect of human cytomegalovirus on the production and biologic action of interleukin-1. J. Infect. Dis. 162, 381.

Mutter, W., Reddehase, M.J., Busch, F.W., Bühring, H-J. & Koszinowski, U.H. (1988): Failure in generating hemopoietic stem cells is the primary cause of death from cytomegalovirus disease in the immunocompromised host. J. Exp. Med. 167, 1645.

Quinnan, G.V. Jr, Kirmani, N., Rook, A.H., Manischewitz, J.F., Jackson, L., Moreschi, G., Santos, G.W., Saral, R. & Burns, W.H. (1982): Cytotoxic T cells in cytomegalovirus infection. N. Engl. J. Med. 307:7.

Rakusan, T.A. & Juneja, H.S. (1989): Inhibition of hemopoietic colony formation by human cytomegalovirus in vitro. J. Infect. Dis. 159, 127.

Reed, E.C., Bowden, R.A., Dandliker, P.S., Lilleby, K.E. & Meyers, J.D. (1988): Treatment of cytomegalovirus pneumonia with ganciclovir and intravenous cytomegalovirus immunoglobulin in patients with bone marrow transplants. Ann. Intern. Med. 109, 783.

Reusser, P., Fisher, L.D., Buckner, C.D., Thomas, E.D. & Meyers, J.D. (1990): Cytomegalovirus infection after autologous bone marrow transplantation: occurrence of cytomegalovirus disease and effect on engraftment. Blood 75, 1888.

Reusser, P., Riddell, S.R., Meyers, J.D. & Greenberg, P.D. (1991): Cytotoxic T-lymphocyte response to cytomegalovirus after human allogeneic bone marrow transplantation: Pattern of recovery and correlation with cytomegalovirus infection and disease. Blood 78, 1373.

Rice, G.P.A., Schrier, R.D. & Oldstone, M.B.A. (1984): Cytomegalovirus infects human lymphocytes and monocytes: virus expression is restricted to immediate-early gene products. Proc. Natl. Acad. Sci. USA 81, 6134.

Rodgers, B.C., Scott, D.M., Mundin, J. & Sissons, J.G.P. (1985): Monocyte-derived inhibitor of interleukin 1 induced by human cytomegalovirus. J. Virol. 55, 527.

Simmons, P., Kaushansky, K. & Torok-Storb, B. (1990): Mechanisms of a cytomegalovirus-mediated myelosuppression: Perturbation of stromal cell function versus direct infection of myeloid cells. Proc. Natl. Acad. Sci. USA 87, 1386.

Sing, G.K. & Ruscetti, F.W. (1990): Preferential suppression of myelopoiesis in normal human bone marrow cells after in vitro challenge with human cytomegalovirus. Blood 75, 1965.

Stanier, P., Taylor, D.L., Kitchen, A.D., Wales, N., Tryhorn, Y. & Tyms, A.S. (1989): Persistence of cytomegalovirus in mononuclear cells in peripheral blood from blood donors. Br. Med. J. 299, 897.

Torok-Storb, B., Simmons, P., Khaira, D., Stachel, D. & Myerson, D. (1992): Cytomegalovirus and marrow function. Ann. Hematol. 64, A128.

Verdonck, L.F., de Gast, G.C., van Heugten, H.G., Nieuwenhuis, H.K. & Dekker, A.W. (1991): Cytomegalovirus infection causes delayed platelet recovery after bone marrow transplantation. Blood 78, 844.

Wingard, J.R., Chen, D.Y-H., Burns, W.H., Fuller, D.J., Braine, H.G., Yeager, A.M., Kaiser, H., Burke, P.J., Graham, M.L., Santos, G.W. & Saral, R. (1988): Cytomegalovirus infection after autologous bone marrow transplantation with comparison to infection after allogeneic bone marrow transplantation. Blood 71, 1432.

Résumé

Les travaux présentés dans cet article ainsi que des observations antérieures suggèrent que les différentes souches de CMV sont capables d'infecter des cellules cibles différentes et que des variations parmi les patients infectés par des types cellulaires spécifiques de CMV peuvent contribuer de façon significative à l'évolution clinique. Ces observations indiquent que définir exactement comment le CMV interfère avec la fonction médullaire nécessite 1) de déterminer les types cellulaires infectés dans les cas de myélosuppression, 2) de caractériser la souche de CMV responsable de l'infection et 3) de déterminer comment il interfère avec les fonctions cellulaires. Ces informations sont indispensables pour établir des stratégies permettant d'éviter la myélosuppression et pouraient peut-être aider à comprendre les tropismes et la pathogénie qui pourait être codés génétiquement pour les différentes souches de CMV.

Inhibition of hematopoietic stem/progenitor cells by human immunodeficiency virus type 1

Brian R. Davis*, Giorgio Zauli**,*** Dazhi Cen*, Vincent P. Antao*, Maria Carla Re**, Giuseppi Visani** and Michele La Placa**

*California Pacific Medical Center Research Institute, 2330 Clay St., San Francisco, CA, USA; **Institute of Microbiology, University of Bologna, Via Massarenti 9, Bologna, Italy; ***Institute of Human Anatomy, University of Ferrara, Via Fossato di Mortara 66, 44100 Ferrara, Italy

Summary

Although hematopoietic cytopenias occur frequently in HIV-1 induced disease, the mechanisms by which HIV-1 contributes to these cytopenias are presently not known. In this report, we focus on mechanisms involving the direct or indirect inhibition of stem/progenitor cells by HIV-1. Stem/progenitor cells are defective in a significant subset of persistently cytopenic or ARC/AIDS HIV-1 infected individuals. These defects occur with little or no productive HIV-1 infection of the hematopoietic stem/progenitor cell compartment. Significantly, direct *in vitro* exposure of normal CD34+ marrow cells to HIV-1 induces stem/progenitor cell defects resembling those in HIV-1 (+) cytopenic or ARC/AIDS patients. In order to elucidate the mechanisms responsible for this direct induction of defects, we have established experimental systems involving either primary CD34+ cells or leukemic CD34+ cell lines. We have also investigated an indirect mechanism by which HIV-1 may inhibit stem/progenitor cells through inducing negative regulatory cytokine expression in accessory cells. An increased understanding of both direct and indirect mechanisms may potentially lead to the development of appropriate anti-viral or cytokine strategies successful in reversing HIV-1 associated cytopenias.

HIV-1 induced disease is primarily characterized by the specific deficit in CD4+ T lymphocytes, which leads to an immune deficient status. However, cytopenias in other hematopoietic lineages (anemia, neutropenia, leukopenia, thrombocytopenia, or pancytopenia) are also frequently observed in HIV infected individuals, with the frequency of these cytopenias generally increasing with the stage of disease (Zon et al., 1987; Scadden et al., 1989). Multiple mechanisms are likely responsible for the occurrence of cytopenias in HIV-1 infected individuals. A number of proposed mechanisms invoke the direct involvement of HIV-1 in inducing cytopenias. These mechanisms include the direct action of HIV-1 on a) hematopoietic stem/progenitor cells (inhibiting their ability to produce mature blood cells); b) more mature marrow or blood cells (inhibiting their ability to produce mature blood cells, e.g. the production of platelets by megakaryocytes); c) marrow accessory cells, such as T-lymphocytes or macrophages (affecting the balance between positive and negative regulatory hematopoietic factors); or d) the immune system (leading to the induction of antibodies which target and destroy platelets, neutrophils, etc, and/or release of cytokines upon immune stimulation). A direct involvement for HIV-1 in induction of cytopenias is suggested by 1) cytopenias occurring most frequently during times of increased viremia (e.g. during symptomatic acute primary infection and in the AIDS stage of disease, occurring least frequently during the low viremia asymptomatic stage), and 2) the ability of anti-viral agents, such as AZT, to reverse some cytopenias, such as thrombocytopenia. Other possible mechanisms for cytopenias in HIV-1 indfected individuals, which do not involve HIV-1 directly, include suppression of hematopoiesis by certain opportunistic infections (e.g. cytomegalovirus or parvovirus) or marrow toxic anti-viral agents (e.g. AZT).

Several *in vivo* defects in hematopoiesis have been identified in HIV-1 infected individuals. The decreased frequency of hematopoietic progenitor cells (CFU-GM, BFU-E [burst forming unit - erythroid], CFU-Meg [colony forming unit - megakaryocyte], and CFU-TL [colony forming unit - T lymphocyte]) in the marrow and blood of HIV-infected individuals with AIDS (Acquired Immunodeficiency Syndrome) (Carlo Stella et al., 1987; Leiderman et al., 1987, Bagnara et al., 1990; Ganser et al., 1990; Lunardi-Iskandar et al., 1989; Louache et al., 1992) or with isolated thrombocytopenia (Zauli et al., 1992a) indicates progenitor cell disfunction. Prior studies have evaluated the colony forming cells in the context of total bone marrow or adherent cell depleted marrow. Such analysis does not permit distinguishing between 1) intrinsic defects in stem/progenitors, 2) reduced colony formation due to selected depletion of progenitors *in vivo*, or 3) inhibition of stem/progenitors by accessory cells also present in the bone marrow sample. In two studies, the defective colony formation could be partially, but not completely, corrected by either T-cell depletion (Carlo Stella et al., 1987) or treatment of marrow cells with anti-sense oligonucleotides to tat or rev regulatory gene sequences (Louache et al., 1992). These data are consistent with intrinsically defective stem/progenitors being further suppressed by accessory cells. Our laboratory has shown that purified CD34 (+) cells (the CD34 antigen is expressed by all known hematopoietic stem/progenitor cells but is absent on all mature hematopoietic cells) from symptomatic, but not asymptomatic, HIV-1 (+) individuals are defective in colony formation. The frequency of CFU-GM in the marrow of HIV-1 (+) ARC or AIDS patients was significantly less ($p<.01$) per CD34 (+) cells, with the mean CFU-GM frequency reduced 60% with respect to normal controls. These results are significant in demonstrating that the stem/progenitor cells are themselves defective, and that the defective colony growth is not simply a consequence of either inhibition by accessory cells, or a decrease in the number of CD34 (+) cells. Similar results have been obtained in another study of HIV-1 (+) symptomatic individuals (Stanley et al., 1992) and HIV-1 (+) thrombocytopenic individuals (Zauli et al., 1992a). In this last study, we demonstrated that there was a correlation between the severity of defect in the CD34 (+) cell population and the level of gag p24 antigen in the bone marrow plasma. These findings suggested that increased viral replication in the marrow environment, either acting through direct or indirect mechanisms, may be responsible for induction of these defects.

Presented in Figure 1 are potential mechanisms by which HIV, directly or indirectly, could inhibit hematopoietic stem/progenitor cells. In Fig. 1a, HIV-1 productively infects stem/progenitors, resulting in synthesis of proviral DNA, synthesis of viral proteins, expression of env proteins on the cell surface, and release of new virus particles. Presumably such productive infection could either lead to lytic cell death or, in combination with antibodies to envelope viral proteins, could result in antibody mediated suppression or antibody dependent complement mediated cytotoxicity. Fig. 1b shows non-productive infection, leading either directly to death of stem/progenitors (perhaps via an apoptotic mechanism), or to the induced expression of inhibitory cytokines resulting in eventual suppression or cell death. In Fig. 1c, virions, or envelope gp120, bound to the surface of uninfected stem/progenitors, are fixed by anti-viral antibodies, leading to eventual cell death (again perhaps through apoptosis). Finally, Fig. 1d reflects the ability of HIV infected accessory cells (T lymphocytes, macrophages) to suppress hematopoietic stem/progenitors by the release of inhibitory cytokines. Also shown in Fig. 1d is the ability of extracellular viral regulatory proteins, specifically tat, to upregulate inhibitory cytokines in T cells or macrophages, with a subsequent inhibition of stem/progenitors.

Prior to initiation of these studies, much emphasis had been placed on the possibility that HIV-infected stem/progenitor cells played a major role in the hematopoietic suppression observed in HIV-1(+) individuals. These suggestions followed from the work of Donahue et al. (1987), who reported a suppression in colony formation of 50-90% when marrow from ARC/AIDS patients was plated in the presence of serum immunoglobulin from HIV-seropositive individuals or rabbit anti-env gp120 serum. These data suggested that a significant percentage of bone marrow progenitors in ARC/AIDS individuals were infected with HIV and that the suppression of hematopoiesis in HIV-infected individuals might be an immune-mediated phenomenon involving antibodies directed agains virus-infected progenitors (as diagrammed in Fig. 1a). Furthermore, Folks et al. (1988) reported that purified normal CD34 (+) bone marrow cells could be infected *in vitro* with HIV-1. In two separate studies examining the extent of HIV infection (by DNA PCR) directly in CD34 (+) stem/progenitor cells from HIV-1 infected individuals, we demonstrated that the stem/progenitors are rarely infected *in vivo* (Davis et al., 1991; Zauli et al., 1992a). For numerous CD34 (+) cell samples exhibiting defective behavior, there was no evidence for HIV-1 infection. Even in the rare cases in which there was perhaps some detectable infection, it was at a level insufficient to explain the defective biological activity of the stem/progenitor cells (Zauli et al., 1992a). Similar results have also been obtained by three other laboratories (Louache et al., 1992; Von Laer et al., 1990; Molina et al., 1990); an additional group reported evidence of low level CD34 (+) cell infection *in vivo*, but again at levels insufficient to explain the observed defects (Stanley et al., 1992). The absence of significant HIV infection in CD34 (+) cells from ARC/AIDS or cytopenic patients is inconsistent with the immunological model of hemopoietic suppression (Fig. 1a). In spite of the *in vivo* evidence indicating that HIV-1 infection of CD34 (+) stem/progenitor cells occurs rarely, significant effort has been invested in recent years seeking to demonstrate productive *in vitro* infection of these cells. There is still significant controversy in this field, with some groups reporting infectability (Folks et al., 1988; Steinberg et al., 1991; Kitano et al., 1991) and other groups reporting an absence of infectability (Molina et al., 1990; Zauli et al., 1991; Zauli et al., 1992b; Cen et al., submitted). The fact that production of viral antigen does not start until several weeks following virus exposure (Folks et al., 1988; Kitano et al., 1991) may indicate that infection actually occurs after the *in vitro* maturation of CD34 (+) cells.

More relevant, we believe, to mechanisms of cytopenias in HIV-infected individuals is the ability of HIV-1 to directly induce defects in stem/progenitor cells (Steinberg et al., 1991; Zauli et al., 1991, Zauli et al., 1992b; Cen et al., submitted), defects which resemble those in CD34 (+) cells from HIV-1 (+) patients with cytopenias or advanced disease (Zauli et al., 1992a; Stanley et al., 1992). Exposure of normal CD34 (+) cells to certain HIV-1 isolates induces defects in CFU-GM, BFU-E, CFU-Meg, and CFU-TL derived colony formation; furthermore, the same isolates inhibit the ability of exposed CD34 (+) cells to proliferate/survive in IL-3 containing liquid culture. Induction of defects in primary CD34 (+) cells occurred with minimum viral doses in the range of 0.1 - 1.0 ng/ml gag p24, a level of virus frequently observed in the plasma of patients with early symptomatic or late stage HIV-1 induced disease (Ho et al., 1989). The ability of anti-gp120 antibodies to neutralize the induction of CD34 (+) cell defects by the virus inoculum indicates that HIV-1 is directly responsible for this effect. One group (Molina et al., 1990) did not observe any reduction in colony formation by marrow mononuclear cells exposed to HIV-1. The exact reason for this discrepancy is not known. However, since the induction of CD34 (+) cell defects by HIV-1 is dose responsive, it is possible that an amount of virus insufficient to induce stem/progenitor defects was used in these experiments. The *in vitro* induction of these defects does not require productive infection of the stem/progenitor cells with HIV-1 (Zauli et al., 1991, 1992b; Cen et al., submitted). These data, together with the rare infection of CD34 (+) cells *in vivo*, efffectively rule out the productive infection model in Fig. 1a, but are consistent with the non-productive infection model for induction of stem/progenitor defects, pictured in Fig. 1b. In collaborative studies with Zauli et al., we have demonstrated that direct exposure of primary CD34 (+) cells, or the CD34 (+) TF-1 erythromyeloid progenitor cell line, to HIV-1 leads to an increase in apoptosis (Zauli et al., submitted), suggesting that this may be a critical event in defect induction (Fig. 1b, upper panel). Addition of anti-gp120 antibody, following pre-exposure of normal TF-1 cells to gp120 or HIV-1, significantly enhanced the induction of apoptosis in these cells. This mechanism for induced cell death of stem/progenitors is pictured in Fig. 1c. It may be operative *in vivo* where free gp120, HIV-1, and anti-HIV-1 antibodies are present, and is consistent with the previously described findings of Donahue et al. (1987). In our laboratory, we have examined the HIV-1 induced inhibition of the CD34 (+) KG-1 lymphomyeloid progenitor cell line (Antao et al., manuscript in preparation). Interestingly, the induction of defects by HIV-1 in KG-1 cells appears to occur by the stimulated production of an inhibitory cytokine, which acts back on KG-1 cells to inhibit or even kill them (Fig. 1b, lower panel). Naturally, we are interested in determining whether a similar mechanism is operative for primary CD34 (+) cells.

In an effort to examine the effect of HIV-1 infection on hematopoiesis, in a context as close as possible to the *in vivo* marrow microenvironment, we studied three HIV-1 isolates in long term bone marrow cultures. These experiments provide the first evidence for productive HIV-1 infection of LTBMCs. Furthermore, we observed a difference in the ability of the HIV-1 isolates to inhibit myelopoiesis in the LTBMCs, which, for at least two of the three isolates examined, correlated with the differential ability of the isolates to directly inhibit CD34 (+) cells in liquid culture and semi-solid colony assays (Cen *et al.*, submitted). These data suggest that inhibition of stem/progenitors may be an important mechanism for HIV-1 induced suppression of hematopoiesis. Furthermore, these data suggest that HIV-1 isolates may have differential ability to induce cytopenias *in vivo*, and the development of cytopenias may depend on the presence of the appropriate isolates. Furthermore, the dose responsiveness of the inhibition relative to the virus load suggests that the the magnitude of the viral burden *in vivo* may also affect the occurrence of cytopenias.

Virus induced defects in stem/progenitor cells constitute only one potential mechanism whereby HIV may suppress hematopoiesis. Another possible mechanism would be through direct infection of other cell types in the marrow, leading to decreased production of stimulatory cytokines or increased production of inhibitory cytokines, which themselves act on stem/progenitor cells. We have demonstrated that T-lymphocytes are the primary marrow reservoir for HIV *in vivo*, with marrow macrophages also infected at a lower level with HIV-1 (Davis *et al.*, 1991). Furthermore, we have observed the ability of the HIV-1 tat protein (at concentrations between 10 ng/ml and 10 μg/ml) to induce marrow macrophages to increase their expression of TGF-β-1, a potent inhibitory of stem/progenitor cells (Zauli *et al.*, 1992c). These data indicate that virus present in the marrow, or in the blood which permeates the marrow, may inhibit stem/progenitor cells through such an indirect mechanism. This mechanism is pictured in Fig. 1d.

Although cytopenias occur frequently in HIV-1 induced disease, the exact mechanisms by which HIV-1 contributes to these cytopenias are presently not known. Our prior approach has been to combine an examination of the *in vivo* situation (i.e. what cells in the bone marrow are HIV infected, what defects are observed in cell types purified from the bone marrow of individuals with cytopenias) with the establishment of *in vitro* experimental systems designed to directly investigate the role of HIV-1 in inducing hematopoietic suppression. In future studies, we will focus primarily on the mechanisms pictured in Fig. 1b and 1c, using previously developed experimental systems. We believe that these and other studies are important in that they seek to dissect, in a focused manner, the mechanisms responsible for defective hematopoiesis in HIV infected individuals. It is very possible that an increased understanding of relevant mechanisms may lead to the development of appropriate anti-viral or cytokine strategies which will be successful in reversing these cytopenias. Furthermore, since T-lymphopoiesis also originates in the marrow with multipotent stem cells and progenitors committed to the T-cell lineage, it may be that some of these mechanisms are also relevant to the inability of the marrow and thymus to restore normal CD4 cell counts during the course of HIV-1 induced disease.

References

Bagnara GP *et al.* (1990): Early loss of circulating hemopoietic progenitors in HIV-1 infected subjects. *Exp Hematol* 18, 426-430.

Carlo Stella C *et al.* (1987): Defective *in vitro* growth of the hemopoietic progenitor cells in the acquired immunodeficiency syndrome. *J. Clin. Invest.* 80, 286-293.

Cen D *et al.* (submitted): Effect of different human immunodeficiency virus type-1 isolates on long-term bone marrow hematopoiesis.

Davis BR et al. (1991): Absent or rare human immunodeficiency virus infection of bone marrow stem/progenitor cells in vivo. J. Virol. 65, 1985-1990.

Donahue RE et al. (1987): Suppression of in vitro haematopoiesis following human immunodeficiency virus infection. Nature 326, 200-203.

Folks TM et al. (1988): Infection and replication of HIV-1 in purified progenitor cells of normal human bone marrow. Science 242, 919-922.

Ganser A et al. (1990): Changes in the haematopoietic progenitor cell compartment in the acquired immunodeficiency syndrome. Res. Virol. 141, 185-193.

Ho DD et al. (1989): Quantitation of human immunodeficiency virus type 1 in the blood of infected persons. N. Eng. Jour. Med. 321, 1621-1625.

Kitano K et al. (1991): Macrophage-active colony-stimulating factors enhance human immunodeficiency virus type 1 infection in bone marrow stem cells. Blood 77, 1699-1705.

Leiderman IZ et al. (1987): A glycoprotein inhibitor of in vitro granulopoiesis associated with AIDS. Blood 70, 1267-1272.

Louache F et al. (1992): Role of human immunodeficiency virus replication in defective in vitro growth of hematopoietic progenitors. Blood 80, 2991-2999.

Lunardi-Iskandar Y et al. (1989): Impaired in vitro proliferation of hemopoietic precursors in HIV-1 infected subjects. Leuk. Res. 13, 573-581.

Molina JM et al. (1990): Lack of evidence for infection of or effect on growth of hematopoietic progenitor cells after in vivo or in vitro exposure to human immunodeficiency virus. Blood 76, 2476-2482.

Scadden DT et al. (1989): Pathophysiology and management of HIV associated hematologic disorders. Blood 74, 1455-1463.

Stanley SK et al. (1992): CD34 (+) bone marrow cells are infected with HIV in a subset of seropositive individuals. J. Immunol. 149, 689-697.

Steinberg HN et al. (1991): In vitro suppression of normal human bone marrow progenitor cells by human immunodeficiency virus. J. Virol 65, 1765-1769.

Von Laer D et al. (1990): CD34 (+) hematopoietic progenitor cells are not a major reservoir of the human immunodeficiency virus. Blood 76, 1281-1286.

Zauli G et al. (1991): Evidence for an HIV-1 mediated suppression of in vitro growth of enriched (CD34 (+)) hematopoietic progenitors. J. Acquir. Immune Defic. Syndromes 4, 1251-1253.

Zauli G et al. (1992a): Impaired in vitro growth of purified (CD34 (+)) hematopoietic progenitors in human immunodeficiency virus-1 seropositive thrombocytopenic individuals. Blood 79, 2680-2687.

Zauli G et al. (1992b): Evidence for an HIV-1 mediated suppression of uninfected hematopoietic (CD34 (+)) progenitor cells in vivo. J. Infect. Diseases 1666, 710-716.

Zauli G et al. (1992c): Tat protein stimulates production of transforming growth factor-$\beta 1$ by marrow macrophages: a potential mechanism for HIV-1 induced hematopoietic suppression. *Blood* 80, 3036-3043.

Zauli G et al. (submitted): Hematopoietic progenitor (CD34 (+) cells are driven to apoptotic death by membrane interaction with human immunodeficiency virus type 1.

Zon, LI et al. (1987): Haematologic manifestation of the human immune deficiency virus (HIV). *Brit. Jour. Haematology* 66, 251-256.

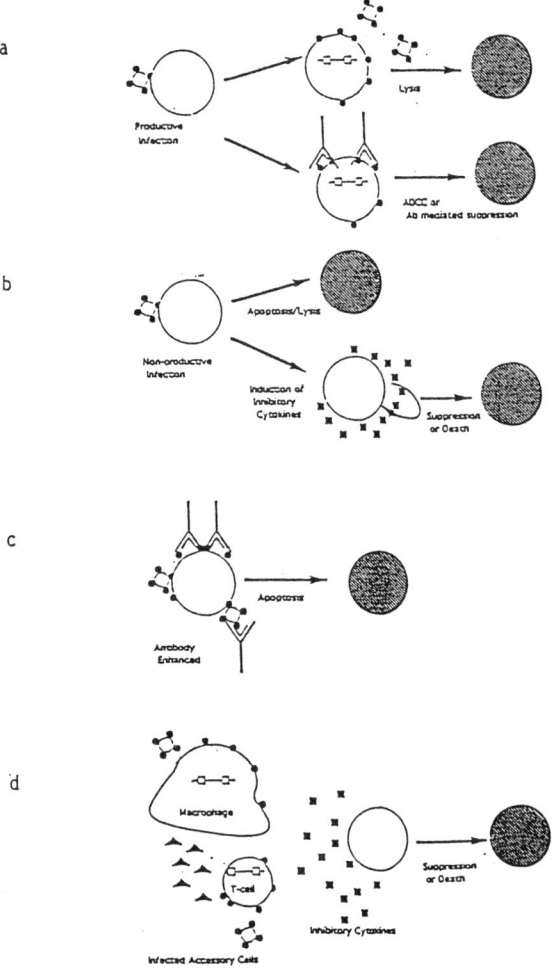

Fig. 1. Potential Mechanisms for HIV-induced Stem/Progenitor Defects. Large open circles: CD34+ stem/progenitors; large shaded circles: suppressed or dying stem/progenitors; small dark circles: env gp120; small dark squares: cellular-produced inhibitory cytokines; small dark triangles: viral regulatory proteins (e.g. tat). Virions, antibody molecules, and integrated proviral DNA are also shown.

Résumé

Bien que des cytopénies surviennent fréquemment dans les maladies induites par le VIH-1, le mécanisme par lequel le VIH-1 contribue à ces cytopénies n'est pas encore connu. Cet article est centré sur les mécanismes impliquant une inhibition directe ou indirecte des cellules souches/progéniteurs par le VIH-1. Les cellules souches/progéniteurs sont défectifs dans un sous-groupe significatif d'individus à cytopénie persistante ou infectés par ARC/AIDS/VIH-1. Ces anomalies se produisent avec peu ou pas d'infection productive de VIH-1 du compartiment cellules souches/progéniteurs. De façon significative, l'exposition directe in vitro de cellules $CD34^+$ de moelle normale au VIH-1 induit des anomalies des cellules souches/progéniteurs ressemblant à celles observés chez les patients cytopéniques $VIH-1^+$, atteints d'ARC ou de SIDA. Afin d'élucider les mécanismes responsables de cette induction directe d'anomalies, nous avons établi des systèmes expérimentaux soit avec des cellules $CD34^+$ soit avec des lignées leucémiques $CD34^+$. Nous avons aussi étudié un mécanisme indirect par lequel le VIH-1 pourrait inhiber les cellules souches/progéniteurs en induisant l'expression de cytokines régulatrices négatives dans les cellules accessoires. Une meilleure compréhension des mécanismes directs et indirects pourrait conduire au développement de stratégies appropriées utilisant des antiviraux ou des cytokines afin de prévenir les cytopénies induites par le VIH-1.

HIV exerts a dual effect on hematopoietic progenitor cells: direct infection and modulatory activities on their growth

Valérie Calenda and Jean-Claude Chermann

INSERM U 322, Unité des Rétrovirus et Maladies Associées, Campus Universitaire de Luminy, BP 33, 13273 Marseille Cedex 9, France

A variety of hematological alterations in AIDS patients have already been reported. Several possible explanations have been proposed : virus infection of myeloid precursors, immune mechanisms and *in vitro* suppression of progenitor cells after HIV infection (as reviewed by Calenda et al. 1992a). To date, however, the real causes of impaired hematopoiesis in AIDS patients are still not clear. We performed long term liquid cultures of bone marrow cells from healthy seronegative donors after incubation with HIV-1 or HIV-2 CEM conditoned medium.

Cell viability, HIV p24, reverse transcriptase activity in liquid culture and clonogenic assays for erythroid and granulomonocytic progenitor cells were assessed during all experimental observation period. No effect on cell viability and no viral replication was observed in HIV infected liquid cultures. The virus was rescued from highly purified progenitor cells by cocultivation with HIV permissive cells, indicating a non productive infection (Calenda *et al.*, 1992b ; Calenda & Chermann 1992c). PCR experiments and in situ hybridization confirmed the presence of viral DNA in CD34+ cells after incubation with HIV (Calenda *et al.*, 1992d).

Clonogenic assays revealed an impaired capacity of progenitor cells to give rise to colonies of differentiated progeny. Moreover we found an indirect effect of HIV on progenitor cells by way of regulatory proteins (Tat, Nef) able to modulate their growth in semi solid assays. These data indicate that HIV associated impaired hematopoiesis, can be explained by a direct infection of hematopoietic progenitor cells and a modulatory effect on progenitor cells growth linked to HIV regulatory Tat and Nef proteins.

REFERENCES

Calenda, V., Sebahoun, G., and Chermann J.C. (1992b): Modulation of normal human erythropoietic progenitor cells in long term liquid cultures after HIV-1 infection. Aids Res. Hum. Retroviruses 8: 61-67.

Calenda, V., and Chermann, J.C. (1992a): The effects of HIV on hematopoiesis. Eur. J. Haematol. 48: 181-186.

Calenda, V., and Chermann, J.C. (1992c): Severe in vitro inhibition of erythropoiesis and transient stimulation of granulopoiesis after bone marrow infection with eight different HIV-2 isolates. AIDS 6: 943-948.

Calenda, V., Tamalet, C., and Chermann, J.C. (1992d): Transient stimulation of granulopoiesis and drastic inhibiton of erythropoiesis in HIV-2-infected long-term liquid bone marrow cultures. J. of AIDS 5: 1148-1157.

Donahue, R.E., Johnson, M.M., Zon, L.I., Clark, S.C., and Groopman, J.E. (1987): Suppression of in vitro haematopoiesis following human immunodeficiency virus infection. Nature 326: 200-203.

Human immunodeficiency virus type-1 triggers apoptotic cell death of hematopoietic progenitor (CD34+) cells

Giorgio Zauli[1,3], Marco Vitale[2], Brian R. Davis[3], Maria Carla Re[4], Davide Gibellini[4], Silvano Capitani[1] and Michele La Placa[4]

[1]Institute of Human Anatomy, University of Ferrara, Via Fossato di Mortara 66, 44100 Ferrara, Italy; [2]Institute of Human Anatomy, University of Brescia, Italy; [3]Medical Research Institute, San Francisco, CA, USA; [4]Institute of Microbiology, University of Bologna, Italy

At flow cytometry analysis, a significant ($p<0.05$) increase in the frequency of apoptotic cell death was observed in both HIV-1-treated bone marrow purified CD34+ cells and interleukin-3 dependent TF-1 cell line with respect to mock-treated controls after 72 hours of culture (**Table 1**). Similar results were obtained when apoptosis was evaluated at electron microscopy and gel electrophoresis analyses.

Table 1: analysis of the percentage of TF-1 cells and CD34+ bone marrow progenitors in apoptosis after exposure to HIV$_{IIIB}$. *statistically significant differences ($p<0.05$) between HIV-1-treated and mock-treated cells. Mean±S.D. of five separate duplicated experiments are reported.

IL-3 (ng/ml)	Mock-treated TF-1 cultures	HIV$_{IIIB}$-treated TF-1 cultures	ICR-3 TF-1 cultures
0	51.0±6.0	52.5±5.3	55.0±6.0
0.02	30.5±5.5	*41.0±4.0	*49.0±8.6
0.2	18.5±4.3	*29.5±3.0	*34.5±6.0
1	16.0±3.6	*27.5±7.0	*28.5±6.5
2	11.5±3.6	14.0±4.0	14.5±4.6
10	9.5±3.2	10.5±2.7	11.0±4.3
	Mock-treated CD34+ cell cultures	HIV$_{IIIB}$-treated CD34+ cell cultures	
1	5.5±2.0	*13.5±3.5	

Nevertheless, no signs of productive or latent virus replication were ever observed in HIV-1 treated CD34+ cells up to 16 days of liquid culture. The HIV-1-induced apoptosis of CD34+ cells was likely triggered by the simple interaction of HIV-1 envelope glycoprotein gp120 with CD4 receptor, expressed at low level on the surface of hematopoietic progenitors. In fact, in cross-linking experiments, treatment of TF-1 cells with recombinant gp120 plus a polyclonal anti-gp120 antibody or with anti-CD4 monoclonal antibody plus rabbit anti-mouse immunoglobulin G significantly ($p<0.05$) increased the percentage of apoptotic death.

These data strongly suggest that HIV-1, and also free gp120 in the presence of anti-gp120 antibody, could play a direct role in the pathogenesis of peripheral blood cytopenias of AIDS patients, affecting the growth of hematopoietic progenitor cells.

Negative regulators may mediate hematopoietic suppression in HIV-infected bone marrow long term cultures

G.N. Schwartz[1], S.W. Kessler[2], M.L. Francis[3], D.L. Hoover[4], S.W. Rothwell[5], L.M. Burrell[6] and M.S. Meltzer[3]

[1]Transplantation Therapy Section, National Cancer Institute, Bethesda, MD, [2]Immune Cell Biology Program, Naval Medical Research Institute, Bethesda, MD, Departments of [3]Cellular Immunology, [4]Bacterial Diseases, and [5]Hematology, Walter Reed Army Institute of Research, Washington, DC, and [6]Hematology-Oncology, Walter Reed Army Medical Center, Washington, DC, USA

INTRODUCTION

Anemia, neutropenia, and thrombocytopenia are some of the complications seen in patients with AIDS (Ganser et al., 1990; Lunardi-Iskandar et al., 1989). In addition to the low numbers of blood cells, AIDS patients have low numbers of the progenitor cells that produce mature blood cells. Lunardi-Iskandar et al., (1989) observed that in bone marrow from AIDS patients the frequency of progenitor cells that formed colonies derived from colony-forming units for granulocytes and macrophages (CFU-GM), and burst-forming units for erythroid cells (BFU-E) was only 34% and 21%, respectively, of the frequency detected in bone marrow from HIV-1 seronegative individuals. Some patients with early stages of HIV-infection were also cytopenic and had low numbers of progenitor cells. The causes for the low numbers of blood cells and hematopoietic progenitor cells in HIV-infected individuals are not known.

The bone marrow microenvironment is a close association of accessory cells that provides the niches, adhesion molecules, and regulatory factors necessary for the production and differentiation of hematopoietic progenitor cells (Allen & Dexter, 1984). Some of the accessory cells that make up the marrow microenvironment include fibroblasts, macrophages, endothelial cells, and T cells. Some populations of these accessory cells express CD4, the receptor for HIV-1 and can become infected with HIV-1 (McElrath et al., 1989; Scadden et al., 1990; Sun et al., 1989). Transforming growth factor-β (TGF-β), interferon-α (IFN-α), tumor necrosis factor-α (TNF-α), and interleukin-4 (IL-4) are multifunctional cytokines that are increased in cultures of cells from HIV-infected individuals or cells exposed to HIV-1 or HIV-viral proteins (Clerici & Shearer, 1993; Eyster et al., 1983; Roux-Lambard et al., 1989; Zauli et al., 1992) and suppress in vitro hematopoiesis (Broxmeyer et al., 1983; Moore, et al 1991; Sonada et al., 1990). One possible explanation for the low numbers of blood cells and CFC in HIV-infected individuals is that production of negative regulators of hematopoiesis is dysregulated when accessory cells of the bone marrow microenvironment are infected with HIV-1. Stromal cell layers, an adherent layer of cells that forms in bone marrow long term cultures (LTC), simulates the marrow microenvironment (Allen & Dexter, 1984). This report presents the results of studies using LTC of human bone marrow cells to investigate the effect of HIV-1 on factors that regulate hematopoiesis.

METHODS

Stromal cell layers were established by culturing low density marrow cells from HIV-seronegative donors for 4 weeks in 1.5 cm^2 cultures wells with 0.5 ml media. The confluent stromal cell layers were exposed to 9.0 Gy γ-radiation to eliminate residual progenitor cells and then infected with HIV-1$_{ADA}$, a monocytotropic strain of

HIV-1 (Gendleman et al., 1988). Two or 24 hours later the stromal cell layers were washed 3 times and then cocultured with 5×10^3 highly purified autologous CD34+ bone marrow cells (Folks et al., 1988). At days 0, 2, 4, 8, and 11 neutralizing antibodies to TGF-β, TNF-α, INF-α, and IL-4 were added to noninfected and HIV-infected LTC. Nonadherent and adherent cells from LTC were subcultured in methylcellulose containing PHA-LCM and erythropoietin for the detection of colonies derived from BFU-E and CFU-GM. The mean ± SEM for multiple studies was calculated from the mean number of CFC from 2 to 4 LTC per study.

An enzyme immunoassay for the detection of HIV-1 p24 antigen was used to monitor infection in LTC. IFN-α activity in culture media from LTC was assayed by protection of MDBK cells against cytopathicity of vesicular stomatitis virus. Cytokine mRNA was estimated by reverse transcription PCR from total RNA of noninfected and HIV-infected stromal cell layers. RNA was prepared by phenol/chloroform/isoamyl alcohol extraction from cells lysed in quanidium isothiocyanate, and first strand cDNA was synthesized using an oligo$(dT_1)_6$ primer. Equivalent cDNA synthesis from RNA preparations was verified by amplification of β-actin.

RESULTS AND CONCLUSIONS

Within the first week after exposure of the stromal cell layers to HIV-1, the concentration of HIV-1 p24 antigen increased from 0.2 ng/ml to 3 ng/ml. The concentration of HIV-1 p24 antigen continued to increase throughout the 2 to 4 weeks of culture and was 2 to 20-fold higher when stromal cell layers were cocultured with CD34+ cells. A progressive increase in HIV-1 p24 demonstrated that there was a productive infection. RT-PCR analysis of stromal cell layers 2, 8, and 19 days after infection with HIV-1 demonstrated that noninfected and HIV-infected stromal cell layers expressed equivalent levels of mRNA for GM-CSF, IL-6, and IFN-α.

One week after the coculture of CD34+ cells with HIV-infected stromal cell layers, the number of CFU-GM in 7 out of 10 experiments was reduced compared to noninfected control LTC. In those 7 experiments, the number of CFU-GM was 54 ± 4% of the number in noninfected LTC. The number of BFU-E in HIV-infected LTC was only 39 ± 8% of the number in noninfected LTC (n = 6). There were fewer BFU-E in HIV-infected LTC, whether or not there was a reduced number of CFU-GM.

One week after coculture of stromal cell layers with CD34+ cells, the concentration of IFN-α increased from < 1 IU/ml in noninfected LTC to between 4 and 16 IU/ml, in HIV infected LTC. No IFN-α was detected in LTC infected with HIV-1 and then incubated with anti-IFN-α. The number of CFU-GM in HIV-infected LTC incubated with anti-IFN-α was similar to the number in untreated infected LTC (Table 1). The number of BFU-E, however, was 1.9 ± 0.3-fold greater (n = 3) in HIV-infected LTC incubated with anti-IFN-α (Table 1).

Table 1. Effect of Anti-IFN-α on the Number of BFU-E and CFU-GM in HIV-Infected LTC

Colony Assay	Percentage of Number in Noninfected Long Term Cultures	
	Infection + No Antibody	Infection + Anti-IFN-α
BFU-E (n = 3)[a]	46 ± 1	90 ± 15
CFU-GM (n = 1)	41	58

[a]Values are the mean ± S.E.M. of the mean values from 3 experiments for BFU-E

Suppression of BFU-E numbers in HIV-infected LTC was not prevented by either anti-TNF-α or anti-IL-4. Similarly, anti-TGF-β did not prevent the suppression of CFU-GM in HIV-infected LTC (Table 2, Study A). One and 2 weeks after the addition of CD34+ cells, the number of CFU-GM in HIV-infected LTC incubated with anti-IL-4 was 2 to 4-fold greater than in untreated LTC (Table 2, Study A & B). Anti-TNF-α required a longer period of time to exert an effect. At 1 week there was no difference between control LTC and HIV-infected LTC treated with anti-TNF-α. By 2 weeks, however, there were 2.9-fold more CFU-GM in LTC incubated with anti-TNF-α than in control infected LTC. Neither anti-IL-4 nor anti-TNF-α promoted an increase in the number of CFU-GM in noninfected or HIV-infected LTC in experiments where CFU-GM numbers were similar in infected and noninfected LTC. These results suggest that anti-IL-4 and anti-TNF-α are not potentiating new production of CFU-GM, but are overcoming negative regulatory effects caused by an increase in the levels of growth inhibitory cytokines after infection of the stromal cell layers with HIV-1.

Table 2. Effect of anti-TNF-α and anti-IL-4 on the number of CFU-GM in HIV-infected LTC

Treatment	Day 0[b]	Number of CFU-GM per LTC[a] 1 Week	2 Weeks
Study A.			
Infection + No Ab	81 ± 10	147 ± 40	71 ± 1
Infection + Anti-TGF-β		144 ± 32	81
Infection + Anti-IL-4		358 ± 62	120 ± 21
Study B			
No Infection	115 ± 21	211 ± 41	237 ± 50
Infection + No Ab		88 ± 25	118 ± 18
Infection + Anti-TNF-α		100 ± 14	347 ± 79
Infection + Anti-IL-4		203 ± 53	437 ± 79

[a] Values are the mean ± S.E.M. of the mean values of 2 LTC each. The sum of the number of CFU-GM in the nonadherent and adherent fractions was used to calculate the mean number per LTC.

[b] Number of CFU-GM in 5×10^3 CD34+ marrow cells added to irradiated stromal cell layers

REFERENCES

Allen, T.D., and Dexter, T.M. (1985): The essential cells of the hemopoietic microenvironment. *Exp. Hematol.* 12, 517.

Clerici, M. and Shearer, G.M. (1993): A T_H1 to T_H2 switch is a critical step in the etiology of HIV-infection. *Immunology Today* 14, 107.

McElrath, M.J., Pruett, C., and Cohn, Z.A. (1989): Mononuclear phagocytes of blood and bone arrow; comparative roles as viral reservoirs in human immunodeficiency virus type infection. *Proc Nat Acad Sci (USA)* 86, 675.

Eyster, M.E., Goedirt, J., Poon, M.C., and Preblo, O.T. (1983): Acid-labile alpha interferon: a possible preclinical marker of the acquired immune deficiency syndrome in hemophilia. *New Eng. J. Med.* 309, 583.

Folks, T.M., Kessler, S.W., Orstein, J.M., Justement, S., Jaffe, E.S., and Fauci, A.S. (1988): Infection and replication of HIV-1 in purified progenitor cells of normal bone marrow. *Science* 242, 919.

Ganser, A. Ottman O.G., Van Briesen, H., Rubsamen-Waigmann, H., and Hoelzer, D. (1990): changes in hematopoietic progenitor cell compartment in the acquired immunodeficiency syndrome. *Res. Virology* 141, 185.

Gendelman, H.E., Orenstein, J.M., Martin, M.A., Ferrua, C., Mitra, R., Phipps, T., Wahl,L.A., Lane, H.C., Fauci, A.S., Burke, D.S., Skillman, D., and Meltzer, M.S. (1988): Efficient isolation and propagation of human immunodeficiency virus on recombinant colony-stimulating factor-1 treated monocytes. *J. Exp. Med.* 167, 1428.

Lunardi-Iskandar, Y., Georgoulias, V., Bertoli, A.M., Augery-Bourget, Y., Ammar, A., Vittecoq, D., Rosenbaum, W., Meyer, P., and Jasmin, C. (1989): Impaired in vitro proliferation of hemopoietic precursors in HIV-1 infected subjects. *Leukemia Res.* 13, 573.

Moore, M.A.S. (1991): Clinical implications of positive and negative hematopoietic stem cell regulators. *Blood* 78, 1.

Roux-Lombard, P., Modoux, C., Cruchaud, A., and Dayer, J.M. (1989): Purified blood monocytes from HIV 1-infected patients produce high levels of TNF-α and IL-1. *Clin. Immunol. & Immunopathol.* 50, 374.

Scadden, D.T., Zeira, M., Woon, A., Wang, Z., Schieve, L., Ikeuchi, K., Lim, B.,and Groopman, J.E. (1990): Human immunodeficiency virus infection of human bone marrow stromal fibroblasts. *Blood* 76, 317.

Sonoda, Y, Okuda, T., Yokota, S., Maekawa, T., Shizumi, Y., Nishigaki, H., Misawa, S., Fujii, H., and Abe, T. (1990): Actions of human interleukin-4/B-cell stimulatory factor-1 on proliferation and differentiation of enriched hematopoietic progenitor cells in culture. *Blood* 75, 1615.

Sun, N.C.J., Shapshak, P., Lachant, N.A., Hsu, M.Y., Lieger, L., Schmid, P., Beall, G., and Imagawa, D.T. (1989): Bone marrow examination in patients with AIDS and AIDS-related complex (ARC): morphologic and in situ hybridization studies. *A.J.C.P.* 92, 589.

Zauli, G., Davis, B.R., Re, M.C., Visani, G., Furlini, G., and La Placa, M. (1992): tat protein stimulates production of transforming growth factor-β1 by marrow mcrophages: a potential mechansims for human immunodeficiency virus-1-induced hematopoietic suppression. *Blood* 80, 3036.

This research was supported in part by Naval Medical Research and Development Command work unit # 63105A DH29 AD 010. Portions of this work were done while G.N. Schwartz, Ph.D. held a Senior National Research Council Research Associateship at Walter Reed Army Institute of Research. The opinions or assertions contained herein are the private views of the authors and are not to be construed as official or reflecting the views of the Departments of the Army or Navy or the Department of Defense.

Retrovirus induced inhibition of hematopoiesis: failure to establish long-term marrow cultures using LP-BM5 murine retrovirus infected marrow cultures

Vincent S. Gallichio, Nedda K. Hughes and Kam-Fai Tse

Hematology/Oncology Division, Departments of Internal Medicine and Clinical Sciences, University of Kentucky Medical Center and Department of Veterans Affairs, Lexington, KY, USA

Hematological abnormalities are associated with retrovirus infections. In human HIV-infection by the time full-blown symptoms develop, patients usually present with hypoplastic marrows and are pancytopenic. Although these observations have been a well documented phenomenon, the mechanism(s) upon which pancytopenia develops clinically is not well understood. We report here studies incorporating the use of LP-BM5 retrovirus infected immunodeficient mice and their ability to form long-term marrow cultures (LTMC) *in vitro*. Normal C57BL6 mice were infected with LP-BM5 (MuLV) immunodeficiency virus (10 µg total protein) i.p. Five-weeks post-viral infection, mice were sacrificed and marrow harvested from normal and virus-infected mice. LTMC were established in the presence of dose-escalation zidovudine (AZT) i.e., 10^{-6} M, 5×10^{-7} M, and 10^{-7} M *in vitro*. Compared to controls prepared from normal marrow, LTMC prepared from virus-infected cells failed to establish and subsequently release supernatant derived mononuclear cells from virus-infected cultures. AZT was ineffective in either establishing LTMC or consistently producing mononuclear cells. BFU-E, CFU-GM, and CFU-Meg were all reduced and none observed post-five weeks of culture. AZT failed to reverse this decrease in progenitor cells. Microscopic examination of cultures at 10-weeks failed to observe the establishment of a stromal layer using LP-BM5 infected marrow cells. The use of the LTMC system allowed for the evaluation of the effect of AZT on several classes of hematopoietic progenitors, i.e., CFU-GM, CFU-Meg, and BFU-E. The results described here demonstrate the importance of the effect of AZT on the marrow derived stroma and its ability to support hematopoietic stem cells. These results further indicate that in this *in vitro* system, AZT overtime, i.e., 15-weeks, produced an inhibition of BFU-E to 20% of control; however, this was observed only at the highest concentration tested, 10^{-6}M. In fact inhibition, although not as dramatic compared to BFU-E, was also observed for CFU-GM and CFU-Meg and at lesser concentrations. There has been considerable controversy as to the reported differences in toxicity profile on erythropoiesis and the induction of anemia, when comparing *in vivo* versus *in vitro* studies. The data reported here attempts to clarify this problem. In comparison of the *in vivo* versus *in vitro* data, these results differed from the observations described for the effects of AZT *in vivo* where the reduction in erythroid progenitor stem cell was markedly specific for CFU-E rather than BFU-E. In fact, depending upon the organ tested, e.g., bone marrow in spleen, BFU-E following AZT exposure, are elevated from the spleen. These data indicate LP-BM5 infection inhibits the formation of a normal stroma and therefore LTMC. LP-BM5 infection serves as a useful model to understand answers to questions related to interactions between retroviral infection, human AIDS and reduced stromal support of hematopoiesis.

VII. Bone marrow protection

VII. *Protection de la moelle osseuse*

MDR gene transfer into human hematopoietic cells

Charles Hesdorffer*, Maureen Ward**, Sylvio Podda**,
Christine Richardson**, Lauren Michel**, Larry Smith**,
Michael Gottesman***, Ira Pastan*** and Arthur Bank*,**

*Department of Medicine and **Genetics and Development, College of Physicians and Surgeons of Columbia University, New York; ***The National Cancer Institute, Washington DC, USA

ABSTRACT

This paper reviews the principles of gene therapy and its applications. The construction and safety of the unique packaging line created in our laboratory is described. Its usefulness in neomycin gene transfer is discussed. We explain why the MDR gene is a useful candidate for gene transfer in humans and discuss how we have been able to show expression of this gene eight months after gene transfer in our in vivo experiments in mice. Experiments are detailed involving the transfer of the MDR gene into human hematopoietic cells in which we have shown 10-50% of the colonies contain and express the transduced MDR gene. Finally, the implications of these findings on future human protocols is examined.

1. PRINCIPLES OF GENE TRANSFER:

There is considerable interest in the technology of gene transfer as a potential treatment of various diseases with the replacement or addition of normal genes. There are three requirements for gene therapy:

(1) Efficient gene transfer - the ability to insert a normal gene into a cell.
(2) Efficient gene expression - the proper transcription of the transferred gene to produce its protein product at an effective level.
(3) Safety - the gene transfer technology must be safe for the recipient.

2. GENE TRANSFER:

Retroviruses have been shown to be the most effective means of inserting new genetic material into cells (Mann et al, 1983; Joyner et al, 1983; Cone & Mulligan, 1984; Miller 1990). Retroviruses have the unique ability to enter cells with high efficiency and to integrate their DNA into the cell's genome. The major problem associated with the use of retroviruses is the formation of replication-competent viruses which as they spread could lead to multiple random integrations in the host genome. These integrations could lead to the activation of oncogenes with consequent tumor formation (Neel et al, 1981; Varmus et al, 1981). In order to prevent this complication modified retroviral packaging lines have been constructed to prevent the transfer into cells of replication-competent retroviruses.

Several years ago, a psi 2 packaging line was used. This line has a deleted psi sequence, the signal necessary for packaging viral RNA. When this line is transfected with a retroviral vector carrying the gene to be transferred, a single recombinational event can lead to the production of wild-type virus. Miller and Buttimore (1986), attempted to circumvent this problem by creating a packaging line with an additional mutation. This packaging line, PA317, in which additional deletions of the 5' LTR are employed, has been used extensively in various gene transfer studies in humans. Although two recombinational events are needed to produce wild-type virus with the use of this packaging line, intact retrovirus has been documented (Donahue et al, 1992).

Our laboratory has constructed retroviral packaging lines in which three mutational events are necessary for the production of wild-type virus (Markowitz et al, 1988). These packaging lines were created using the Moloney murine leukemia virus present in the plasmid 3PO with the psi mutation. Two plasmids were made one of which had the gag and pol genes of 3PO and the other, the env gene. Cell lines expressing the gag-pol and env regions from the different plasmids were made co-transfecting them into NIH 3T3 cells. Recipient cells were selected for those containing and expressing the desired plasmids using selectable markers. In this way safe efficient ecotropic and amphotrophic packaging lines were created. GP+E86 is the optimal ecotropic line, and GP-AM12, the best amphotrophic line. Over the past 4-5 years these lines have been tested extensively in our laboratory and others and used by over 300 laboratories worldwide and no incidence of wild-type viral breakout has been demonstrated. Stringent viral studies including the use of the 3T3 amplification assays as well as the evaluation of focus forming units in the sarcoma positive, leukemia negative (S+L-) assay (Haapala et al, 1985) have been performed indicating the in vitro safety of these retroviral packaging lines.

3. NEOMYCIN GENE TRANSFER IN MICE:

During the past decade a number of groups have shown that the neomycin resistance (neo^R) gene in a retroviral vector can be transferred to live irradiated mice and can be demonstrated in the majority of the recipient's bone marrow cells (Mann et al, 1983; Joyner et al, 1983; Cone & Mulligan, 1984; Miller et al, 1984; Williams et al, 1985; Dick et al, 1985; Keller et al, 1985; Eglitis et al, 1985). Most workers showed a 20% success rate in integrating the neo^R gene into lethally irradiated mouse recipients. Using our unique retroviral vector GP+E86 we noted 50% efficiency of neo^R gene transfer (Hesdorffer et al, 1990). In addition, we were able to use this system to define aspects of stem cell kinetics which may be important in the process of gene transfer in humans. We showed that at least a certain number of stem cells must be infused in order to allow long-term reconstitution of the mouse marrow. In experiments in which successive generations of mice were given transduced hematopoietic cells, we were able to determine that approximately 30 stem cells seemed to be the limiting number of stem cells that would allow long-term marrow reconstitution (Hesdorffer et al, 1990). This figure parallels that of Spangrude and co-workers (1988) who used a different approach in which the mouse hematopoietic precursor cells were selected with a series of monoclonal antibodies directed against various cell surface antigens.

4. THE MDR GENE - why use it?

The successful transfer of the neo^R gene is enhanced if the cells exposed to the retroviral vector are preselected with G418, a neomycin analogue, prior to infusion into a mouse. We can infer from this that a selectable gene transferred along with a co-transferred non-selectable gene on the same retroviral vector would enrich the population of infused marrow cells containing both genes. The selection would insure that if only a small number of cells are retrovirally infected (or transduced), the marrow could be enriched for these cells. If there is no selection, then untransduced cells compete with the transduced cells for growth in recipients. The selectable marker G418 cannot be used in animals or humans because of its toxicity, but the human multiple drug resistance (MDR) gene is useful as a potential in vivo selectable marker.

The MDR gene is a 600kb piece of DNA located on the long arm of chromosome 7 which encodes for the production of a transmembrane glycoprotein. This glycoprotein called P-glycoprotein acts as an active energy dependant pump, increasing drug efflux and decreasing cell membrane permeability in various cancer cell lines (Dano, 1973; Stein et al, 1992; Skovsgaard, 1978; Inaba et al, 1979; Fojo et al, 1985). Table 1. indicates the list of anti-cancer drugs whose mechanism of resistance has been shown to depend upon the MDR gene. Bone marrow cells express low levels of the MDR gene. Therefore, these cell are particularly susceptible to drugs which depend on the presence of MDR for their inactivation.

TABLE 1.

Vinca Alkaloids - vinblastine
Anthracyclines - daunorubicin
Epipodophyllotoxins - etoposide
Antibiotics - actinomycin D
Others - taxol, mitomycin C, topotecan, mithramycin

5. MDR GENE TRANSFER IN MICE:

A retroviral vector was constructed using an MDR cDNA which contains the entire coding sequence of the MDR gene and in which the expression of the MDR gene is driven by Harvey leukemia virus LTRs enhanced by a Harvey LTR enhancer. The MDR cDNA alone when injected into mouse fertilized eggs resulted in a line of mice expressing increased MDR in its bone marrow cells. This line of mice was made less leukopenic by daunomycin than normal mice similarly treated. The MDR cDNA in a retroviral vector was shown to transduce mouse and human cells. We transfected this retroviral vector into our amphotropic and ecotropic packaging lines. The highest titer producer lines containing and expressing the intact MDR gene (based upon the survival of cells exposed to colchicine) were isolated: 5×10^5 viral particles/ml for the ecotropic GP+E86 line and 5×10^4 viral particles/ml for the amphotropic AM12 cell line.

In initial experiments we co-cultured the ecotropic producer line with mouse bone marrow cells and transfused these transduced cells into lethally irradiated mice (Podda et al, 1992). Using PCR analysis with human MDR-specific oligonucleotide primers 50 days after transplantation, we demonstrated the presence of the human MDR gene in the peripheral blood of 90% of the transplanted mice. Additionally, at eight months posttransplantation, 50% of the transplanted mice demonstrate the human MDR gene in the peripheral blood. Southern blot analysis of the marrow of sacrificed animals corroborates the presence of the human MDR gene. Furthermore, in order to define the expression of the MDR gene we used a mouse monoclonal antibody, 17F9, which recognizes an external epitope of the MDR P-glycoprotein on the surface of cells. Using cells treated with this antibody, and then exposed to a fluorescent mouse anti-IgG2b in fluorescent-activated cell sorting (FACS) analysis, we showed that a significant number of the bone marrow cells from the mice sacrificed at eight months posttransplantation express high levels of MDR protein. To determine the cell type containing and expressing the MDR gene, cell gating on the basis of size and morphology was used to exclude lymphocytes from this analysis. Approximately 14% of the granulocyte-macrophage cells in the bone marrow samples analyzed showed markedly increased amounts of MDR P-glycoprotein on their surface as compared to controls. These data indicate that the cells containing and expressing the MDR gene are derived from bone marrow stem cells.

In subsequent experiments we demonstrated our ability to select in vivo for MDR-transduced cells in the bone marrow. Four transduced mice, initially positive by PCR for the presence of the MDR gene at 50 days posttransplantation, subsequently lost their MDR-PCR signal at eight months. These mice received a single dose of 140mgm of taxol in an attempt to enrich for bone marrow cells containing and expressing the human MDR gene and to prove that MDR expression allows for cell selection. Seven days after taxol, the PCR signal in all four mice reappeared. In addition, FACS analysis of the peripheral blood of two of these mice showed 5-8% MDR-positive granulocytes (Podda et al, 1992).

These experiments demonstrate a number of important principles in mice prior to beginning human studies using the MDR gene: (1) bone marrow stem cells can be stably transduced with the human MDR gene for long periods of time; (2) MDR-transduced cells express high levels of MDR P-glycoprotein such that they may be protected from the toxicity of MDR-responsive drugs such as taxol; and (3) MDR-transduced cells can be selected in vivo using MDR-responsive drugs.

6. MDR GENE TRANSFER INTO HUMAN HEMATOPOIETIC CELLS:

More recently we have transduced human bone marrow cells with our highest titer amphotropic MDR producer line. In these studies supernatants from the producer line were used instead of co-cultivation with the MDR producer cells to avoid contamination of the bone marrow with producer cells and mimic the conditions to be used in human clinical trials. Ficoll-separated nucleated bone marrow cells (NBMC) were cultured for 24 hours in media containing 10% fetal calf serum, 10 units/ml IL-3, 200 units/ml IL-6, and 50 units/ml human stem cell factor (SCF). The NBMC were subsequently exposed to supernatants from the amphotropic MDR producer line 1-4 times, each for 8-12 hours, in order to determine whether retroviral transfer could be optimized by providing an excess ratio of virus to NBMC (Ward et al, 1992). PCR analysis with MDR primers from different exons yielded a unique 157 basepair band from transduced integrated MDR cDNA, while the endogenous human MDR gene containing introns gave either no signal or a band of larger size. The results of an experiment in which the MDR gene was successfully transduced into human NBMC is shown in fig. 1 (Ward et al, 1992). Analysis of both BFU-E and CFU-GEMM from transduced marrows indicates that the in vitro proliferative potential of the cells is preserved and that 10-50% of the colonies contain and express the transduced MDR gene (Ward et al, 1992).

The use of ficoll-separated whole marrow in potential human studies is problematic since the volume of virus containing the number of viral particles required to successfully transduce human marrow cells would be, however, economically and practically unfeasible. It is now well documented that the putative early marrow progenitor ($CD34^+$) cell can be separated from whole marrow and are enriched 50-100 fold for reconstituting early hematopoietic precursors including stem cells. The CD34+ cells can successfully reconstitute the marrow of an individual following exposure to high doses of marrow ablative chemotherapy (Kessinger and Armitage, 1991; Berenson et al, 1991). Using the same protocol of gene transduction, Fig. 2. indicates that marrow enriched for $CD34^+$ cells, using either the Applied Immune Sciences plates (Lebkowski et al, 1992) or the Ceprate^{TM}SC stem cell concentrator (Berenson et al, 1991), can be successfully transduced with the MDR gene.

7. IMPLICATIONS FOR HUMAN PROTOCOLS:

Human bone marrow stem cells express higher than normal levels of MDR (Chaudhary & Roninson, 1991). However, this higher level of MDR expression has no functional meaning for animals or patients. Normal mice become severely neutropenic after exposure to daunomycin, while mice, whose bone marrow contains an exogenous human MDR gene under the influence of a strong promoter, are unaffected (Galski et al, 1989; Mickisch et al, 1991); MDR-transduced bone marrow cells in live mice are preferentially protected from toxicity and are enriched after exposure to taxol (Sorrentino et al, 1992; Podda et al, 1992). Furthermore, it is well known that the primary toxicity of MDR-responsive drugs, when used to treat patients with cancer, is marrow ablation.

The addition of the MDR gene with a strong promoter (as in the retroviral vector we are using in our studies) should lead to the protection of transduced human marrow cells exposed to MDR-responsive drugs such as the anthracyclines, vincaalkaloids, etoposides and taxol. We are planning studies in which patients undergoing bone marrow harvesting and autologous bone marrow transplantation for breast, brain and ovarian cancer have the MDR gene transduced into a portion of their harvested marrow. The gene transduced marrow will subsequently be reinfused into these patients after marrow ablation with high doses of chemotherapy. The ultimate goal of these studies is to treat patients with residual or recurrent disease with higher than normally tolerable doses of MDR-responsive drugs such as taxol and topotecan (an etoposide). The success of MDR gene therapy may add to the clinicians armamentarium of growth factors and peripheral stem cell support in the management of patients with incurable malignancies as well as provide insights into the process of gene therapy for patients more generally.

FIGURE 1: NBMC incubated for varying periods with the MDR producer cell line. Lysates were assayed for the integration of the human MDR cDNA by PCR. Lane 1: PCR buffer. Lane 2: Lysis buffer. Lane 3: MDR retroviral plasmid. Lane 4: MDR amphotropic producer cell DNA. Lane 5: untransduced NBMC. Lanes 6-9: MDR-transduced NBMC; 1,2,3 and 4 rounds of infection respectively. The MDR signal increases with the number of rounds of infection.

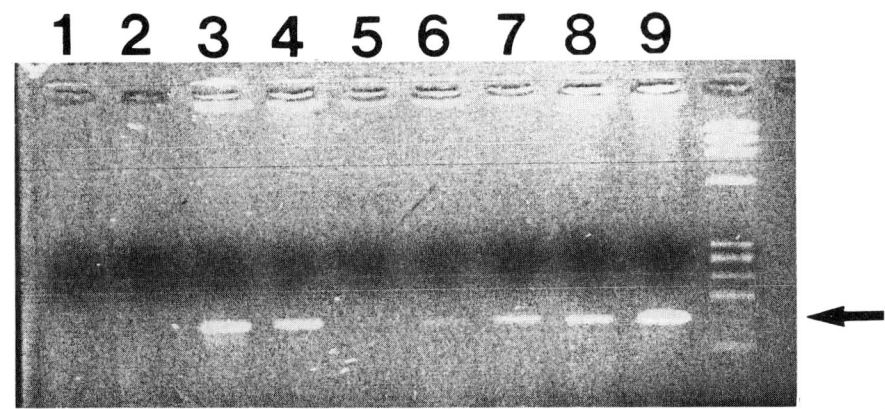

FIGURE 2: MDR-PCR analysis of $CD34^+$ cells exposed to multiple rounds of infection. Lane 1: $CD34^+$ transduced cells. Lane 2: dH_2O. Lane 3: Lysis buffer. Lane 4: $CD34^+$ untransduced cells. Lane 5: MDR amphotropic producer cell DNA. Lane 6: MDR retroviral plasmid.

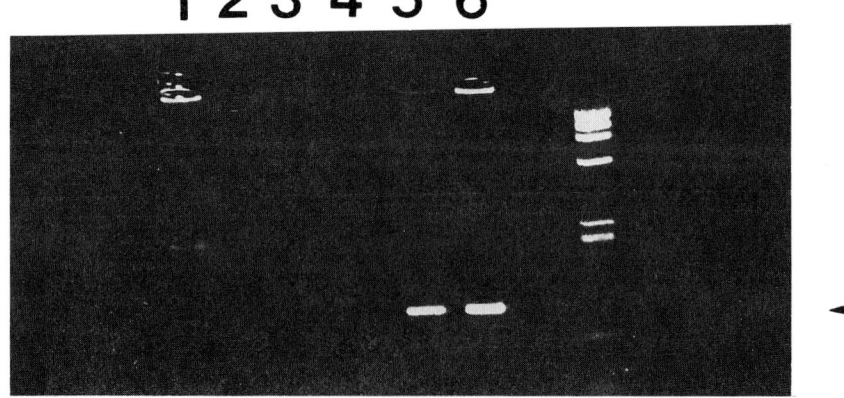

REFERENCES

Berenson RJ. et al. (1991): Blood 77,1717-1722.
Chaudhary PM & Roninson IB. (1991): Cell 66,85-94.
Cone R. & Mulligan RC. (1984): Proc. Natl. Acad. Sci. 81,6349-6353.
Dano K. (1973). Biochem. Biophys. Acta 323,466-483.
Dick JE. et al. (1985): Cell 42,75-79.
Donahue RE. et al. (1992): J. Exp. Med. 176,1125-1135.
Eglitis MA. et al (1985): Science 230,1395-1398.
Fojo A. et al. (1985) Cancer Res. 45,3002-3007.
Galski H. et al. (1989): Mol. Cell Biol. 9,4357-4363.
Haapala et al. (1985): J. Virology 53,827-833.
Hesdorffer CS. et al. (1990): DNA & Cell Biol. 9,717-723.
Inaba M. et al. (1979) Cancer Res. 39,2200-2203.
Joyner A. et al. (1983): Nature 305,556-558.
Keller G. et al (1985): Nature 318,149-155.
Kessinger and Armitage. (1991): Blood 77,211-213.
Lebkowski JS. et al. (1992): Transplantation 53,1011-1019.
Mann R. et al. (1983): Cell 33,153-159.
Markowitz D. et al. (1988): J. Virol. 62,1120-25.
Markowitz D. et al. (1988): Virology 167,400-405.
Mickisch G. et al. (1991): Proc. Natl. Acad. Sci. 88,547-551.
Miller AD. et al. (1984): Science 225,630-632.
Miller AD. and Buttimore C. (1986) Mol. Cell. Biol. 6,2895-2902.
Miller AD. (1990): Blood 76,271-78.
Neel BG. et al. (1981): Cell 23,323-334.
Podda S. et al. (1992): Proc. Natl. Acad. Sci. 89,9676-9680.
Skovsgaard T. (1978) Cancer Res. 38,4722-4727.
Sorrentino B. et al. (1992): Science 257,99-103.
Spangrude GJ. et al. (1988): Science 241,58-62.
Stein W. et al. (1992) Trends in Biochemical Sci. (Submitted).
Varmus HE. et al. (1981) Cell 25,23-36.
Ward M. et al. (1992): Blood suppl 80,239A.
Williams D. et al. (1985): Nature 310,476-480.

Résumé

Cet article passe en revue les principes de la thérapie génique et ses applications. La construction et les caractéristiques de la lignée d'empaquetage unique créée dans notre laboratoire y sont décrites. Son utilisation pour le transfert du gène néomycine est discutée. Nous expliquons pourquoi le gène MDR est un candidat intéressant pour le transfert de gènes chez l'homme et discutons comment nous avons été capables de montrer son expression huit mois après son transfert dans des expériences in vivo chez la souris. Nous rapportons en détail des expériences de transfert du gène MDR dans les cellules hématopoïétiques humaines dans lesquelles nous avons montré que 10-50% des colonies contiennent et expriment le gène MDR transduit. L'implication de ces résultats dans les futurs protocoles cliniques est examinée.

ns. Colloque INSERM/John Libbey Eurotext Ldt.

Cytokine interactions in protection from lethal irradiation: synergy of IL-1 and c-kit ligand

R. Neta*, J.M. Wang**, J.J. Oppenheim**, N. Davis* and C.M. Dubois***

*Department of Experimental Hematology, Armed Forces Radiobiology Research Institute, Bethesda, MD;
**Laboratory of Molecular Immunoregulation, National Cancer Institute, Frederick, MD and
***Immunology Division, Faculty of Medicine, University of Sherbrooke, Sherbrooke, Quebec, Canada

ABSTRACT

Administration of LPS, IL-1, or TNF prior to whole body ^{60}Co γ-irradiation protects mice from lethal hematopoietic syndrome. However, to confer such protection, each one of these agents requires endogenous production and interaction of IL-1, TNF, and IL-6, since neutralizing antibodies to each of the three cytokines blocked IL-1 and TNF-induced radioprotection. Neutralizing antibodies to c-kit ligand (KL) also blocked LPS and IL-1-induced radioprotection indicating that endogenously produced KL is also required for protection from lethal hematopoietic syndrome. Conversely, radioprotection induced by KL was reduced by anti-IL-1R antibody, indicating that IL-1 is required for KL radioprotection. In contrast to IL-1 which induces IL-6, Colony Stimulating Factors (CSFs) and MnSOD, KL administration did not result in induction of circulating IL-6, CSF or MnSOD gene expression in bone marrow cells, suggesting distinct mechanisms for protection by these cytokines. The combined administration of IL-1 and KL resulted in synergistic radioprotective effect. As both the mRNA for c-kit and KL binding on the bone marrow cells were elevated within hours of IL-1 administration to mice, this synergy may be based on IL-1-induced upregulation of KL receptors . Thus, interaction of several endogenously produced cytokines: IL-1, IL-6, TNF, and KL are responsible for protection from radiation-induced lethal hematopoietic syndrome. This interaction may in part depend on receptor upregulation by and for these cytokines.

INTRODUCTION

Death resulting from exposure to $LD_{100/30}$ doses of ionizing radiation can be prevented by transplantation of undamaged bone marrow cells and has therefore been referred to as lethal hematopoietic syndrome (Bond et al., 1965, van Bekkum, 1969). Alternatively, the death of animals receiving $LD_{100/30}$ doses of radiation can be prevented by pretreatment with immunomodulatory agents or proinflammatory cytokines (Ainsworth and Hatch, 1958, Neta et al., 1986; 1988). We have demonstrated that in the case of immunomodulatory LPS, this effect is mediated by the endogenously produced proinflammatory cytokines, IL-1 and TNF, since antibodies to these cytokines block LPS-induced radioprotection (Neta et al., 1991). Radioprotection with IL-1 and TNF, in turn, can be abrogated with anti-IL-6 antibody as well as anti-TNF and anti-IL-1R antibody (Neta et al., 1991; 1992), providing evidence that obligatory interaction of these three endogenously produced cytokines is required for protection from lethal hematopoietic syndrome.

Recently, the ligand for c-kit (KL), a hematopoietic cytokine, has been identified and cloned (Williams et al., 1990, Zsebo et al., 1990). KL deficient mice (Steel) have a defect in stromal cells which produce the ligand for c-kit. The availability of KL in a recombinant form permitted numerous studies aimed at elucidating hematopoietic regulatory interactions. KL synergizes with a number of cytokines to stimulate the growth of hematopoietic progenitors in vitro as well as stimulates hematopoiesis in vivo (Broxmeyer et al., 1991, de Vries et al., 1991, Muench et al., 1992, Ogawa et al., 1991). Kit receptor is expressed in an adult animal on mast cells and hematopoietic early and more committed progenitor cells, but not on mature neutrophils, lymphocytes or macrophages (Broxmeyer et al., 1991). Mice receiving antibody to c-kit receptor displayed a loss of hematopoietic progenitor cells (Ogawa et al., 1991), whereas administration of KL prior to lethal irradiation protected mice from death from hematopoietic failure (Zsebo et al., 1992).

In this study we have first evaluated the contribution of endogenously produced KL in radioprotection induced by IL-1 and LPS. The antibody to KL abrogated IL-1 and LPS-induced radioprotection. This suggested that endogenously present KL may be the primary mediator required for radioprotection. To test this hypothesis we have also examined the effect of anti-IL-1R antibody on KL radioprotection. Our results show that endogenously produced IL-1 is contributing to KL radioprotection, that IL-1 and KL synergize in radioprotection and that this synergy may be based on upregulation of c-kit receptors by IL-1.

MATERIALS AND METHODS

Mice. B6D2F1 female mice, 8-10 weeks old, were purchased from Jackson Laboratories (Bar Harbor, MA). Mice were handled as previously described (Neta et al., 1988).

Antibodies. A rat monoclonal IgG1, anti-IL 1 receptor antibody (35F5), was kindly provided by Dr. Chizzonite (Hoffmann-La-Roche). Chromatographically purified rat IgG (Sigma) was used as control. Polyclonal anti KL antibody (P2) was raised in rabbit against purified recombinant yeast derived murine KL and was a generous gift from Dr. Douglas Williams (Immunex). As a control equivalent concentrations of normal rabbit serum or pre-immune serum was employed.

Cytokines. Recombinant human interleukin 1 (rHu IL 1α 117-271 Ro 24-5008 lot IL 1 2/88 activity 3×10^8 units/mg) was kindly provided by Dr. Peter Lomedico, Hoffmann-La Roche (Nutley, NJ). Rat KL-PEG (Zsebo et al., 1992) was a generous gift from Dr. Ian McNiece (Amgen, Thousand Island, CA).

Irradiation. Mice were randomized, placed in Plexiglass containers and were bilaterally irradiated using ^{60}Co γ-photons at a mid-line tissue dose rate of 40 cGy/min. The mice received whole body irradiation and the field was uniform within ±2%. The number of surviving mice was recorded daily for 30 days.

KL binding on Bone Marrow Cells. Human recombinant KL was iodinated by chloramine T method which yielded ^{125}I-KL preparations with specific activities at about 20 uCi/ug protein. 5×10^6 BMC were used for each sample and the assay was performed as described (Dubois et al., 1991).

Northern blot analysis. Bone marrow cells collected from 6 femurs/group were peleted and total cellular RNA was isolated by acid guanidium thiocyanate-phenol-chloroform extraction. RNA was separated by electrophoresis on 1% agarose and transferred onto a Hybond-N (Amersham, Oakville, Ontario) membrane for Northern analysis. The murine c-kit probe was a 3710 bp EcoRI-HindIII insert of the c-kit cDNA clone pGEM3 c-kit3 and the human superoxide dismutase probe was a 0.1 Kb PstI insert of the SODI cDNA clone pSP 64 cSOD. As a control for RNA integrity, blots were rehybridized with a 1 kB Pat1 cDNA probe (ATCC) of the housekeeping gene glyceraldehyde phosphate dehydrogenase (GAPDH).

RESULTS

The effect of anti-cytokine antibody on LPS and IL-1-induced radioprotection. To assess the contribution of KL to IL 1- and LPS-enhanced survival from lethal irradiation, mice were treated with the antibodies or control immunoglobulin prior to administration of LPS or IL 1 and $LD_{85/30}$ irradiation. The results (Table I) indicate that as with anti-IL-1R antibody, anti-KL antibody blocked IL 1-induced protection from radiation lethality. Furthermore, this antibody also completely blocked LPS-induced radioprotection.

Table 1. The Effect of Anti-KL Antibodies on Survival of Mice Radioprotected with LPS or IL 1

Treatment	LPS	%survival	IL-1	%survival
saline	−	17	−	15
rat Ig	+	60	+	58
aIL 1R	+	25*	+	12.5*
aKL	+	5*	+	10*
rab. Ig	+	−	+	83

CD2F1 or B6D2F1 mice, received intraperitoneally 100 μg of anti-IL 1R antibody, 1:10 dilution of polyclonal rabbit anti-KL serum or preimmunized rabbit or saline in 0.5 ml total volume. Twenty hours later mice received 100 ng IL 1 or 1 μg of LPS intraperitoneally, and 1 day later were given 950 cGy γ-radiation.
* Different ($p < 0.05$) from the treatment groups, but not different from the control, saline only treated mice.

The effect of anti-IL-1R antibody on KL radioprotection.
To examine whether endogenously produced IL-1 contributes to KL-induced radioprotection mice received 100 ug/mouse of anti-IL-1R antibody, rat immunoglobulin in an equivalent dose or saline, and 6 hours later a single dose of 5 ug PEG-KL. Whereas at a supralethal dose of 1050 cGy of radiation 63% of saline pretreated and 64% of immunoglobulin pretreated mice were protected from death by KL, only 25% of mice receiving anti-IL-1R antibody and KL survived irradiation. Thus, although anti-IL-1R antibody did not entirely abolish KL radioprotective effect, it reversed its protective effect in 40% of treated animals indicating that IL-1 contributes to KL-induced radioprotection.

The effect of combinations of IL-1 and KL in protection from lethal irradiation. The observed co-dependency of KL and IL-1 in their radioprotective effect led us to examine the effect of their combined administration on survival of irradiated mice. Single injection of the two cytokines in combination, 18 hours before irradiation, resulted in synergistic radioprotection apparent at 1200 cGy to 1300 cGy doses of radiation. Whereas all of IL-1 treated mice were dead following exposure to 1200 cGy, and only 4% of KL-treated mice survived irradiation with this dose, 83% (n=61) of IL-1/KL treated mice survived irradiation with 1200 cGy and 38% (n=21) of mice survived 1300 cGy dose. The radioprotective effect of KL was not associated with induction of circulating hematopoietic growth factors, such as IL-6 or CSF or upregulation of MnSOD mRNA expression in bone marrow cells (data not shown), all of which are induced with IL-1.

<u>The effect of IL-1 on KL binding and c-kit gene expression by bone marrow cells</u>. Our previous results established that IL-1 treatment of mice resulted in upregulation of IL-1 binding due to expression of increased numbers of IL-1 receptors on hematopoietic progenitor cells. It was therefore plausible that IL-1 may similarly lead to upregulation of KL binding on bone marrow cells. Indeed, treatment of mice with 1 ug of IL-1, a dose employed in a combined IL-1/KL radioprotective treatment, resulted in 2 to 3 fold upregulation of radiolabeled KL binding, beginning at 4 hours after treatment and still detectable at 18 to 20 hours post IL-1 injection. This increase in KL binding correlated with an increase in c-kit mRNA expression within 2 hours following IL-1 treatment.

DISCUSSION

It was recently reported that, as for IL-1, KL treatment of mice prior to whole body exposure to ^{60}Co γ-irradiation increases the numbers of surviving mice concomitant with hematopoietic recovery that occurs by 6 to 10 days post irradiation (Zsebo et al., 1992). The mechanism for the radioprotection by KL is presently unknown. KL administration does not result in induction of cytokine cascade including CSFs and IL-6 or in upregulation of mRNA for MnSOD, all of which are induced by IL-1 and probably contribute to radioprotection. Our studies employing antibodies to KL and to IL-1R have established that endogenous production of both cytokines is required to protect mice treated with LPS or IL-1 from lethal effects of radiation (Table 1) and that radioprotection with KL also requires participation of IL-1. Combined administration of the two cytokines results in the ability of treated mice to survive doses of radiation over 40% greater than the original lethal dose (DRF=1.47) and at 1200 through 1300 cGy doses of radiation these cytokines had synergistic effect. The above findings suggested that IL-1 and KL cooperate in promoting survival after lethal irradiation. This cooperation may be based in part on IL-1 upregulating receptors for KL as indicated by our results showing an increased ^{125}I-KL binding as well as expression of c-kit mRNA by bone marrow cells.

Although numerous reports document synergy of IL-1, IL-3, IL-6, G-CSF, GM-CSF and KL in promoting in vitro growth of colonies of various lineages, the basis for this synergy have not been determined. Recent report demonstrated that hematopoietic growth factors IL-3 and GM-CSF down-regulate the mRNA for c-kit on mast cells and stem cell progenitor cell lines (Welham and Schrader, 1991). IL-4 in that study did not affect c-kit expression, however in another study with human mast cells and progenitor cells IL-4 downregulated c-kit expression (Sillaber et al., 1991). In contrast, TNF enhanced c-kit expression on acute myelogenous leukemia blasts (Brach et al., 1992). Whether TNF has this effect on hematopoietic progenitors and whether this upregulation is in part the basis for TNF-induced radioprotection awaits further studies. Our results provide the first evidence that IL-1/KL synergy may be based on direct or indirect effect of IL-1 in enhancing expression of c-kit on the bone marrow cells.

Previous studies from our as well as other laboratories demonstrated that administration of even minute (0.1 ug/mouse) doses of IL-1 has profound hematopoietic effects. These effects include increased cycling of hematopoietic progenitor cells and proliferative expansion of myeloid progenitor cells in the marrow. These were first detected within 6 hours and increased progressively for 48 hours (Neta et al., 1987, Johnson et al., 1989, Schwartz et al., 1989). A more recent study (Hestdal et al., 1992) demonstrated that IL-1 treatment of mice results within 24 hours in a 5 fold upregulation of HPP-CFC and 2-3 fold upregulation of CFU-c grown in the presence of GM-CSF or IL-3. This cell expansion was thought to be based on IL-1 induction of hematopoietic CSFs and their receptors upregulation. Our current results present an additional mechanism for expansion of progenitor cells through increased expression of c-kit. It remains to be established whether the IL-1-induced upregulation affects the c-kit bright (more primitive) or c-kit dull (less primitive) progenitor cells and whether this effect is direct or requires secondary mediators. Whereas IL-1-induced hematopoietic growth factors such as G-CSF and IL-6 synergise in vitro with KL in the expansion of hematopoietic cells, their administration does not result in upregulation of the expression of c-kit (Neta et al., unpublished results). Thus, these factors do not mediate c-kit upregulation by IL-1. This unique effect of IL-1 may contribute to its radioprotection.

ACKNOWLEDGEMENTS

We thank Mrs. Francine Grondin for her assistance with Northern blots. This work was supported by the Armed Forces Radiobiology Research Institute, Defense Nuclear Agency, under Research Work Unit 00129. The opinions or assertions contained herein are the private views of the authors; no endorsement by the Defense Nuclear Agency has been given or should be inferred. The research was conducted according to the principles enunciated in the <u>Guide for the Care and Use of Laboratory Animals</u> prepared by the Institute of Laboratory Animal Resources, National Research Council.

REFERENCES

Ainsworth EJ and Hatch MH: (1958): Decreased x-ray mortality in endotoxin-treated mice. Radiat Res 9:96.

Bond, V.P., T.M. Fliedner, and J.O. Archambeau. 1965. "Mammalian Radiation Lethality. A Disturbance in Cellular Kinetics." Academic Press, New York

Brach M.A., Buhring H.J., Grub H.J. et al. (1992): Functional expression of c-kit by acute myelogenous le ukemia blasts is enhanced by tumor necrosis factor through posttranscriptional mRNA stabilization by a labile protein. Blood, 80, 1224-1230.

Broxmeyer HE, Maze R, Miazawa K, et al. (1991): The kit receptor and its ligand, steel factor, as regulators of hemopoiesis. Cancer Cells, 3: 480-487.

de Vries P, Brasel KA, Eisenman JR et al. (1991): The effect of recombinant mast cell growth factor on purified murine hematopoietic stem cells. J. Exp. Med. 173, 1205-1211.

Dubois C.M., Ruscetti F.W., Keller J.R. et al. (1991): In vivo interleukin 1 (IL-1) administration indirectly promotes type II IL-1 receptor expression on hematopoietic bone marrow cells: Novel mechanism for the hematopoietic effects of IL-1.

Hestdal K., Jacobson S.E.W., Ruscetti F.W. et al. (1992): In vivo effect of interleukin 1a on hematopoiesis: Role of colony stimulating factor receptor modulation. Blood 80, 2486-2494.

Johnson C.S., Keckler D.J., Topper M.I. et al. (1989): In vivo hematopoietic effects of recombinant interleukin 1a in mice: Stimulation of granulocytic, monocytic megakaryocytic and early erythroid progenitors, suppression of late stage erythropoiesis, and reversal of erythroid suppression with erythropoietin. Blood, 73, 678-683.

Muench MO, Schneider JG, and Moore MAS (1991): Interactions among colony-stimulating factors, IL-1b, IL-6, and kit-ligand in the regulation of primitive murine hematopoietic cells. Exp. Hematol. 20, 339-349.

Neta R, Douches SD, and Oppenheim JJ. (1986): Interleukin-1 is a radioprotector. J. Immunol., 136:2483-2485.

Neta R, Oppenheim JJ, and Douches SD. (1988): Interdependence of the radioprotective effects of human recombinant IL-1, TNF, G-CSF, and murine recombinant GM-CSF. J.Immunol., 140:108-111.

Neta R, Oppenheim JJ, Schreiber RD, et al. (1991): Role of cytokines (interleukin 1, tumor necrosis factor, and transforming growth factor b in natural and lipopolysaccharide-enhanced radioresistance. J. Exp.Med., 173, 1177-1182.

Neta R., Perlstein R., Vogel S.N. et al. (1992): Role of IL 6 in protection from lethal irradiation and in Endocrine responses to IL 1 and TNF. J. Exp. Med. 175, 689 - 694.

Neta R, Sztein MB, Oppenheim JJ et al. In vivo effects of IL-1. I. Bone marrow cells are induced to cycle following administration of IL-1. J. Immunol., 139:1861-1866 (1987).

Ogawa M, Matsuzaki Y, Nishikawa S et al. (1991): Expression and function of c-kit in hematopoietic progenitor cells. J. Exp. Med. 174, 63-71.

Sillaber C, Strobl H, Bevec D, et al. (1991): IL-4 regulates c-kit proto-oncogene product expression in human mast and myeloid progenitor cells. J. Immunol. 147, 4224-4228.

Schwartz G.N., Patchen M.L., Neta R. et al. (1989): Radioprotection of mice with interleukin 1. Relationship to number of spleen colony forming units. Radiat. Res., 119:101-112.

Van Bekkum, D.W., (1969). Bone marrow transplantation and partial body shielding for estimating cell survival and repopulation. In "Comparative Cellular and Species Radiosensitivity", eds. V.P. Bond and T. Sugahara, Igaku Shoin, Tokyo, p.175.

Welham M.J. and Schrader J.W. (1991): Modulation of c-kit mRNA and protein by hematopoietic growth factors. Mol. and Cell. Biol. 11, 2901-2904.

Williams DE, Eisenman J, Baird A, et al. (1990): Identification of a ligand for c-kit proto-oncogene. Cell 63, 167-174.

Zsebo K.M., Wypych J., McNiece I.K., et al. (1990) Identification, purification and biological characterization of hematopoietic stem cell factor from Buffalo rat-conditioned medium. Cell, 63, 195-201.

Zsebo KM, Smith KA, Hartley CA, et al. (1992): Radioprotection of mice by recombinant stem cell factor. Proc. Natl. Acad. Sci. USA, 89, 9464-9468.

Résumé

L'administration de LPS, IL-1 ou TNF avant une irradiation gamma totale par le ^{60}Co protége les souris du syndrome hématopoiétique létal. Cependant, pour conférer une protection, chacun de ces agents nécessite la production endogène et l'interaction de l'IL-1, TNF et IL-6, car des anticorps neutralisant chacune de ces cytokines bloquent la radioprotection induite par le TNF et l'IL-1. Des anticorps neutralisant le c-kit ligand (KL) bloquent aussii la radioprotection induite par le LPS et l'IL-1, indiquant que le KL produit de façon endogène est aussi nécessaite pour la protection de l'hématopoièse. Inversement, la radioprotection induite par le KL est diminuée par l'anticorps contre l'IL-1R, ce qui indique que l'IL1 est nécessaire à la radioprotection par KL. Contrastant avec l'IL-1 qui induit de l'IL-6, des CSFs et la MnSOD, l'administration de KL n'entraine pas d'induction d'IL-6 circulante ni d'expression des gènes des CSFs et de MnSOD dans les cellules de moelle, ce qui suggère des mécanismes de protection différents pour ces cytokines. L'administration combinée d'IL-1 et de KL entraine une effet radioprotecteur synergique. Comme à la fois le mRNA de c-kit et la liaison de KL aux cellules médullaires sont augmentés dans les heures qui suivent l'administration d'IL-1 aux souris, cette synergie est peut-être basée sur l'induction par l'IL-1 d'une augmentation de l'expression des récepteurs de KL. Ainsi, les interactions entre plusieurs cytokines produites de façon endogène, IL-1, IL-6, TNF et KL, sont responsables de la protection contre le syndrome hématopoiétique létal induit par les radiations. Cette interaction dépendrait en partie de l'"augmentation de l'expression des récepteurs par et pour ces cytokines.

Protective effects of TNF and IL-1

James R. Zucali and Jan Moreb

Department of Medicine, Division of Medical Oncology, University of Florida, Gainesville, FL 32610, USA

ABSTRACT: Treatment of neoplastic conditions often involves the use of radiation therapy or cytotoxic drugs. Unfortunately, hematopoietic tissue is known to be highly sensitive to the damaging effects of such agents resulting in decreased bone marrow and immune function. Protection of bone marrow stem cells from these insults may provide a possible means of reducing the myelotoxicity associated with radio and chemotherapy while hopefully not affecting the susceptibility of the tumor cells to this therapy. Interleukin-1 (IL-1) and tumor necrosis factor alpha (TNF) have been shown to be beneficial in protecting the hematopoietic system from radiation and chemotherapy. In this report, we give an overview of studies using IL-1 and TNF as protective agents and discuss possible mechanisms involved in their protective action.

Dose intensity is emerging as a crucial determinant of success in cytotoxic cancer therapy; however, myelosuppression presents as one of the major complications encountered. The administration of biological response modifiers, or more specifically cytokines, either before or after cytotoxic therapy is being used to overcome this problem. Interleukin 1 (IL-1) and tumor necrosis factor alpha (TNF) are two well characterized cytokines that have been successfully cloned (1,2). Both have been shown to play a major role in the mobilization of the immune response, the activation of inflammatory processes, and the regulation of hematopoiesis (3,4). Because of their diverse and overlapping effects on hematopoiesis, IL-1 and TNF are being investigated for their usefulness in protection from the lethal myelosuppression and the neutropenic effects that accompany the use of chemotherapy and ionizing radiation.

A number of studies have described the beneficial effects of IL-1 and TNF in protecting mice from both irradiation and chemotherapy. In vivo, IL-1 is able to stimulate neutrophil egress from the bone marrow (5) and accelerate hematopoietic recovery in mice treated with chemotherapeutic agents (6). IL-1 administration has also been shown to protect normal mice from the acute lethal effects of cyclophosphamide, mafosphamide and 5-fluorouracil by accelerating hematopoiesis in these myelosuppressed animals (7). IL-1 in combination with G-CSF has been shown to accelerate recovery of hematopoietic progenitors and decrease the severity and duration of neutropenia associated with the administration of 5-fluorouracil to mice (8). In addition, IL-1 has been reported to prevent the marked depletion of myeloid and erythroid elements seen in the bone marrow after treatment with doxorubicin (9).

In our laboratory, we have shown that pretreatment of bone marrow cells with IL-1 and TNF results in the protection of very early hematopoietic progenitor cells from the lethal effects of the chemotherapeutic agent 4-hydroperoxycyclophosphamide (4-HC) (10,11). Although all types of colonies in culture are protected by IL-1 and TNF from 4-HC, the most interesting are the undifferentiated blast colony forming cells. These colonies grow quickly into large colonies and give rise to secondary undifferentiated blast cell and mixed colonies upon replating (10,11). Because such features point to the primitive nature of blast cell colonies, we decided to determine whether these progenitors were indeed true pluripotent stem cells. Human bone marrow cells were preincubated with IL-1 and TNF prior to a lethal dose of 4-HC. These cells were then plated on top of long-term irradiated bone marrow stroma. Reconstitution was observed for 5-6 weeks in these long-term stromal cultures indicating that long-term reconstituting cells, which may correlate to those cells responsible for the engraftment of transplanted bone marrow in vivo, were protected (12). In addition, we also injected bone marrow cells pretreated with or without IL 1 prior to 4-HC treatment into lethally irradiated mice and recorded their survival. We found that marrow cells pretreated with IL-1 were able to reconstitute 90% of the irradiated animals by 30 days as compared to 40% survival in animals transplanted with cells not pretreated with IL-1 (Fig 1). These studies suggested that hematopoietic stem cells were being protected by IL-1 and TNF.

Several previous studies by different investigators (13,14) have suggested that the reason for the relative resistance to 4-HC by stem cells when compared to other more differentiated hematopoietic cells or tumor cells is the high levels of cytosolic aldehyde dehydrogenase (ALDH) present. This enzyme, which is responsible for the inactivation of 4-HC, is present in elevated levels in CD 34 positive cells (14). In our studies, the protection from 4-HC by IL-1 and TNF was shown to require at least 20 hours of incubation with the cytokines. This fact, together with the prevention of protection by cycloheximide addition, suggested that upregulation of protein synthesis was involved in the protective actions of IL-1 and TNF (15). Therefore, we investigated the possibility that ALDH was induced by IL-1 and TNF. Using a specific inhibitor for cytosolic ALDH, diethylaminobenzaldehyde (DEAB), we completely prevented the protection observed by preincubation with IL-1 and TNF (15). Furthermore, we have also been able to demonstrate about a three fold enhancement of ALDH mRNA in bone marrow cells incubated with IL-1 and TNF. These results suggest that elevated levels of ALDH may be responsible for the protection observed.

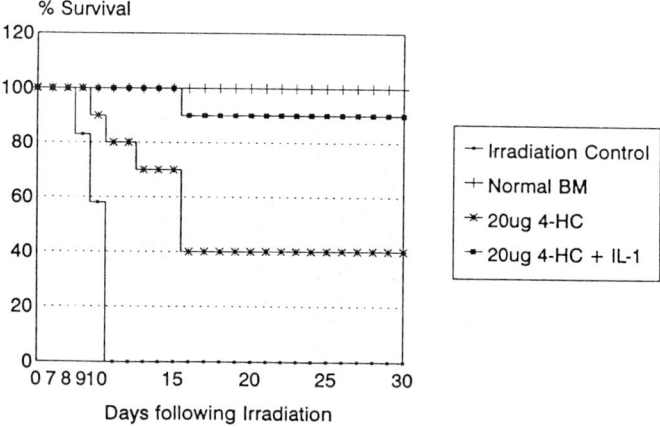

Fig. 1. Survival of mice injected with bone marrow cells treated for 20 hours with or without 100 ng/ml IL-1 prior to 30 minutes exposure to 4-HC.

Administration of IL-1 or TNF before irradiation will protect mice from doses of radiation that would be fatal to untreated animals. IL-1 will also promote the survival of previously untreated mice if administered within hours of the radiation (16). However, this approach requires higher doses of the cytokine and it is only effective when a sublethal dose of irradiation has been administered (16). Presumably, IL-1 after irradiation induces repair processes that correct established radiation damage; whereas, IL-1 given before irradiation may, in addition, prevent some of the deleterious effects of irradiation. IL-1 and TNF are the only cytokines reported to demonstrate optimal protection when administered prior to irradiation (16,17). These results suggest that IL-1 and TNF act to protect and aid in the recovery of host bone marrow cells from an irradiation insult.

Because all of the studies concerning protection of mice by IL-1 and TNF were done in vivo, it is not at all clear whether IL-1 and TNF act directly on hematopoietic stem cells. We have incubated male bone marrow cells with IL-1 prior to subjecting these cells to 8.0 Gy irradiation. These male cells were then injected into lethally irradiated female mice and survival recorded. In addition, animals were sacrificed at monthly intervals and the bone marrow, spleen and thymus analyzed for the presence of male cells using a Y chromosome specific probe and Southern analysis. We found an increase in survival of mice which received IL-1-treated male cells when compared to media-treated cells (Fig. 2). In addition, we found that male cells were present at all time intervals only in all three tissues of mice which had received IL-1-treated bone marrow cells. Furthermore, cells obtained from IL-1-treated primary recipients could repopulate secondary and tertiary recipients and male cells persisted in all three tissues of these animals. These results suggest that IL-1 treatment could protect long-term repopulating cells capable of bringing about reconstitution from radiation.

Since ionizing radiation produces a variety of free radical species capable of damaging intracellular targets, much attention has focused on the characterization and quantitation of endogenous antioxidant enzymes. The primary antioxidant enzymes are glutathione peroxidase, which acts on H_2O_2 and organic peroxides; catalase, which acts on H_2O_2; and superoxide dismutase, which converts superoxide O_2 to H_2O_2 (18). The importance of superoxide dismutase in the protection against oxidative stress has been suggested by a number of investigators (19,20) Petkau et al. (21) have demonstrated that the intravenous injection of mice with superoxide dismutase protects hematopoietic stem cells from irradiation as measured by spleen colony formation.

We have studied the radioprotective effects of IL-1 and TNF on a number of cell lines and normal murine bone marrow cells and found that IL-1 and TNF can induce radioprotection in normal hematopoietic bone marrow cells and in a human melanoma cell line A375 (22). In contrast, IL-1 and TNF were not radioprotective for the leukemic cell lines HL-60, K562, or Mo7e. We also observed that IL-1 and TNF were capable of inducing elevated mRNA and protein for manganese superoxide dismutase in both normal hematopoietic cells and the A375 cell line while no increase in MnSOD mRNA was seen in HL-60 or K562 leukemic cell lines. These results suggest that IL-1 and TNF may exert their radioprotective effects by increasing the amounts of MnSOD in cells that display protection. Further proof will require inhibiting the induction of MnSOD in cells and correlating the inhibition with the loss of radioprotection or alternatively, upregulating the production of MnSOD in cells that do not express it and demonstrating that there is a corresponding increase in radioresistance. More recent work has utilized cells transfected with sense MnSOD cDNA (23). These authors have shown that overexpression of MnSOD allowed for increased recovery from treatment with cytostatic and cytotoxic doses of IL-1 and TNF. In addition, cells transfected with the sense MnSOD cDNA showed increased survival after treatment with doxorubicin, mitomycin C and gamma irradiation in vitro.

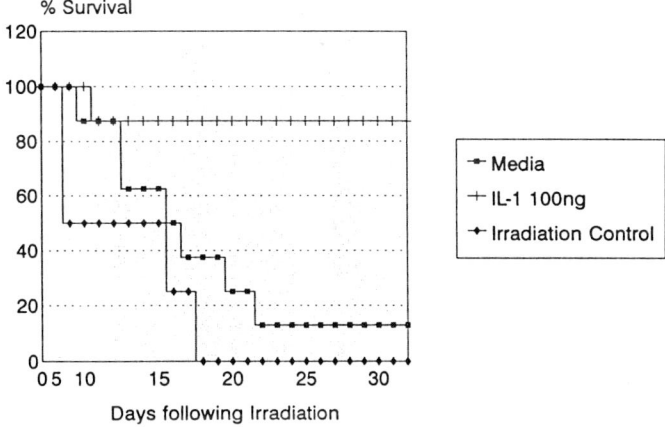

Fig. 2. Survival of mice transplanted with bone marrow cells treated for 20 hours with or without 100 ng/ml IL-1 prior to receiving 8.0 Gy irradiation.

CONCLUSIONS

A significant amount of data has accumulated in the literature regarding the protective ability of IL-1 and TNF against chemotherapy and irradiation. The mechanism(s) responsible for such protection are being investigated. Because IL-1 and TNF induce a number of different enzymes, it would seem reasonable that different mechanism might be involved in the protection observed from different therapeutic modalities. So far, two enzymes have been implicated; ALDH in the protection from 4-HC and MnSOD in the protection from irradiation. There are still many questions which need to be answered. For example, are there more than one mechanism working in the same cells at the same time? Are the mechanisms of resistance unique to normal hematopoietic cells or will tumor cells also be affected by IL-1 and TNF? Although we have not yet encountered protection of leukemic cells from 4-HC by IL-1 and TNF, we have observed radioprotection of a human melanoma cell line A375 induced by IL-1. More studies will be needed to clarify this observation. Future studies will probably involve gene therapy as a means of assessing the importance of specific enzymes which may be involved in the protection of hematopoietic cells from various chemotherapeutic agents or irradiation.

REFERENCES
1. Auron, P.E. et al (1984): Nucleotide sequence of human monocyte interleukin 1 precursor cDNA. Proc. Natl. Acad. Sci. USA 81, 7907-7911.
2. Pennica, D.G. et al (1984): Human tumor necrosis factor: precursor structure expression and homology to lymphotoxin. Nature 312, 724-729.
3. Walker, R.I. (1988): Acute radiation injuries. Pharmacol. Therapeutics 39, 9-12.
4. Dinarello, C.A. (1989): Interleukin-1 and its biologically related cytokines. Adv. Immunol. 44, 153-205.
5. Stork, L.C. et al (1989): Interleukin-1 accelerates murine granulocyte recovery following treatment with cyclophosphamide. Blood 73, 938-944.
6. Moore, M.A.S. & Warren, D.J. (1987): Synergy of interleukin 1 and granulocyte colony-stimulating factor: In vivo stimulation of stem cell recovery and hematopoietic regeneration following 5-fluorouracil treatment of mice. Proc. Natl. Acad. Sci. USA 84:7134-7138.
7. Futami, H. et al (1990): Chemoprotective effects of recombinant human IL-1α in cyclophosphamide-treated normal and tumor-bearing mice. J. Immunol. 145, 4121-4130.
8. Moore, M.A.S. et al (1990): Hematologic effects of interleukin-1β, granulocyte colony-stimulating factor, and granulocyte-macrophage colony-stimulating factor in tumor-bearing mice treated with 5-fluorouracil. J. Natl. Cancer Inst. 82, 1031-1037.
9. Epstein, D.A. et al (1989): Prevention of doxorubicin-induced hematotoxicity in mice by interleukin 1. Cancer Res. 49, 3955-3960.
10. Moreb, J. et al (1989): Protective effects of IL-1 on human hematopoietic progenitor cells treated in vitro with 4-hydroperoxycyclophosphamide. J. Immunol. 1423, 1937-1942.
11. Moreb, J. et al (1990): The effects of tumor necrosis factor-α on early human hematopoietic progenitor cells treated with 4-hydroperoxycyclophosphamide. Blood 76, 681-689.
12. Zucali, J.R. et al (1992): Protection of cells capable of reconstituting long-term bone marrow stromal cultures from 4-hydroperoxycyclophosphamide by interleukin-1 and tumor necrosis factor-α. Exp. Hematol. 20, 969-973.

13. Kohn, F.R. & Sladek, N.E. (1985): Aldehyde dehydrogenase activity as the basis for the relative insensitivity of murine pluripotent hematopoietic stem cells to oxazaphosphorines. Biochem. Pharmacol. 34, 3465-3471.

14. Kastan, M.B. et al (1990): Direct demonstration of elevated aldehyde dehydrogenase in human hematopoietic progenitor cells. Blood 75, 1947-1950.

15. Moreb, J. et al (1992): Role of aldehyde dehydrogenase in the protection of hematopoietic progenitor cells from 4-hydroperoxycyclophosphamide by interleukin-1ß and tumor necrosis factor. Cancer Res. 52, 1770-1774.

16. Neta, R. & Oppenheim, J.J. (1988): Cytokines in therapy of radiation injury. Blood 72:1093-1095.

17. Neta, R. et al (1988): Interdependence of the radioprotective effects of human IL-1, TNF-α, G-CSF, and murine recombinant G-CSF. J. Immunol. 140, 108-111.

18. Kumar, K.S. et al (1988): Radioprotection by antioxidant enzymes and enzyme mimetics. Pharmacol. Therapeutics 39,301-309.

19. Wong, G.H.W. & Goeddel, D.V. (1988): Induction of manganese superoxide dismutase by tumor necrosis factor: Possible protective mechanisms. Science 242, 941-944.

20. St. Clair, D.K. et al (1991): Overproduction of human Mn-superoxide dismutase modulates paraquat-mediated toxicity in mammalian cells. Fed. Eur. Biochem. Soc. 293, 199-203.

21. Petkau, A. et al (1975): Radioprotection of bone marrow stem cells by superoxide dismutase. Biochem. Biophys. Res. Commun. 67, 1167-1174.

22. Eastgate, J. et al (1993): A role for manganese superoxide dismutase in radioprotection of hematopoietic stem cells by interleukin-1. Blood 81,639-646.

23. Hirose, K. et al (1993): Overexpression of mitochondrial manganese superoxide dismutase promotes the survival of tumor cells exposed to interleukin-1, tumor necrosis factor, selected anticancer drugs, and ionizing radiation. FASEB 7, 361-368.

Résumé

Le traitement des néoplasies implique souvent l'utilisation de radiothérapie ou de drogues cytotoxiques. Malheureusement, le tissu hématopoïétique est connu pour être très sensible aux effets délétères de ces agents qui entrainent une diminution des fonctions médullaire et immune. La protection des cellules souches hématopoïétiques pourrait fournir un moyen de réduire la myélotoxicité associée à la radio/chimiothérapie sans affecter la sensibilité des tumeurs à ces traitements. L'interleukine-1 (IL-1) et le Tumor Necrosis Factor α (TNF α) paraissent être bénéfiques pour protéger le système hématopoïétique des radiations et de la chimiothérapie. Danc cet article, nous présentons une revue des études utilisant l'IL-1 et le TNFα comme agents protecteurs et nous discutons les mécanismes possibles de leur action protectrice.

Protection of non-self-renewing repopulating stem cells by the pentapeptide pEEDCK

Walter R. Paukovits, Marie-Hélène Moser and Johanna B. Paukovits

Laboratory of Growth Regulation, Institute for Tumor Biology-Cancer Research, University of Vienna, Vienna, Austria

Self-renewal and differentiation are thought to be characteristic for hematopoietic stem cells (HSC's). Investigating reconstituting HSC's we found an irreversible decimation after repeated injection of cytosine arabinoside. Suicide experiments revealed that these pre-CFU-S were non-proliferating in normal mice, and stayed quiescent after ara-C treatment, indicating that the most primitive HSC are a non-mitotic storage population, devoid of self-renewal potential. They maintain hematopoiesis by progression into the CFU-S compartment, subject to CFU-S population size. Protection of CFU-S against ara-C by inhibiting their proliferation with pGlu-Glu-Asp-Cys-Lys (pEEDCK) also reduced drainage of the stem cell pool, providing a new way for protection of hematopoiesis against drug induced aplasia. Our results show a dissociation of the "prime source" function of stem cells from the functions of self-renewal and differentiation, which are properties of the CFU-S. This dissociation may require a reformulation of traditional stem cell concepts.

All hematopoietic cells produced in an adult organism are thought to arise from a population of primitive stem cells (HSC), which are capable of differentiating into the major lineages of blood cells, which then multiply, mature, and ultimately become functional erythrocytes, granulocytes, etc.. During this process a series of cell divisions leads to a multiplicative population increase at each maturation step. In this way a high output of mature cells is sustained by a slowly dividing HSC population. Their number is thought to remain relatively constant over prolonged times, despite continuous losses by differentiation-accompanied cell divisions. To explain this constancy, the assumption has been made that each HSC has (in principle) the choice to either divide and become more differentiated, or to divide with both daughter cells remaining in the stem cell compartment (self renewal divisions). The balance between these alternative routes is thought to be determined by the requirements of the organism in such a way that the number of stem cells is kept constant.

For many years the most stem-cell-like experimentally accessible members of the hematopoietic system were the spleen colony forming units, CFU-S (Till and McCulloch, 1961) and major features of the stem cell concept were developed with the CFU-S as a model population. They are assayed by injecting bone marrow cells into lethally irradiated recipient mice. Some of them settle in the spleen where they grow to form macroscopic nodules of hematopoietic cells. Each spleen colony originates as a clone from a single CFU-S, and contains further (secondary) CFU-S as well as more mature cells of various lineages. The original single CFU-S seems thus to have self-renewed as well as differentiated. Since the CFU-S were thought to represent the HSC population, the capacities for self renewal and differentiation were considered as the defining properties of hematopoietic stem cells.

More recently it has become clear (Ploemacher and Brons, 1989; Harrison et al., 1990; Jones et al., 1990; Iscove, 1990) that the roots of the hematopoietic system are to be found beyond the CFU-S population, rendering the CFU-S merely the progeny of a still more basic pre-CFU-S population, responsible for the long term maintenance of hematopoiesis. Pre-CFU-S cells have been identified recently (Ploemacher and Brons, 1989; Harrison et al., 1990; Lerner and Harrison, 1990) by their potential for long term repopulation of the hematopoietic system after damage. They can be determined by the two-step method of Ploemacher and Brons (1989) or by competitive repopulation assays as described by Harrison et al. (1990). It appears, that pre-CFU-S and CFU-S are physically distinct and can be separated from each other (Ploemacher and Brons, 1989; Jones et al., 1990), although the extent of a possible overlap of the two populations is still an open question (Iscove, 1990; Spangrude et al., 1988). A similar cell type seems to be responsible for the sustained production of progenitor cells (CFU-GM) in long term bone marrow cultures (Fraser et al., 1990). The physical properties of these LTC replacing cells (Ploemacher and Brons, 1989; Eaves, 1991), in particular small size, low density, $CD34^+$, $My-10^{++}$, $HLA-DR^{low}$, low forward light scattering, and low fluorescence with rhodamine, are very similar to the in vivo pre-CFU-S, but different from the CFU-S especially with respect to the last parameter. The CFU-S, which can no longer be regarded as the ultimate hematopoietic stem cells, nevertheless seem to constitute a highly important bottleneck population, being the first and most immature transit population in the production line of the hemopoietic system. Their role appears to be (a) to undergo differentiation and to feed differentiated cells into the amplification compartments, while (b) at the same time maintaining their own population size by performing self-renewal divisions, and (c) to increase their proliferation rate and progenitor output in respon-se to a demand situation.

We were interested in the stem cell properties of the pre-CFU-S, in particular, if the essential properties defining a stem cell, namely self-renewal and differentiation, could also be demonstrated in the pre-CFU-S population. Since both, pre-CFU-S and CFU-S, are known to be proliferatively quiescent under normal conditions (Ploemacher and Brons, 1989; Lajtha et al., 1969) functional differences might become more clearly evident under circumstances leading to active proliferation of early hematopoietic cells, e.g. regeneration after treatment with cytostatic drugs. Earlier studies performed in our laboratory (Paukovits et al., 1990b, 1991b) have allowed to exactly determine the proliferative behavior of the CFU-S population after repeated injections of ara-C. Based on such results we could specify conditions leading to maximal proliferative activity (percentage of cells in S-phase) in this normally relatively quiescent population. For modulating CFU-S proliferation in a specific and reversible way we have used the CFU-S inhibitory pentapeptide pGlu-Glu-Asp-Cys-Lys (pEEDCK, Paukovits et al., 1990c, 1991b; Moser and Paukovits, 1991). By using Ploemacher's method (Ploemacher and Brons, 1989) we have then investigated the proliferative behavior of the pre-CFU-S under conditions forcing the CFU-S into maximal activity, unless inhibited by pEEDCK.

In this paper we describe the results of these experiments showing a) that the pre-CFU-S do not proliferate normally, b) that they remain quiescent even under conditions leading to their irreversible numerical decimation, c) that protection of CFU-S by pEEDCK reduces the compensatory progression of pre-CFU-S into active hematopoiesis. This surprising result indicates that, contrary to expectations based on the traditional stem cell concept, compensatory self-renewal is not a property of hematopoietic stem cells. They seem to constitute an exhaustible reservoir of quiescent primordial cells, which can (upon demand) progress into the CFU-S compartment where they express a high potential for proliferation and thus initiate the production of blood cells. It seems that the traditional picture of hematopoietic stem cells (and possibly our general concepts about what stem cells are and what they do) may require some modification.

Proliferative activation of quiescent CFU-S

Injection of ara-C leads to the intensive proliferation of the initially quiescent, and thus ara-C resistant, CFU-S within 12 to 16 hrs (Guigon et al., 1978; Paukovits et al., 1990b). In our experiments two ara-C injections were thus given 12 hrs apart. This results in a numerically reduced (Paukovits et al., 1991b) but actively cycling CFU-S population devoid of S-phase cells. During the following hours surviving non-S-phase CFU-S progress into the empty S-phase, detectable by suicide assay. After about 8 hrs a plateau is reached (Paukovits et al., 1991b), where 67% of the entire CFU-S population are synthesizing DNA (table 1). The application of pEEDCK (30 µg/kg) after the second ara-C injection reduced the S-phase population to about half of the plateau value (table 1), making them less sensitive to a third ara-C injection given at this time.

Table 1

CFU-S proliferation after 2 injections of ara-C, and its inhibition by pEEDCK. Groups of 5 female Balb/c mice were injected twice with ara-C (300 mg/kg, i.p.) with a 12 hour interval between the two drug applications. Control mice received isotonic saline. pEEDCK (30 µg/kg, i.p.) was applied 2,4, and 6 hrs after the second ara-C injection. Another two hours later the proliferative activity of the CFU-S was determined by the suicide method, using ara-C as the in vitro suicide agent (Paukovits et al., 1990b). CFU-S were determined by the spleen colony technique as described previously (Paukovits et al., 1991b). pEEDCK was synthesized, purified, and activated as described previously (Paukovits et al., 1990c). The values given ($\pm$ SEM) are the results of independent experiments, each with N evaluable recipient mice. The last line of the table gives the means of all experiments, together with the inter-experimental variation ($\pm$SD).

Percent CFU-S in S-phase					
2 x NaCl	N	2 x ara-C	N	2 x ara-C + pEEDCK	N
4.5±1.8	8	68.3±5.7	10	26.6±5.0	10
3.0±5.9	13	54.0±4.4	9	30.1±5.7	10
13.3±6.8	7	53.0±3.7	12	33.3±4.6	12
14.0±6.9	7	82.6±6.0	13	30.7±4.3	12
		65.2±5.5	14	31.3±1.5	15
		77.6±3.1	11		
		75.0±4.4	13		
		63.8±7.1	6		
8.7±5.8	35	67.4±10.7	88	30.4±2.4	59

Long term effects of repeated ara-C treatment on the repopulating capacity of the bone marrow.

To ensure complete hematopoietic recovery the animals were then left until day 67 without further treatment. Then the bone marrow cellularity, the femoral content of CFU-GM and day-11 CFU-S, and the marrow repopulating ability (pre-CFU-S) were determined. Table 2 gives the results obtained. They show that, down to the day-11 CFU-S level, all parameters had returned to normal. The repopulating ability of the marrow, as expressed by the content of pre-CFU-S cells was, however, severely reduced (table 2). While mice, which had received control (solvent) treatment only, showed a high pre-CFU-S level of about 190 per femur of primary recipient, the ara-C treatment had reduced this value to about 4. On the contrary, those mice, which had received the CFU-S inhibiting pEEDCK peptide before the third ara-C injection, had considerably improved pre-CFU-S numbers of about 90 (table 2). In other experiments we repeatedly measured pre-CFU-S numbers in drug treated mice until one year after treatment. No recovery of pre-CFU-S numbers was observed (table 3).

Relationship of pre-CFU-S losses to CFU-S proliferation.

The extent of the pre-CFU-S loss (table 2) is directly proportional to the cycling activity (i.e. drug sensitivity) of the CFU-S (table 1), with a regression parameter of $r^2 = 0.987$, indicating a possible causal relationship between the two parameters. Other experiments indicated, that this relationship remains exactly the same, independent of the nature of the agent used to induce CFU-S recruitment and killing.

In a similar way excessive rebound overshoots after leukodepression are the consequence of excessive proliferation of surviving CFU-S. We were thus interested if the observed pre-CFU-S losses (table 2) were correlated with the excess leukocyte production during rebound. The average daily increase in leukocyte counts (i.e. the excess production over steady state) during the rebound phase between day 5 and day 10 was 2020 ± 360 WBC/mm^3/ day in the 3 x ara-C treated mice, and 830 ± 170 WBC/mm^3/day in mice protected by pEEDCK. Regression analysis shows that the reduction of pre-CFU-S numbers is linearly correlated with these values.

Table 2
Hematopoietic status of mice 2 months after treatment with 3 doses of ara-C w/o inhibitory pEEDCK, or w/o stimulatory (pEEDCK)$_2$-dimer. Groups of 5 mice were treated with 3 x 300 mg/kg ara-C and 3 x 30 µg/kg pEEDCK (or solvent). The stimulatory (pEEDCK)$_2$-dimer (Paukovits, 1990; Paukovits et al., 1990c, 1991b) was applied as a continuous infusion at a dose rate of 1.4 µg/kg/day starting 4 hours and continued until day 7. All animals were subsequently left without further treatment until day 67. At this time the femoral bone marrow was obtained and assayed for its content of nucleated cells, committed progenitors CFU-GM, pluripotent spleen colony forming units day-11 CFU-S, and marrow repopulating pre-CFU-S.

	numbers per femur ± SD			
	nucleated cells (x10^{-6})	CFU-GM	day11-CFU-S	pre-CFU-S
control	13.7±4.5	32300±1800	6300± 800	192±34
3 x ara-C	14.2±2.5	24300±5200	6200±1300	4±5
3 x ara-C + pEEDCK	13.0±2.1	31800±6500	6700± 400	93±20
3 x ara-C + dimer	16.0±5.0	39200±6600	7400±1900	19±12

Table 3
Femoral pre-CFU-S content of mice after treatment with three doses of ara-C (300 mg/kg, as in table 2).

days after treatment	treatment	
	3 x ara-C	3 x NaCl
pre-treatment value	192 ± 34	192 ± 34
1	62 ± 11	
19	30 ± 22	200 ± 17
67	4 ± 5	
360	10 ± 7	195 ± 27

Table 4:
Determination of the proliferative status of the pre-CFU-S population 9 hrs after repeated ara-C treatment. Mice were treated with 3 x 300 mg/kg ara-C or 3 x isotonic saline. Nine hours after the last ara-C injection the bone marrow was obtained, and subjected to "suicide" by incubation with ara-C. After washing the cells, both aliquots were transplanted into irradiated recipients and the pre-CFU-S assay was performed as described above. Due to technical differences pre-CFU-S values cannot be compared with those of table 2.

	in vivo treatment	
	3 x NaCl	3 x ara-C
cells per femur (x10^{-6})	13.0 ± 2.5	3.3 ± 0.5
day-11 CFU-S per femur	2150 ± 420	190 ± 9
percent in S-phase	21.3 ± 10.8	91.7 ± 7.6
pre-CFU-S	90.5 ± 15.1	30.0 ± 10.5
percent in S-phase	0.0 ± 9.1	-7.7 ± 21.6

Determination of pre-CFU-S in S-phase by suicide.

The complete absence of proliferative activity in the pre-CFU-S population was more directly shown by suicide experiments in which we determined the number of pre-CFU-S in S-phase in mice injected three times with ara-C. Table 4 shows that 9 hours after the last ara-C injection the bone marrow cellularity was strongly decreased (75%), most day-11 CFU-S had been eliminated (91%), and almost all of the survivors were synthesizing DNA (92%). The femoral pre-CFU-S content of these mice was reduced to one third, but no pre-CFU-S were killed in the suicide experiment. This triple ara-C treatment, which was shown above to lead to long term depletion (2% left after 2 months) of the HSC population, thus did not induce any proliferative activity of the pre-CFU-S at a time when day-11 CFU-S were proliferating at a maximal rate.

Effects of triple ara-C treatment on long-term-culture initiating cells (LTC-IC)

The pre-CFU-S like long-term-culture initiating cells (LTC-IC) responsible for maintaining CFU-GM production in long term bone marrow cultures (Eaves, 1991) were assayed in the femoral marrows of mice treated with 3 x ara-C or isotonic NaCl. This assay measures the number of CFU-GM present in the second of two consecutive long-term bone marrow cultures, assuming that all CFU-GM present at this time (about 6-8 weeks after the initial inoculation of the first feeder layer with the test marrow) were produced from LTC-IC, which have similar physical properties as the pre-CFU-S. Table 5 gives the results obtained, showing that the number of CFU-GM present in and above the adherent layer at the end of the second culture period was strongly reduced in marrow cultures from 3 x ara-C treated mice as compared with cultures from control animals.

===

Table 5:
Reduced content of long-term-culture initiating cells (LTC-IC) in marrow from ara-C treated mice. Production of CFU-GM was measured in the second of two consecutive long term bone marrow cultures. Bone marrow cells from 3 x ara-C (or NaCl) treated mice were cultivated (see experimental section) for 19 days on a feeder layer of normal bone marrow cells. After enzymatic disruption of the adherent layer (containing the LTC-IC) the cell suspension was transferred onto a second feeder layer and further incubated for 21 days. Then the content of CFU-GM was determined in aliquots of the adherent and non-adherent layers.

in vivo treatment	culture layer[1]	CFU-GM per culture (mean ± SD)
3 x NaCl	N	489 ± 30
3 x ara-C	N	240 ± 27
3 x NaCl	A	921 ± 226
3 x ara-C	A	227 ± 44
3 x NaCl	N+A	1410
3 x ara-C	N+A	467

[1].......N = non-adherent, A = adherent
===

Discussion

HSC's have been conceptually endowed with certain properties thought to be general characteristics of stem cells. Two of these properties have been regarded as being most important, firstly the ability to differentiate and, secondly, the capacity for self-renewal ensuring replacement of stem cells which had been lost from the compartment. This picture was developed when CFU-S were thought to represent the hemopoietic stem cell population. It is now known, that CFU-S are not the ultimate HSC's, but when the "leadership" of the hemopoietic system was transferred from the CFU-S to their predecessors, the pre-CFU-S, their "stem cell properties" (self renewal and differentiation) were automatically co-transferred. However, nobody has ever seen a real HSC proliferating or self-renewing. What has been observed is, that HSC produced progeny (CFU-S and other), but it remains unknown if this process has anything to do with HSC proliferation. It has become customary to associate stem cell maintenance (e.g. in culture, Fraser et al., 1990) with stem cell proliferation, but that HSC thrive under certain conditions does not necessarily mean that they proliferate under these conditions. We have tried to avoid such uncertainties by directly measuring DNA-synthesis in the pre-CFU-S population with suicide techniques. By combining Ploemacher's method for the detection of pre-CFU-S with Dresch's suicide technique (Dresch et al., 1983), utilizing the strictly phase-specific cytotoxicity of ara-C, we determined the percentage of HSC in S-phase. As our results show, the complete insensitivity of the pre-CFU-S to suicide conditions, their irreversible decimation, as well as the exactly linear relationship between CFU-S proliferation and pre-CFU-S expenditure indicate, that the pre-CFU-S did not self-renew in order to compensate the drastic reduction of their number (table 2 and 3) brought about by repeated ara-C treatment (Paukovits et al, 1993). A similar conclusion must be drawn from the reduction of LTC-IC after in vivo treatment with ara-C (table 4).

The putative self-renewal capacity of HSC's seems thus not to apply to the pre-CFU-S. This view is also supported by the observation that pre-CFU-S do not readily form colonies in the spleens of irradiated hosts (Ploemacher and Brons, 1989) indicating that they remain quiescent even under such extremely stimulatory conditions. Furthermore a rapid exhaust of the repopulation potential has been observed (Jones et al., 1989, Harrison et al., 1990; Brecher et al., 1990) upon serial transplantation, which has been interpreted as the dominance of excessive differentiation stimuli overbalancing the stimuli for pluripotent HSC's to replenish themselves (Harrison et al., 1990). Our results, however, suggest that proliferation in the stem cell compartment is absent and not merely overbalanced by other influences. Only after their transition into the next compartment (CFU-S) the potential of these cells becomes activated.

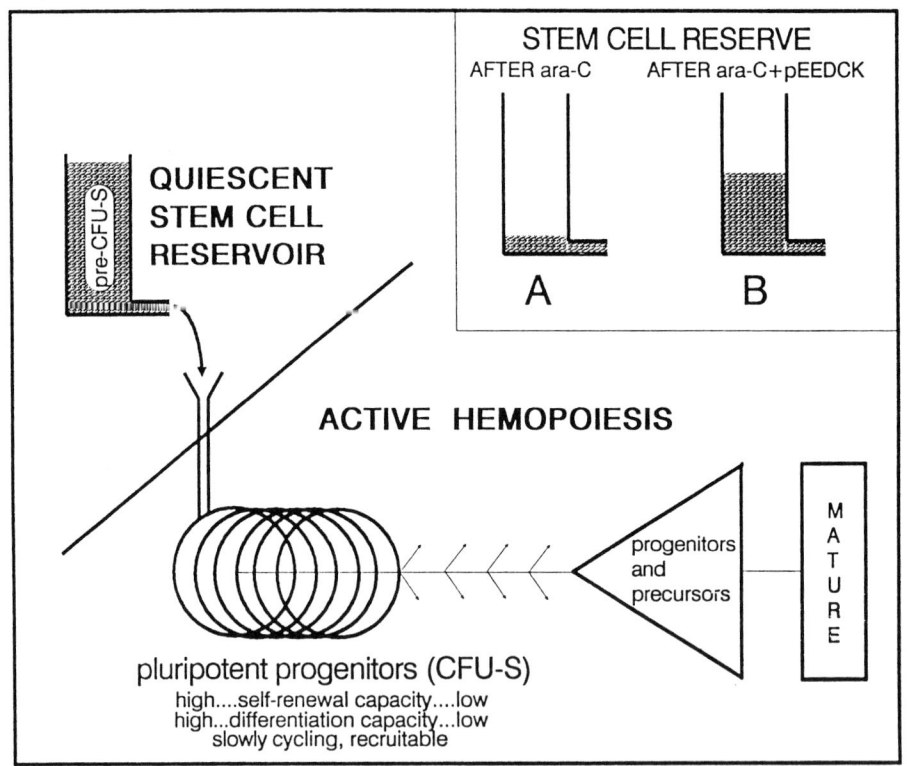

Fig. 1
Schematic representation of Hematopoiesis with quiescent, not-self-renewing, and exhaustible stem cells.
A permanently quiescent population of hematopoietic stem cells (HSC) slowly feeds new clones into active hemopoiesis. Self-renewal, differentiation and recruitability are properties of pluripotent progenitors (e.g. CFU-S). Hemopoietic stress leads to an excessive influx of stem cells (pre-CFU-S) into active hemopoiesis, resulting in a diminished HSC-reserve (insert A). The inhibitory action of pEEDCK makes post-stem-cell compartments more resistant against cytostatic drugs, resulting in a reduced need for compensatory allocation of HSC into active hemopoiesis and better preservation of the stem cell reservoir (insert B).

It seems thus, that the HSC-like pre-CFU-S population functions as a limited storage pool (fig.1) of primitive hematopoietic cells, which under normal conditions only occasionally feed a new clone into the production line of the system. In this respect the hematopoietic system shows a certain similarity to other organs, e.g. the ovary, where a certain number of non-replaceable primary cells is waiting for their turn to enter the differentiation and proliferation pathway. Pre-CFU-S numbers are large enough to maintain hematopoiesis for many lifetimes (Harrison et al., 1990) under normal circumstances, but being devoid of the ability to proliferate and self-renew, they are rapidly consumed under conditions of hematopoietic stress. As long as at least some pre-CFU-S remain available, a superficially normal hematopoietic activity is maintained, albeit with a severely restricted capacity to withstand further stress situations.

The term "stem cell" has usually involved two different notions. Firstly, a stem cell is that cell which functions as the prime source of all other cells in a given tissue, and secondly, stem cells are (by definition) capable of differentiation and self-renewal. It was tacitly assumed that both notions described the same cell population. Our results now indicate that this is not necessarily the case. The functions of self renewal and differentiation seem to be dissociated from the prime source function. Observations like the one reported here may require a reformulation of traditional stem cell concepts.

<u>Practical implications of the use of pEEDCK for long term protection of hematopoiesis against cytostatic damage.</u>

The experiments described in this paper also have another, more practically important, aspect. It is known that many of the hematological problems associated with the clinical use of cytostatic drugs are related to the high proliferative activity of many cells in the bone marrow. Once recruited by the first drug application also the initially quiescent and thus relatively resistant CFU-S become highly sensitive to subsequent doses, resulting in hematopoietic damage down to the level of primitive pluripotent cells. We have shown (Paukovits et al., 1990b, 1991b) that keeping CFU-S proliferation as low as possible is helpful in preventing hematotoxic side effects of cancer chemotherapy. For this purpose we use the CFU-S and CFU-GM specific leukocyte derived pentapeptide pEEDCK (Paukovits and Laerum, 1982; Paukovits et al.,1990c,1991b). This peptide lacks the pleiotropism and/or toxicity exerted by other inhibitors (TGF-ß, TNF-α, MIP-1-α) also proposed for this purpose (for review see Moser and Paukovits, 1991). Using clinically relevant doses of ara-C, we have shown (Paukovits et al., 1990a, 1991a, 1991b) that proliferation inhibition of early hematopoietic cells during critical phases of chemotherapy leads to considerable improvement of the hematological situation (Moser and Paukovits, 1991; Paukovits et al., 1990a, 1991a, 1991b), which could make this new approach highly interesting for clinical use.

In contrast to the post-chemotherapeutic use of hematopoietic stimulatory factors which accelerate regeneration but do not avoid damage (Murphy and Rizzoli, 1990), the use of the inhibitory pEEDCK peptide reduces the extent of damage in the earliest hematopoietic compartments (Moser and Paukovits, 1991). In this way pEEDCK improves the starting conditions for subsequent regeneration, which proceeds rapidly and smoothly without further help. Keeping the CFU-S population at low proliferative levels also has important benefits for the long term reconstitution of the hematopoietic system (Paukovits et al., 1993). We have shown here that the extensive killing of recruited CFU-S by repeated drug application results in the (compensatory) progression of large numbers of pre-CFU-S into the CFU-S compartment. The resulting drastic drainage of the HSC pool is however not compensated by self-renewal divisions of the remaining pre-CFU-S resulting in the irreversible loss of repopulating capacity. The use of pEEDCK preserves most of the CFU-S (Paukovits et al, 1991b) and as a consequence reduces the pressure exerted on the pre-CFU-S to proceed into CFU-S in order to compensate for the losses. Under the conditions used in our experiments about half of the pre-CFU-S remain available. Our results also show that post-chemotherapeutic stimulator treatment has no beneficial effect on the preservation of the HSC pool. Our results on the use of pEEDCK for short term amelioration and/or prevention of neutropenia and on the reduction of long term damage after cytostatic treatment may be of great practical importance.

Acknowledgement: We are grateful for the support of this work by the Austrian Foundation for Scientific Research (Grant P8086-med) and by the Anton Dreher Stiftung of the Medical Faculty of the University of Vienna.

References

Brecher, G., Neben, S., Yee, M., (1990), Bone marrow proliferation after passage through an irradiated host. In: The Biology of Hematopoiesis (Wiley-Liss), p. 449-458.
Dresch, C., El Kebir, N., Metral, J., Karsdorf, A., (1983), Exp. Hematol. 11: 187-192.
Eaves, C., Cashman J.D., Sutherland H.J., et al., (1991), Ann. N.Y. Acad. Sci. 628: 298-306
Fraser, C.C., Eaves, C.J., Szilvassy, S.J., Humphries, R.K., (1990), Blood 76: 1071-1076.
Guigon, M., Sainteny, F., Dumenil, D., Lepault, F., Frindel, E., (1978), Exp. Hematol. 6: 270-272.
Harrison, D.E., Stone, M., Astle, C.M., (1990), J.Exp.Med. 172: 431-437.
Iscove, N., (1990), Nature 347: 126-127.
Jones, R.J., Celano, P., Sharkis, S.J., Sensenbrenner, L.L., (1989), Blood 73: 397-401.
Jones, R.J., Wagner, J.E., Celano, P., Zicha, M.S., Sharkis, S.J., (1990), Nature 347: 188-189.
Lajtha, L.G., Pozzi, L.V., Schofield, R., Fox, M., (1969), Cell Tissue Kinet. 2: 39-49.
Lerner, C., Harrison, D.E., (1990), Exp.Hemat. 18: 114-118.
Moser, M.-H., Paukovits, W.R., (1991), Trends in Pharmacol.Sci (TIPS) 12: 304-310
Murphy, M.J., Rizzoli, V. (eds), (1990), Int. J. Cell Cloning 8, Suppl 1, pp. 1-399 .
Paukovits, W.R., (1990), Stem cell (CFU-S) inhibitory peptides: Biological properties and application of pEEDCK as hemoprotector in cytostatic tumor therapy. In: Exp. Hematol. Today-1989 (eds: Gorin, N.C., Douay, L.) pp. 72-80, Springer Verlag (New York).
Paukovits, W.R., Guigon, M., Binder, K.A., Hergl, A., Laerum, O.D., Schulte-Hermann, R., (1990b), Cancer Res. 50: 328-332.
Paukovits, W.R., Hergl, A., Schulte-Hermann, R., (1990c), Molec. Pharmacol. 38: 401-409.
Paukovits, W.R., Laerum, O.D., (1982), Zeitschr. Naturforschung (c), 37: 1297-1300.
Paukovits, W.R., Moser, M.H., Binder, K.A., Paukovits, J.B., (1990a), Cancer Treatment Rev. 17: 347-354.
Paukovits, W.R., Moser, M.H., Rutter, R., Paukovits, J.B., (1991a), Ann. N.Y. Acad. Sci. 628: 92-104
Paukovits, W.R., Moser, M.H., Binder, K.A. Paukovits, J.B., (1991b), Blood 77: 1313-1319
Paukovits, W.R., Moser, M.H., Paukovits, J.B., (1993), Blood, April 1993, in press
Ploemacher, R.E. & Brons, N.H.C., (1989), Exp.Hematol. 17: 263-266.
Spangrude, G.J., Heimfeld, S. & Weissmann, I.L., (1988), Science 241: 58-62.
Till, J.E., McCulloch, E.A., (1961), Radiat. Res. 14: 213-222.

Résumé

L'auto-renouvelement et la différenciation sont des caractéristiques des cellules souches hématopoiétiques. En étudiant les cellules souches capables de reconstitution, nous avons trouvé une destruction irréversible de cette population après des injections répétées de cytosine arabinoside. Des expériences de suicide montrent que ces pré-CFU-S ne prolifèrent pas chez les souris normales et demeurent quiescentes après traitement par l'Ara-C, ce qui indique que les cellules souches les plus primitives sont une population de réserve non mitotique, dépourvue de capacité d'autorenouvelement. Elles assurent le maintien de l'hématopoièse par progression dans le compartiment des CFU-S, en fonction de la taille de cette population. La protection des CFU-S contre l'Ara-C en inhibant leur prolifération avec p-Glu-Glu-Asp-Cys-Lys (pEEDCK) réduit aussi le drainage du pool des cellules souches, représentant une nouvelle voie de protection de l'hématopoièse contre l'aplasie induite par les drogues. Nos résultats montrent une dissociation entre la fonction de "primitivité" des cellules souches et les fonctions d'auto-renouvelement et de différenciation. Cette dissociation rend nécessaire une reformulation des concepts traditionnels de cellule souche.

Protective effects of AcSDKP (Seraspenide) against the *in vitro* toxicity of Asta Z to hematopoietic and stromal cells

M. Guigon, C. Hamilton, D. Bonnet, C. Jiang, F.M. Lemoine, F. Isnard and A. Najman

Department of Hematology, CHU St-Antoine, Paris, France

SUMMARY

The tetrapeptide AcSDKP is an inhibitor of murine CFU-S entry into cell cycle, which also inhibits the cycling of normal human progenitors in a reversible way, without affecting the growth of leukemic cells. In animal models, it has been shown to protect mice against the toxicity of high doses of chemotherapy. In this paper, we report for the first time that AcSDKP can protect CFU-GM progenitors, primitive stem cells (LTC-ICs) as well as stromal cells from the toxicity of high doses of Mafosfamide (Asta Z) used in the purging of marrows from AML and NHL patients in complete remission before autograft.

INTRODUCTION

The efficacy of potent cytotoxic drugs used in chemotherapy of malignancies is often limited by their toxicity for normal tissues, especially the bone marrow. Our approach to reduce the myelotoxicity of chemotherapeutic agents has been to try to preserve hematopoietic stem cells by maintaining them out of the cell cycle at the time of the treatment. More than 15 years ago, we have shown that a dialysable extract of fetal calf bone marrow inhibited the entry into S phase of murine pluripotent stem cells (CFU-S) after radiation or chemotherapy (Frindel and Guigon 1977, Guigon and Frindel 1978). This crude inhibitor was able to increase the survival of mice given lethal doses of Cytosine Arabinoside (Ara-C) (Guigon et al. 1982,1987). The inhibitor was further identified as a tetrapeptide Acetyl-Ser-Asp-Lys-Pro (AcSDKP) by Lenfant et al.(1989 a); it is now obtained by chemical synthesis (Lenfant et al. 1989b) and termed Seraspenide. As the native one, the synthetic peptide increased the survival of mice given Ara-C or Cyclophosphamide, without modifying the response of malignant cells to treatment (Bogden et al. 1991).

In view of its potential clinical application, it was important to know whether this peptide also acted on human hematopoietic cells. Using the synthetic molecule kindly provided by Ipsen-Biotech (Paris, France), we have shown that the tetrapeptide AcSDKP blocks temporarily and reversibly the cycling of normal human progenitors of the granulo-macrophagic (CFU-GM) and erythroid (BFU-E) lineages (Guigon et al. 1990, Bonnet et al.1992a). Experiments carried out on purified populations indicate that AcSDKP inhibits both the clonogenic progenitors present in the CD34+HLA-DR^{high} population and the most primitive stem cells present in the CD34++HLA-DR^{low} population. The inhibition, obtained with nanomolar concentration of the peptide, is direct and reversible (Bonnet et al.1993 and this book). Interestingly, AcSDKP has no effet on leukemic cell lines (Guigon et al. 1991) and does not modify the cycling of leukemic progenitors from chronic myeloid leukemia CML patients (Bonnet et al.1992b,Cashman et al.1992a), thus appearing to have a differential effect on normal and leukemic cells. Furthermore, we did not find any effect of the peptide on the proliferation of blast cells from AML patients, even after stimulation by growth factors (Bonnet et al. this book), although our data cannot exclude an effect on AML clonogenic cells which is now being investigated. The properties of this tetrapeptide strongly suggest that it could be used to protect normal marrow stem cells/progenitors during chemotherapy. Indeed the first results of a clinical trial in which AcSDKP has been administered to patients in association with monochemotherapy are encouraging (Carde et al. this book).

The potential protective effect of Seraspenide was further investigated in different models. We recently reported that AcSDKP may protect normal human progenitor cells against the in vitro toxicity of Zidovudine (AZT), used in AIDS treatment (Grillon et al. 1993). Data reported in this book indicate that AcSDKP can protect normal hemopoietic cells during hyperthermia (Wierenga and Konings this book) and phototofrin marrow purging (Coutton et al. this book) without protecting leukemic cells. We have been interested to know whether AcSDKP can protect normal hematopoietic stem cells/progenitors during Mafosfamide (Asta Z) purging before autograft. Indeed, the selective, reversible inhibitory effect of AcSDKP could result in protection of normal progenitors and potentially allow the use of higher doses of purging agent, enhancing the chance of completely purging an autologous bone marrow before reinfusion. In this paper, we report the effect of AcSDKP incubation prior to Mafosfamide treatment on the hematopoietic and stromal cells of bone marrows from patients with acute myeloid leukemia (AML) and non-Hodgkin's lymphoma (NHL) in complete remission.

MATERIAL AND METHODS

Reagents
Studies were carried out with the synthetic tetrapeptide AcSDKP (Ipsen-Biotech, Paris, France) and with Mafosfamide-L-Lysine (Asta-Z 7654) from Asta Pharma, Bielefeld, Germany. Recombinant growth factors were kindly provided by Genetics Institute, Cambridge, USA (GM-CSF and IL-3), Boehringer, Paris, France (Epo), Rhone-Poulenc, Paris, France (G-CSF) and Amgen, Thousand Oaks, USA (SCF). All these factors were used at concentrations determined to be optimal in preliminary experiments.

Cells
Bone marrow cells were obtained after informed consent either from AML patients or from NHL patients in complete remission before autograft from the Department of Hematology of St Antoine Hospital.

Protocol
10^7 fresh mononuclear cells/ml were first incubated with or without AcSDKP (at concentrations ranging from 0.05-50 ng/ml) in Iscove modified Dulbecco medium (IMDM) containing 2% heat-inactivated fetal calf serum (FCS) for 20hrs. The conditions of incubation with the peptide are critical: 1) a low concentration of serum is required because the peptide is rapidly degraded at high concentrations of serum

(Grillon et al.1993) and 2) a 20 hour-incubation appeared necessary to reduce the cell cycling of normal progenitors (Guigon et al. 1989). Then, Asta Z (100 µg/ml) was added for another 30 minutes at 37°C in a waterbath with frequent agitation. Chilled IMDM containing 10% FCS was then added and the cells were washed two more times, resuspended in IMDM and counted. Then, cells were plated in methylcellulose in order to determine the number of clonogenic progenitors. In some cases, long term cultures were established in order to study more primitive progenitors and the stroma formation.

Clonogenic assay
Cells were plated in 35-mm Petri dishes in 1-ml aliquots of IMDM containing 30% FCS, 5×10^{-5} M 2-mercaptoethanol (2-ME), 0.8% (wt/vol) methylcellulose and a mixture of the following recombinant growth factors: IL-3 (100 U/ml), GM-CSF (200 U/ml), G-CSF (1 ng/ml), SCF (100 U/ml) and Epo (3 U/ml). Triplicate cultures were incubated for 14 days at 37°C in a humidified atmosphere containing 5% CO_2 in air. Granulocyte-macrophage colonies (CFU-GM) containing at least 50 cells were scored on day 14-18.

Long Term Bone Marrow Culture (LTBMC)
LTBMCs were initiated in long-term culture medium (LTCM; Terry Fox Laboratory,Vancouver, Canada) with Asta Z-treated cells previously incubated with or without AcSDKP. At weekly intervals, half of the medium and non-adherent cells were removed and an equivalent volume of fresh medium was added. The non-adherent cells were counted using the trypan blue exclusion method. The establishment and the appearance of a stromal cell layer was monitored weekly by examination under an inverted microscope.

Long-Term-Culture Initiating Cell (LTC-IC) Assay
1×10^6 cells in 2 ml LTCM were seeded on a preestablished irradiated stromal layer and were incubated at 33°C for 5 weeks with a weekly half-medium change. Then, the adherent layer was detached and prepared as a single cell suspension using trypsin. Cells were counted using the trypan blue exclusion method and plated for in methylcellulose to determine the total clonogenic content of each LTCM. Division by 4 of the total number of progenitors at 5 week gives the number of LTC-ICs present in the initial population (Sutherland et al. 1989).

RESULTS

Effect of the preincubation with AcSDKP on the number of CFU-GM progenitors from Asta Z-treated cells
Table1 shows that under our culture conditions i.e. in presence of 5 growth factors, the number of CFU-GM in the marrow from 11 AML and 9 NHL patients in complete remission varied widely which is probably related to the nature of the disease and/or previous treatments. Asta Z treatment (100µg/ml) decreased drastically the number of colonies from 85 to 99,7%.
The preincubation of cells with AcSDKP before Asta Z increased significantly the number of CFU-GM in 4 out of 10 evaluable AML and 2 out of 9 NHL. It should be mentioned that the preincubation of cells with AcSDKP alone resulted in a significant decrease of CFU-GM number in 4 AML (UPN 1, 7, 10 and 11) and in 5 NHL (UPN 2a,3a,4a,6a, 8a) respectively, whereas in the other cases, no inhibitory effect of the peptidfe was observed (data not shown).

Table 1
Effect of the preincubation with AcSDKP on the toxicity of Asta Z to CFU-GM

Number of CFU-GM/10E5 cells

UPN	Untreated	Asta Z	AcSDKP + Asta Z
AML			
1	290 ± 32	26 ± 4	10 ± 1
2	233 ± 16	16 ± 2	20 ± 2
3	327 ± 12	35 ± 4	32 ± 3
4	358 ± 25	32 ± 3	48 ± 3*
5	430 ± 28	10 ± 2	nd
6	198 ± 9	18 ± 2	29 ± 2*
7	307 ± 12	18 ± 1	23 ± 3
8	892 ± 18	3 ± 1	35 ± 4*
9	280 ± 22	13 ± 2	10 ± 2
10	357 ± 14	2 ± 1	9 ± 2*
11	120 ± 15	1 ± 1	2 ± 1
NHL			
1a	582 ± 6	87 ± 2	97 ± 8
2a	658 ± 12	81 ± 4	137 ± 9*
3a	730 ± 10	84 ± 2	88 ± 13
4a	310 ± 8	29 ± 3	17 ± 2
5a	377 ± 44	47 ± 3	52 ± 4
6a	645 ± 37	73 ± 5	123 ± 13*
7a	185 ± 30	17 ± 2	20 ± 2
8a	203 ± 13	7 ± 2	6 ± 1
9a	22 ± 2	1 ± 1	2 ± 1

Results are given as mean ± SEM of triplicates
UPN : unique patient number
* p<0,05

Table 2
Effect of the preincubation with AcSDKP (0.05ng/ml) on the toxicity of Asta Z (100ug/ml) to LTC-ICs

	Number of LTC-IC/10E6 cells		
UPN	Untreated	Asta Z	AcSDKP + Asta Z
AML			
5	44,5 ± 2	10 ± 1 (78%)	15 ± 1*
7	41 ± 5	7 ± 2 (83%)	11 ± 4
11	220 ± 7	12 ± 3 (95%)	45 ± 7*
NHL			
5a	23 ± 10	12 ± 5 (48%)	30 ± 9*
6a	42 ± 10	53 ± 12	37 ± 4
8a	47 ± 14	14 ± 12 (70%)	6 ± 5

Results are given as mean ± SEM of triplicates
UPN : unique patient number
(%) percentage of toxicity
* $p<0.05$

Effect of the preincubation with AcSDKP on the number of LTC-ICs from Asta Z-treated cells

We have also investigated the effect of AcSDKP on the toxicity of Asta Z to more primitive stem cells, using the LTC-IC assay. Indeed, although we observed a large variability in the number of LTC-ICs in these marrows harvested from patients who previously received intensive chemotherapy, Asta Z at 100 µg /ml was toxic for these cells in 5 out of 6 cases. Table 2 shows that the toxicity ranged from 49 to 95%. Interestingly, the preincubation of cells with AcSDKP resulted in a significant increase of LTC-IC number in 3 cases out of 5 cases in which Asta Z was toxic. Furtherrmore, in UPN 11, the preincubation of cells with AcSDKP alone decreased by 53% the LTC-IC number, indicating a inhibitory effect of the peptide for this population.

Effect of the preincubation with AcSDKP on the stromal layer from Asta Z-treated cells

The establishment and organization of the stromal layer in long-term cultures from 4 AML and 3 NHL patients were monitored weekly under an inverted microscope. In untreated cultures, the stromal layer formed normally and a confluent layer of adherent cells was seen by week 3. By contrast, cultures with Asta Z- treated cells never reached confluence and upon trypsinization and numeration, the number of cells remained very low, being about 8% of the matched untreated culture at week 5. Interestingly, in the cultures established with AcSDKP-Asta Z treated cells, a significantly increase of the stromal cellularity was observed, with a threefold increase of the number of cells as compared to the matched Asta Z- treated cultures. This difference in the cellularity of the stromal layer is clearly shown by the macroscopic observation of the dishes. Figure 1 presents an example of the aspect of the stromal layer after a 4 week-culture of Asta Z- treated cells without (a) or with a preincubation with AcSDKP (b).

DISCUSSION

The present study demonstrates for the first time the effectiveness of AcSDKP in reducing mafosfamide (Asta Z) toxicity to mature and primitive hematopoietic progenitors as well as stromal cells in bone marrows from AML and NHL patients in complete remission before autograft.

Based on preliminary data, we chose to test AcSDKP at concentration of 0.05ng and 1ng/ml. These doses were previously reported to inhibit the cell cyle and reduce the number of progenitors from human normal marrow (Guigon et al.1990). whereas high concentrations did not show any inhibitory effect (Bonnet et al.1993). We have assumed that the protective effect of AcSDKP was related to its reversible inhibitory effect on progenitor growth. However, in 3 cases of AML where AcSDKP increased CFU-GM number (UPN 4,6,8) , there was no inhibitory effect of AcSDKP as assessed by colony counts. Indeed, we have noticed previously that AcSDKP can reduce cell cycling without inducing a concomitant decrease of the number of colonies. The absence of inhibitory effect of the peptide could also be related to the disease and/or the previous treatment; indeed, in our series of patients, it turns out that an inhibitory effect of AcSDKP was found more frequently in NHL than in AML marrow samples.Furthermore, one cannot excluded individual suceptibility already noticed in normal subjects. In this respect, the evaluation of the seric level of AcSDKP by enzymatic immuno-assay (Pradelles et al.1990) would certainly be of interest.

The results presented on table 1 shows a protective effect of CFU-GM in 4/10 AML and 2/9 NHL cases only. It should be emphasized that the evaluation of a potential protective effect of an inhibitory molecule for clonogenic progenitors is limited by the experimental technique. Indeed, the number of colonies observed in the AcSDKP+ Asta Z-treated group can be the result of both the inhibitory effect of AcSDKP - which blocks temporarily and reversibly progenitor growth (Bonnet et al. 1993) - and the toxicity of Asta Z, since neither inhibited nor killed progenitor cells will give rise to

Figure 1

Aspect of the stromal layer in a 4-week-culture of Asta Z-treated bone marrow without (a) or with (b) a preincubation with AcSDKP. (40x)

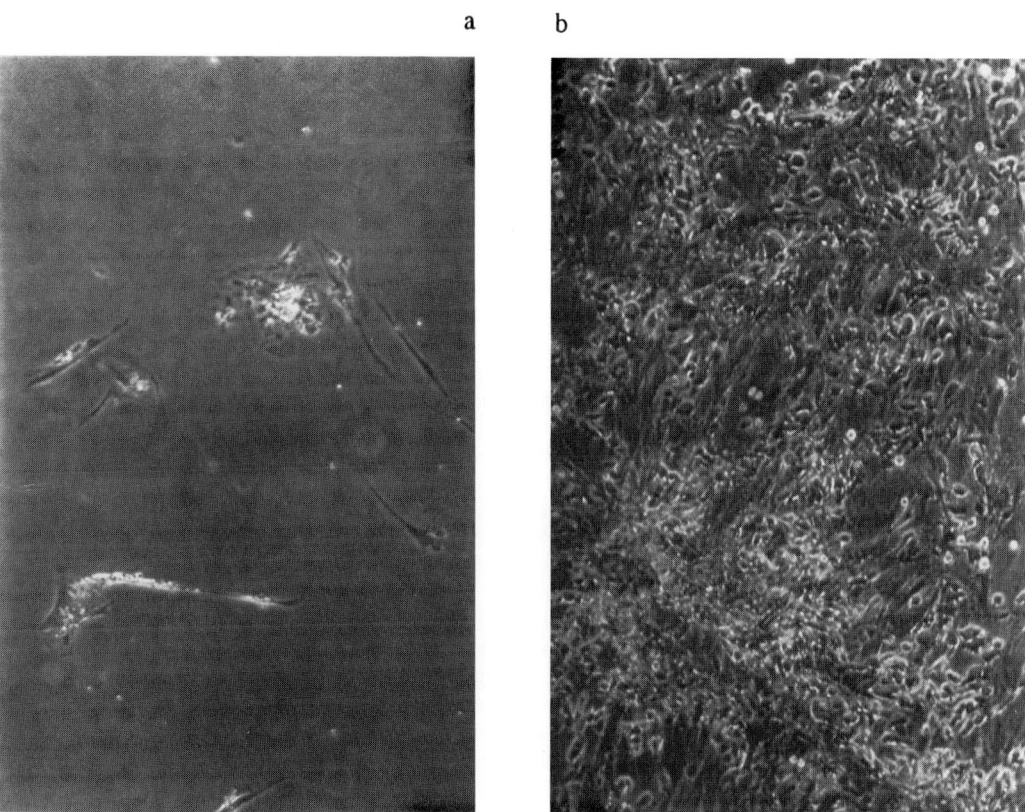

colonies. Therefore, as described elsewhere (Grillon et al. 1993), the protective effect has a minimal estimate P_{min}, if we assume that AcSDKP-inhibited cells were then killed by Asta Z treatment, and a maximal estimate P_{max}, if all AcSDKP-inhibited cells were assumed to be preserved from Asta Z treatment, the real protective effect being between the two values P_{min} and P_{max}. Therefore, our results, which represent the P_{min}, underestimate the protective effect of AcSDKP and it is likely that the beneficial effect of AcSDKP does occur in a higher number of cases.

Primitive stem cells, which are normally quiescent, have been reported to be less sensitive to Asta Z (Siena et al.1985; Carlo-Stella et al. 1992). The rather elevated toxicity (48-95%) observed in our experiments may reflect the increased cycling of the primitive progenitors due to the proximity of ablative chemotherapy rmaking them more vulnerable to the drug. There is now evidence that AcSDKP acts also on primitive stem cells (Cashman et al. 1992b; Bonnet et al. 1993; this book) and the inhibitory effect on LTC-IC that we have observed in UPN 11 is in agreement with these data. The fact that iAcSDKP can protect this population is of high interest for long term engraftment.

The deleterious effect of Asta Z on the stromal layer has been already reported (Siena et al 1985; Kalechman et al. 1992). We have observed a severe impairment of the stromal layer which might be due to the high dose of Asta Z and/or to the fact that we used bone marrow from patients (AML or NHL) who had just recently completed a cycle of chemotherapy. Interestingly, the preincubation with AcSDKP ameliorates the formation of the stromal layer which however never reached confluence. Surprisingly, such an increase of stromal cells with AcSDKP was found in marrows from NHL patients only. Here again the role of the disease and/ or previous treatments should be evoked. These observations confirm previous data showing that AcSDKP has no adverse effect on stromal cells (Bonnet et al. 1992c) and on the contrary might be beneficial; however whether its acts directly or via the induction of cytokines is not yet known.

Several mechanisms can be involved in the radio-and/or chemoprotection (Kalechman et al.1991, Moreb et al. 1992) and are discussed in this book (Neta et al.; Moreb et al., Sredni et al).Although the mechanism of the protective effect of AcSDKP remained to be elucidated, the fact that this small peptide of low cost, devoid of toxicity and easy to produce can protect progenitors, primitive stem cells and stromal cells in the marrow of a number of patients requiring autografting allow to consider it of interest in the purging setting, inasmuch as it has been shown to have no effect on leukemic cell growth (Bonnet et al.1992b and this book). Furthermore, its effect on stromal cells may be an additional advange because of the importance of microenvironment for optimal development of newly grafted bone marrow.

REFERENCES

- Bogden AE, Carde P, Deschamps de Paillette E, Moreau JP, Tubiana M, Frindel E.(1991)
Amelioration of chemotherapy induced toxicity by cotreatment with a hematopoiesis-inhibiting tetrapeptide AcSDKP. Ann NY Acad Sci, 628 : 126-139
- Bonnet D, Fabrega S, Lemoine F.M, Aoudjhane M, Najman A, Guigon M.(1992a)
Differential effect of the tetrapeptide AcSDKP on human normal and leukemic blood progenitors. Int J Cell Cloning 10 : Supp 1, 185-187
- Bonnet D, Césaire R, Lemoine F.M, Aoudjhane M, Najman A, Guigon M.(1992b)
The tetrapeptide AcSDKP, an inhibitor of the cell cycle status for normal human bone marrow progenitors, has no effect on leukemic cells. Exp Hematol 20 : 251-255
- Bonnet D, Lemoine F.M, Khoury E, Pradelles P, Najman A, Guigon M. (1992c)
Reversible inhibitory effects and absence of toxicity of the tetrapeptide Acetyl-N-Ser-Asp-Lys-Pro (AcSDKP) in human long-term bone marrow culture. Exp. Hematol, 20 : 1165
- Bonnet D, Lemoine F.M, Pontvert-Delucq S, Baillou C, Najman A, Guigon M. (1993)
Direct and reversible inhibitory effect of the tetrapeptide AcSDKP (Seraspenide) on the growth of human $CD34^+$ subpopulations in response to growth factors. Blood in press

- Carlo-Stella C, Mangoni L, Almici C, Garau D, Craviotto L, Piovani G, Caramatti C, Rizzoli V. (1992). Differential sensitivity of adherent CFU-blast, CFU-mix, BFU-E, and CFU-GM to mafosfamide: Implications for adjusted dose purging in autologous bone marrow transplantation. Exp. Hematol. 20:328-333,
- Cashman JD, Eaves AC, Eaves CJ.(1992a)
Primitive neoplastic hematopoietic cells from patients with chronic myeloid leukemia can ignore the reversible indirect inhibitory action of the tetrapeptide AcSDKP. Exp. Hematol. I20, 782
- Cashman JD, Eaves AC, Eaves CJ. (1992b)
Evidence for an indirect mechanism mediating the inhibitory effect of the tetrapeptide AcSDKP on primitive human hematopoietic cell proliferation. J.Cell.Biochem, Suppl 16C, 95 (abstr)
- Frindel E, Guigon M. (1977)
Inhibition of CFU entry into cycle by a bone marrow extract.
Exp Hematol 5 : 74-76
- Grillon C, Bakala J, Lenfant M. (1993)
Optimization of cell culture conditions for the evaluation of biological activities of AcSDKP, a natural inhibitory molecules. Growth factors : in press.
- Grillon C, Bonnet D, Mary JY, Lenfant M, Najman A, Guigon M (1993).
The tetrapeptide AcSerAspLysPro (Seraspenide) an hematopoietic inhibitor, may reduce the in vitro toxicity of 3'-azido-3'deoxythymidine to human hematopoietic inhibitors. Stem Cells : in press
- Guigon M, Frindel E (1978)
Inhibition of CFU-S entry into cell cycle after irradiation and drug treatment. Biomedicine29: 176-178
- Guigon M, Mary JY, Enouf J, Frindel E. (1982)
Protection of mice against lethal doses of 1-β-arabinofuranosyl-cytosine by pluripotent stem cell inhibitors. Cancer Res 42 : 638-642
- Guigon M.(1987) Biological properties of low molecular weight pluripotent stem cell (CFU-S) inhibitors. In : Najman A, Guigon M, Gorin NC, Mary JY. (eds) : The inhibitors of Hematopoiesis, vol.162. John Libbey Eurotext, 241-251
- Guigon M, Bonnet D, Lemoine F.M, Kobari L, Parmentier C, Mary JY, Najman A .(1990)
Inhibition of human bone marrow progenitors by the synthetic tetrapeptide AcSDKP. Exp Hematol 18 : 1112-1115
- Guigon M, Bonnet D, Césaire R, Lemoine F.M, Najman A.(1991)
Effects of small peptidic inhibitors of murine stem cells on human normal and malignant cells. Ann NY Acad Sci, 628 : 105-114
- Kalechman Y,Sotnik-Barkai I, Albeck M, Horwith G, Sehagl SN, Sredni B. (1991)
Use and mechanism of action of AS101 in protecting bone marrow colony forming units-granulocyte-macrophage following purging with Asta-Z 7557. Cancer Research 51; 5614-5620
- Lenfant M, Wdzieczak-Bakala J, Guittet E, Promé JC, Sotty D, Frindel E. (1989a)
Sequence determination of an inhibitor of hemopoietic pluripotent stem cell proliferation. Proc Natl Acad Sci USA 86 : 779-782
- Lenfant M, Itoh K, Sakoda M, Sotty D, Sasaki NA, Wdzieczak-Bakala J, Mori KJ.(1989b)
Enhancement of the adherence of hematopoietic stem cells to mouse bone-marrow derived stromal cell line MS-1-T by the tetrapeptide acetyl-N-Ser-Asp-Lys-Pro. Exp Hematol 17 : 898-902
- Moreb J, Zucali JR, Zhang Y, Colvin MO, Gross MA. (1992)
Role of aldehyde dehydrogenase in the protection of hematopoietic progenitor cells for 4-hydroperoxycyclophosphamide by interleukin 1 and tumor necrosis factor. Cancer Research 52, 1770-1774
- Pradelles P, Frobert Y, Créminon C, Liozon E, Massé A, Frindel E. (1990)
Negative regulator of pluripotent hematopoietic stem cell proliferation in human white blood cells and plasma as analysed by enzyme immunoassay. Biochem Biophys Res Communic 170 : 986-993
- Siena S, Castro-Malaspina H, Gulati SC, Lu L, Colvin MO, Clarkson BD, O'Reilly RJ, Moore MAS. (1985).
Effects of in vitro purging with 4-hydro-peroxycyclophosphamide on the hematopoietic and microenvironmental elements of human bone marrow. Blood 65(3):655-662,
- Sutherland HJ, Eaves CJ, Eaves AC, Dragowska W, Lansdorp PM. (1989)
Characterization and partial purification of human marrow cells capable of initiating long-term hematopoiesis in vitro. Blood 74 : 1563-1570

ACKNOWLEDGMENTS

D.Bonnet is a recipient from CIFRE award. We are grateful to Dr E.Deschamps de Paillette and Dr F.Thomas (Ipsen-Biotech, France) for supporting this work and providing AcSDKP and to Amgen (USA) for providing stem cell factor. We thank Dr J.Stachowiak for providing the marrow samples.

Résumé

Le tétrapeptide AcSDKP (Seraspenide) inhibe l'entrée en synthèse d'ADN des CFU-S de la souris; il inhibe aussi de façon réversible la mise en cycle des progéniteurs humains normaux, sans modifier la croissance de cellules leucémiques. Dans des modèles animaux, AcSDKP protège les souris de la toxicité de doses élevées de chimiothérapie. Nous présentons ici pour la première fois des résultats qui montrent qu'AcSDKP protège les progéniteurs CFU-GM, les cellules souches primitives (LTC-IC) et les cellules stromales de la toxicité de fortes doses Mafosfamide (Asta Z) utilisé dans la purge in vitro de moelles de patients atteints de LAM et de LNH en rémission complète avant autogreffe.

Effect of the tetrapeptide AcSDKP on the hyperthermic sensitivity of the murine hematopoietic stem cell compartment

P.K. Wierenga and A.W.T. Konings

Department of Radiobiology, State University Groningen, Groningen, The Netherlands

Current achievements in the research on the heat sensitivity of normal and malignant hematopoietic cells has encouraged proposals for the application of hyperthermia as a purging agent in the treatment of leukemia prior to autologous bone marrow transplantation.

Hyperthermic characteristics of the various subsets of the murine hematopoietic stem cell compartment were investigated in our laboratory. It could be demonstrated that there is an increase in hyperthermic sensitivity from the primitive stem cells (LTRA and MRA) towards the most differentiated progenitors (CFU-GM, BFU-E and CFU-E) and that the relationship between the stem cell hierarchy and heat sensitivity is predominantly based on the differences in proliferative activity [1-4].

To increase the selectivity of a hyperthermic purging protocol towards killing of leukemic cells, the normal hematopoietic stem cell compartment should be protected by decreasing the proliferative activity of the normal subsets without affecting the leukemic cells. For this reason bone marrow cells ($2-3 \times 10^6$/ml) from CBA/H mice were incubated in the presence of 10^{-9} M AcSDKP (kindly provivded by Laboratoires Beaufour, Paris) for 8 and 24 hours at 37°C, respectively. After the incubation period, the cell suspensions were heat-treated for 60 minutes at 43°C. In the absence of AcSDKP this hyperthermic regimen results in ±3% survival of the CFU-GM. The results on the murine CFU-GM demonstrate that after an incubation period of 8 hours with AcSDKP the amount of cells in S-phase decreased to $14 \pm 3\%$, versus $33 \pm 2\%$ in the control situation. The absolute number of CFU-GM was decreased by <15%. Similar results were obtained with the CFU-S-8: a decrease in the amount of S-phase cells to 12% versus 32% in the control situation and a comparable minor loss in absolute number of CFU-S-8. The decrease in proliferative activity of both subsets will lead to a threefold increase in survival after the hyperthermic treatment. The proliferative activity of the CFU-S-12 (13% of cells in S-phase) was unaffected by the AcSDKP treatment while again the absolute number decreased <15%. The effect of a 24 hours incubation period with AcSDKP on the number of CFU-GM in S-phase was comparable with the 8 hours period but the beneficial effect of the addition of AcSDKP on the survival of the CFU-GM in the hyperthermic purging protocol, however, is adversely affected by an almost 50% decrease in CFU-GM numbers.

Based on these results, it can be concluded that an incubation with 10^{-9} M AcSDKP for 8 hours might have a possitive effect on the outcome of a hyperthermic purging protocol.

1) Wierenga PK and Konings AWT, Int. J. Hyperthermia (1990), 6, 793-801.
2) Wierenga PK and Konings AWT, Int. J. Hyperthermia (1991), 7, 785-793.
3) Wierenga PK and Konings AWT, Exp. Hematol (1993), in press.
4) Wierenga PK, Down JD and Konings AWT, submitted.

Photoprotection of the hematopoietic progenitors by the tetrapeptide N-AcSDKP

C. Coutton, A. Bohbot, M. Guigon* and F. Oberling

*Centre de Recherche en Immunologie, Laboratoire d'Hématologie et d'Immunothérapie des Cancers, 1, place de l'Hôpital, 67091 Strasbourg Cedex, France; *Laboratoire d'Hématologie, Faculté de Médecine Saint-Antoine, 27, rue de Chaligny, 75012 Paris, France*

Phototherapy could be used as an *in-vitro* bone marrow purging procedure before autologous grafting in acute leukemia. In the best conditions leukemic cells are reduced by 4 to 5 log, but in the same manner only 30% of normal CFU-GM progenitors are preserved (Atzpodien et al., 1987). In an autologous bone marrow purpose it would be necessary to increase normal progenitor survival without decreasing the purging effect. In this way we studied the photoprotective effect of the tetrapeptide N-AcSDKP which is a reversible inhibitor of normal hematopoiesis (Guigon et al., 1990).

Normal mononuclear bone marrow cells (NMC) obtained from heathly volonteers (after informed consent) were separated by centrifugation on Ficoll gradient, and aliquots were frozen at -196°C in nitrogen liquid. NMC were incubated for 20 hr with or without N-AcSDKP (10^{-9} M and 10^{-10} M) at 37°C before sensitization for 1hr with photofrin II (2.5 µg/ml) at 37 °C. Cells were irradiated by 2, 10 or 25 J/cm2 with an argon laser (λ=514 nm). The CFU-GM recovery was evaluated by clonogenic assay in methylcellulose (0.8%) supplemented with GM-CSF (200 U/ml) and FCS (24%). In comparison we tested the effect of N-AcSDKP (10^{-7} M to 10^{-12} M) on the leukemic cell lines HL-60 and K-562 by using the same protocol.

NMC's incubation with N-AcSDKP before photosensitization induce an increase of the percentage of CFU-GM recovery when the cells were irradiated by high energy doses (10 or 25 J/cm2). Similar results were obtained either with fresh or thawed NMC (p values ≤ 0.05 by dunett's test). This photoprotective effect was more pronounced for 10^{-10} M than 10^{-9} M as shown in Fig.1.
In opposition, photokilling of the leukemic cells HL-60 and K-562 was unmodified by the incubation with N-AcSDKP at 10^{-7} M to 10^{-12} M ($p \geq 0.1$ by student's test).
Our results suggest that N-AcSDKP is a photoprotector of normal CFU-GM progenitors.

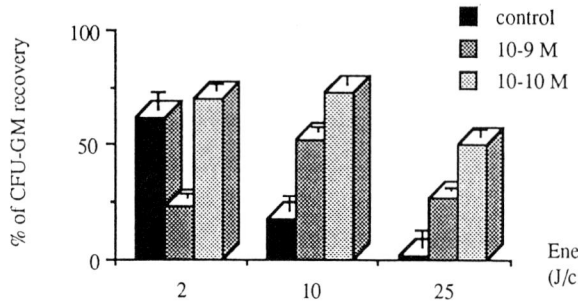

Fig.1; representative experiment. The % of CFU-GM recovery after photosensitization were calculated in comparison with unirradiated cells incubated with the same N-AcSDKP concentration ($p \leq 0.05$).

REFERENCES :-Atzpodien,J.,Gulati,S.C.,Strife,A.,Clarkson,B.D.(1987): Photoradiation models for ex-vivo treatment of autologous bone marrow grafts. In Blood 70: 484-489.
-Guigon,M.,Bonnet,D.,Lemoine,F.,Kobari,L.,Parmentier,C.,Mary,J.Y.,Najman,A. (1990): Inhibition of human bone marrow progenitors by the tetrapeptide AcSDKP. In Exp. Hematol. 18: 1112-1115.

Phase I-II trial of Seraspenide (AcSDKP): a suppressor of myelopoiesis protects against chemotherapy myelotoxicity

P. Carde, N. Mathieu-Tubiana, E. Vuillemin, V. Delwail, O. Corbion,
A. Vekhoff, F. Isnard, J.-M. Ferrero, E. Garcia-Giralt, J.-F. Gimonet,
A.-M. Stoppa, E. Leger-Picherit, E. Fadel, L. da Costa, C. Domenge,
D. Khayat, F. Guilhot, A. Monnier, R. Zittoun, B. Brun, D. Maraninchi,
C. Chastang, E. Goncalves, J.-Ph. Monpezat, F. Beaujan,
E. Deschamps de Paillette, E. Frindel, M. Guigon and A. Najman

French Cooperative Group, Coordinator : P. Carde, Institut Gustave-Roussy, 39, rue Camille-Desmoulins, 94805 Villejuif, France

ABSTRACT

Seraspenide, a synthetic tetrapeptide, inhibits cell cycle entry of normal hematopoietic stem cells. In mice it protects hemopoiesis against the damage caused by cytarabine, cyclophophamide and carboplatin. Seraspenide has been given to 53 cancer patients undergoing monochemotherapy. Patients underwent 2 consecutive cycles of monochemotherapy (cytarabine or ifosfamide) in a double-blind cross-over randomized study: first cycle: chemotherapy + Seraspenide or placebo; second cycle: chemotherapy + reverse order. Treatment groups were: - CYTARABINE 0.75 to 2g/m² i.v. q. 12 h. x 4 or- IFOSFAMIDE 1.5 to 3.0 g/m²/d i.v. 1 h. x4 d. + Seraspenide (or placebo) 0.5 to 156,25 mg/kg continuous i.v. infusion A protective effect of Serapenide has been demonstrated (Wilcoxon matched-paired rank test) on the comparison of 6 parameters concerning the blood leucocytes and neutrophils during the 2 cycles: area under the curve (AUC) from day 8 to day 22 & depth of the nadir of leucocytes ($p < 0.07$) and neutrophils and duration of leucopenia & neutropenia < 500/ml (5.7 vs 7.1 days, $p < 0.04$). Seraspenide has been devoided of toxicity. These results reflect the advantage observed in the Seraspenide arm for each individual chemotherapy. Seraspenide, as an inhibitor of cell cycle entry of hemopoietic cells, protects from drug-induced hemopoietic damage. Its assessment for long term protection is underway.

INTRODUCTION

The Colony Forming Units in the Spleen (CFU-S), pluripotent stem cells in mice, are normally quiescent and it has been demonstrated that they are therefore relatively protected from the toxicity of drugs until they are triggered into cycle through a feedback mechanism (Vassort et al., 1971). The existence of low molecular weight endogenous negative regulators was demonstrated using dialysable fetal calf bone marrow extracts that are able to inhibit the entry into cell cycle of the CFU-S pluripotent stem-cells challenged by Cytarabine & varied stimulations (Frindel & Guigon, 1977).

The protective activity of Seraspenide against myelotoxicity has been assessed on the comparison of 6 parameters concerning the blood *leukocytes* and neutrophils: the integration of their area under the curve (AUC) from day 8 to day 22, the depth of their nadir and the durations of hypoplasia < 1000 *leukocytes* /mm^3 and < 500 neutrophils /mm^3. The demonstration of a stastical difference providing evidence for a protective effect of Seraspenide has been performed using the Wilcoxon matched-paired rank test.

Thirty-one patients had unevaluable data for myelotoxicity assessment: 27 of them had only 1 cycle of monochemotherapy in combination with placebo (18 patients) or Seraspenide (9 patients), mostly due to tumor progression prior to the initiation of cycle 2. Four patients had major protocol violations.

Treatment groups were:

- CYTARABINE 0.75 to 2g/m² i.v. 1 hour infusion q. 12 hours x 4 (= 2 days)+ Seraspenide 0.5 to 62.5 µg/kg i.v. 12 hours perfusion for 48 hours (4 dose levels: 0,5 to 62,5 mg/kg/administration of chemotherapy). Seraspenide infusion was initiated 2 hours prior cytarabine and continued 10 hours after the ending of the last cytarabine infusion.

- IFOSFAMIDE 1.5 to 3.0 g/m²/d i.v. 1 hour infusion x 4 days +Seraspenide i.v. continuous infusion for 96 hours (4 dose levels: 2,5 to 156,25µg/kg/administration). Seraspenide infusion was initiated simultaneously with ifosfamide and continued 23 hours after the ending of the last ifosfamide infusion.

RESULTS

The tolerance of Seraspenide either alone (n= 13 patients) or combined with monochemotherapy (n= 53 patients) has been so far excellent. The maximal tolerated dose has not been reached.

The myeloprotective effect was apparent in the series of the first 53 evaluable patients. In the overall series (n=53) as judged on the AUC, depth of the nadir of leukocytes (p < 0,07) and neutrophils and duration of neutropenia (5,7 vs 7,1 days, p < 0,04). The results were more impressive at the Seraspenide doses of 2.5 and 12.5 µg/kg (levels 2 and 3) overall (n=30): when the Seraspenide was associated with the chemotherapy the nadirs were less profound for the leukocytes (p < 0,004) as well as for the neutrophils (p< 0,008). The AUC for the leukocytes as well as for the neutrophils were greater in combination with the Seraspenide (p< 0,04 et < 0,01, respectively). Finally, the duration of the neutropenia was 1,8 days shorter in combination with Seraspenide (5,9 vs 7,7 days, p< 0,05) (Table II).

Fetal calf bone marrow and liver extracts were proved, through multistep purification, to contain the tetrapeptide acetyl-SDKP. A myeloprotection from myelotoxic drugs was anticipated to occur, using acetyl-SDKP as preventing a proportion of the stem cell pool to be prompted into cycle and theerfore to be destroyed by the next application of the cytotoxic drug (Guigon et al. ,1982; Bogden et al. ,1991). Acetyl-SDKP (INN= Seraspenide) has been identified as a small peptide (4 amino-acids = Acetyl-Ser-Asp-Lys-Pro =Acetyl SDKP) and synthesized (Lenfant et al. 1989). It plays a physiological hemoregulatory role in mice as suggested for instance by the observation that the CFU-S are induced into cycle when mice are given anti-acSDKP monoclonal antibodies Volkov et al. 1992; Frindel & Monpezat, 1989; Frindel et al., 1992). It has been isolated from human placenta and it is found in man, where it is particularly aboundant in lymphopoietic and hemopoietic organs (Pradelles et al. ,1990). It has been possible to quantify the concentration of acetyl-SDKP in the blood of normal volunteers and of patients with varied malignancies: myeloproliferative disorders appear to demonstrate high circulating levels of acetyl-SDKP. Acetyl-SDKP is synthetized in long-term marrow cultures (Charbord, personal communication) and it inhibits in a reversible way the proliferation of human progenitors (Guigon et al., 1990; Bonnet et al., 1992). The potential for inhibiting hemopoietic stem cells in human beings appear to be more pronounced for the more primitive stem cells (CD 34/DR low progenitors rather than CD 34/DR high or later progenitors). The mechanism of action of AcetylSDKP may call for intermediary cells (Cashman et al. 1992; Robinson et al. 1992) but there is also evidence for a direct action on stem cells (Bonnet et al. 1993).

PATIENTS AND METHODS

A phase I-II study was undertaken in 1991 in cancer patients (Table I), mostly heavily pretreated, undergoing 2 consecutive cycles of monochemotherapy (cytarabine or ifosfamide).

Table I: Cancer type distribution in Seraspenide phase I-II study (n = 53)

HEMATOLOGICAL TUMORS n= 14	SOLID TUMORS n= 39	
6 Non Hodgkin lymphoma	10 metastatic breast ca.	3 metastatic hepatic ca.
3 acute lymphoid leukaemia	5 sarcoma	2 rhino-pharyngeal ca.
2 acute myeloid leukaemia	4 HEENT	2 endometrial ca.
3 misceallenous	3 ovarian ca.	1 colon ca.
	3 esofagus ca.	6 misceallenous

The Phase I-II study has been designed so as to combine Seraspenide to chemotherapy at the 1st or the 2d cycle with a double-blind cross-over:

> 1st cycle: chemotherapy + Seraspenide or placebo
> 2d cycle: chemotherapy + reverse order

Table II: RESULTS: MYELOPROTECTION AGAINST BOTH MONOCHEMOTHERAPIES on the comparison of the hematological data in the *30 patients at the 2.5 & 12.5 µg/kg Seraspenide dose levels*

PATIENT GROUPS & END-POINTS		PLACEBO		SERASPENIDE		
		mean	st. error	mean	st. error	p*
Leucocytes	A.U.C.	44111	3734	47561	3842	0,038
	Nadir /mm³	1622	187	2078	256	0,004
	Duration <1000 /mm³	3,8	0,9	3,9	0,9	ns
Neutrophils	A.U.C.	23623	2234	27757	2752	0,01
	Nadir /mm³	518	97	796	145	0,008
	Duration<500/mm³ **	7,7	1,2	5,9	1,3	<0,05

* Wilcoxon matched-paired rank test
** 1.8 days shorter in the cycle with Seraspenide

These results reflect the advantage that has observed individually in the Seraspenide arm for each chemotherapy :
- in patients treated with cytarabine (n=17) as judged on the nadir of the leucocytes and neutrophils (Table III):

Table III: Myeloprotection against **CYTARABINE** asssessed on the comparison of the hematological data in all the 17 patients & in the 11 patients at the 2,5 et 12,5 µg/kg Seraspenide dose levels

PATIENT GROUPS & END-POINTS		PLACEBO		SERASPENIDE		
		mean	st. error	mean	st. error	p*
CYTARABINE (n=17) nadir leucocytes /mm³	all dose levels	1540	242	2062	345	<0,05
CYTARABINE nadir leucocytes (n=11)	dose levels 2,5 & 12,5 µg/kg	1453	274	2173	484	<0,05
nadir neutrophils /mm³		81	53	504	220	<0,01

*Wilcoxon matched-paired rank test

The impact on peripheral blood for the Cytarabine patients is translated in an improvement of the WHO index of toxicity for leukocytes and neutrophils (also perhaps with a trend for platelets) as suggested by the comparison of the grade 3 & 4 toxicity levels (Table IV).

Table IV - WHO INDEX FOR HEMATOLOGICAL TOXICITIES IN THE 2.5 & 12.5 DOSE LEVELS - (n = 17)

PARAMETERS	GRADE WHO*	PLACEBO (n)	SERASPENIDE (n)
Erythrocytes ($\times 10^3/mm^3$) < 2.5	4	4	4
2.5 à < 3.0	3	6	8
3.0 à < 3.5	2	5	3
3.5 à < 4.0	1	2	1
> = 4.0	0	0	1
Leucocytes ($\times 10^3/mm^3$) < 1.0	4	8	5
1.0 à < 2.0	3	3	4
2.0 à < 3.0	2	3	3
3.0 à < 4.0	1	3	3
> = 4.0	0	0	2
Neutophils ($\times 10^3/mm^3$) < .5	4	14	12
5 à < 1.0	3	2	1
10 à < 1.5	2	1	2
15 à < 2.0	1	0	1
> = 2.0	0	0	1
Platelets ($\times 10^3/mm^3$) < 25.0	4	9	8
25.0 à < 50.0	3	7	4
50.0 à < 75.0	2	1	1
75.0 à < 100.0	1	0	4
> = 100.0	0	0	0

*WHO index exept for erythrocytes

- in patients treated with ifosfamide (n=36), as judged on the AUC of the leucocytes and neutrophils (Table V):

(TABLE V):

Myeloprotection against **IFOSFAMIDE** assessed on the comparison of hematological data in all the 36 patients
and in the 19 patients at the 2,5 et 12.5 µg/kg Seraspenide dose levels

PATIENT GROUPS & END-POINTS	PLACEBO		SERASPENIDE		
	mean	st. error	mean	st. error	p*
IFOSFAMIDE all dose levels (n=36)					
A.U.C. leucocytes /mm^3	43864	3223	46505	3060	<0,07
A.U.C. neutrophils /mm^3	27755	2640	29585	2343	<0,06
IFOSFAMIDE dose levels (n=19) 2.5 & 12.5µg/kg					
A.U.C. leucocytes & 12.5µg/kg (n=19)	41271	4046	47145	4603	<0,05
A.U.C. neutrophils /mm^3	23622	2516	29710	3623	<0,01

*Wilcoxon matched-paired rank test

DISCUSSION

Seraspenide is the first isolated and characterized inhibitor that inhibits the entry of stem cells into cycle (Frindel & Guigon, 1977) and protects from cytotoxic drugs in-vivo (Guigon et al. ,1982; Bogden et al. ,1991). Unlike growth factors it may preserve the stem cell pool (Hornung & Longo, 1992). At the difference of TGFb or TNFa, the Seraspenide is remarkably well tolerated, but no comparison is possible with other inhibitors, such as the MIP-1a (Dunlop et al., 1992) and the pentapeptide pEEDCK (Paukovits et al., 1991), which have not been tested in the clinic. This trial is on-going and in addition the protective effect of Seraspenide in patients undergoing aggressive polychemotherapy for HEENT carcinoma has been launched. Up to now Seraspenide appears to inhibit reversibly and specifically the normal progenitors (Guigon et al., 1990; Bonnet et al., 1992), thus Seraspenide should enhance the therapeutic index.

CONCLUSION

Seraspenide is non-toxic in patients undergoing chemotherapy. Preliminary data in the first 53 patients who undergone chemotherapy with and without Seraspenide suggest that it protects from drug-induced hemopoietic damage. Only one modality of administration and few dose levels have been explored till now. The optimal chronology of administration in relation to chemotherapy initiation and discontinuation is yet to be explored. The potential for hemoprotection has only been tested over 1 cycle of chemotherapy; it may be improved with the repetition of the cycles. Further clinical investigations are required to assess for long term protection, which is not warranted by current methods, especially positive growth factors, except for bone marrow transplants. In this setting the potential benefits from negative bone marrow regulators like the Seraspenide may be even greater than anticipated.

REFERENCES

Bogden, A.E., Carde, P., Deschamps De Paillette, E., Moreau, J.P., Tubiana, M., Frindel, E. (1991): Amelioration of chemotherapy induced toxicity by co-treatment with AcSDKP, a tetrapeptide inhibitor of hematopoietic stem cell proliferation. Ann. N.Y. Acad.Sci. : 628: 126-139

Bonnet, D., Lemoine, F., Khoury, E., Pradelles, P., Najman, A., Guigon, M. (1992): Reversible inhibitory effects and absence of toxicity of the tetrapeptide Acetyl N - Ser-Asp-Lys-Pro (AcSDKP) in human long-term bone marrow culture. Exp. Hematol. ; 20: 1165-1169

Bonnet D., Lemoine F.M., Pontvert -Delucq S., Baillou C., Najman A., Guigon M. Direct and reversible inhibitory effect of the tetrapeptide acetyl-N-Ser-Asp-Lys-Pro (Seraspenide) on the growth of human CD34[+] subpopulations in response to growth factors. Blood submitted 1993

Cashman J.D., Eaves A.C., Eaves C.J. (1992): Evidence for an indirect mechanism mediating the inhibitory effect of the tetrapeptide AcSDKP primitive human hematopoietic cell proliferation. J. Cell. Biochem.; Suppl. 16C;

Dunlop, D.J., Wright, E.G., Larimore, S., Graham, G.J., Holycake, T., Kerr, D.J., Wolpe, S.D., Pragnell, I.B. (1992): Demonstration of stem cell inhibitor and myeloprotective effects of SCI/rh MIP 1 a in vivo. Blood ; 79: 2221-2225

Frindel, E., And Guigon, M. (1977) Inhibition of CFU entry into cycle by a bone marrow extract.Exp. Hematol. ; 5:p.74-76

Frindel, E., Masse, A., Pradelles, P., Volkov, L., Rigaud, M. (1992): The correlation of endogenous plasma levels in mice and the kinetics of CFU-S entry into cycle after ARA-C treatment: fundamental and clinical aspects. Leukemia ; 6:579-601

Frindel, E., And Monpezat, J.P. (1989): The physiological role of the endogenous colony forming units-spleen (CFU-S) inhibitor Acetyl-N-Ser-Asp-Lys-Pro (AcSDKP). Leukemia ; 3: 753-754

Guigon, M, Bonnet, D., Lemoine, F., Kobari, L., Parmentier, C., Mary, J.Y., Najman, A. (1990): Inhibition of human bone marrow progenitors by the synthetic tetrapeptide AcSDKP. Exp. Hematol. ; 18: 1112-1115

Guigon, M., Mary, J.Y., Enouf, J., Frindel, E. (1982): Protection of mice against lethal doses of 1b-D-arabinofuranosylcytosine by pluripotent stem cell inhibitors.Cancer Res. ; 42: 638-641

Hornung, R.L., And Longo, D.L. (1992): Hematopoietic stem cell depletion by restorative growth factor regimens during repeated high dose cyclophosphamide therapy. Blood ; 80: 77-83

Lenfant, M., Wdzieczak-Bakala, J., Guittet, E., Prome, J.C., Sotty, D, Frindel, E. (1989): Isolation and structure determination of an inhibitor of hematopoietic pluripotent stem cell proliferation. Proc. Natl. Acad. Sci. USA ; 86: 779-782

Paukovits, W.R., Moser, M.H., Binder, K.A., Paukovits, J.B. (1991): Protection from arabinofuranosyl cytosine and N-mustard induced myelotoxicity using hemoregulatory peptide p-Glu-Glu-Asp-Lys-Lys monomer and dimer. Blood ; 77: 1313-1319

Pradelles, P., Frobert, Y., Creminon, C., Liozon, E., Masse, A., Frindel, E. (1990): Negative regulator of pluripotent hematopoietic stem cell proliferation in human white blood cells and plasma as analysed by enzyme immunoassay. Biochem. Biophys. Res. Com. ; 170: 986-993

Robinson S., Lenfant M., Wdzieczack-Bakala J. , Melville J., Riches A. (1992): The mechanism of action pf the tetrapeptide Acetyl-N-Ser-Asp-Lys-Pro (AcSDKP) in the control of haematopoietic stem cell proliferation. Cell Prolif.; 25: 623-632

Vassort, F., Frindel, E., Tubiana, M (1971): Effect of hydroxyurea on the kinetics of colony forming units of bone marrow in the mouse. Cell Tissue Kinet. ; 4: 423-431

Volkov, L., Pradelles, P., Conte, L., Frindel, E. (1992): L'étude du rôle d'un régulateur négatif (Ac-Ser-Asp-Lys-Pro) dans les phénomènes de prolifération cellulaire. C. R. Acad. Sci. Paris 315, série III, 499-504

AKNOWLEDGEMENTS: our gratitude to Emilia FRINDEL, Martine GUIGON & Maurice TUBIANA, who conceived the role of negative regulators of hemopoiesis and searched for the substance, to Maryse LENFANT who pioneered the search for the active peptide and its mechanism of action, to & to Johanna WDZIECZAK-BAKALA and all co-workers and clinician colleagues, to Arthur BOGDEN, François THOMAS for stimulating discussions. Funds Suzanne AXEL, Jacques et Monique ROBOH contributed to the realisation of preclinical & clinical studies

SERASPENIDE (INN)- manufactured by Ipsen-Biotech / Beaufour (Paris)
Patent = Institut Gustave Roussy / INSERM / CNRS. Granted by the ANVAR.

Résumé

Le Séraspenide (acétylSDKP) est un tétrapeptide synthétique, mais naturellement présent chez l'homme, qui inhibe l'entrée en cycle des cellules souches hématopoiétiques, in vivo et in vitro, chez la souris comme chez l'homme. In vivo, chez la souris il protège l'hématopoïèse et la survie après doses léthales de chimiothérapies du type cytarabine, cycloposphamide, carboplatine, adriamycine. In vitro, chez l'homme il a également une activité protectrice vis-à-vis de différents types de progéniteurs hématopoiétiques exposés à la cytarabine, au mafosfamide, à l'AZT et à l'hyperthermie. Le Séraspénide seul a été administré en phase I à 13 patients porteurs de cancer. Aucun effet biologique ou clinique n'a été mis en évidence sur 4 paliers de doses. En phase I-II il a été associé à l'un ou l'autre (au hasard) de deux cycles successifs de monochimiothérapie strictement identiques (dose, modalités d'administration) chez 53 patients porteurs de cancer et on a comparé l'évolution hématologique de ces patients (test des paires appariées de Wilcoxon). Deux cycles identiques de CYTARABINE, se sont succédé à 4 semaines d'intervallechez 17 patients, à des doses qui variaient individuellement pour chacun d'eux de 0,75 à 2g/m² i.v. toutes les 12 h. x 4. Le Séraspenide a été associé en parallèle à l'un des deux cycles, en perfusion continue, à 4 doses, de 0,5 à 62,5 mg/kg/administration de chimiothérapie. Une protection sur le nadir des leucocytes et sur celui des neutrophiles a été observée ($p < 0,05$ & $0,01$). De même, deux cycles identiques d'IFOSFAMIDE se sont succédé à 4 semaines d'intervalle chez 36 patients, à des doses qui variaient individuellement pour chacun d'eux de 1,5 à 3g/m²/jour i.v. en 1 h. x 4 jours. de 0,75 à 2g/m² i.v. toutes les 12 h. x 4. Le Séraspenide a été associé en parallèle à l'un des deux cycles, à 4 doses, de 2,5 à 156,25 mg/kg/administration de chimiothérapie. Une protection sur l'aire sous la courbe des leucocytes et sur celui des neutrophiles a été observée ($p < 0,05$ & $0,01$). Quand on groupe les patients CYTARABINE et IFOSFAMIDE aux 2ème et 3ème paliers de Séraspénide la durée de la neutropénie est également plus brève ($p < 0,05$). Aucun effet adverse n'a été détecté. Les observations précédentes, suggérant que le Séraspénide peut conférer une protection hématologique vis-à-vis de deux chimiothérapies différentes, sans effet secondaire, incitent à poursuivre les investigations cliniques et l'exploration des mécanismes sous-jacents.

WR 2721 (Ethiofos) protects normal progenitor/stem cells from cyclophosphamide derivatives toxicity, with preservation of their anti-leukemic effects: application to *in vitro* marrow purging

Luc Douay*,**, Chen Hu*, Marie-Catherine Giarratana*, Sandrine Bouchet*, Dominique Bardinet* and Norbert-Claude Gorin*

*Formation Associée Claude-Bernard, Unité de Recherche sur les greffes de cellules souches hématopoïétiques, Laboratoire de Cultures Cellulaires, CHU Saint-Antoine, 27, rue de Chaligny, 75012 Paris, France; **Laboratoire d'Hématologie, Hôpital d'enfants Armand-Trousseau, 26, avenue du Docteur-Arnold-Netter, 75571 Paris Cedex 12, France

ABSTRACT

We have investigated on WR 2721, an organic thiophosphate, to evaluate the protection of the normal progenitor/stem cell pool from cyclophosphamide derivatives (CYD) toxicity, in conjunction with the preservation of their anti-leukemic effect. Late and early normal progenitor cells were studied in presence of placenta conditioned medium (PCM- CFU-GM) and in presence of G-CSF, GM-CSF, IL3, EPO and SCF (5R- CFU-GM). Leukemic progenitor cells (CFU-L) were grown in presence of PHA, leukocyte feeder layer ± 25U IL2. A significant ($p<0.05$) protecting effect was observed on PCM- CFU-GM in 6/11 cases, as well as on 5R-CFU-GM in 5/8 cases. The WR 2721 protection yielded a increase of late and early progenitor recovery of 4 fold (2 to 8) and 2 fold (1 to 5) respectively. In contrast, in no case (0/5), WR 2721 exhibited any protection effect upon leukemic CFU-L. In conclusion, the data show that WR 2721 can protect human normal progenitor/stem cells from CYD toxicity, with preservation of their anti-leukemic effect. The therapeutic of such a protecting agent is obvious in the setting of ABMT with purged marrow.

In vitro chemotherapy for elimination of residual leukemic cells from human autologous grafts has been extensively investigated since the initial report by Sharkis et al (1980) on the tumoricidal effect of cyclophosphamide derivatives on rat BNML leukemic cells. Numerous preclinical and clinical studies have shown the efficiency in humans of 4-hydroperoxycyclophosphamide (4-HC) and mafosfamide for in-vitro purging in acute leukemia (AL) (Douay et al, 1982 ; Siena et al, 1985). In addition to killing tumor cells, however, these drugs are toxic to normal progenitor/stem cells. A linear correlation between the CFU-GM content of purged marrow and time to hemopoietic reconstitution has been demonstrated (Rowley et al, 1987 ; Rowley et al, 1989). The lower the CFU-GM recovery, the longer time to engraftment. Since delays in marrow recovery increase a patient's risk of infection and bleeding, especially in acute myeloblastic leukemia (AML) (Douay et al, 1987), a method which might reduce cyclophosphamide derivatives toxicity to normal marrow without compromising the anti-leukemic effect has to be investigated. The protection of normal hematopoietic progenitor cells from ionizing radiation and chemotherapeutic agents has been a subject of recent interest, and several studies have shown the capacity of certain biological response modifiers to confer radio-or chemoprotection to early bone marrow cells in vivo (Neta et al, 1986 ; Moore et al, 1987). Recently, Moreb et al.(1990) have demonstrated the ability of interleukin-1 (IL1) and tumor necrosis factor alpha (TNF-alpha) to protect in vitro human hematopoietic precursors from the toxicity of 4-HC. Lemoli et al (1992) have reported similar data with TGF β. Frindel and Guigon (1977) have reported the tetrapeptide acSDKP to inhibit the entry into DNA synthesis of murine pluripotent stem cells and the growth of human progenitor cells (Guigon et al, 1990), suggesting a possible use for protection against chemotherapy toxicity.

In the present study we have investigated on a different approach for bone marrow protection against 4-HC and mafosfamide. WR 2721 (Ethiofos, Amifostine) is a phosphorylated sulfhydryl compound which has been shown to protect bone marrow and other normal tissues from damage produced by radiotherapy or alkylating agent chemotherapy (Capizzi et al, 1993 ; Stemmer et al, 1992). Available evidence suggest that conversion of parent drug to WR 1065, it's free thiol metabolite, is required for activity. This conversion is catalyzed by alkaline phosphatases which in most tissues have an alkaline pH optimum. The selective activation of WR 2721 may be due to the higher level of alkaline phosphatase as well as a higher pH in normal bone marrow, compared to the more acidic environment and lower alkaline phosphatase levels of most tumors, where conversion to WR 1065 will be slower. Once WR 1065 is produced, it covalently binds to alkylating species and neutralizes their effects. In the setting of in-vitro purging for autologous bone marrow transplantation, in patients with hematological malignancies, we have investigated on WR 2721 to evaluate the preservation of antileukemic effects of cyclophosphamide derivatives, in conjonction with protection of the normal progenitor cells.

MATERIALS AND METHODS

Rationale for the definition of the maximun tolerable dose of drug : the CFU-GM LD 95.

1- In a phase 1 study, Kaiser et al (1985) have reported that engraftment may occur with marrow containing no residual granulocyte-macrophage colony-forming units (CFU-GM), as assessed by conventional semisolid assay. Nevertheless, failures of engraftment have occured in some patients grafted with marrow totally depleted of CFU-GM.

2- In an in-vitro preclinical study (Douay et al, 1982 ; Gorin et al, 1986), we have shown that normal CFU-GM express individual sensitivity to Mafosfamide that might be retained in favor of an individual adjustment of the dose.

3- Using long-term marrow culture, we have shown (Douay et al, 1982) that the importance of the pluripotential stem cell pool is parallel to the residual CFU-GM content of the treated marrow ; consequently, doses of Mafosfamide reducing the CFU-GM content to 0% could severely injure the primitive pool.

4- As a consequence of these observations, we have defined the optimal dose for the treatment of marrow in-vitro as the dose sparing a sufficient fraction of CFU-GM that would guarantee engraftment : a residual amount of 5% CFU-GM (CFU-GM LD95).

Bone marrow cells were obtained from the iliac crest of 11 donors - normal donors for allogeneic transplantation (n=4) ; non-Hodgkin's lymphoma (NHL ; n=2); acute non lymphoblastic leukemia (ANLL ; n=4) ; acute lymphoblastic leukemia (ALL ; n=1) ; -at the time of marrow collection in complete remission for ABMT or for allograft and processed as described below.

Bone marrow leukemic cells were obtained from 5 patients at time of diagnosis : ALL n=1 , AML n= 3 , APCML n=1. The mean blast contamination was 90%.

Cell separation, in-vitro protection and treatment : light density mononucleated cells (MNC) were separated by centrifugation on a Ficoll-isopaque density gradient (d=1.077 g/ml) at 400 g for 20 min. Interface cells were washed, suspended in TC 199 medium (Gibco, France) supplemented with 20% autologous plasma and adjusted to 2.10^7/ml. MNCs were then incubated with WR 2721 (3mg/ml) for 15 min at 37°C, and then washed once in cold Iscove's modified Dulbecco's medium (IMDM ; J. Bio France), counted and readjusted to 10^7/ml. MNCs were subsequently incubated with concentrations of freshly diluted mafosfamide (ASTA Z 7654, ASTA Pharma, Bielefeld, Germany) or 4-hydroperoxycyclophosfamide (Scios Nova Inc, Mountain view, USA), ranging from 20 to 200µg/ml in steps of 10 or 20µg. Incubation was performed for 30 min at 37°C in a waterbath with frequent agitations. Cells were then washed once in cold IMDM and cultured for residual various normal progenitor cells simultaneously, or leukemic progenitors.

Progenitor cell studies

The assay for growth factors-stimulated CFU-GM (5R-CFU-GM) was performed as follows. 5×10^4 untreated and mafosfamide treated MNCs were plated in 35-mm petri dishes in 1-ml aliquots of IMDM containing 30% foetal calf serum (FCS) (Gibco), detoxified bovine serum albumin 10 mg/ml (Sigma) and 0.92% methylcellulose (Flucka, Buchs, Switzerland). Cultures were stimulated with a combination of human recombinant growth factors (GF) : Interleukin 3 (IL3 ; 10ng/ml), Granulocyte growth factor (G-CSF ; 10ng/ml), granulocyte-macrophage growth factor (GM-CSF ; 10ng/ml), erythropoietin EPO ; 3U/ml) and Mast Cell Growth Factor (MGF ; 100 ng/ml). EPO (specific activity 173000 U/mg protein) was kindly provided by Boehringer (Germany), GM-CSF (specific activity $1-5 \times 10^7$ CFU/mg protein) by Behring (Marburg, Germany), MGF (specific activity 10^5U/mg) by Immunex (Washington, USA), IL3 (specific activity $1-3.5 \times 10^8$U/mg protein) by Genzyme (Boston, USA), G-CSF by Rhone Poulenc.
All growth factors were used at optimal concentration, as determined in preliminary experiments. Dishes were incubated at 37°C in a humidified atmosphere supplemented with 5% CO2. Three dishes were plated for each individual data point per experiment. Granulocyte-macrophage colonies (CFU-GM) containing more than 50 cells were scored on day 14.

The assay for placental conditioned medium (PCM) stimulated CFU-GM (PCM-CFU-GM) was performed in agar. The basic medium was McCoy's 5A medium without serum (GIBCO-Biocult) supplemented with 30% FCS (GIBCO-Biocult). Colony-stimulating activity (CSA) was supplied by 10% human PCM. Cells were seeded at 5×10^4/ml of medium containing equal volumes of 0.6% agar and 2xMcCoy's 5A medium to achieve a final serum concentration of 15%. Three 35-by 10-mm petri dishes (Greiner) were plated for each assay. The cultures were incubated for 10 days at 37°C in a humidifed 5% CO2 atmosphere. The colonies (CFU-GM) containing more than 50 cells were scored on day 10.

The assay for leukemic progenitor cells (CFU-L) was performed according to our extensively published technique (x). Briefly two hundred thousand cells were resuspended in McCoy's 5A medium containing 0.72% méthylcellulose (Fluka), 5% FCS, 2,5% PHA, ± 25U recombinant human (rh) interleukin 2 (IL2) (Genzyme), and 7.5×10^{-5}M alpha-thioglycerol. The mixture was overlaid onto a 0.5% agar feeder layer containing 2×10^6 irradiated leukocytes in 35x10-mm petri dishes. Several dishes were incubated in an atmosphere of 5% CO2. Colonies (>50 cells) were scored after 5-7 days, harvested with a Pasteur pipette, pooled, and washed twice in phosphate-buffered saline (PBS). After cytocentrigugation (Cytospin ; Shandon), slides were stained with May-Grünwald-Giemsa and studied for immunological markers by flow cytometry analysis.

Analysis of data

As extensively previously reported (Gorin et al, 1986 ; Douay et al, 1989), sensitivity of progenitor cells to mafosfamide was studied by measuring their remaining proportion and comparing it to untreated marrow. Because the evolution of this proportion in relation to the dose of mafosfamide generated a sigmoid curve, a probit transformation was used to linearize the relationship taking into account the number of replicates used for each dose. Each linear curve was characterized by its slope and the dose necessary to achieve a 95% (LD 95) reduction of the initial progenitor cell pool. Statistical comparison were performed using the Fisher exact test.

RESULTS AND DISCUSSION

We demonstrated that pre-incubation of bone marrow cells with WR 2721 followed by washing and then incubation with CY derivatives (CYD) produced a significant increase of normal progenitor cell recovery in 8 out 11 experiments compared to drug alone, without reducing the elimination of leukemic progenitor cells (CFU-L) in 5/5 cases tested.

In details (Table 1), a significant ($p<0.05$) protecting effect was observed on late (PCM-CFU-GM) and earlier normal progenitor cells (5R-CFU-GM) in 6/11 and 5/8 cases tested respectively. No protection was observed in 3/11 cases only 1 normal marrow, 2 AML marrow in CR.

Overall data settled that protected cells exhibited a lower sensitivity to CYD as shown by the increase value of the LD95 of 37% (8 to 65) for PCM CFU-GM and 30% (11 to 55) for 5R-CFU-GM in parallel to a decrease of the slope of the dose-response curve (fig 1).

When considering the maximum tolerable dose (MDT), the WR 2721 protection yielded an increase of progenitor recovery of 4 fold (2 to 8) and 2 fold (1 to 5) for PCM-CFU-GM and 5R CFU-GM respectively.

Table 1 : Protection of late and early progenitor cells by WR 2721 against cyclophosphamide derivatives

UPN	Diagnosis	Drug	LD 95 (1)			
			PCM CFU-GM(2)		5R CFU-GM(3)	
			WR-	WR+	WR-	WR+
1	NL	Mafosfamide	40	40	150	150
2	NL	Mafosfamide	23	24	80	93
3	NL	Mafosfamide	63	61	106	209
4	NL	Mafosfamide	39	53	-	-
5	NHL	4-HC	25	35	-	-
6	NHL	Mafosfamide	42	58	-	-
7	ALL	4-HC	28	32	28	38
8	AML	4-HC	23	38	35	55
9	AML	Mafosfamide	65	70	96	107
10	AML	Mafosfamide	34	33	68	67
11	AML	Mafosfamide	36	37	90	78

(1) estimated dose (µg/ml) allowing a 5% survival of CFU-GM ; (2) CFU-GM grown in agar in presence of PCM and (3) in methylcellulose in presence of MGF + IL3 + GM-CFS + G-CSF + EPO.

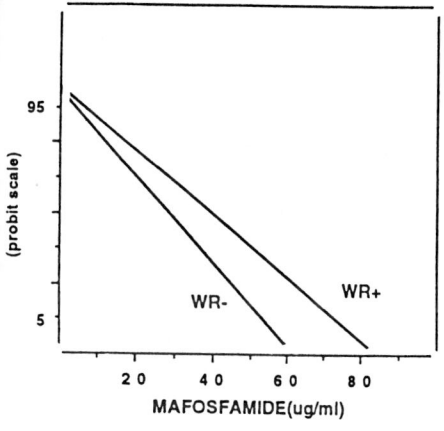 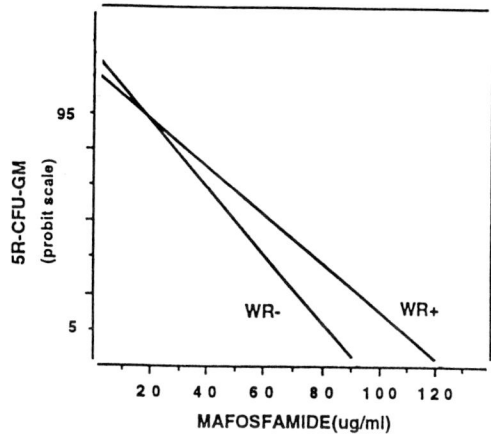

fig 1 : Dose-response curve of late (PCM - CFU-GM) and early (5R - CFU-GM) progenitor cells treated by CYD, in presence (WR+) or not (WR-) of ethiofos (WR 2 72 77) . $p<0.05$.

We next examined the response of CFU-L. Table 2 indicates a decrease of the LD95 in all experiments ($p<0.05$) testifying in favor of a non protecting effect of WR 2721 upon leukemic cells, but also suggesting a possible potentiating effect on the cytotoxicity of CYD against these tumor cells. Moreover, the cloning efficiency of CFU-L was decreased by a mean level of 20% after incubation with WR 2721, in absence of CYD, in 3/5 cases (data not shown).

Fig 2 illustrates the survival of CFU-L exposed either to mafosfamide alone or to WR 2721 before mafosfamide treatment.

Table 2 : effect of WR 2721 on leukemic progenitor cells (CFU-L)

	CFU-L LD 95 ($\mu g/ml$)	
Diagnosis	WR-	WR+
AML	42	38
AML	34	18
AML	50	45
ALL	36	28
AP-CML	64	51

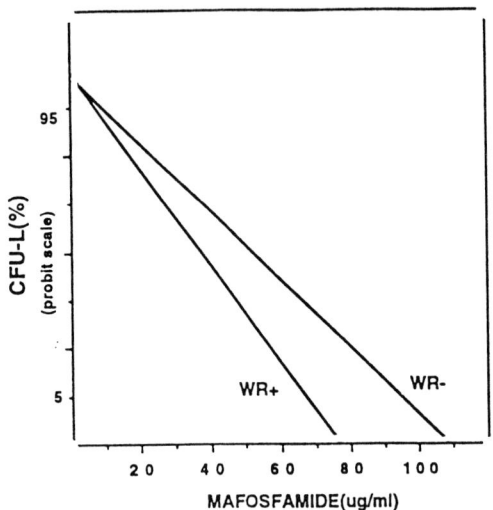

Fig 2 : Dose-response curve of CFU-L treated by CYD, in presence (WR+) or not (WR-) of ethiofos (WR 2721) p<0.05.

In conclusion, the data presented here clearly show that WR 2721 protects human normal progenitor/stem cells from CYD toxicity, with preservation of their antileukemic effect confirming the previous data of Valeriote and Tolen (1982) in mice with nitrogen mustard. The therapeutic use of such a protecting agent is obvious in the setting of ABMT with purged marrow.

REFERENCES

ALLIERI MA, FABREGA S, OZSAHIN H, DOUAY L, BARBU V, GORIN NC. (1992). Detection of BCR/ABL translocation by polymerase chain reaction in leukemic progenitor cells (ALL-CFU) from patients with acute lymphoblastic leukemia (ALL). Exp. Hematol. 20 : 312-314.

CAPIZZI RL, SCHEFFLER BJ, SCHEIN PS. (1993). WR 2721 mediated protection of normal bone marrow from cytotoxic chemotherapy. In Press, Cancer 15-20.

CASTELLI MP, BLOCK PL, SCHNEIDER M, PENNINGTON R, ABE F, TALMADGE JE (1988). Protective, restorative and therapeutic properties of recombinant human IL-1 in rodent models. J. Immunol. 140 : 3830.

DOUAY L, GORIN NC, GEROTA I, NAJMAN A, DUHAMEL G. (1982). In vitro treatment of leukemic bone marrow for autologous transplantation. Exp. Hematol. (suppl. 12) 10: 113.

DOUAY L, LAPORTE JP, MARY JY. (1987). Differences in kinetics of hematopoietic reconstitution between ALL and ANLL after autologous bone marrow transplantation with marrow treated in-vitro with mafosfamide (ASTA Z 7557). Bone Marrow Transplant. 2 : 33-43.

DOUAY L, MARY JY, GIARRATANA MC, NAJMAN A, GORIN NC (1989). Establishment of a reliable experimental procedure for bone marrow purging with Mafosfamide (ASTA Z 7557). Exp. Hematol. 17 : 429-432.

FRINDEL E, GUIGON M (1977). Inhibition of CFU entry into cycle by a bone marrow extract. Exp. Hematol. 5 74.

GORIN NC, DOUAY L, LAPORTE JP, MARY JY, NAJMAN A, SALMON C, AEGERTER P, STACHOWIAK J, DAVID R, PENE F, KANTOR G, DELOUX J, DUHAMEL E, VAN DEN AKKER J, GEROTA J, PARLIER Y, DUHAMEL G (1986). Autologous bone marrow transplantation using marrow incubated with ASTA Z 7557 in adult acute leukemia. Blood 67 : 1367.

GUIGON M, BONNET D, LEMOINE F, KOBARI L, PARMENTIER C, MARY JY, NAJMAN A (1990). Inhibition of human bone marrow progenitors by the synthetic tetrapeptide AcSDKP. Exp. Hematol. 18 : 1112.

HAGENBEEK A, MARTENS ACM (1982). Autologous bone marrow transplantation in acute leukemia : separation of hematopoietic stem cells from clonogenic leukemia cells by 4-hydroperoxycyclophosphamide. Exp. Hematol. (suppl 11) 10 : 14.

KAISER H, STUART RK, BROOKMEYER R, BESCHORNER WE, BRAINE HG, BURNS WH, FULLER DJ, KORBLING M, MANGAN KF, SARAL R, SENSENBRENNER L, SHADDUCK RK, SHENDE AC, TUTSCHKA PJ, YEAGER AM, ZINKHAM WH, COLVIN OM, SANTOS GW (1985). Autologous bone marrow transplantation in acute leukemia : a phase 1 study of in-vitro treatment of marrow with 4-HC to purge marrow cells. Blood 65 : 1504.

LEMOLI RM, STRIFE A, CLARKSON BD, HALEY JD, GULATI SC. (1992). TGF-beta 3 protects normal human hematopoietic progenitor cells treatment with 4-hydroperoxycyclophosphamide in-vitro. Exp. Hematol. 20 : 1252-1256.

MOORE MAS, WARREN DJ (1987). Synergy of interleukin-1 and granulocyte-macrophage colony stimulating factor : in vivo stimulation of stem-cell recovery and hematopoietic regeneration following 5-fluorouracile treatment in mice. Proc. Natl. Acad. Sci. USA 84 : 7134.

MOREB J, ZUCALI JR, RUETH S (1990). The effects of tumor necrosis factor alpha on early human hematopoietic progenitor cells treated with 4-hydroperoxycyclophosphamide. Blood 76 : 681.

NETA R, DOUCHES SD, OPPENHEIM JJ (1986). Interleukin-1 is a radio protector. J. Immunol 136 : 2483.

NETA R, OPPENHEIM JJ, DOUCHES SD (1988). Interdependence of the radioprotective effects of human recombinant interleukin-1, tumor necrosis factor,

ROWLEY SD, JONES RJ, PIANTADOSI S. (1989). Efficacy of ex vivo purging for autologous bone marrow transplantation in the treatment of acute nonlymphoblastic leukemia. *Blood*, 74 : 501-506.

ROWLEY SD, ZUEHLSDORF M, BRAINE HG. (1987). CFU-GM content of bone marrow graft correlates with time to hematologic reconstitution following autologous bone marrow transplantation with 4-hydroperoxycyclophosphamide-purged bone marrow. *Blood,* 70 : 271-275.

SHARKIS SJ, SANTOS GW, COLVIN OM (1980). Elimination of acute myelogenous leukemic cells from marrow and tumor suspension in the rat with 4-hydroperoxycyclophosphamide. *Blood* 55 : 521.

SIENA S, CASTRO-MALASPINA H, GULATI SC, LU L, COLVIN OM, CLARKSON BD, O'REILLY RJ, MOORE MAS. (1985). Effect of in vitro purging with 4-hydroperoxycyclophosphamide on the hematopoietic and microenvironmental elements of human bone marrow. *Blood* 65 : 655.

SPHALL EJ, JONES RB, JOHNSTON C, HAMI L, MORGAN MJ, STEMMER S, BUNN PA, WALTER J, McCULLOCH W, SCHEIN PS. (1991). Amifostine (WR2721) shortens the engraftment period of 4-HC purged bone marrow in breast cancer patients receiving high-dose chemotherapy with autologous bone marrow support. (ABMS). *Blood* 78 (suppl. 1) : 192A.

STEMMER SM, SHPALL EJ, JONES RB, HAMI L, BEARMAN SI, MYERS SE, TAFFS S, SHAW L, CAPIZZI RL, SCHEIN P. (1992). Amifostine (WR-2721) shortens the engraftment time of 4-hydroperoxycyclophosphamide (4-HC) purged bone marrow in lymphoma patients receiving high dose chemotherapy (HDC) with autologous bone marrow support (ABMS). *Blood* 80: 269.

VALERIOTE F, TOLEN S. (1982). Protection and potentiation of nitrogen mustard cytotoxicity by WR 2721. Cancer research 42 : 4330-4331.

YUHAS JM, STORER JB. (1969). Differentital chemoprotection of normal and malignant tissues. *J. Natl. Cancer Inst.* 42 : 331-335.

Résumé

Nous avons étudié le WR 2721, un thiophosphate organique, pour évaluer la protection des progéniteurs hématopoïétiques normaux contre la toxicité des dérivés de l'endoxan, en association avec le maintien de leur effet antileucémique. Les progéniteurs normaux tardifs et précoces ont été étudiés en présence de milieu conditionné placentaire (PCM-CFU-GM) et en présence de G-CSF, GM-CSF, IL3, EPO et SCF (5R CFU-GM). Les progéniteurs leucémiques (CFU-L) ont été cultivés en présence de PHA, sous-couche leucocytaire ± 25U/IL2. Un effet protecteur significatif ($p<0,05$) a été observé sur les PCM-CFU-GM dans 6/11 cas ainsi que sur les 5R CFU-GM dans 5/ 8 cas. La protection par le WR 2721 a permis d'augmenter la récupération en progéniteurs tardifs et primitifs de 4 fois (2 à 8) et de 2 fois (1 à 5) respectivement. En revanche, le WR 2721 n' a jamais protégé les progéniteurs leucémiques dans 5 cas étudiés. En conclusion, nos résultats montrent que le WR 2721 peut protéger les progéniteurs hématopoïétiques normaux de la toxicité des dérivés de l'endoxan avec conservation parallèle de leur effet anti-leucémique. L'intérêt thérapeutique d'un tel agent protecteur est évident dans le cas des AGMO avec une moelle purgée.

Immunomodulation by AS101: mode of action and clinical applications in cancer therapy

Benjamin Sredni, Michael Albeck and Yona Kalechman

C.A.I.R. Institute, Department of Life Sciences, Bar Ilan University, Ramat Gan, 52900 Israel

SUMMARY

AS101 (ammonium trichloro[dioxyethylene-O,O']tellurate) is a new synthetic compound with immunomodulating properties and minimal toxicity. The compound can induce the secretion of a variety of cytokines in vitro and in vivo both in humans and in animals. AS101 has been found to protect mice from adverse effects of radiation and chemotherapy caused to early progenitor cells in the bone marrow, to the granulocytic-monocytic lineage, as well as to the T cell and thymocytic compartment. Moreover, the compound has been shown to increase survival of mice subjected to lethal doses of radiation or chemotherapy. AS101 has also been shown to increase survival of mice bearing solid tumors or metastases when given alone or in conjunction with chemotherapy. The augmentation of the number of Colony Forming Units-Granulocyte-Macrophage (CFU-GM) in the bone marrow (BM), as well as the increase in the number of spleen and BM cells, and the secretion of Colony Stimulating Factor (CSF) by these cells in mice treated with AS101 before being subjected to irradiation or Cyclophosphamide (CYP) treatment (Kalechman et al., 1990; Kalechman et al., 1991a), suggests AS101's potential for accelerating the recovery of hemopoiesis after cytotoxic chemotherapy. On the basis of these studies, a phase II randomized clinical trial utilizing AS101 in combination with chemotherapy has been initiated on non-small cell lung cancer (NSCLC) patients. Preliminary results show significant hair and bone marrow preservation of platelets and granulocytes (Tichler et al., 1993) thus allowing for dose intensification, improved self-image and compliance during chemotherapy treatment.

HISTORICAL BACKGROUND

In the past decade there has been growing interest in the potential of synthetic compounds to modify immune responses by imitation of cytokine action. The restorative properties of immunomodulators spring from their ability to induce the differentiation and proliferation of particular groups of cells. In cancer patients, immunotherapy is one of a variety of modalities proposed to either prevent or eliminate tumor metastases. The administration of various natuarlly occurring or synthetic biological response modifiers has been shown to enhance host immune reactivity and to elicit antitumor activity. However, the maximum tolerated dose of biological response modifiers, such as interleukins, IL-1, IL-2, gamma-interferon (γ-IFN) and tumor necrosis factor (TNF), is lower than the dose required to exhibit maximal biological activity.

Developing strategies to overcome the toxicity of biological response modifiers and, alternatively, developing and characterizing new agents that do not exhibit such side effects, could greatly improve the prospects of effective immunotherapy in treating human patients.

We have developed a new synthetic compound, AS101, a low molecular weight organic tellurium compound soluble in organic solvents but only slightly soluble in water.

IMMUNOMODULATING EFFECTS IN VITRO: ANIMAL CELLS

Effects on IL-1, IL-2, CSF and TNF Secretion

AS101 was found to stimulate mouse spleen cells to proliferate and to produce lymphokines such as IL-2 and CSF, which are regulators of lymphopoiesis and myelopoiesis (Sredni et al., 1987; Sredni et al., 1988). AS101 at 0.05-1µg/ml induced optimal CSF production by mouse spleen cells, and optimal IL-2 production was found at 0.1-0.5µg/ml AS101. The ability of AS101 to stimulate the secretion of CSF was not restricted to spleen cells. Much higher levels of CSF were secreted by BM cells stimulated by AS101 (Kalechman et al., 1990). The levels of CSF secreted by 0.5x10⁶ BM cells were three times higher than the levels secreted by 5x10^6 speen cells/ml (Table 1).

Table 1. Cytokine secretion by mouse cells stimulated by AS101

AS101 µg/ml	IL-2 Production u/ml	CSF Production by spleen cells (colonies/10^5 cells)	CSF Production by BM cells (colonies/10^5 cells)	IL-1 Production By PEC cells (u/ml)	TNF Production by PEC cells (u/ml)
1.0	4.0±0.3	2.0±0.1	36±2	–	–
0.5	9.0±2.0	2.0±0.1	78±6	22±3	6±0.5
0.1	12.0±2.0	35.0±3.7	102±6	9±2	3±0.3
0.05	4.0±0.3	38.0±3.2	20±2	–	–
0.01	0.5±0.1	15.0±2.0	0	–	–

AS101 was also able to induce the secretion of IL-1. The secretion of IL-1 was analyzed in peritoneal exudate cells from BALB/c and C3H-HeJ mice injected with thioglycolate. AS101 at 0.5µg/ml induced the secretion in vitro of very large amounts of IL-1 from both BALB/c and C3H/HeJ peritoneal exudate cells (PEC).

The LPS contamination of the substance was established by a Limulus assay and shown to contain less than 2u/ml both in the AS101 preparation and in the peritoneal exudate-conditioned medium. The effect of AS101 induced supernatants could be ablated by anti-IL-1 antibodies. The effect of AS101 on macrophages was also demonstrated by the ability of the compound to stimulate mouse PEC cells, isolated by peritoneal puncture 72h after IP thioglycolate injection, and to secrete TNF (Sredni et al., 1990) (Table 1).

IMMUNOMODULATING EFFECTS IN VITRO: HUMAN CELLS

Cytokine Secretion

AS101 can stimulate normal human mononuclear cells (MNC) to express IL-2 receptors, to produce IL-2 and to proliferate (Sredni et al, 1987; Sredni et al, 1988). Various concentrations of AS101 were tested in parallel for induction o f all three functions.Concentrations of 0.1-0.5µg/ml were found to be optimal.Approximately 25% MNC from normal donors stained positively with monoclonal anti-IL-2 receptor antibodies after culture with 0.5µg/ml of AS101 for 48h, compared to only 2% of cells incubated in medium alone (Table 2). AS101 has also the ability to stimulate human MNC to secrete γ-IFN and TNF when incubated with the cells for 96h (Sredni et al., 1990).

Table 2. Cytokine secretion and induction of IL-2 receptors by human cells stimulated by AS101

AS101 µg/ml	IL-2 production u/ml	IL-1 production by PEC cells u/ml	TNF production by MNC u/ml	γ-IFN production by MNC u/ml	Induction of IL-2R (stimulation index)
1	1.8±0.4	1±0.5	–	–	7.0±1.5
0.5	12.2±5.3	13±2	5.5±1	37.5±7.5	35.7±8.6
0.1	14.3±4.6	21±3	2.0±0.3	0	32.9±12.9
0.05	1.1±0.2	6±2	–	–	2.9±1.4
0.01	1.5±0.4	1±0.4	–	–	1.1±0.3

PROTECTION FROM ADVERSE EFFECTS OF RADIATION

The hemopoietic system is the primary critical tissue for radiation-induced mortality. It may therefore be of vital importance that AS101 was found to offer protection against the adverse effects of radiation by means of accelerating the restoration of functional hematopoietic cells. We hypothesized that AS101 could exert radioprotective effects via induction of cytokines, such as IL-1 which has been previously reported to have strong radioprotective properties (Neta et al., 1986), or other cytokines such as GM-CSF or TNF which were also reported to synergize with IL-1 to exert their radioprotective effects. We found that injection of AS101, every other day for two weeks prior to a sublethal dose of 450 rads, accelerated the rate of recovery of BM or spleen cellularity. The number of BM cells in the AS101-injected mice reached almost that of normal non-irradiated mice (Kalechman et al., 1990).

The same protocol of AS101 injections showed an increase in CSF production by BM or spleen cells. Following exposure to 450 rads, there was a dramatic decrease in the number of CFU-GM. Pretreatment of mice with AS101 increased twelve-fold the number of precursor cells (Table 3).

The recovery of spleen colony-forming units (CFU-S) in the BM and spleen of AS101 treated mice was significantly faster than that of control PBS treated mice. The effect of AS101 on the recovery of endogenous CFU-S was also very significant. The number of these endogenous CFU-S has been found to have a close correlation to the survival rate of mice treated with radioprotective drugs. AS101 was found to increase the number of endogenous colonies either when it was injected prior to irradiation or immediately after irradiation. Each endogenous spleen colony from irradiated AS101 injected mice was found to include a higher number of CFU-S in AS101 treated mice.

AS101 was found to exert its radioprotective effects also on the T cell compartment. T cell and thymocytes proliferation in the presence of Con A was significantly higher in AS101 treated irradiated mice than in PBS treated irradiated mice. IL-2 secretion by mouse spleen T cells was also significantly higher in AS101 treated mice (Table 3).

Table 3. Effect of AS101 on the recovery of hemopoietic cells from irradiated mice and their functioning

| | Irradiated | | | Non-irradiated |
| | PBS treated | AS101 treated | | control |
		10 μg	30 μg	
spleen cells $\times 10^7$	1.8±0.3	4.3±0.5	3.8±2	9±1
BM cells $\times 10^6$	14.0±1	33.0±3	25.0±2	45±3
CSF secretion by BM cells (colonies/10^5 cells)	60.0±2	82.0±3	59.0±3	93±4
No. BM-CFU-C	10.0±2	51.0±3	13.0±3	120±6
IL-2 secretion by spleen cells (u/ml)	6.0±0.5	26.0±3	24.0±2	29±3
thymocytes proliferation (stimulation index)	12.3±2	36.8±3	34.6±4	46±5
spleen cells proliferation (proliferation index)	15.7±2	53.0±6	48.0±4	66±7
BM CFU-S	11.0±2	42.0±3	14.0±2	59±3
spleen CFU-S	3.0±0.5	18.0±3	32.0±3	41±4

These results provided sufficient grounds for reasoning that AS101, by enhancing the hematopoietic functioning in sublethal irradiated mice, could also protect mice against lethal doses of irradiation. Indeed, we found that in mice treated with varying doses of AS101 prior to a 840 rad irradiation, there was a significant increase in survival rate (80% vs. 30% in control PBS injected mice) (Fig. 1). Similar results were obtained after oral pretreatment with AS101, with 50μg/mouse being the optimal dose (Sredni et al., 1992).

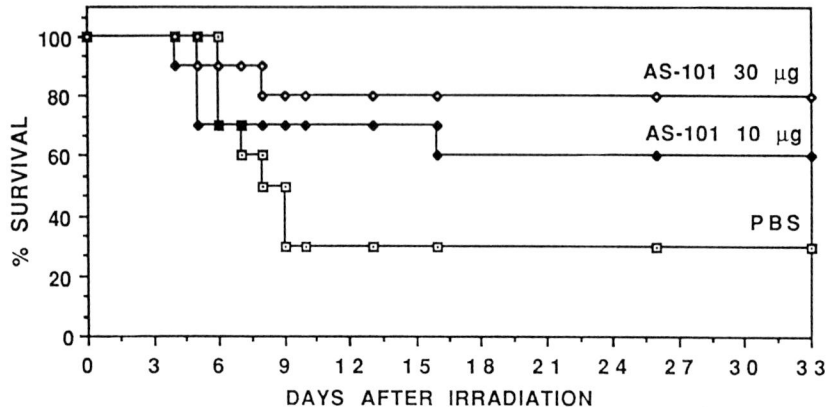

Fig. 1. Radioprotective effect of AS101 on lethally irradiated mice. BALB/c mice were irradiated with 840-rad irradiation after being injected with different concentrations of AS101 or PBS for 2 weeks, every other day. The rate of survival was monitored during 33 days. Results represent three experiments with a total of 30 mice.

CHEMOTHERAPY

Protection from side effects of chemotherapy

Chemically-induced cytoreduction is used therapeutically for the treatment of neoplasms and also to ablate bone marrow prior to bone marrow transplantation. The efficacy of chemotherapy against advanced tumor depends on the dose intensity of the drugs utilized. Treatment with such agents is limited, however, by myelosuppression, immune suppression and other toxic effects on normal cells. Immunotherapy with cytokines such as IL-1 and G-CSF may help to minimize the myelosuppressive effects of DNA-damaging agents. Since AS101 has been found to release these cytokines, we assumed that this agent might exert chemoprotective effects via induction of cytokines. We found that injection of AS101 every other day for two weeks prior to a sublethal dose of 250mg/kg of CYP accelerated the rate of recovery of BM cellularity, and BM colony forming units granulocyte-macrophage (CFU-GM) progenitor cells, and the secretion of CSF (Kalechman et al.,1991a; Kalechman et al., 1991b). AS101 was found to exert its chemoprotective effects also on the T cell compartment (Kalechman et al., 1992). T cell and thymocyte proliferation in the presence of Con-A was significantly higher in AS101 treated mice subjected to CYP at 250mg/kg than in control PBS and CYP injected mice. IL-2 secretion by mouse spleen T cells was also significantly higher in AS101 treated mice (Table 4).

Table 4. Effect of AS101 on the recovery of hemopoietic cells from cyclophosphamide treated mice and their functioning

	CTX treated			Non-CTX treated control
	PBS treated	AS101 treated		
		2.5μg	10μg	
BM cells x10^6	33±3	50.0±4	46.0±4	53±6
CSF secretion by spleen cells (colonies/10^5 BM cells)	54.3±9.1	156.5±25.7	137.3±27.6	180±16
No. BM-CFU-C	14±2.5	40.0±3	42.0±3	120±11
IL-2 secretion by spleen cells (u/ml)	17.0±3	31.0±4	26.0±3	39±5
spleen cell proliferation (stimulation index)	11.0±2	42.0±5	36.0±4	62±6
thymocyte proliferation (stimulation index)	9.0±1	39.0±3	37.0±4	45±6

Effect of AS101 on hematopoiesis

In order to examine the effects of AS101 administration on suppression of hematopoiesis induced by cytotoxic agents, CYP at a sublethal dose of 250mg/kg was administered IP to normal mice after AS101 injection. The number of WBC, neutrophils, platelets and lymphocytes was significantly decreased (p<0.05). Injection of 10µg AS101 to mice before treatment with CYP produced a significant protection and a faster recovery of hematopoiesis (Fig. 2).

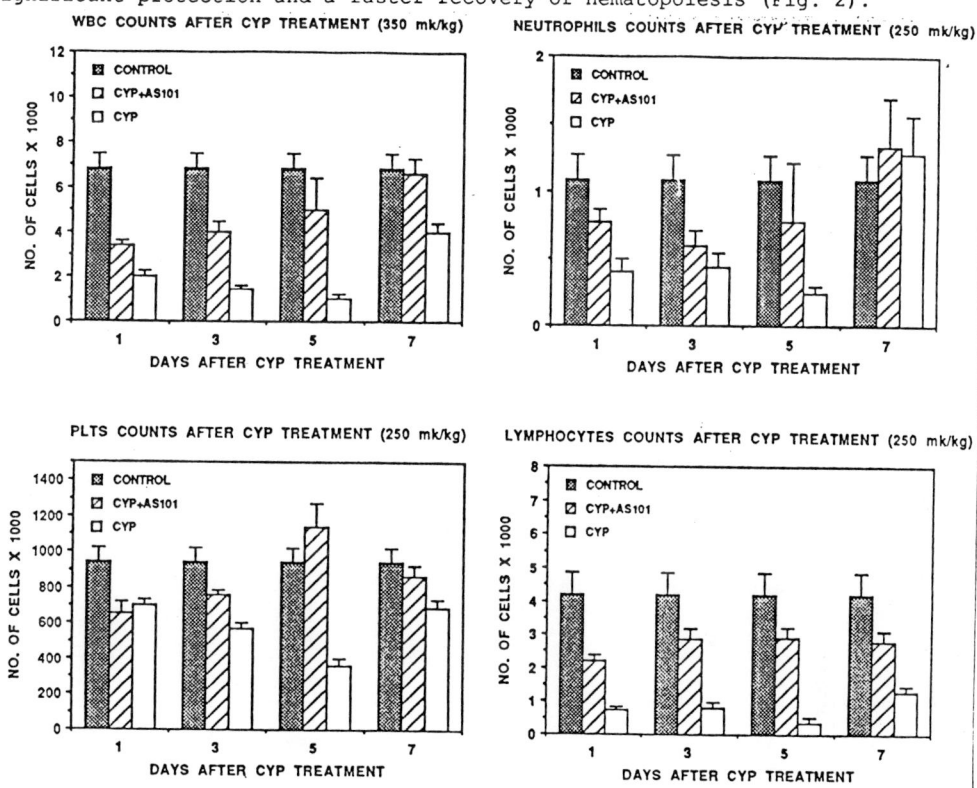

Fig. 2. Effect of AS101 on blood cell counts. CYP (250mg/kg) was administered IP to normal mice with and without pretreatment with AS101 (10µg IP).

Protective effects of single IP doses of AS101 24h prior to chemotherapy

A two-week pretreatment with oral or IP AS101 (every other day) has previously been shown to protect the hematopoietic system from damage caused by CYP (Sredni et al., 1992; Kalechman et al., 1991a). These studies have been extended by assessing the ability of AS101 to protect stromal cells in the BM from toxic effects of either CYP or other chemotherapeutic drugs, when given as a single IP dose to mice prior to chemotherapy (Kalechman et al.,1993). This protection could be of utmost importance as stromal cells in the BM control the initial phases of hemopoiesis by regulating stem cell self-renewal and differentiation.

CYP was injected to mice with or without AS101. Stromal cells were prepared from their BM and the following parameters were examined: CFU-F (Colony Forming Unit-Fibroblasts - representing the number of stromal colonies), secretion of CSF by stromal layers, secretion of IL-6 by stromal layers, and the ability of stromal layers to sustain BM committed stem cells (CFU-GM).

It was concluded that all above mentioned stromal functions were significantly decreased following CYP treatment. Treatment with AS101 before CYP injection significantly increased stromal functions above levels of CYP treated mice (Table 5).

Table 5. Protective effect of AS101 against cyclophsphamide-induced bone marrow toxicity
Effect of AS101 on fibroblast colony forming cells (CFU-F), CSF secretion, IL-6 and CFU-GM in the bone marrow of mice treated with CYP. Mice were injected with AS101 (10μg) 24h before treatment with CYP (250mg/kg). At various time points mice were sacrificed and their BM cells were tested. Results represent mean±SEM of 3 different experiments including 3 mice/group/experiment.

	CFU-F	CSF	IL-6 (U/ml)	CFU-GM
Control	53±4	38±3	32	62±5
CYP 250mg/kg	12±2	13±2	15	14±2
AS101 10μg+CYP	36±3	43±4	26	37±3

AS101 was also found to protect mice from hemopoietic damages caused by sublethal doses of various chemotherapeutic drugs and to increase the rate of survival of mice treated with lethal doses of these agents. These included cyclophosphamide, 5-FU, Doxorubicin, Cyclohexylchloroethylmitrosourea (CCNU) or etoposide (VP-16) (Kalechman et al., 1993 (submitted)). The protective properties of AS101 enabled the treatment of tumor bearing mice with combined therapy of the compound and very high doses of chemotherapy, thereby significantly increasing their survival rate. On the basis of these results, phase II clinical trials on cancer patients treated with AS101 in combination with chemoterhapy have been initiated.

Fifty-six patients with non-small cell lung cancer were randomized to receive either chemotherapy alone (CT) using Carboplatin, 300 mg/m^2/IV day 1, VP-16 200mg/m^2 po days 3,5 and 7, or CT plus AS101 3mg/m^2/IV 3 times a week (AS101). AS101 was administered for 2 weeks prior to initiating CT. Postchemotherapy results are shown in Table 6.

Table 6. The effect of AS101 on hair and bone marrow preservation

Alopecia	#pts	Grade 0	I	II	III	IV	Mantel-Haenszel Chi-Square
CT	30	13%	7%	30%	43%	7%	
AS101	26	23%	23%	35%	19%	0%	(p=0.02)
Granulocytopenia		Grade 0-II			III-IV		
CT	30	47%			53%		
AS101	26	85%			15%		(p=0.002)
Thrombocytopenia		Grade 0-II			III-IV		
CT	30	81%			19%		
AS101	26	96%			4%		(p=0.077)
Colony Stimulating Factor		Increase			Decrease		
CT	30	17%			83%		
AS101	26	77%			23%		(p=0.001)

AS101 added minimal toxicity with 28% of AS101 having Grade I-II postinjection drug fever compared to 3% in CT group. Garlic halitosis was universal in the AS101 group and absent in the CT group.

The results showed a significant increase in hair and bone marrow preservation as reflected by a significant decrease in thrombocytopenia and neutropenia induced by chemotherapy (Tichler et al., 1993). Furthermore, an increase in IL-1, GM-CSF and IL-6 secretion was observed in patients treated with the combined therapy. On the basis of the these effects of AS101 we hypothesized that this compound might exert its chemoprotective effect indirectly, via the enhanced production of cytokines.

In order to define the factor by which AS101 may mediate its protection, studies are being designed to determine whether the in vivo treatment with antibodies directed against murine IL-1a, IL-6 and various CSFs would abrogate AS101-induced chemoprotectiveness. These properties of AS101 make it a promising candidate for chemoimmunotherapy protocols in cancer patients for minimizing adverse cytotoxicity resulting from a variety of drugs. Phase II clinical trials with AS101 in combination with chemotherapeutic drugs, including Carboplatin, VP-16 or Doxorubicin are currently in progress.

REFERENCES

Kalechman, Y., Albeck, M., Oron, M., Sobelman, D., Gurwith, M., Sehgal, S.N., and Sredni, B. (1990): The radioprotective effects of the immunomodulator AS101. *J. of Immunol.* 145, 1512-1517.

Kalechman, Y., Albeck, M., Oron, M., Sobelman, D., Gurwith, M., Horwith, G., Kirsch, T., Maida, B., Sehgal, S.N., and Sredni, B.(1991a): The protective and restorative role of AS101 in combination with chemotherapy. *Cancer Research* 51(5), 1499-1503.

Kalechman, Y., Sotnik Barkai, I., Albeck, M., Horwith, G., Sehgal, S.N., and Sredni B.(1991b): The use and mechanism of action of AS101 in protecting bone marrow CFU-GM after purging with ASTA-Z. *Cancer Research* 51(20), 5614-5620.

Kalechman, Y., Sotnik Barkai, I., Albeck, M., and Sredni, B. (1992): The effect of AS101 on the reconstitution of T-cell reactivity following irradiation or cyclophosphamide treatment. *Exp. Hematol.* 20, 1302-1308.

Kalechman, Y., Sotnik Barkai, I., Albeck, M., and Sredni, B. (1993): Protection of bone marrow stromal cells from toxic effects of Cyclophosphamide in vivo, or ASTA-Z 7557 and Etoposide (VP-16) in vitro by Ammonium trichloro(dioxyethylene-O,O')tellurate (AS101). *Cancer Research* (in press).

Kalechman, Y., Sotnik Barkai, I., Albeck, M., and Sredni, B. (1993): The protective role of AS101 in combination with several cytotoxic drugs acting by different mechanisms of action. *Blood* (submitted).

Neta, R., Douches, S., and Oppenheim, J.J. (1986): Interleukin-1 is a radioprotector. *J. of Immunol.* 136, 2483-2487.

Sredni, B., Caspi, R.R., Klein, A., Kalechman, Y., Danzinger, Y., Ben Ya'akov, M., Tamari, T., Shalit, F., and Albeck, M. (1987): A new immunomodulating compound (AS-101) with potential therapeutic application. *Nature* 330, 173-176.

Sredni, B., Caspi, R.R., Lustig, S., Klein, A., Kalechman, Y., Danziger, Y., Ben Ya'akov, M., Tamari, T., Shalit, F., and Albeck, M.(1988): The biological activity and immunotherapeutic properties of AS101, a synthetic organotellurium compound. *Nat. Immun. Cell Growth Regul.* 7, 163-168.

Sredni, B., Kalechman, Y., Albeck, M., Gross, O., Aurbach, D., Sharon, P., Sehgal, S.N., Gurwith, M.J., and Michlin, H. (1990): Cytokine secretion effected by synergism of the immunomodulator AS101 and the protein kinase-C inducer bryostatin. *Immunology* 70(4), 473-477.

Sredni, B. Albeck, M., Kazimirsky, G., and Shalit, F. (1992): The immunomodulator AS101 administered orally as a chemoprotective and radioprotective agent. *Int. J. Immunopharmac.* 14(4), 613-619.

Tichler, T., Shapiro, Y., Shani, A., Braderman, Y., Catane, R., Brenner, H., Kaufman, B., Pavlotsky, F., Kalechman, Y., Farbstein, M., Albeck, M., and Sredni, B. (1993): Hair preservation and platelet/granulocyte sparing effect using AS101 with chemotherapy - a randomized study. (Abstract) *ASCO*, Miami, USA, 1993.

Résumé

L'AS101 (ammonium trichloro(dioxyethylene-o-o') tellurate) est un nouveau composé synthétique possédant des propriétés immunomodulatrices et une toxicité minimale. Ce composé peut induire la secrétion d'un certain nombre de cytokines in vitro et in vivo tant chez l'animal que chez l'homme. L'AS101 peut protéger les souris des effets adverses des radiations et de la chimiothérapie sur les progéniteurs précoces de la moelle, sur la lignée granulo-macrophagique aussi bien que sur les compartiments des cellules T et des thymocytes. En outre il a été montré que ce composé augmente la survie de souris soumises à des doses létales d'irradiation ou de chimiothérapie. L'AS101 augmente également la survie de souris porteuses de tumeurs solides ou de métastases lorsqu'il est administré seul ou associé à la chimiothérapie. L'augmentation du nombre des CFU-GM dans la moelle ainsi que l'augmentation du nombre de cellules médullaires et spléniques et la secrétion de CSF par ces cellules chez les souris traitées par l'AS101 avant irradiation ou cyclophosphamide (Kalechman et al. 1990,1991a) suggèrent la capacité de l'AS101 à accélérer la récupération de l'hématopoïèse après chimiothérapie cytotoxique. Sur la base de ces études, un essai clinique randomisé phase II utilisant l'AS101 associé à la chimiothérapie a été entrepris chez des patients atteints de cancer du poumon (autres qu'à petites cellules) Les résultats préliminaires montrent une préservation significative des plaquettes et des granulocytes ainsi que des cheveux des cheveux (Tichler et al. 1993), ce qui permet une intensification de la dose et une meilleure acceptation de la chimiothérapie.

A new macrophage-associated activity protects mice from radiation induced death

Vladimir Kravtsov and Ina Fabian

Department of Histology and Cell Biology, Sackler School of Medicine, Tel Aviv University, Tel Aviv, Israel

We have found that crude conditioned medium (CM) from murine histiocytic J-774 cell line increased survival rate of γ-irradiated mice. At the first step of purification CM was subjected to ultrafiltration (Amicon Diaflo Membranes) and when a fraction of Mr about 50kDa (Active Fraction, AF) was administrated to C57Bl/6 mice (12-14 week old, female) twice, 24 and 48h after irradiation (800cGy of total body irradiation, LD90/60) only 20% of AF-treated mice died within 60 days after irradiation as compared to 90% of saline-treated mice. AF-treatment (24 and 48h post irradiation) of sublethally irradiated mice (550 cGy) resulted in a 2.5 fold increase in endogenous D10-CFU-S (E-CFU-S). No changes were found in the number of D12-CFU-S formed in lethally irradiated recipients transplanted with bone marrow cells (BMC) of non-irradiated AF-treated mice. However, the number of myeloid progenitors (CFU-C) in the bone marrow of AF-treated mice was reduced dramatically. In agar cultures of BMC AF failed to support colony formation when was added alone and inhibited in a dose dependent manner colony formation in the presence of GM-CSF or IL-3. AF inhibited also ^{3}H-TdR incorporation by PHA-stimulated peripheral blood lymphocytes. The inhibitory effects of AF as well as its ability to increase the number of E-CFU-S were sensitive to heat inactivation at 80°C but not at 60°C. Western Blot Analysis of AF revealed the absence of detectable amounts of IL-6, GM-CSF and IL-1α and in the thymocyte co-stimulation assay AF did not enhance incorporation of ^{3}H-TdR by PHA-stimulated murine thymocytes. We suggest that J-774 cells are a source of a new cytokine(s) that accelerates the multiplication of cycling, but not dormant CFU-S and inhibits proliferation of committed myeloid progenitors (CFU-C). The ability to accelerate accumulation of pluripotent stem cells can partly explain the efficacy of AF in the therapy of γ-irradiated mice. Purification of the molecules responsible for the described effects of AF is being performed now.

Chemotherapy bone marrow toxicity: stimulation or/and protection of stem cells

Albert Najman

La cigale ou la fourmi
(Jean de La Fontaine)

Department of Hematology, CHU Saint-Antoine, Paris, France

Chemotherapy provides the most effective treatment for disseminated malignant diseases by reducing substantial number of abnormal cells. The efficacy of this treatment depends upon the proper utilisation of antineoplastic agents in accordance with biological factors concerning tumor resistance and tumor growth.

Taking these rules into consideration, the most effective mean of decreasing the development of tumor resistance is through a combination of cytotoxic drugs rather than single drug regimens. Furthermore dose intensity is fundamentally important for a successful antitumoral effect (1). Based on the concept of a first order cytokinetic effect, several courses of chemotherapy are needed to kill a sufficient number of malignant cells (2). As cytotoxic drugs are more effective on dividing cells than on resting cells, the consequences are usually harmful for the hematopoietic system which is one of the most proliferative tissue in the body. Indeed normal cells may be killed in the same proportion as the malignant ones as the activity of cytotoxic drugs is not specifically orientated to the abnormal cells. Depletion of the differentiated cell pool causes the cycling of resting stem cells in order to facilitate bone marrow recovery (3). Newly formed cells became now sensitive to chemotherapy and the same sequence happened again. Thus myelotoxicity is a usual side effect of chemotherapy which may become limiting and preclude therapy. The dose as well as the schedule of delivery are major points which may be hindered by cytopenia. In addition, myelotoxicity is the main impediment for increasing treatment intensity which may be instrumental to improve the outcome of patients with high risk cancers.

The use of GM-CSF and G-CSF, hematopoietic growth factors which stimulate mainly the production of granulocytes, was extensively studied these last years, with the purpose of reducing post chemotherapy neutropenia.

Several clinical trials with GM-CSF and G-CSF have clearly shown that the importance and the duration of neutropenia can be significantly reduced after chemotherapy (4,5,6). While G-CSF acts primarily on the production of neutrophils, GM-CSF stimulate also eosinophils and monocytes. In both cases, accelerated neutrophil recovery is completed by a reduced number of febrile days although GM-CSF can cause by itself in some cases fever which has to be correctly interpretated. Randomized trials with G-CSF have shown that it facilitates the delivery of chemotherapy by reducing the need to delay administration and also by allowing to maintain the adequate dose (7,8). Hematopoietic growth factors (HGF) allow also the administration of high dose chemotherapy. Neidhart et al (9) reported the successful use of G-CSF following high dose cyclophosphamide, cisplatin and etoposide regimen. Laporte et al (10) have treated five patients with advanced non Hodgkin lymphomas and bone marrow involvement by a myelo-ablative combination of high dose carmustine, etoposide, ara-c and melphalan followed by GM-CSF. Recovery was observed in 3/5 patients without any support of hematopoietic stem cells.

Indeed high dose myelo-ablative chemotherapy is usually followed by the reinfusion of autologous hematopoietic stem cells harvested from the bone marrow or from peripheral blood. GM-CSF and G-CSF improve the collection of peripheral blood progenitor cells (11, 12). They clearly improve neutrophil recovery after autologous bone marrow transplant (13,14). When peripheral blood progenitors cells are added, platelet recovery may be also partially improved (15).

Hematopoietic growth factors offer with obvious evidence a way to face chemotherapy myelosuppression. The results achieved on the granulocyte lineage will be extended to platelets when a specific HGF will become available. One have however to consider the shortcomings of the clinical use of HGF. Delivered at pharmacological doses, their effect on bone marrow and on other tissues are not yet clearly known. Long term effects on the microenvironnment have to be evaluated. Unpredictable consequences on the natural cytokines cascade cannot be excluded. Above all, their efficacy depends on the presence of sufficient amount of stem cells.

Preservation of the hematopoietic system may be the alternative way to reduce or avoid myelosuppression.

- Till now, the real number of hematopoietic stem cells is largely unknown. Likewise, their ability to overcome iatrogenic damage has to be more precisely evaluated. Encouraging data were recently reported in mice transplanted with ex vivo expanded bone marrow (16) but the long term outcome of such manipulations cannot be asserted. In untreated normal animals, the pluripotent stem cells (CFU-S) renewal capacity appears to be preserved during the life time. However, when the hematopoietic system is damaged by radiation or chemotherapy, the self renewal of stem cells seems to be reduced. When CFU-S are serially transplanted in mice, a decline in the repopulating ability is observed whatever the interval between two transplants may be (17). In the patients treated with allogenic bone marrow transplantation, Messner and al (18) observed that the number of bone marrow progenitors remains, even after 4 years, lower than it was in the donor. In the case of autologous bone marrow transplant, a successful engraftment is clearly related to the number of collected CFU-GM (19).

- The proportion of normal stem cells killed by the cytotoxic drugs or used for differentiation to repopulate the bone marrow is another very approximate parameter. The limiting myelotoxicity of chemotherapy suggests that the bone marrow reconstitution can really be profoundly altered, either by a too great reduction of the number of stem cells, or by a damage to the regulatory network. This last point is probably a critical one which remains largely underevaluated.

Therefore, a policy based more on saving than on the uncontrolled use of our stem cells pool could be the right answer. To allow the killing of the lowest number of normal cells and to limit the differentiation of stem cells to the right level needed to repopulate the bone marrow after each course of chemotherapy are the two goals of such a policy. The optimal conditions to protect the hematopoietic system require an early start of the procedure. The protective agent has to be orientated towards the normal cells and to be inactive on the malignant cells. In addition, it should not interfere with the antitumoral effect of chemotherapy.

Different ways to protect the bone marrow were presented in this meeting which have to be selected on the above criteria. Studies have to go further to choose the best candidate for this new approach of therapy of malignant diseases. The optimal procedure could be a first phase of bone marrow sparing using inhibitory molecules and then a second phase of adapted and controlled stimulation to achieve a complete recovery, opening the way to new treatments of malignancy.

REFERENCES

1- FREI E III., CANELLOS G.P. (1980)
   Dose : a critical factor in cancer chemotherapy.
   Am. J. Med. 69 585-594

2- SKIPPER H.E. (1967)
   Criteria associated with destruction of leukemia and solid tumor cells in animals.
   Cancer Research 27 (1) 2636-2645

3- FRINDEL E., DUMENIL D., SAINTENY F. (1980)
   Role of pluripoietins in murine bone marrow stem cell differentiation.
   Leukemia Research 4 287-299

4- GROOPMAN J.E., MOLINA M., SCADDEN D.T. (1989)
   Hematopoietic growth factors.
   New Engl. J. Med. 321 1449-1459

5- BIESMA B., DEVRIES E.G.E., WILLEMSE P.H.B., SLUITER W.J. et al. (1990)
   Efficacy and tolerability of recombinant human granulocyte macrophage colony stimulating factor in patients with chemotherapy related leukopenia and fever.
   European J. Cancer 26 932-936

6- MORSTYN G., CAMPBELL L., LIESCHKE G. et al. (1989)
   Treatment of chemotherapy induced neutropenia by subcutaneously administered granulocyte colony stimulating factor with optimization of dose and duration of therapy.
   J. Clin. Oncol. 7 1554-1562

7- GABRILOVE J.L., JAKUBOWSKI A., SCHER H. et al. (1988)
   Effect of granulocyte colony stimulating factor on neutropenia and associated morbidity due to chemotherapy for transitional cell carcinoma of the urothelium.
   N. Engl. J. Med. 318 1414-1422

8- DAVIS I., MORSTYN G. (1991)
The role of granulocyte colony stimulating factor in cancer chemotherapy.
Seminars in Haematology 28 suppl. 2 25-33

9- NEIDHART J., MANGALIK A., KOHLER W. et al. (1989)
Granulocyte colony stimulating factor stimulates recovery of granulocytes in patients receiving dose intensive chemotherapy without bone marrow transplantation.
J. Clin. Oncol. 7 1685-1692

10- LAPORTE J.Ph., FOUILLARD L., DOUAY L. et al. (1991)
Autologous bone marrow transplantation after the BEAM regimen.
Lancet 338 601-602

11- GIANNI A.M., SIENA S., BREGNI M. et al. (1989)
Granulocyte macrophage stimulating factor to harvest circulating hematopoïetic stem cells for autotransplantation.
Lancet ii 580-585

12- MOLINEUX G., POJDA Z., HAMPSON I.N., LORD B.I., DEXTER T.M. (1990)
Transplantation potential of peripheral blood stem cells induced by granulocyte colony stimulating factor.
Blood 76 2153-2158

13- GORIN N.C., COIFFIER B., HAYAT M. et al. (1992)
Recombinant human granulocyte macrophage colony stimulating factor after autologous bone marrow transplantation with unpurged and purged marrow in non Hodgkin's lymphoma : a double blind placebo controlled trial.
Blood 80 1149-1157

14- TAYLOR K.M., JAGANNAHTH S.M., SPITZER G. et al. (1989)
Recombinant granulocyte colony stimulating factor hastens granulocyte recovery after high dose chemotherapy and autologous bone marrow transplantation in Hodgkin's disease.
J. Clin. Oncol. 7 1791-1799

15- SHERIDAN W.P., BEGLEY C.G., JUTTNER C.A. et al. (1992)
Effect of peripheral blood progenitor cells mobilised by filgrastin on platelet recovery after high dose chemotherapy.
The Lancet 339 640-644

16- MUENCH M.O., FIRPO M.T., MOORE M.M. (1993)
Bone marrow transplantation with IL1 plus kit ligand ex vivo expanded bone marrow accelerates hematopoietic reconstitution in mice without the loss of stem cell lineage and proliferative potential.
Blood 81 3463-3470

17- SCHOFIELD R. (1979)
The pluripotent stem cell.
Clinics in Haematology 8 221-238

18- MESSNER H.A., CURTIS J.E., MINDEN M.D., et al (1987).
Clonogenic hematopoietic precursors in bone marrow transplantation.
Blood 70 1425-1432

19- GORIN N.C. (1989)
Indications for autologous bone marrow transplantation of stem cells in hematology.
Biol. Clin. Hemat. 11 63-74

Author index
Index des auteurs

Adamson J.W., 155
Albeck M., 477
Almendral J.M., 381
Amiral J., 235
Antao V.P., 391

Baatout S., 239
Baillou C., 37, 169
Ballerini P., 259
Banchereau J., 163
Bank A., 411
Bardinet D., 469
Batard P., 283
Baudier J., 103
Bauvois B., 263
Beaujan F., 459
Belloc F., 305
Benninger L., 141
Bernard P., 305
Bhatnagar P., 193
Bi S., 111
Billard C., 255
Blackwood E.M., 75
Blanchet J.P., 187
Bohbot A., 457
Boiron J.M., 305
Bondurant M.C., 213
Bonnet D., 169, 177, 445
Bosmans E., 361
Bouchard C., 257
Bouchet S., 469
Bourette R., 187
Bréchot C., 91
Bron D., 361
Brotherton D., 221
Broustet A., 305
Broxmeyer H.E., 141
Brûlet P., 315, 325
Brun B., 459
Brunet A., 55
Brosh N., 243, 251
Brown E., 283
Brown K.E., 375
Bueren J.A., 381
Bujan W., 267
Burrell L.M., 403

Caen J.P., 225, 235
Calenda V., 399
Capitani S., 401
Carde P., 183, 191, 459
Cardoso A., 283
Carlo-Stella C., 349
Carmi P., 243
Caroll M.P., 291
Cashman J.D., 123
Caux C., 163
Cavadore J.C., 65
Cayre Y.E., 259
Cen D., 391
Chambard J.C., 55
Chastang C., 459
Chatelain C., 239
Chatelain C.B., 239
Chatelain B., 239
Chenivessel X., 91
Chermann J.C., 399
Chevillard C., 237
Choquet D., 257
Clark S.C., 283
Coffman T., 335
Cohen A., 341
Conquet F., 315
Cook D., 335
Cooper S., 141
Corazza F., 267
Corbion O., 459
Cornetta K., 141
Coutton C., 457
Craig S., 161, 221
Czaplewski L., 221

Da Costa L., 459
Dainiak N., 201, 301
Davis B.R., 391, 401
Davis N., 419
Deacon D., 133
Delforge A., 361
Delwail V., 459
De Marsh P., 193
Deschamps de Paillette E., 189, 191, 459
Devault A., 65
Dexter T.M., 161

Dinarello C.A., 41
Domenge C., 459
Dorval C., 361
Dorée M., 65
Douay L., 469
Du D.L., 269
Dubois C.M., 23, 419
Dumenil D., 315, 325
Dunlop D.J., 159
Duperray C., 237
Durand B., 155
Dvilansky A., 271

Eaves A.C., 123
Eaves C.J., 123
Eisenman R.N., 75
Escary J.L., 315, 325
Evans S., 221
Ezan E., 191
Ezine S., 325

Fabian I., 485
Fadel E., 459
Fardoun D., 265, 359
Ferrand N., 15
Ferrero J.M., 459
Fesquet D., 65
Fisher D.E., 307
Fisher J., 221
Fitzsimmons E., 159
Fondu P., 267
Francis M.L., 403
Freedman M.H., 341
Frey C., 193
Freshney M.G., 159
Fridman W.H., 257
Fries B., 383
Frindel E., 459

Gallichio V.S., 407
Gandrillon O., 15
Garcia M., 15
Garcia-Giralt E., 459
Gewirtz A.M., 275
Giarratana M.C., 469
Gibellini D., 401
Gilbert R., 221
Gimonet J.F., 459
Gluckman E., 365
Godden J., 185
Goldman J.M., 111
Goncalves E., 459
Gorin N.C., 469
Gorodinsky M., 271
Gothelf Y., 243

Gottesman M., 411
Graham G.J., 185, 223
Grenier N., 189
Grillon C., 183
Grognet J.M., 191
Grunberger T., 341
Guha A., 201
Guigon M., 37, 169, 177, 187, 445, 457, 459
Guilhot F., 459

Hague N., 141
Hamilton C., 445
Hamood M., 267
Han Z.C., 225, 235
Hatzfeld A., 283
Hatzfeld J., 283
Hendrie P., 141
Henglein B., 91
Hérodin F., 189, 191
Hesdorffer C., 411
Hestdal K., 23
Heyworth C.M., 161
Hilton D.I., 3
Holyoake T.L., 159
Honigwachs-Shaanani J., 243, 253
Hoover D.L., 403
Howard L., 221
Hu C., 469
Hughes N.K., 407
Hunter M.G., 161

Irvine A.E., 223
Isnard F., 191, 445, 459
Ivanović Z., 49

Jacobsen S.E.W., 23
Jiang C., 445
Jiang C.Y., 269
Jiang F.Z., 269

Kalechman Y., 477
Kato J., 83
Keller J.R., 23
Kelley L.L., 213
Kessler S.W., 403
Khayat D., 459
Khoury E., 37
King A., 193
Kirby S., 335
Konings A.W.T., 455
Koury M.J., 213
Kravtsov V., 485
Kretser de T., 241
Kretzner L., 75

Labbaye C., 259

Labbé J.C., 65
Lagneaux L., 361
L'Allemain G., 55
Lallemand Y., 315
Lamas E., 91
Lanza F., 111
Laouar A., 263
La Placa M., 391, 401
Lasfar A., 255
Lebeurier I., 225, 235
Lebrun-Texeira G., 265
Lee B.C., 243
Leger-Picherit E., 459
Lemoine F.M., 37, 169, 177, 445
Lenfant M., 179, 181, 183
Lenormand B., 265
Lenormand P., 55
Levesque J.P., 283
Levy A., 243, 253
Li M.L., 283
Li X.S., 269
Liu L., 169
Lodish H.F., 3
Longmore G.D. 3
Lorca T., 65
Lord B.I., 161
Lowry P., 133
Lu L., 141
Lu H., 301
Lukić M., 49

Mac J., 75
Mahon F.X., 305
Mathieu-Tubiana N., 459
Mangoni L., 349
Mantel C., 141
Mao N., 269
Maraninchi D., 459
Martyré M.C., 51
Matsushime H., 83
Meltzer M.S., 403
Mestries J.C., 189
Michel L., 411
Migliaccio A.R., 155
Migliaccio G., 155
Milenković P., 49
Mire-Sluis A.R., 119
Monnier A., 459
Monpezat J.P., 459
Moran D., 115
Moreb J., 429
Morgan P., 221
Morge X., 191
Morris T.C.M., 223
Moser M.H., 437
Mouchiroud G., 187

Myerson D., 383

Najman A., 37, 169, 177, 445, 459, 487
Nathan I., 271
Neta R., 419

Oberling F., 457
Oppenheim J.J., 419
Owen-Lynch J., 161

Pagès G., 55
Pastan I., 411
Patel S., 221
Paukovits J., 437
Paukovits W.R., 437
Pearson M.A., 161
Peled A., 243, 251
Pelus L.M., 193
Perreau J., 315, 325
Peulvé P., 359
Plumb M., 335
Podda S., 411
Pontvert-Delucq S., 37, 169
Pouysségur J., 55
Pradelles P., 191
Pragnell I.B., 159, 185, 335
Preston Mason R., 201

Quesenberry P., 133

Randall L., 119
Re M.C., 391, 401
Reiffers J., 305
Rice A., 305
Richardson C., 411
Riches A., 179, 183, 185, 357
Rieger K.J., 181
Rizzoli V., 349
Robinson S., 179, 183, 357
Rogalsky V., 115
Rousseau A., 181
Roussel M.F., 83
Ruscetti F.W., 23

Saez-Servent N., 181
Samal B., 155
Samarut J., 15
Sanders M., 301
Sariban E., 267
Sarris A., 141
Sautès C., 257
Schwartz G.N. 403,
Segovia J.C., 381
Sherr C.J., 83
Smith L., 411
Smithies O., 335

Socié G., 365
Sookdeo H., 283
Sredni B., 477
Staquet P., 239
Sternberg D., 243, 253
Stevenson E.C., 223
Steward W.P., 159
Stoppa A.M., 459
Stošić-Grujičić S., 49
Stratford May W., 291
Stryckmans P., 361
Sumereau-Dassin E., 265, 359

Tang P.H., 269
Teixeira-Lebrun G., 359
Thomas F., 189, 191
Thorpe R., 119
Tiret L., 315
Todorov G., 115
Toledo J., 243
Torok-Storb B., 283
Tse K.F., 407

Vadhan-Raj S., 141
Vanek W., 341
Vannier J.P., 265, 359
Vekhoff A., 459

Visani G., 391
Vitale M., 401
Vittet D., 237
Vuillemin E., 459
Voelter W., 181

Wadhwa M., 119
Walker F., 301
Wang J., 91
Wang J.M., 419
Ward M., 411
Watowich S.S., 3
Wdzieczak-Bakala J., 179, 181, 183
Whetton A.D., 161
Wierenga P.K., 455
Wietzerbin J., 31, 51, 255
Woods N., 221

Young N.S., 375

Zauli G., 391, 401
Zhang M.W., 269
Zindy F., 91
Zipori D., 243, 251, 253
Zittoun R., 459
Zucali J.R., 429

Colloques **INSERM**
ISSN 0768-3154

Other *Colloques* published as co-editions by John Libbey Eurotext and INSERM

133 Cardiovascular and Respiratory Physiology in the Fetus and Neonate. *Physiologie Cardiovasculaire et Respiratoire du Fœtus et du Nouveau-né.*
Scientific Committee : P. Karlberg,
A. Minkowski, W. Oh and L. Stern;
Managing Editor : M. Monset-Couchard.
ISBN : John Libbey Eurotext 0 86196 086 6
   INSERM 2 85598 282 0

134 Porphyrins and Porphyrias. *Porphyrines et Porphyries.*
Edited by Y. Nordmann.
ISBN : John Libbey Eurotext 0 86196 087 4
   INSERM 2 85598 281 2

137 Neo-Adjuvant Chemotherapy. *Chimiothérapie Néo-Adjuvante.*
Edited by C. Jacquillat, M. Weil and D. Khayat.
ISBN : John Libbey Eurotext 0 86196 077 7
   INSERM 2 85598 283 7

139 Hormones and Cell Regulation (10th European Symposium). *Hormones et Régulation Cellulaire (10ᵉ Symposium Européen).*
Edited by J. Nunez, J.E. Dumont and R.J.B. King.
ISBN : John Libbey Eurotext 0 86196 084 X
   INSERM 2 85598 284 7

147 Modern Trends in Aging Research. *Nouvelles Perspectives de la Recherche sur le Vieillissement.*
Edited by Y. Courtois, B. Faucheux, B. Forette,
D.L. Knook and J.A. Tréton.
ISBN : John Libbey Eurotext 0 86196 103 X
   INSERM 2 85598 309 6

149 Binding Proteins of Steroid Hormones. *Protéines de liaison des Hormones Stéroïdes.*
Edited by M.G. Forest and M. Pugeat.
ISBN : John Libbey Eurotext 0 86196 125 0
   INSERM 2 85598 310 X

151 Control and Management of Parturition. *La Maîtrise de la Parturition.*
Edited by C. Sureau, P. Blot, D. Cabrol, F. Cavaillé and G. Germain.
ISBN : John Libbey Eurotext 0 86196 096 3
   INSERM 2 85598 311 8

Colloques INSERM
ISSN 0768-3154

153 Hormones and Cell Regulation (11th European Symposium). *Hormones et Régulation Cellulaire (11ᵉ Symposium Européen).*
Edited by J. Nunez and J.E. Dumont.
ISBN : John Libbey Eurotext 0 86196 104 8
INSERM 2 85598 324 X

158 Biochemistry and Physiopathology of Platelet Membrane. *Biochimie et Physiopathologie de la Membrane Plaquettaire.*
Edited by G. Marguerie and R.F.A. Zwaal.
ISBN : John Libbey Eurotext 0 86196 114 5
INSERM 2 85598 345 2

162 The Inhibitors of Hematopoiesis. *Les Inhibiteurs de l'Hématopoïèse.*
Edited by A. Najman, M. Guigon, N.C. Gorin and J.Y. Mary.
ISBN : John Libbey Eurotext 0 86196 125 0
INSERM 2 85598 340 1

164 Liver Cells and Drugs. *Cellules Hépatiques et Médicaments.*
Edited by A. Guillouzo.
ISBN : John Libbey Eurotext 0 86196 128 5
INSERM 2 85598 341 X

165 Hormones and Cell Regulation (12th European Symposium). *Hormones et Régulation Cellulaire (12ᵉ Symposium Européen).*
Edited by J. Nunez, J.E. Dumont and E. Carafoli.
ISBN : John Libbey Eurotext 0 86196 133 1
INSERM 2 85598 347 9

167 Sleep Disorders and Respiration. *Les Evénements Respiratoires du Sommeil.*
Edited by P. Lévi-Valensi and D. Duron.
ISBN : John Libbey Eurotext 0 86196 127 7
INSERM 2 85598 344 4

169 Neo-Adjuvant Chemotherapy. *Chimiothérapie Néo-Adjuvante.*
Edited by C. Jacquillat, M. Weil, D. Khayat.
ISBN : John Libbey Eurotext 0 86196 150 1
INSERM 2 85598 349 5

171 Structure and Functions of the Cytoskeleton. *La Structure et les Fonctions du Cytosquelette.*
Edited by B.A.F. Rousset.
ISBN : John Libbey Eurotext 0 86196 149 8
INSERM 2 85598 351 7

Colloques INSERM
ISSN 0768-3154

172 The Langerhans Cell. *La Cellule de Langerhans.*
Edited by J. Thivolet, D. Schmitt.
ISBN : John Libbey Eurotext 0 86196 181 1
INSERM 2 85598 352 5

173 Cellular and Molecular Aspects of Glucuronidation. *Aspects Cellulaires et Moléculaires de la Glucuronoconjugaison.*
Edited by G. Siest, J. Magdalou, B. Burchell
ISBN : John Libbey Eurotext 0 86196 182 X
INSERM 2 85598 353 3

174 Second Forum on Peptides. *Deuxième Forum Peptides.*
Edited by A. Aubry, M. Marraud, B. Vitoux
ISBN : John Libbey Eurotext 0 86196 151 X
INSERM 2 85598 354 1

176 Hormones and Cell Regulation (13th European Symposium). *Hormones et Régulation Cellulaire (13e Symposium Européen).*
Edited by J. Nunez, J.E. Dumont, R. Denton
ISBN : John Libbey Eurotext 0 86196 183 8
INSERM 2 85598 356 8

179 Lymphokine Receptors Interactions. *Interactions Lymphokines-récepteurs.*
Edited by D. Fradelizi, J. Bertoglio
ISBN : John Libbey Eurotext 0 86196 148 X
INSERM 2 85598 359 2

191 Anticancer Drugs (1st International Interface of Clinical and Laboratory responses to anticancer drugs). *Médicaments anticancéreux (1re Confrontation internationale des réponses cliniques et expérimentales aux médicaments anticancéreux).*
Edited by H. Tapiero, J. Robert, T.J. Lampidis
ISBN : John Libbey Eurotext 0 86196 223 0
INSERM 2 85598 393 2

193 Living in the Cold (2nd International Symposium). *La Vie au Froid (2e Symposium International).*
Edited by A. Malan, B. Canguilhem
ISBN : John Libbey Eurotext 0 86196 234 9
INSERM 2 85598 395 9

Colloques INSERM
ISSN 0768-3154

194 Progress in Hepatitis B Immunization. *La Vaccination contre l'hépatite B.*
Edited by P. Coursaget, M.J. Tong
ISBN : John Libbey Eurotext 0 86196 249 4
INSERM 2 85598 396 7

196 Treatment Strategy in Hodgkin's Disease. *Stratégie dans la maladie de Hodgkin.*
Edited by P. Sommers, M. Henry-Amar,
J.H. Meezwaldt, P. Carde
ISBN : John Libbey Eurotext 0 86196 226 5
INSERM 2 85598 398 3

198 Hormones and Cell Regulation (14th European Symposium). *Hormones et Régulation Cellulaire (14ᵉ Symposium Européen).*
Edited by J. Nunez, J.E. Dumont
ISBN : John Libbey Eurotext 0 86196 229 X
INSERM 2 85598 400 9

199 Placental Communications : Biochemical, Morphological and Cellular Aspects. *Communications placentaires : aspects biochimique, morphologique et cellulaire.*
Edited by L. Cedard, E. Alsat, J.C. Challier,
G. Chaouat, A. Malassiné
ISBN : John Libbey Eurotext 0 86196 227 3
INSERM 2 85598 401 7

204 Pharmacologie Clinique : Actualités et Perspectives. (6ᵉ Rencontres Nationales de Pharmacologie clinique).
Edited by J.P. Boissel, C. Caulin, M. Teule
ISBN : John Libbey Eurotext 0 86196 225 7
INSERM 2 85598 454 8

205 Recent Trends in Clinical Pharmacology (6th National Meeting of Clinical Pharmacology).
Edited by J.P. Boissel, C. Caulin, M. Teule
ISBN : John Libbey Eurotext 0 86196 256 7
INSERM 2 85598 455 6

206 Platelet Immunology : Fundamental and Clinical Aspects. *Immunologie plaquettaire : aspects fondamentaux et cliniques.*
Edited by C. Kaplan-Gouet, N. Schlegel,
Ch. Salmon, J. McGregor
ISBN : John Libbey Eurotext 0 86196 285 0
INSERM 2 85598 439 4

Colloques INSERM
ISSN 0768-3154

207 Thyroperoxidase and Thyroid Autoimmunity. *Thyroperoxydase et auto-immunité thyroïdienne.*
Edited by P. Carayon, T. Ruf
ISBN : John Libbey Eurotext 0 86196 277 X
INSERM 2 85598 440 8

208 Vasopressin. *Vasopressine.*
Edited by S. Jard, R. Jamison
ISBN : John Libbey Eurotext 0 86196 288 5
INSERM 2 85598 441 6

210 Hormones and Cell Regulation (15th European Symposium). *Hormones et Régulation Cellulaire (15e Symposium Européen).*
Edited by J.E. Dumont, J. Nunez, R.J.B. King
ISBN : John Libbey Eurotext 0 86196 279 6
INSERM 2 85598 443 2

211 Medullary Thyroid Carcinoma. *Cancer Médullaire de la Thyroïde.*
Edited by C. Calmettes, J.M. Guliana
ISBN : John Libbey Eurotext 0 86196 287 7
INSERM 2 85598 440 0

212 Cellular and Molecular Biology of the Materno-Fetal Relationship. *Biologie cellulaire et moléculaire de la relation materno-fœtale.*
Edited by G. Chaouat, J. Mowbray
ISBN : John Libbey Eurotext 0 86196 909 1
INSERM 2 85598 445 9

215 Aldosterone. Fundamental Aspects. *Aspects fondamentaux.*
Edited by J.P. Bonvalet, N. Farman, M. Lombes, M.E. Rafestin-Oblin
ISBN : John Libbey Eurotext 0 86196 302 4
INSERM 2 85598 482 3

216 Cellular and Molecular Aspects of Cirrhosis. *Aspects cellulaires et moléculaires de la cirrhose.*
Edited by B. Clément, A. Guillouzo
ISBN : John Libbey Eurotext 0 86196 342 3
INSERM 2 85598 483 1

217 Sleep and Cardiorespiratory Control. *Sommeil et contrôle cardio-respiratoire.*
Edited by C. Gaultier, P. Escourrou, L. Curzi-Dascalora
ISBN : John Libbey Eurotext 0 86196 307 5
INSERM 2 85598 484 X

Colloques INSERM
ISSN 0768-3154

218 Genetic Hypertension. *Hypertension génétique.*
Edited by J. Sassard
ISBN : John Libbey Eurotext 0 86196 313 X
INSERM 2 85598 485 8

219 Human Gene Transfer. *Transfert de gènes chez l'homme.*
Edited by O. Cohen-Haguenauer, M. Boiron
ISBN : John Libbey Eurotext 0 86196 301 6
INSERM 2 85598 497 1

220 Medicine and Change: Historical and Sociological Studies of Medical Innovation. *L'innovation en médecine : études historiques et sociologiques.*
Edited by Ilana Löwy
ISBN : John Libbey Eurotext 2 7420 0010 0
INSERM 5 85598 508 0

221 Structures and Functions of Retinal Proteins. *Structures et fonctions des rétino-protéines.*
Edited by J.L. Rigaud
ISBN : John Libbey Eurotext 0 86196 355 5
INSERM 2 85598 509 9

222 Cellular and Molecular Biology of the Adrenal Cortex. *Biologie cellulaire et moléculaire du cortex surrénal.*
Edited by J.M. Saez, A.C. Brownie, A. Capponi, E.M. Chambaz, F. Mantero
ISBN : John Libbey Eurotext 0 86196 362 8
INSERM 2 85598 510 2

223 Mechanisms and Control of Emesis. *Mécanismes et contrôle du vomissement.*
Edited by A.L. Bianchi, L. Grélot, A.D. Miller, G.L. King
ISBN : John Libbey Eurotext 0 86196 363 6
INSERM 2 85598 511 0

224 High Pressure and Biotechnology. *Haute pression et biotechnologie.*
Edited by C. Balny, R. Hayashi, K. Heremans, P. Masson
ISBN : John Libbey Eurotext 0 86196 363 6
INSERM 2 85598 512 9

Colloques INSERM
ISSN 0768-3154

228 Non-Visual Human-Computer Interactions. *Communication non visuelle homme-ordinateur.*
Edited by D. Burger, J.C. Sperandio
ISBN : John Libbey Eurotext 2 7420 0014 3
INSERM 2 85598 540 4

230 From Research in Oncology to Therapeutic Innovations. *De la recherche oncologique à l'innovation thérapeutique.*
Edited by P. Tambourin, M. Boiron
ISBN : John Libbey Eurotext 2 7420 0016 X
INSERM 2 85598 542 0

231 Human Ochratoxicosis and its pathologies. *Ochratoxicose humaine et ses pathologies.*
Edited by E.E. Creppy, M. Castegnaro, G. Dirheimer
ISBN : John Libbey Eurotext 2 7420 0017 8
INSERM 2 85598 543 9

232 Anxiety : Neurobiology, Clinic and Therapeutic Perspectives. *Anxiété : Neurobiologie, Clinique et Perspectives Thérapeutiques.*
Edited by M. Hamon, H. Ollat, M.-H. Thiébot
ISBN : John Libbey Eurotext 2 7420 0018 6
INSERM 2 85598 544 7

LOUIS-JEAN
avenue d'Embrun, 05003 GAP cedex
Tél. : 92.53.17.00
Dépôt légal : 873 — Novembre 1993
Imprimé en France